Biofuel and Biorefinery Technologies

Volume 15

This book series provides detailed information on recent developments in biofuels & bioenergy and related research. The individual volumes highlight all relevant biofuel production technologies and integrated biorefinery methods, describing the merits and shortcomings of each, including cost-efficiency. All volumes are written and edited by international experts, academics and researchers in the respective research areas.

Biofuel and Biorefinery Technologies will appeal to researchers and post-graduates in the fields of biofuels & bioenergy technology and applications, offering not only an overview of these specific fields of research, but also a wealth of detailed information.The ideas and technologies presented in this book series contribute to the UN Sustainable Development Goal 7: *Affordable and Clean Energy.*

Satinder Kaur Brar •
Carlos Saul Osorio Gonzalez •
Carlos Riccardo Soccol • Rahul Saini
Editors

The Microbiology of the Drop-in Biofuel Production

Springer

Editors
Satinder Kaur Brar
Department of Civil Engineering
Lassonde School of Engineering - York University
Toronto, ON, Canada

Carlos Saul Osorio Gonzalez
Department of Civil Engineering
Lassonde School of Engineering - York University
Toronto, ON, Canada

Carlos Riccardo Soccol
Department of Bioprocess Engineering and Biotechnology
Federal University of Paraná
Curitiba, Paraná, Brazil

Rahul Saini
Department of Civil Engineering
Lassonde School of Engineering - York University
Toronto, ON, Canada

ISSN 2363-7609 ISSN 2363-7617 (electronic)
Biofuel and Biorefinery Technologies
ISBN 978-3-031-61636-5 ISBN 978-3-031-61637-2 (eBook)
https://doi.org/10.1007/978-3-031-61637-2

This Springer imprint is published by the registered company Springer Nature Switzerland AG
The registered company address is: Gewerbestrasse 11, 6330 Cham, Switzerland

To the research community working collaboratively on sustainable development goals.

Preface

In the pursuit of sustainable energy sources, the exploration of green and alternative fuels has gained wide attention. One of these examples is drop-in biofuels, which have emerged as a potential solution not only to fossil fuel-based energy sources but also to conventional biofuels such as biodiesel and bioethanol. This is so because, along with being renewable as well as ability to be produced through microbial systems, drop-in biofuels do not require blending and can be used directly in existing vehicle engines. In this sense, the book, "The Microbiology of Drop-in Biofuel Production" delves into the intricate world of microbial-driven processes that enable the production of drop-in biofuels, examining opportunities, challenges, and the sustainable prospects of this innovative approach.

The book is structured into sixteen comprehensive chapters, each contributing with a unique perspective to the overarching theme of drop-in biofuel production. In the initial chapters, readers will be introduced to the fundamentals and breakdown of several chapters in "Chapter 1: Biofuels: Opportunities, Risks, and Sustainability," specifically extrapolating toward drop-in biofuel. In addition, this chapter provides a foundational understanding of the challenges, importance, and sustainability aspects of drop-in biofuels.

As the book progresses, the focus narrows down to specific biochemical routes for drop-in biofuel production. "Chapter 2: The Current Prospect of Drop-in Biofuel Production Through Biochemical Route" offers an in-depth analysis of the current state-of-the-art on drop-in biofuel production, paving the way for subsequent chapters to explore the diverse feedstocks utilized in the process. "Chapter 3: Renewable Residues as Feedstocks for Drop-in Biofuel Production" highlights the potential of utilizing renewable residues, while "Chapter 4: Microalgal-Based Drop-in Biofuel Production" and "Chapter 5: Yeast-Based Drop-in Biofuel Production" delve deeper into the microbial-based drop-in biofuel synthesis. Similarly, Chapters 6 and 7 extend the exploration to bacterial and fungal contributions, elucidating "Chapter 6: Bacterial-Based Drop-in Biofuel Production" and "Chapter 7: The Role of Fungi in Drop-in Biofuel Production." The following chapters provide information about specialized areas such as sustainable aviation fuel production, microbial isoprenoids, and microbial alkane/alkene production (Chaps. 8, 9, 10), offering a nuanced understanding of the versatility of microbial involvement in drop-in biofuel production.

Addressing the practical aspects, "Chapter 11: Downstream Processing of Drop-in Biofuels" provides crucial insights into biofuel synthesis steps, while "Chapter 12: Recent Advances in Genetic and Metabolic Engineering for Drop-in Biofuel Production" explores the cutting-edge technologies shaping the future of microbial biofuel engineering.

Moreover, environmental considerations are thoroughly examined in "Chapter 13: Life Cycle Analysis of Drop-in Biofuels Production," providing a comprehensive view of the ecological footprint associated with drop-in biofuel processes using different case studies. Lastly, the book concludes with a look toward the future, exploring current and future legislation in drop-in biofuel production (Chap. 14) and projecting "Chapter 15: The Future of Microbial-Based Drop-in Biofuel Commercialization."

This book thus aims to serve as a comprehensive resource for researchers, academics, and industry professionals seeking a deeper understanding of the microbiological particulars involved in drop-in biofuel production. By bringing together diverse standpoints and the latest advancements in the field, the editors hope to contribute to the collective knowledge that will drive the sustainable evolution of biofuels while endorsing renewable biorefineries.

Key Features:

- Covers processes and technologies to produce drop-in biofuel.
- Explores different types of feedstocks for drop-in biofuel production.
- Provides role of different microbial species in drop-in biofuel production.
- Presents the different types of drop-in biofuels and their applications from motor vehicles to aviation fuel.
- Provides policies and regulations aspect of drop-in biofuel.

Toronto, ON, Canada
Curitiba, Paraná, Brazil

Satinder Kaur Brar
Carlos Saul Osorio Gonzalez
Carlos Riccardo Soccol
Rahul Saini

Acknowledgments

We acknowledge the relentless support of all students and researchers who contributed chapters in the book and made the content cohesive and appealing to the readers.

Financial support from Natural Sciences and Engineering Research Council of Canada Collaborative Research and Training Experience Program (CREATE, CREAT/543238-2020 "Training in Applied Biotechnology for Environmental Sustainability (TABES)"), Discovery program (RGPIN-2020-06067, RGPIN-2021-03628), and James and Joanne Love Chair in Environmental Engineering are acknowledged.

Contents

Contributors

Marcela M. Albuquerque Department of Bioprocess Engineering and Biotechnology, Federal University of Paraná, Curitiba, Brazil

Kerthika Devi Athiyappan National Institute of Technology Rourkela, Rourkela, India

Bapan Banik Drug Delivery Research Laboratory, Department of Pharmaceutical Sciences, Dibrugarh University, Dibrugarh, India

Rafael N. Barros Department of Bioprocess Engineering and Biotechnology, Federal University of Paraná, Curitiba, Brazil

Gaurav Bharadwaj Department of Civil Engineering, Lassonde School of Engineering, York University, Toronto, ON, Canada

Gustavo Amaro Bittencourt Department of Bioprocess Engineering and Biotechnology, Federal University of Paraná, Curitiba, Brazil

Satinder Kaur Brar Department of Civil Engineering, Lassonde School of Engineering, North York, ON, Canada

Walter José Martinez Burgos Department of Bioprocess Engineering and Biotechnology, Federal University of Paraná, Curitiba, Brazil

Júlio C. de Carvalho Department of Bioprocess Engineering and Biotechnology, Federal University of Paraná, Curitiba, Brazil

Ishani Chakrabartty Learning and Development Solutions, Indegene Pvt. Ltd., Manyata Embassy Business Park, Bengaluru, India

Frank William Rowley Chaplen Department of Biological and Ecological Engineering, Oregon State University, Corvallis, OR, USA

Naorem Bela Devi Department of Biosciences and Bioengineering, Indian Institute of Technology Guwahati, Guwahati, India

Luis Fernando Flores Escuela de Producción, Tecnología y Medio Ambiente, Universidad Nacional de Río Negro, Viedma, Argentina

Qidou Gao College of Enology, Northwest A&F University, Xianyang, People's Republic of China

Khalid Rehman Hakeem Department of Biological Sciences, Faculty of Science, King AbdulAziz University, Jeddah, Saudi Arabia

Sharareh Harirchi Swedish Centre for Resource Recovery, University of Borås, Borås, Sweden
Department of Biotechnology, Iranian Research Organization for Science and Technology, Tehran, Iran

Alonso G. Hernández-Mendoza Instituto de Biotecnología y Ecología Aplicada, Universidad Veracruzana, Xalapa, Veracruz, Mexico

Leonardo Wedderhoff Herrmann Department of Bioprocess Engineering and Biotechnology, Federal University of Paraná, Curitiba, Brazil

Susan Grace Karp Department of Bioprocess Engineering and Biotechnology, Federal University of Paraná, Curitiba, Brazil

Guneet Kaur School of Engineering, University of Guelph, Guelph, ON, Canada

Mandana Lak Department of Microbiology, Faculty of Biological Sciences, Alzahra University, Tehran, Iran

Luiz Alberto Junior Letti Department of Bioprocess Engineering and Biotechnology, Federal University of Paraná, Curitiba, Brazil

Chandana Malakar Department of Animal Biotechnology, College of Veterinary Sciences, Assam Agricultural University, Guwahati, India

Maria Clara Manzoki Department of Bioprocess Engineering and Biotechnology, Federal University of Paraná, Curitiba, Brazil

Ariel Marchlewicz Faculty of Natural Science, Institute of Biology, Biotechnology and Environmental Protection, University of Silesia in Katowice, Katowice, Poland
Faculty of Mathematics and Science, Department of Biological and Environmental Science, University of Jyväskylä, Jyväskylä, Finland

Ricardo Hernández Martínez CONAHCYT- Colegio de Postgraduados Campus Córdoba, Mpio. de Amatlán de los Reyes, Veracruz, Mexico

Sergio Martínez-Hernández Instituto de Biotecnología y Ecología Aplicada, Universidad Veracruzana, Xalapa, Veracruz, Mexico

Adriane B. P. Medeiros Department of Bioprocess Engineering and Biotechnology, Federal University of Paraná, Curitiba, Brazil

Yugal Kishore Mohata Department of Applied Biology, School of Biological Sciences, University of Science and Technology Meghalaya, Ri-Bhoi, India

Makary Nasser School of Engineering, University of Guelph, Guelph, ON, Canada

Seyed Ali Nojoumi Microbiology Research Center, Pasteur Institute of Iran, Tehran, Iran
Department of Mycobacteriology and Pulmonary Research, Pasteur Institute of Iran, Tehran, Iran

Amilia Nongbet Department of Botany, School of Biological Sciences, University of Science and Technology Meghalaya, Ri-Bhoi, India

Priscilla Zwiercheczewski de Oliveira Department of Bioprocess Engineering and Biotechnology, Federal University of Paraná, Curitiba, Brazil

Carlos S. Osorio-Gonzalez Department of Civil Engineering, Lassonde School of Engineering, York University, Toronto, ON, Canada

Kannan Pakshirajan Department of Biosciences and Bioengineering, Indian Institute of Technology Guwahati, Guwahati, India

Balasubramanian Paramasivan National Institute of Technology Rourkela, Rourkela, India

Bridget Signe Price School of Chemical, Biological, and Environmental Engineering, Oregon State University, Corvallis, OR, USA

Mohadasseh Ramezani Microorganisms Bank, Iranian Biological Resource Centre (IBRC), ACECR, Tehran, Iran

Rahul Saini Department of Civil Engineering, Lassonde School of Engineering, York University, Toronto, ON, Canada

Saurabh Jyoti Sarma Department of Biotechnology, School of Engineering and Applied Sciences, Bennett University, Greater Noida, India

Thamarys Scapini Department of Bioprocess Engineering and Biotechnology, Federal University of Paraná, Curitiba, Brazil

Suparna Sen Department of Biosciences and Bioengineering, Indian Institute of Technology Guwahati, Guwahati, India

Malvika Sharma School of Engineering, University of Guelph, Guelph, ON, Canada

Sumit Sharma Department of Biotechnology, School of Engineering and Applied Sciences, Bennett University, Greater Noida, India

Shikha Singh Department of Biotechnology, School of Engineering and Applied Sciences, Bennett University, Greater Noida, India

Carlos Ricardo Soccol Department of Bioprocess Engineering and Biotechnology, Federal University of Paraná, Curitiba, Brazil

Luciana Porto de Souza Vandenberghe Department of Bioprocess Engineering and Biotechnology, Federal University of Paraná, Curitiba, Brazil

Chloe Tan School of Engineering, University of Guelph, Guelph, ON, Canada

Bikash Ranjan Tiwari Institut national de la recherche scientifique - Centre Eau Terre Environnement, Université du Québec, Quebec City, QC, Canada

Lachi Wankhede Department of Civil Engineering, Lassonde School of Engineering, York University, Toronto, ON, Canada

Adenise L. Woiciechowski Department of Bioprocess Engineering and Biotechnology, Federal University of Paraná, Curitiba, Brazil

Xiaobing Yang College of Enology, Northwest A&F University, Xianyang, People's Republic of China

Xiaochun Zheng College of Enology, Northwest A&F University, Xianyang, People's Republic of China

List of Abbreviations

AAR	Acyl-ACP reductase
ACCT	Acetyl-CoA C-acetyltransferase
ABE	Acetone Butanol Ethanol
ABNT	Brazilian National Standards Organization
AcAc-CoA	Acetoacetyl-CoA
ACP	Acyl Carrier Protein
ADH	Alcohol-dependent hemiterpene
ADO	Fatty aldehyde deformylating oxygenases
AibAB	Glutaconyl-CoA decarboxylase
ANP	Brazilian National Agency for Petroleum, Natural Gas and Biofuels
ANSI	American National Standards Institute
AR	Aldehyde reductase
ASTM	American Society for Testing and Materials
AT	Acyltransferase
ATF1	Alcohol acetyltransferase
ATJ	Alcohol-to-Jet
ATP	Adenosine triphosphate
BCAP	Biomass Crop Assistance Program
BioCCS or BECCS	Bioenergy with Carbon Capture Solutions
$BioH_2$	Biohydrogen
BMD	MVA-3, 5-PP decarboxylase
BSA	Bovine serum albumin
CAPEX	Capital expenditures
CAR	Carboxylic acid reductase
cbjALD	acyl-CoA reductase
CDP-ME	4-diphosphocytidyl-2-C-methyl-D-erythritol
CDP-MEP	4-diphosphocytidyl-2-C-methyl-D-erythritol 2-phosphate
CEPA	Canadian Environmental Protection Act
CER1	Aldehyde decarbonylases

CK	Choline kinase
CMK	CDP-ME kinase
CNPE	Brazilian National Council of Energy Policy
CO2	Carbon dioxide
CoA	Coenzyme A
DAG	Diacylglycerol
DCW	Dry cell weight
DES	Deep eutectic solvents
DGAT	Diacylglycerol acyltransferase
DH	Dehydratase
DHAP	Dihydroxyacetone phosphate
DMA	Dimethylallyl monophosphate
DMAA	Dimethylallyl alcohol
DMAP	Dimethylallyl pyrophosphate
DMAPP	Dimethylallyl diphosphate
DX	1-deoxy-D-xylulose
DX	DXP synthase
DXP	1-deoxy-D-xylulose-5-phosphate
DXR	DXP reductase
DXR	DXP reductoisomerase
DXS	1-deoxy-D- xylulose-5-phosphate synthase
EBP	Ethanol Blended Petrol
ECCC	Environment and Climate Change Canada
EDTA	Ethylenediaminetetraacetic acid
EIZS	Epi-isozizaene synthase
ER	Enoylreductase
ERG10	Acetoacetyl-CoA thiolase
ERG12	Mevalonate kinase
ERG13	Hydroxymethylglutaryl-CoA synthase
ERG19	Mevalonate 5-pyrophosphate decarboxylase
ERG20	Farnesyl diphosphate synthase
ERG8	Mevalonate 5-phosphate kinase
FA	Fatty acid
FAEE	Fatty acid ethyl ester
FAME	Fatty acid methyl esters
FAR	Fatty acyl-ACP/CoA reductase
FFA	Free fatty acids
FFVs	Flex-fuel vehicles
FPP	Farnesyl pyrophosphate
FPP	Farnesyl diphosphate
FPPS	FPP synthase
FQD	Fuel Quality Directive
FS	Farnesene Synthase
FSA	Fructose-6-phosphate aldolase

FucA	Fuculose-l-phosphate aldolase
FucI	Fucose isomerase
FucK	Fuculokinase
G3P	Glyceraldehyde-3-phosphate
GA	Glycolaldehyde
GA-HA	Glycolaldehyde-Hydroxyacetone
GAP	Glyceraldehyde 3-phosphate
GESs	Geraniol synthases
GFPP	Geranylfarnesyl pyrophosphate synthase
GFT	Gasification with consequent Fischer–Tropsch
GGPP	Geranylgeranyl diphosphate
GGPPS	Geranylgeranyl pyrophosphate synthase
GHG	Greenhouse gases
GHG	Greenhouse gas
GMO	Genetically Modified Organisms
GPAT	Glycerol 3-phosphate acyltransferase
GPP	Geranyl diphosphate
GPPS	Geranyl pyrophosphate synthase
GPPSs	GPP synthases
HA	Hydroxyacetone
H/C	Hydrogen/Carbon
H_2O_2	Hydrogen peroxide
HDR	HMB-PP reductase
HDRD	Hydrogenation-derived renewable diesel
HDS	HMB-PP synthase
HEFA	Hydrotreated/hydroprocessed ester and fatty acid
HHV	Higher heating value
HMB-PP	4-hydroxy-3-methyl-butenyl 1-diphosphate
HMG-CoA	3-hydroxy-3-methylglutaryl-CoA
HMGR	Hydroxy-methyl-glutaryl-CoA reductase
HMGR	HMG-CoA reductase
HMGS	HMG-CoA synthase
HPOs	Hydro-treated pyrolysis oils
HRT	Hydraulic retention time
HTL	Hydrothermal liquefaction
HVO	Hydrotreated/hydrogenated vegetable oil
IDI	isopentenyl-diphosphate isomerase
IEA	International Energy Agency
ILUC	Indirect land-use change
IP	Isopentenyl monophosphate
IPA	Isoprenoid alcohol
IPK	Isopentenyl phosphate kinase
IPP	Isopentenyl pyrophosphate
ISO	International Standards Organization

IspD	2-C-methyl-D-erythritol 4-phosphate cytidylyltransferase
IspE	4-diphosphocytidyl-2-C-methyl-D-erythritol kinase
IspF	2-C-methyl-D-erythritol 2,4-cyclodiphosphate synthase
IspG	HMB-PP synthase
IspH	HMB-PP reductase
IUP	Isopentenol utilization pathway
JP5	Jet Propellant 5
KR	Ketoreductase
KS	Ketosynthase
LCFS	Low-Carbon Fuel Standard
LiuC	Enoyl-CoA hydratase
LPA	Lysophosphatidic acid
LPAT	Lysophosphatidic acid acyltransferase
LS	Limonene synthase
M3K	Mevalonate 3-kinase
MAPA	Brazilian Ministry of Agriculture, Cattle and Supplying
MB	3-methyl-2-butenal
MB-CoA	3-methyl-2-butenoyl-CoA
MCT	MEP cytidylyltransferase
MDH	MVA5P dehydratase
MDS	MEcPP synthase
MEcPP	2-C-methyl-D-erythritol-2,4-cyclodiphosphate
MEP	2-C-methyl-erythritol 4-phosphate
MG	Methylglyoxal
MG-CoA	3-methylglutaconyl-CoA
MGS	MG synthase
MOF	Metalloorganic framework
MPD	Phosphomevalonate decarboxylase
MrTPS2	α-isocomene synthase
MVA	Mevalonic acid
MVA-3, 5-PP	3, 5-bisphosphomevalonate
MVA-3-P	MVA-3-phosphate
MVA-5-P	MVA-5-phosphate
MVA-PP	MVA-5-diphosphate
MVD	Diphosphomevalonate decarboxylase
MVK	MVA kinase
NADH	Nicotinamide adenine dinucleotide
NADHP	Nicotinamide adenine dinucleotide phosphate
NOx	Nitrogen oxides
NPPS	Neryl pyrophosphate synthase
NPQ	Non-photochemical quenching
NREL	United States National Renewable Energy Laboratory
OleTje	Cytochrome P450 fatty acid decarboxylase
Ols	Olefins

OPEX	Operational expenditures
OYE2	NADPH oxidoreductase
PA	Phosphatidic acid
PAP	Phosphatidic acid phosphatase
PBR	Photobioreactor
PDAT	Phospholipid:diacylglycerol acyl transferase
PL	Phospholipid
PSII	Photosystem II
PDH	Pyruvate dehydrogenase
PDHC	Pyruvate dehydrogenase complex
pentS	Pentalenene synthase
PhoN	Nonspecific acid phosphatase
PKSs	polyketide synthases
PMVK	Phosphomevalonate kinase
PPM	Parts per million
PSA	Pressure swing absorption
RED	Renewable Energy Directive
REN21	Renewable Energy Policy Network for the 21st Century
RFS	Renewable Fuel standard
RSB	Roundtable on Sustainable Biomaterials
RTFO	Renewable Transport Fuel Obligation
SAF	Sustainable Aviation Fuel
SDG	Sustainable Development Goals
SOx	Sulfur oxides
SPME	Solid phase microextraction
srRNA	Small regulatory RNA
TAG	Tri-acyl-glycerides
tAHMP	Trans-anhydromevalonate 5-phosphate
tAPD	tAHMP decarboxylase
Te	Heterologous fatty acyl-ACP/CoA thioesterase
ThiM	Hydroxyethylthiazole kinase
UndA	Fatty acid oxidase/decarboxylase
UndB	Fatty acid oxidase/decarboxylase
UNEP	United Nations Environment Programme
UPO	Unspecific peroxygenase
USDA	U.S. The Department of Agriculture
VFA	Volatile fatty acid
WTO	World Trade Organization
WWF	World Wildlife Fund
XK	xylulose kinase
YahK	Alcohol dehydrogenase
αDox	Fatty acid α-dioxygenase

Chapter 1
Biofuels: Opportunities, Challenges, Risks, and Sustainability

Amilia Nongbet, Yugal Kishore Mohata, Khalid Rehman Hakeem, and Ishani Chakrabartty

Abstract The global energy and fuel crisis is not unheard of; with fossil fuel depleting at an exponential rate, along with their deleterious effects on human health and environment, there is an urgent need of a suitable alternative. Biofuels are viewed as a possible solution to solve issues relating to greenhouse gas emissions, global warming, and climate change. However, the establishment of biofuels as the possible fuel source of the future can have a negative impact on biodiversity and natural ecosystems, coupled with habitat loss and fragmentation. With the growing awareness of biofuels worldwide, policy makers are designing new laws and strategies, particularly in industrialized nations, for the sustainable use of biofuels. Though numerous life-cycle assessment (LCA) studies have focused on the environmental impacts of biofuel use, the results obtained are often not satisfactory. The aim of the present chapter is to consolidate and clarify the different challenges, risks as well as opportunities that are associated with the sustainability of biofuels. In this chapter, we have highlighted the fact that for the establishment of biofuel as a sustainable fuel source, several important parameters work in collaboration, some of which have been discussed here, and need extensive study in this regard. The role of nanoscience and technology in the biofuel industry has also been briefly highlighted in this chapter.

A. Nongbet
Department of Botany, School of Biological Sciences, University of Science and Technology Meghalaya, Ri-Bhoi, Meghalaya, India

Y. K. Mohata
Department of Applied Biology, School of Biological Sciences, University of Science and Technology Meghalaya, Ri-Bhoi, Meghalaya, India

K. R. Hakeem
Department of Biological Sciences, Faculty of Science, King AbdulAziz University, Jeddah, Saudi Arabia

I. Chakrabartty (✉)
Learning and Development Solutions, Indegene Pvt. Ltd., Manyata Embassy Business Park, Bengaluru, India

S. K. Brar et al. (eds.), *The Microbiology of the Drop-in Biofuel Production*, Biofuel and Biorefinery Technologies 15, https://doi.org/10.1007/978-3-031-61637-2_1

Keywords Biofuels · Environmental impacts · Fossil fuels · Life-cycle assessment · Risks · Sustainability

1.1 Introduction

Fossil fuels are one of the major sources of energy globally. However, there is an energy crisis worldwide owing to the depletion of wealth of fossil fuel. In addition, the greenhouse gas (GHG) emission from transport and other sectors utilizing fossil fuels is increasing at a very fast pace; 15% of the world's GHG emission and 23% of total CO_2 emission is by transport. Through many years of progress, man has utilized many forms of energy ranging from firewood to nuclear energy, from coal to petroleum (Sims et al. 2014). Today, due to the raise in the human population and the exhaustion of fossil fuels with the increase in pollution levels by vehicles, people are worried about global warming (Liang et al. 2016). Hence, renewable sources of energy need to replace conventional fuel sources. Biomass as a resource of energy can play a pivotal role for sustainable development and strengthening energy security. Bioenergy simply refers to energy derived from biomass and biological resources that include thermal energy, electrical energy, and biofuels, i.e. liquid fuels from biomass, particularly ethanol and biodiesel. Biofuels are generally used for transport, though they may also be used for production of electricity (Cowie et al. 2016). Fuels produced from harvested biomass (biofuels) can either be solid, gas or liquid. Anaerobic digestion of biomass like wood pellets, forest waste, etc. are some of the solid biofuel feedstocks for energy generation used primarily for electricity and thermal energy generation, whereas liquid biofuels are primarily used directly in the transport sector as drop-in fuels (Kazamia and Smith 2014). Biofuels are part of a much wider systems, including energy, agriculture, and forestry and do not exist in isolation (Jeswani et al. 2020). Many nations have formulated policies to blend a certain proportion of their transport fuels with biofuels (International Energy Agency (IEA) 2014). Biofuels production is governed mainly by policy measures, investment subsidies, tax exemptions, and their compulsory combination with fossil fuels, as in European Union (EU) and United States of America (USA) (Public Law No. 109-58 2005; European Parliament 2008). Biofuels have become extremely popular as a source of renewable and sustainable energy and have emerged as a prospective substitute to gasoline and diesel transportation fuels derived from petroleum (Cowie et al. 2016). Brazilian bioethanol production had been extremely successful when marketed domestically and even exported. After that, several Latin American countries have set up different biofuel programs to establish an alternative market. The biofuel production has increased with the growing demand for substitute energy sources. However, this demand may have some depleting effect on biodiversity, particularly for exploitation of biofuel crops, but it still remains quite unclear (Núñez-Regueiro et al. 2021). The production and consumption of biofuels increased rapidly worldwide, in an effort to reduce greenhouse gas emissions, expand transportation fuels, promote renewable energy, and generate or retain employment, especially in rural areas and developing countries. The future development of biofuels seems bright, but it will be important to develop and apply biofuels sustainability criteria at the earliest and in a dependable way worldwide (Núñez-Regueiro et al. 2021).

Biomass cultivation would also provide huge opportunities, especially in the development of rural areas and promote diversity in agriculture, and energy obtained from the biomass will be more acceptable to the people as compared to other fuels. In the total estimate of renewable energy generated, biofuel alone contributed 12.83% of the lot (International Energy Agency (IEA) 2014). The significant parameters of bioenergy extraction are the higher conversion efficiencies and lower costs. Globally, India ranks fourth in the emission of GHGs; its transportation domain contributes 13% of the CO_2 emissions. Surprisingly, India is also the fourth-largest importer as well as consumer of crude oil and products of petroleum after USA, China, and Japan (INCCA 2012; International Energy Agency (IEA) 2014; Aslan et al. 2017). After coal, India's second-largest energy source is oil that amounts for around 30.5% of primary consumption (BP 2013). Therefore, alternative energy resources need to be developed effectively and efficiently to substitute the use of fossil oils. Biofuels have significant advantages for national energy security, mitigation of GHGs, generation of employment, and rural development (Purohit and Dhar 2015). Multiple issues like social, environmental, and economic issues related to energy security and management can be addressed by biomass energy (BP 2013). The power developed from the biomass demands more transportation facilities—this is a serious drawback for the biofuel extraction strategies that involve the emission of various gases from the vehicles for fetching feedstock (Jeswani et al. 2020). Many drawbacks are associated with agriculture-derived biofuels such as ecological damages, lower efficiency energy systems in comparison with the conventional fuels (Jeswani et al. 2020; Cowie et al. 2016). The exploitation of fossil resources as fuel has increased due to the increasing demands with fossil resources on the verge of extinction causing serious issues relating to environment and economy (Zeppini and van den Bergh 2020).

The present chapter stresses the need for biofuel which can be a sustainable source of fuel today and in the near future (Fig. 1.1). In the first section, the different methods of biofuel production have been discussed in detail. The following section highlights the environmental impact of biofuels that has been listed under different sub-headings. In the subsequent section, the life-cycle assessment of biofuels has been done extensively. The risks, challenges, and opportunities that exist in the use of biofuels have been highlighted in the following section. Lastly, concluding remarks on the topic have been provided.

1.2 Alternative Methods of Biofuel Production

Multiple pathways are present through which biofuels may be produced. The most common pathways being (i) fermentation to produce ethanol, (ii) trans-esterification to produce biodiesel—both starting with carbohydrates, in the form of starch or

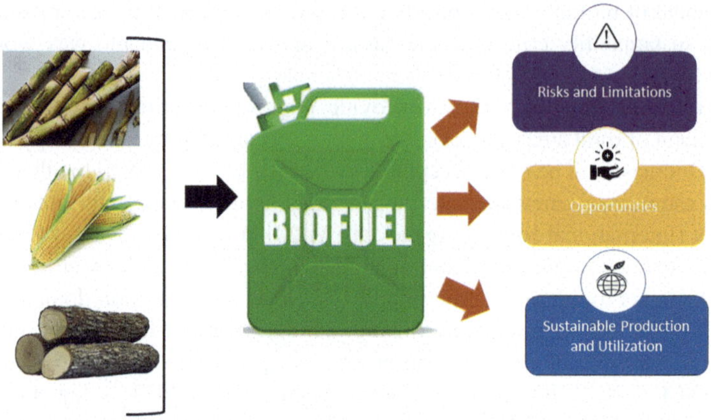

Fig. 1.1 Graphical representation of the futuristic utilization of biofuels

sugar and hydrolyzed into simple sugars. The feedstock for starch crops comprises cereals like maize, wheat, etc. and root crops, e.g., cassava and sugar crops such as sugarcane and sugar beet. After sugar fermentation using yeast to produce ethyl alcohol, the residues may be used as animal feed (Cowie et al. 2016). Second-generation ethanol processes use lingo-cellulosic biomass such as maize stover, perennial grasses such as miscanthus and switchgrass, or short rotation woody crops like willow, poplar. The lignin and cellulose carbohydrates complexes should be subjected to hydrolysis (chemical or enzymatic) to release the sugars for fermentation.

Biodiesel is produced through trans-esterification of vegetable oil (crops like canola, soybean, and sunflower or palm oil) or animal fat (tallow), and the oil reacts with methanol to produce fatty acid methyl ester (FAME). A less common method for producing biofuel is treatment of vegetable oil and tallow into a renewable diesel in the presence of water.

Another way of producing biofuels is from algae which are of great interest. Certain species of algae that grow rapidly can synthesize high amount of lipids that can be readily converted to biodiesel. Two systems for algae production are open raceway ponds used extensively for pharmaceuticals and food additives while the other is known as photo-bioreactor (a closed system) (Cowie et al. 2016). In the photo-bioreactor, algae are supplied with light, nutrients, and CO_2 through plastic or glass tubing. The major drawback for algal biofuel is the high amount of energy required to "harvest" the algae from the water in which they are grown.

The production of biofuels from other sources such as lingo-cellulosic biomass, algal biomass, vegetable oils, and solid municipal waste through hydro-processing (HPR) is still in pilot stage (Furimsky 2013).

1.3 Biofuels: Environmental Impact

The development of life-cycle assessment (LCA) is necessary to comprehend the environmental effects of a product and process to design and implement different policies, in this case relating to biofuels. LCA was developed to assess climate change impacts, water quality to human toxicity, air quality, and a wide range of environmental effects. Multiple LCA studies on biofuels focus only on GHG emissions but sometimes on energy balances also.

1.3.1 Reduction of GHG Emissions

Scientific insights that consider the importance of environment as well as production-related costs further triggered the European Union to offer different regulations to make the biofuels sustainable. These criteria enforce certain limitations to lower GHG emissions in comparison with fossil fuels and prevention of habitat loss and fragmentation, and thus alter biodiversity (Furimsky 2013). We need to extensively diminish GHG emission which primarily comes from the flaming fossil fuels, as it is expected to enlarge by 50%; by 2050, this will be the biggest contributors to climate change (63% in 2000 to 235% by 2050). Many international energy agencies have introduced renewable energy as one of the possible solutions to reduce GHG emission and ensure a stable climate worldwide (Subramaniam et al. 2020).

1.3.2 Use of Land for Biofuels

The amount of land provided to the cultivation of biofuel crops is less than 1% of agriculture land globally as there are concerns over the use of cultivable land for growing feedstock. Critics emphasize that biofuel crops threaten endangered wildlife and cause deforestation besides emission of greenhouse gases. The requirement for land will be increased for growing extra cropland and grazing land, biofuel crops, urban expansion, expansion of industrial forestry (Cowie et al. 2016). The overall agricultural land area is bigger as compared to the bioenergy crops. Further expansion of biofuels may displace production of food for human and animal feed, wood products, and fiber which will lead to a greater risk of indirect land use change. And the demand for land expansion will increase as the land use for growing urban expansion, industrial forestry, biofuel crops, grazing land, protected areas, etc. with the declining productivity will result in deforestation.

1.3.3 Indirect Land Use Change

Indirect Land Use Change (ILUC) causes destruction of biodiversity and threatens the society with loss of food and displaces the poor rural communities (Cowie et al. 2016). For example, cultivation of corn for ethanol in the USA displaces production of animal feed, soybean cultivation in the pastures of Brazil for animal feed generation in turn causes deforestation in the Amazon due to over-grazing.

However, the concept of biofuel-related ILUC is unstable. Their estimates are based on a number of computational models, such as computable general equilibrium models (for example, Global Trade. Analysis Project, GTAP; Modelling International Relationships in Applied General Equilibrium, MIRAGE). These models check the market value by checking the changes in response to demand and supply. But they are criticized for their lack of assurance, and for disappointment to reproduce reality. The estimations for ILUC due to biofuel expansion have been found to lack consistency with the rates of deforestation.

1.3.4 Sustainability of Biofuels

The recent sustainability debate, which was initiated and dominated by stakeholders from developed countries (mainly from Europe), has identified several negative impact of biofuels on the environment and society at large. Such negative impacts include deforestation of the Amazon Forest, degradation of the Cerrado, environmental and social impact of sugarcane harvest, displacement of indigenous communities, and the food-fuel conflict. The Latin American government have undertaken many initiatives for sustainability (established by industry stakeholders) to minimize these negative impacts (Jannsen and Rutz 2011). For sustainable use of resources and reduction of global warming, biofuels have gained massive attention. Harvest of a huge number of additional crops for production of biofuels has a variety of environmental impacts like reduction of the total GHG emissions that typically account for carbon footprints. For assessment, LCA is used, which is a standardized method to assess GHG emissions, land and water use, as well as toxic emissions (Pfister and Scherer 2015). Currently most biofuels are produced from traditional crops that serve as food sources of starch, sugar, or oil and are known as the "first-generation" biofuels. "Second-generation" or "advanced" biofuels utilize lingo-cellulosic crops such as grasses as feedstock and are now focused by the Research and Development. While supplies of the fossil fuels are depleting at a fast pace in many regions, there is no dearth of resources for liquid fuels: deep sea petroleum; shale oil/tar sands, coal to liquid, CNG and gas conversion options are readily available. Though there is a rapid increasing supply from the sources, but concerns over the high environmental costs in terms of carbon footprint, hydrological and biodiversity impacts emitted from these sources are expanding rapidly (Cowie et al. 2016). The massive production of biofuel is expected to increase rapidly in order to meet the growing demands of the future for transport and aviation. Concern over environmental impacts of some biofuel systems has also been raised, whether produced in a first- or second-generation process because the basic "life cycle" of biofuel production involves growing or collecting biomass feedstock, processing feedstock into a liquid fuel product, distributing and combusting fuel (Cowie et al. 2016). The sustainability concept of biofuels is not yet thoroughly fulfilled. The first-generation biofuels failed due to conflict with food and feed to fulfill sustainability. Second-generation biofuels do not satisfy the food-energy-water nexus while the third-generation biofuels affect the nitrogen cycle. The fourth generation are associated with expensive and energy-demanding processes (Abdullah et al. 2019).

1.4 Risks, Challenges, and Opportunities of Biofuels: A Push Toward Sustainability

The impact of biofuels on the reduction of global warming is of a great concern and requires a careful evaluation. With the help of standardized tools such as LCA, comprehensive comparative analysis between biofuels and conventional fossil fuels is necessary to ensure that the associated products and the subsequent market value receive the due importance. In addition to climate change, the impact of biofuel on biodiversity, quality of air, water, and soil need to be analyzed (Cowie et al. 2016). Another concern is the risk of uncontrolled growth of biofuel crops (like weeds) that could affect agricultural production and/or conservation areas. To decrease harmful effects of habitat fragmentation of natural crops and to enhance biodiversity value, plantations of perennial biofuel crops that are native to a particular region can be carried out to design ecological corridors. It is a sad reality that biofuels and cultivation of these crops are posing a big threat to the agriculture/environment nexus (Cowie et al. 2016).

There are three perspectives that raise questions on climate change mitigation value of biofuels—firstly, the imbalance created between the emissions generated to the emissions saved; secondly, there is a claim that importance of emission is nullified by the emissions due to indirect land use, and finally, the most important question is the "rebound effect," i.e. extent of decrease of fossil fuels by the use of biofuels (Cowie et al. 2016).

1.4.1 Sustainable Biofuel Production

For the production of biofuels, a distinguishing feature between plants and algae is their ability to photosynthesize. The production of biofuels by the use of photosynthetic organisms is inexpensive and practical because atmospheric CO_2 supplies carbon and sunlight serves as an energy source. The main photosynthetic pigments involved in the absorption of light are chlorophyll in plants and bacteriochlorophyll in bacteria. Plant biomass serves as a feedstock for synthesis of bio-alcohol, biodiesel, and bio-hydrogen obtained by fermentation (Slade and Bauen 2013). Under certain conditions photosynthetic equipment serves as a source of bio-hydrogen. Although photosynthesis and chloroplast are similar in plants and algae, biofuel production by algae is much more advantageous than higher plants. Based on the present knowledge, microalgae are an attractive feedstock for biofuels production (Slade and Bauen 2013).

It is clear that biofuels can be a possible resource of "futuristic" energy in comparison with the restricted nature, geopolitical insecurity, and detrimental global effects of fossil fuel energy (Rodionova et al. 2017). Emissions of GHG will decrease when compared to fossil fuels (Verdade et al. 2015). Biofuels should be the solutions of future energy, particularly for long distance vehicular transport where other renewable energy options are inadequate. To endorse specific biofuels, decisions should be made on the basis of extensive studies that have huge impacts on all important environmental and socio-economic aspects to avoid obstinate outcomes (Cowie et al. 2016). Biofuels will definitely not be the only way out for a pure and sustainable environment, but incidentally it contributes to the increase in food production. Therefore, significant development and utilization of biofuels could contribute substantially to increment in food security and maintain the environmental quality (Subramaniam et al. 2020).

1.4.2 Novel Biofuel Alternatives

A variety of strategies can be utilized for obtaining biofuel feedstocks, coupled with minimal risk of ILUC. One way can be by using biomass sources that are wastes, such as used cooking oil and secondary processing residues (for example, sawdust, nut shells) and construction and demolition waste; secondly, increasing intensity of production on cropped areas can lessen land requirements and thirdly biofuel crops may be cultivated on minor or "unused" land (Cowie et al. 2016). A fourth strategy is to decrease the menace of indirect land use change by using residues from cropping or forestry systems, such as corn stover, wheat straw, sugarcane trash, forest slash as these lingo-cellulosic biomass types are left to decay in the field.

With the aid of advancements of bioprocessing and fermentation technology, agricultural wastes can be effectively hydrolyzed and transformed into useful bio-products such as vehicular fuels, removal of harmful salts from water, heavy metals and dyes from aqueous solutions, and electricity production. Multiple methods are practiced worldwide that aim to increase the yield better and improve the conversion of biomass into high yield ethanol, butanol, and other improved alcohols. The growth of the biofuel sector can amplify the cost and requirement for feedstock because of its requirement and utilization in biofuel production besides other factors such as competition for land between non-food crops and food crops, irrigation, and nutrition (Subramaniam et al. 2020). At present the technologies are

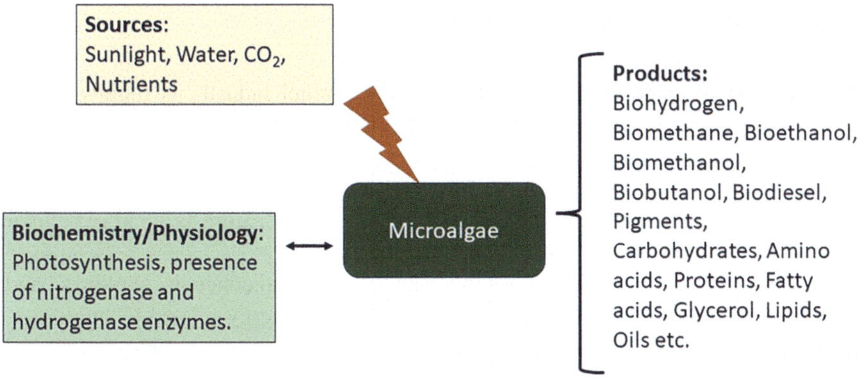

Fig. 1.2 The working mechanism of microalgae in the energy sector

not established enough to prove their competitiveness against fossil fuel, because of high manufacturing costs. However, in the near future, biofuel will be considered as an industry, as it will reduce greenhouse gas emission while providing a politically stable, renewable, and ultimately cost-effective substitute to the fuels usually engaged by countries today (Puri et al. 2012). Biomass will contribute, as estimated, between 15 and 50% of the world's main energy utilization by the year 2050 (Zhang et al. 2014).

Furthermore, microalgae can easily utilize substrates from different waste sources like agricultural runoff, municipal, industrial, or agricultural wastewaters that serve as a source of nitrogen, phosphorus, and minor nutrients as well as water (Kour et al. 2019). Depending on the species used and cultivation method, microalgae can manufacture bio-hydrogen, biomethanol, bioethanol, biodiesel (Poudyal et al. 2016), carbohydrates, proteins, or other compounds that are utilized in different industries particularly in the pharma world (Kour et al. 2019). The best thing is that algal-derived biofuel production requires only sunlight, CO_2, and water and produces a large number of renewable energy co-products (Fig. 1.2).

At present for biofuels, it is still challenging and taxing to be economically viable over fossil fuel. It is proposed that biofuel production can be aimed from nitrogen-rich municipal wastewater as feedstock and CO_2-rich fuel gas treatment—this will contribute to the sustainability and cost reduction of biofuels (Zhang et al. 2014). To deal with the energy crisis, the global joint hard work is necessary for transforming biofuels into our present energy system. Broad investigation biofuel production from "green" sources under varied conditions is crucial as they contribute substantially to optimize the biomass of phototrophic organisms. Based on recent advancements in the field of biofuel production, it is believed that the large-scale production of biofuels is a burning issue and attainable to meet our energy requirement in the future (Rodionova et al. 2017).

1.5 Nanotechnology in Biofuel

With the speedy exhaustion of fossil fuels, the biofuel industry is rapidly a new source of renewable energy along with the ability to deal with environmental challenges. Nanotechnology has tremendous potential to increase the efficiency of biofuel industry. It allows one to produce resources with distinctive properties and modify their structures for precise applications. It exists in nature as a part of cellular systems of different organisms including viruses.

Nature applies the knowledge of molecular architectural concepts, exploiting and mimicking to a large extent, to create new and novel energy-related technologies. With its capacity for molecular fabrication, nanotechnology can provide customized nanomaterials for numerous aspects of energy-efficient biofuel production. Through the use of nanotechnology, sage and clean protocols could be made available for various fields such as geothermal, alternative natural gas, coal and nuclear, solar, wind, and hydraulics (Pattarkine and Pattarkine 2012). The addition of fuel additives to enhance fuel blends performance is another growing application of nanotechnology. Their exceptional properties such as the design, reusability, stability, greater surface area to volume ratio, and catalytic activity make them efficient biofuel additives (Arya et al. 2021). The use of nanoparticles like alumina and hollow carbon nanotubes has revealed to improve the combustion characteristics of biodiesel-operated engines resulting in less harmful emissions (Arya et al. 2021). It has also been reported from earlier studies that the addition of nanomaterials in algal culture system enhanced micro-algal growth as well as induced lipid accumulation and improved extraction efficiency (Mandotra et al. 2018). Nanoparticles (NPs) are excellent solutions to the existing challenges on consumption of biomass in terms of their time and energy efficiency and work-related suitability. Many NPs such as metal and metal oxide are being used in improvement of biofuel production. In addition, nanomaterials such as carbon nanotubes, carbon nanofibers, and nanosheets have been found to be cost efficient as well as stable catalysts for enzyme control, thus improving biofuel synthesis (Arya et al. 2021). Innovative nanomaterials like metallic nanoparticles are manufactured for emitting light at a specific wavelength outside of closed photo-bioreactor which enhanced the algal culture (Pattarkine and Pattarkine 2012). The major improvement of using nanoparticles suspension is to facilitate the growth of specific micro-algal species and to avoid photo inhibition by altering the size and concentration of the nanoparticles (Lee et al. 2015). Their study revealed that nano- and microbubbles in an airlift loop bioreactor (ALB) could be used to boost algal biomass production. Finally, prospective areas of biofuel industry where nanotechnology can potentially bring breakthroughs comprise advanced fermentation, rapid catalytic pyrolysis, jet fuels, catalytic conversion of syngas, hydrocarbon bio-refineries, biofuel cell, biochar, carbon capture and storage, gasification, hydrothermal liquefaction, or carbonization, converting and upgrading of ethanol, volatile products, and tars. In biomass-based energy systems, the use of nano-based precision farming technology could optimize crops production for higher yields of biofuels. Moreover, future research should be focused not only on biofuel sources and energy production, but also on energy competence and storage, transportation, transformation, and end-product use of biofuels. Nanotechnology could extensively contribute to these areas to deal with both the engineering and scientific limitations (Nizami and Rehan 2018).

1.6 Conclusion

Biofuels are probably one of the sustainable resources or substitutes of fossil fuel in the near future. As such there is an utmost necessity to conserve bio-resources that can serve as feedstock or raw material for biofuel production. It is true that in comparison with fossil fuels, the environmental impact of biofuels is a bare minimum because of the lower emissions of hydrocarbons, particulate matter, CO, and other poisonous substances. However, particulate matter is of much smaller dimension and remains suspended freely in the air; thus, they can be inhaled easily and have subsequent harmful effects on human health. But there is no denying to the fact that the use of biofuels has more pros than cons on the human health and environment and hence, it can be termed the fuel of the future.

Life-cycle assessment of biofuels includes a wide range of methodical studies and strategies to outline the environmental impacts of biofuel. This tool lies at the interface of science, technology, and policy (Jeswani et al. 2020), and hence it has multifaceted solutions to the most complex questions related to biofuels. Every aspect related to biofuel is extremely fragile in nature, which adds to its complexity. In addition, sustainability issues related to biofuels encompass a wide range of issues like production cost, competition with fossil fuel, employment provisions, rural development, and others. Also, there is a huge chance of fraud and falsification issues when it comes to trusting the source of biofuels; this can have a huge negative socio-economic impact on the long-term sustainability of biofuels. However, nanotechnology can help to change this scenario to some extent.

All said and done, extensive research in biofuels is still necessary in congruence with wide systems like agriculture, ecosystem, energy, in order to understand its sustainability. It is extremely essential to form an integrated view on biofuels in order to develop futuristic policies on its use, applicability, and subsequently sustainability, without any disadvantage in the near future.

References

Abdullah B, Anuar S, Muhammad S et al (2019) Fourth generation biofuel: a review on risks and mitigation strategies. Renew Sust Energ Rev 107:37–50. https://doi.org/10.1016/j.rser.2019.02.018

Arya I, Poona A, Dikshit PK et al (2021) Current trends and future prospects of nanotechnology in biofuel production. Catalysts 11:1308. https://doi.org/10.3390/catal11111308

Aslan O, Altan A, Hacioglu R (2017) The control of blast furnace top gas pressure by using fuzzy PID. In: Fifth International Conference on Advances in Mechanical and Robotics Engineering. AMRE, pp 22–26

BP (2013) BP statistical review of world energy 2013. In: BP. http://www.bp.com/content/dam/bp/pdf/statistical-review/statistical_review_of_world_energy_2013.pdf. Accessed 6 May 2022

Cowie A, Cowie A, Soimakallio S, Brandão M (2016) Environmental risks and opportunities of biofuels. In: The law and policy of biofuels, pp 3–29. https://doi.org/10.4337/9781782544555.00009

European Parliament (2008) Report on the proposal for a directive of the European Parliament and of the council on the promotion of the use of energy from renewable sources (COM(2008) 0019—C6 0046/2008–2008/0016(COD)). Belgium

Furimsky E (2013) Hydroprocessing challenges in biofuels production. Catal Today 217:13–56. https://doi.org/10.1016/j.cattod.2012.11.008

INCCA (2012) India: Greenhouse gas emissions 2007

International Energy Agency (IEA) (2014) World Energy outlook 2014. In: IEA. https://www.iea.org/reports/world-energy-outlook-2014. Accessed 6 May 2022

Jannsen R, Rutz DD (2011) Sustainability of biofuels in Latin America: risks and opportunities. Energy Policy 39(10):5717–5725

Jeswani HK, Chilvers A, Azapagic A (2020) Environmental sustainability of biofuels: a review. Proc R Soc A Math Phys Eng Sci 476:20200351. https://doi.org/10.1098/rspa.2020.0351

Kazamia E, Smith AG (2014) Assessing the environmental sustainability of biofuels. Trends Plant Sci 19:1–13

Kour D, Rana KL, Yadav N et al (2019) Technologies for biofuel production: current development, challenges, and future prospects. In: Rastegari AA, Yadav AN, Gupta A (eds) Prospects of renewable bioprocessing in future energy systems. Springer, pp 1–52

Lee YC, Lee K, Oh YK (2015) Recent nanoparticle engineering advances in microalgal cultivation and harvesting processes of biodiesel production: a review. Bioresour Technol 184:63–72. https://doi.org/10.1016/j.biortech.2014.10.145

Liang H, Ren J, Gao Z et al (2016) Identification of critical success factors for sustainable development of biofuel industry in China based on grey decision-making trial and evaluation laboratory (DEMATEL). J Clean Prod 131:500–508

Mandotra SK, Kumar R, Upadhyay SK, Ramteke PW (2018) Nanotechnology: a new tool for biofuel production. In: Green nanotechnology for biofuel production, pp 17–28

Nizami AS, Rehan M (2018) Towards nanotechnology-based biofuel industry. Biofuel Res J 18: 798–799. https://doi.org/10.18331/BRJ2018.5.2.2

Núñez-Regueiro MM, Siddiqui SF, Fletcher RJ (2021) Effects of bioenergy on biodiversity arising from land-use change and crop type. Conserv Biol 35:77–87

Pattarkine MV, Pattarkine VM (2012) Nanotechnology for algal biofuels. In: The science of algal fuels. Springer, Netherlands, pp 147–163

Pfister S, Scherer L (2015) Uncertainty analysis of the environmental sustainability of biofuels. Energy Sustain Soc 5:1–12. https://doi.org/10.1186/s13705-015-0058-4

Poudyal RS, Tiwari I, Najafpour MM et al (2016) Current insights to enhance hydrogen production by photosynthetic organisms. In: Hydrogen science and engineering: materials, processes, systems and technology, pp 461–487

Public Law No. 109-58 (2005) Energy Policy Act of 2005. In: H.R.6-109th Congr. Congress.gov

Puri M, Abraham RE, Barrow CJ (2012) Biofuel production: prospects, challenges and feedstock in Australia. Renew Sust Energ Rev 16:6022–6031. https://doi.org/10.1016/j.rser.2012.06.025

Purohit P, Dhar S (2015) Promoting low carbon transport in India biofuel roadmap for India

Rodionova MV, Poudyal RS, Tiwari I et al (2017) Biofuel production: challenges and opportunities. Int J Hydrog Energy 42:8450–8461. https://doi.org/10.1016/j.ijhydene.2016.11.125

Sims REH, Creutzig F, Schaffer R et al (2014) Transport. In: Climate change 2014: mitigation of climate change. Part of the Working Group III contribution to the Fifth Assessment Report of the Intergovernmental Panel on Climate Change, pp 599–670

Slade R, Bauen A (2013) Micro-algae cultivation for biofuels: cost, energy balance, environmental impacts and future prospects. Biomass Bioenergy 53:29–38. https://doi.org/10.1016/j.biombioe.2012.12.019

Subramaniam Y, Masron TA, Azman NHN (2020) Biofuels, environmental sustainability, and food security: a review of 51 countries. Energy Res Soc Sci 68:101549. https://doi.org/10.1016/j.erss.2020.101549

Verdade LM, Piña CI, Rosalino LM (2015) Biofuels and biodiversity: challenges and opportunities. Environ Dev 15:64–78. https://doi.org/10.1016/j.envdev.2015.05.003

Zeppini P, van den Bergh JCJM (2020) Global competition dynamics of fossil fuels and renewable energy under climate policies and peak oil: a behavioural model. Energy Policy 136:110907. https://doi.org/10.1016/j.enpol.2019.110907

Zhang X, Rong J, Chen H et al (2014) Current status and outlook in the application of microalgae in biodiesel production and environmental protection. Front Energy Res 2:1–15. https://doi.org/10.3389/fenrg.2014.00032

Chapter 2
The Current Prospect of Drop-in Biofuel Production Through Biochemical Routes

Sergio Martínez-Hernández, Alonso G. Hernández-Mendoza, and Ricardo Hernández Martínez

Abstract The depletion of fossil fuels and its adverse effects on the environment, such as climate change and global warming, have promoted the search for sustainable alternative energies. One alternative is the use of drop-in biofuels, which are defined as liquid bio-hydrocarbons functionally equivalent to petroleum-derived fuels and are fully compatible with the oil infrastructure and the engines and turbines of existing transportation systems. Drop-in biofuels are produced using three main technologies and their interactions: oleochemical, thermochemical, and biochemical, and hybrids incorporate elements of the three technologies. Biochemical technology has been investigated by exploring various biosynthetic pathways in conventional microorganisms such as *Escherichia coli* and *Saccharomyces cerevisiae*, as well as unconventional ones such as cyanobacteria and algae. Currently, five main biosynthetic pathways have been proposed for the generation of drop-in biofuels: the isoprenoid pathway, the fatty acid pathway, the α-keto acid pathway, the non-decarboxylative Claisen condensation pathway, and the polyketides pathway. The natural production of drop-in biofuels through these pathways is still insufficient to be used for commercial purposes, and thus various strategies have been investigated to increase their productivity and yields. These strategies include metabolic and genetic engineering, upregulating pathway enzymes, deleting or downregulating competing pathways, engineering transcriptional factors, increasing the robustness of host cells, enhancing cell viability, and regulating cell populations. In this chapter, the five biochemical routes for generating drop-in biofuels are described, as well as some strategies implemented to increase titers and yields.

Keywords Isoprenoid pathway · Fatty acid pathway · Keto-acid pathway · Claisen pathway · Polyketide pathway

---

S. Martínez-Hernández (✉) · A. G. Hernández-Mendoza
Instituto de Biotecnología y Ecología Aplicada, Universidad Veracruzana, Xalapa, Veracruz, Mexico
e-mail: sermartinez@uv.mx

R. H. Martínez
CONAHCYT- Colegio de Postgraduados Campus Córdoba, Mpio. de Amatlán de los Reyes, Veracruz, Mexico

S. K. Brar et al. (eds.), *The Microbiology of the Drop-in Biofuel Production*, Biofuel and Biorefinery Technologies 15, https://doi.org/10.1007/978-3-031-61637-2_2

2.1 Introduction

Biofuels have become a compelling substitute for petroleum-based fuels, whose reserves are being depleted and their negative impact on the environment, such as increased climate change and greenhouse gas emissions, is becoming increasingly severe. For example, CO_2 emissions into the atmosphere have shown values of 400 ppm; the last time these levels were this high was more than 15 million years ago (Dahman et al. 2019).

Biofuels are considered renewable, sustainable, efficient, and profitable energy sources that produce less greenhouse gas emissions. Biofuels are derived from organic matter such as plants, agricultural crops and their by-products, algae, municipal solid waste, food waste, animal fats, among others. According to the feedstock source, biofuels have been classified as first-, second-, and third generation. First-generation biofuels are derived from agricultural crops such as corn and sugarcane. Second-generation biofuels are derived from lignocellulosic feedstock, while third-generation biofuels come from algal biomass (Mahapatra et al. 2021). According to their physical state, biofuels can also be classified into solid, gaseous, and liquid. Solid biofuels are also known as primary biofuels, which refers to the direct use of wood and other organic materials. Gaseous biofuels are those such as biohydrogen and methane, while liquid biofuels are those such as biodiesel, biobutanol, and bioethanol.

At present, bioethanol and biodiesel rank as the most abundantly produced biofuels. Data indicate that bioethanol production increased 67% from 2008 to 2018, from 67 to 110.4 billion liters, while biodiesel production increased from 12 to 41 billion liters in the same period (IEA 2019). The main producers worldwide are the USA and Brazil, with 69% of all biofuels, followed by Europe with 9% (REN21 2019; Jeswani et al. 2020). These are mainly first-generation biofuels, where bioethanol is generated by conventional fermentation processes and biodiesel is produced by esterification (fatty acid methyl esters, FAME). The extensive use of these biofuels has been limited, among other factors, because they are chemically and functionally different from fossil fuels. For example, they contain a high oxygen concentration, between 10 and 45%, compared to fossil fuels which lack oxygen. The presence of oxygen limits the compatibility with the existing infrastructure since they are more reactive to the metal in refineries and pipelines. Oxygen also causes increased water attraction, generating a hygroscopic material that can promote biological contamination such as bacterial growth (Karatzos et al. 2017). Furthermore, oxygen reduces energy density, which has an effect on the fuel tank size of transport vehicles and, consequently, on the range of travel. This practically limits their use in aviation. This has also resulted in the so-called blend-wall, which is mainly used mixed with fossil fuels such as gasoline in percentages of 5 to 10% (E5, E10) (Karatzos et al. 2017). Its use with higher mixtures is feasible, but it requires modifications to vehicle engines, service stations, refineries, etc. In this context, an emerging biofuel solution that may overcome these disadvantages are the so-called drop-in fuels.

Drop-in biofuels are defined as liquid bio-hydrocarbons that are functionally equivalent to petroleum-based fuels and are fully compatible with the existing petroleum infrastructure and the engines and turbines of transport systems (Karatzos et al. 2016). Functional equivalence can be determined by evaluating different fuel properties such as carbon number, boiling point, freezing point, flash point, aromatic content, etc. For example, gasoline is composed of a mixture of C4–C12 hydrocarbons with an aromatic content of 20 to 40% and is typically used in spark-ignition engines. Diesel contains C10–C22 hydrocarbons with 25% aromatic content and is mainly used in compression engines. Aviation fuel is a mixture of C8–C16 with a maximum of 25% aromatic content (Karatzos et al. 2017). Drop-in fuels are composed of many different types of hydrocarbons, the mixture of which has functional similarities with fossil fuels.

The following biofuels can be considered within the category of drop-in fuels: renewable diesel, also known as green diesel, hydrotreated/hydrogenated vegetable oil (HVO), hydrotreated/hydroprocessed ester and fatty acids (HEFA), hydrotreated biodiesel or hydrogenation-derived renewable diesel (HDRD) butanol, isobutanol, 1-methyl-1-butanol and 2-methyl-1-butanol, farnesene, bisabolene, pinene, limonene, and fatty acid derivatives with different carbon number such as palmitate (Karatzos et al. 2017).

Drop-in biofuels can be produced using three main technologies and their interaction. Oleochemical technologies use fats and oils as feedstocks to generate products such as HVO and HEFA. Thermochemical technologies include pyrolysis and gasification routes to generate intermediate liquid or gaseous products such as bio-oil and syngas, and subsequently, through different processes, compounds such as alkanes, alcohols, aldehydes, isobutane, hydrogen, and methanol, which can be used as drop-in fuels similar to both gasoline and diesel (Bridgwater 2012; Karatzos et al. 2017). Biochemical technology involves the use of sugar, starch, or cellulose followed by lipid, alcohol, or isoprenoid fermentation. In these advanced biocatalytic processes, sugars are transformed into less oxygenated, more energy-dense molecules, such as longer chain alcohols (butanol, butanediol), and compounds of higher molecular weight, such as isoprenoids and fatty acids. Hybrid technologies, such as fermentation of off-gases into alcohols and subsequent alcohol-to-jet (ATJ) conversion, incorporate elements from the three technologies described above (Karatzos et al. 2017).

In the case of biochemical technology, different biosynthetic pathways have been investigated, exploring the capacities of different host microorganisms, including traditional bacteria and yeasts such as *Escherichia coli* and *Saccharomyces cerevisiae*, as well as unconventional ones such as cyanobacteria and algae (Jiang et al. 2018; Liu et al. 2022). These microorganisms have been evaluated under specialized culture conditions with the purpose of optimizing the processes and obtaining high production yields of biofuels and their intermediates. Metabolic and genetic engineering studies of the various biosynthetic pathways of the microorganisms used for production have also been carried out, including upregulating pathway enzymes, deleting or downregulating competing pathways, engineering transcriptional factors, increasing robustness of host cells, enhancing cell viability,

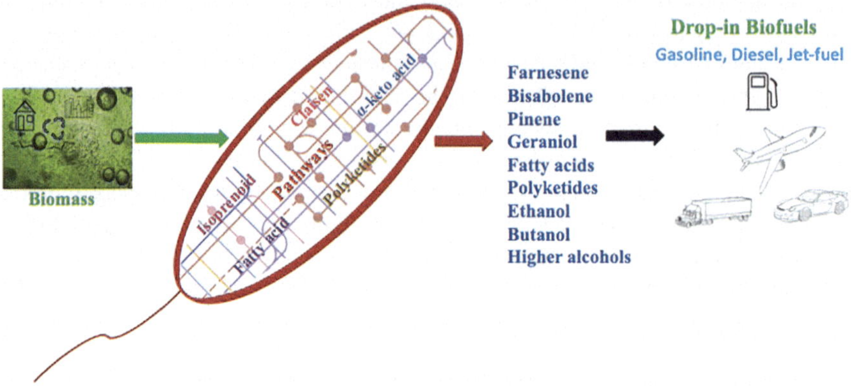

Fig. 2.1 Drop-in biofuel production through biochemical pathways

and regulating cell populations (Jung et al. 2016; Rodriguez and Atsumi 2014; Zhang et al. 2012; Ajjawi et al. 2017; Lim and Jung 2017). Five major biosynthetic pathways have been proposed for the generation of drop-in biofuels: the α-keto acid pathway, the non-decarboxylative Claisen condensation pathway, the fatty acid pathway, the isoprenoid pathway, and the polyketide pathway (Fig. 2.1) (Jiang et al. 2018). Each of these is discussed in the following sections.

2.2 Isoprenoid Pathway

Isoprene (2-methyl-1,3-butadiene) was first discovered as a cellular metabolite in the 1950s (Sanadze 1956). It is the precursor of the family of isoprenoid compounds (Sacchettini and Poulter 1997). Isoprenoids are an extraordinarily chemically and structurally diverse group of compounds. Functionally, they can be used as protective agents and constituents of membranes, reproductive hormones, signal transduction components, among others (Loreto and Velikova 2001; Peñuelas et al. 2005; Terry et al. 1995). These compounds can be naturally synthesized by a diversity of organisms included in the three domains of life, Eukarya, Archaea, and Bacteria (Seemann et al. 2002; Eroglu and Melis 2010; Westfall et al. 2012; Gao et al. 2016). Isoprenoids can also be obtained as by-products of the petrochemical industry and are used as feedstocks for the development of a variety of synthetic chemical products. However, the depletion of fossil sources and their structural complexity make their production difficult or expensive (Khusnutdinov et al. 2023). Terpenes and terpenoids generated via the isoprenoid pathway can be used as fuel substitutes. Monoterpenes (C10), sesquiterpenes (C15), and their derivatives have been identified as diesel and jet-fuel substitutes, which has led to a diversity of studies focused on the biological production of a variety of terpenes and terpenoids with the potential for biofuels such as farnesene, bisabolene, pinene, limonene, geraniol, and linalool, among others (Yang et al. 2012; Jiang et al. 2018; Karatzos et al. 2017; Dickschat 2016).

All isoprenoid pathways begin with the oxidation of glucose to two pyruvate molecules. Subsequently, pathways diverge as pyruvate converts into different intermediates, such as Acetyl-CoA or acetaldehyde. All isoprenoids are biosynthesized from two basic units of five carbons, isopentenyl pyrophosphate (IPP) and dimethylallyl pyrophosphate (DMAPP), which are produced from either the mevalonate (MVA) or 2-C-methyl-erythritol 4-phosphate (MEP) pathway. The carbon chain is elongated by iterative assembly of IPP or DMAPP, generating polyisoprenoid diphosphates, which are then converted to terpenes or terpenoids. The MVA pathway is mainly found in archaea and eukaryotes, in the cytosol and mitochondria of plants and fungi, while the MEP pathway is present in many groups of bacteria, cyanobacteria, green algae, and chloroplasts of higher plants (Rohmer 1999; Bouvier et al. 2005; Sacchettini and Poulter 1997; Jiang et al. 2018; Zargar et al. 2017).

Both pathways consist of several steps and thus various enzymes are involved. The MVA pathway includes seven enzymes and begins with the enzyme acetoacetyl-CoA thiolase (ERG10), which catalyzes the condensation of acetyl-CoA to form acetoacetyl-CoA. Another acetyl-CoA molecule is subsequently incorporated through the activity of the enzyme ERG13 to form HMG-CoA (3-hydroxy-3-methyl-glutaryl-CoA). In the next step, HMGR catalyzes the conversion of HMG-CoA to mevalonate and is considered the most crucial target for terpenoid production. The enzyme mevalonate kinase (ERG12) acts next to generate mevalonate-P, followed by the enzyme ERG8 (mevalonate 5-phosphate kinase) to form mevalonate-PP. In the next step, the enzyme ERG19 (mevalonate 5-pyrophosphate decarboxylase) acts to generate IPP (isopentenyl pyrophosphate), which in turn can be converted to dimethylallyl pyrophosphate (DMAPP) by the enzyme isopentenyl diphosphate isomerase (IDI) (Rohmer 1999; Bouvier et al. 2005; Zargar et al. 2017; Jiang et al. 2018; Liu et al. 2022).

The MEP pathway starts with glyceraldehyde 3-phosphate (GAP) and pyruvate, which are condensed to 1-deoxy-D-xylulose-5-phosphate (DXP) by 1-deoxy-D-xylulose-5-phosphate synthase (DXS). Then, DXP is converted to MEP by DXR (DXP reductase). Subsequently, through a series of reactions, MEP is converted to CDP-ME (4-diphosphocytidyl-2-C-methyl-D-erythritol) by IspD (2-C-methyl-D-erythritol 4-phosphate cytidyltransferase); CDP-ME is transformed into CDP-MEP (4-diphosphocytidyl-2-C-methyl-D-erythritol 2-phosphate) by IspE (4-diphosphocytidyl-2-C-methyl-D-erythritol kinase), which in turn is converted to MEcPP (2-C-methyl-D-erythritol 2,4-cyclodiphosphate) by IspF (2-C-methyl-D-erythritol 2,4-cyclodiphosphate synthase) and subsequently to HMB-PP (4-hydroxy-3-methyl-butenyl 1-diphosphate) by IspG, HMB-PP synthase. The enzyme IspH (HMB-PP reductase) acts next to transform HMP-PP into DMAPP (Yuan and Ching 2014; Jiang et al. 2018; Liu et al. 2022).

The general stoichiometry to produce a molecule of IPP or DMAPP is shown in the following equation (Liu et al. 2022):

$$MVA\ pathway: 1.5\ Glucose + 2\ NADPH + 6\ NAD^{+} \rightarrow IPP + CO_2 + 2\ ADP + 2\ NADP^{+} + 6\ NADH$$

$$MEP\ pathway: Glucose + 2\ ATP + 3\ NADPH + NAD^{+} \rightarrow IPP + CO_2 + 2\ ADP + 3\ NADP^{+} + NADH$$

Molecules of IPP and DMAPP are the basic units for the generation of a diversity of terpenoid compounds that can be used as drop-in fuels such as farnesene, bisabolene, epi-isozizaene, pentalenene, α-isocomene, limonene, pinene, geraniol, among others (Fig. 2.2). To produce farnesene, the enzyme ERG20 catalyzes the sequential condensation of two IPP molecules with one DMAPP molecule to form geranyl pyrophosphate (GPP) and subsequently farnesyl pyrophosphate (FPP), which is transformed into farnesene by the enzyme FS (farnesene synthase). Bisabolene is generated by the activity of the enzyme bisabolene synthase on FPP. Epi-isozizaene, pentalenene, and α-isocomene are produced in a similar way through the enzymes epi-isozizaene synthase (EIZS), pentalenene synthase (pentS), and α-isocomene synthase (MrTPS2), respectively. To produce limonene, pinene, and geraniol, geranyl pyrophosphate synthase (GPPS) converts an IPP and DMAPP molecule to geranyl pyrophosphate (GPP). Subsequently, the enzymes limonene, pinene, and geraniol synthases act on GPP to produce these molecules (Zhao et al. 2017; Liu et al. 2018, 2021; Liu-Jing et al. 2021; Liu et al. 2022; Walid et al. 2023).

The production of these compounds in plants is relatively low, since yield is affected by different factors such as seasonality and regional climate, as well as other factors (Liu et al. 2022). Enhancing production yields can be achieved by employing microorganisms as a viable alternative. However, the natural pathways used by microorganisms continue to exhibit low yields. Therefore, different strategies have recently been investigated in order to improve production and achieve commercial levels. These strategies include novel biosynthetic pathways and their regulation, strain engineering, organelle engineering, multienzyme assembling systems, deleting or downregulating competing pathways, transcriptional engineering, host cell engineering, enhancing cell viability, regulating cell populations, among others (Jung et al. 2016; Rodriguez and Atsumi 2014; Zhang et al. 2012; Ajjawi et al. 2017; Lim and Jung 2017; Jiang et al. 2018).

An increased isoprenoid production has been achieved by overexpressing different enzymes involved in the MVA and MEP pathways. For example, Westfall et al. (2012) found that the overexpression of ERG10 increased the production of isoprenoids in *Saccharomyces cerevisiae*. Gao et al. (2016) engineered the MEP pathway in the cyanobacterium *Synechococcus elongatus* by overexpression of IPP isomerase (IDI), directing about 40% of photosynthetically fixed carbon toward the isoprene biosynthetic pathway, resulting in the production of 1.26 g/L of isoprene from CO_2. Similarly, the overexpression of genes encoding IspD, HMB-PP, and IspG is beneficial in improving terpenoid production (Englund et al. 2018). Furthermore, the synergy between the MEP pathway and the MVA pathway in engineered *E. coli* has demonstrated that the overexpression of both pathways improves isoprene yield approximately 20-fold and three-fold, respectively, compared to

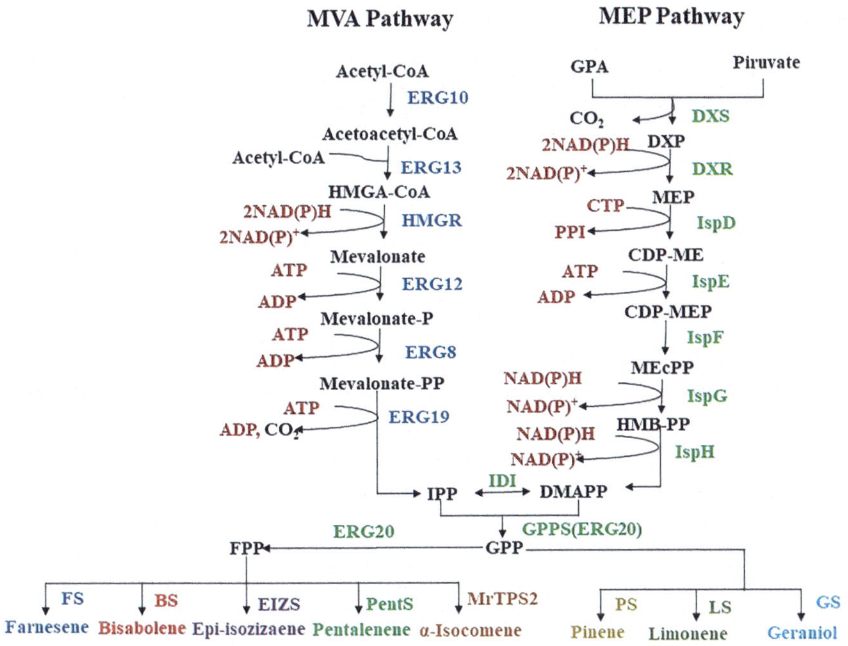

Fig. 2.2 Mevalonate (MVA) and 2-C-methyl-D-erythritol-4-phosphate (MEP) pathways for the production of isoprenoid biofuels (farnesene, bisabolene, epi-isozizaene, pentalenene, α-isocomene, pinene, limonene, geraniol). HMG-CoA, 3-hydroxy-3-methyl-glutaryl-CoA; P-phosphate; PP, diphosphate; IPP, isopentenyl pyrophosphate; GAP, glyceraldehyde 3-phosphate; DXP, 1-deoxy-D-xylulose 5-phosphate; CDP-ME, 4-diphosphocytidyl-2-C- methyl-D-erythritol; CDP-MEP, 4-diphosphocytidyl-2-C-methyl-D-erythritol 2-phosphate; MEcPP, 2-C-methyl-D-erythritol 2,4-cyclodiphosphate; HMB-PP, 4-hydroxy-3-methyl-butenyl 1-diphosphate; DMAPP, dimethylallyl pyrophosphate; GPP, geranyl pyrophosphate. ERG10, acetoacetyl-CoA thiolase; ERG13, HMG-CoA synthase; HMGR, HMG-CoA reductase; ERG12, mevalonate kinase; ERG8, mevalonate 5-phosphate kinase; ERG19, mevalonate 5-pyrophosphate decarboxylase; DXS, 1-deoxy-D- xylulose-5-phosphate synthase; DXR, DXP reductase; IspD, 2-C-methyl-D-erythritol 4-phosphate cytidyltransferase; IspE, 4-diphosphocytidyl-2-C-methyl-D-erythritol kinase; IspF, 2-C-methyl-D-erythritol 2,4-cyclodiphosphate synthase; IspG, HMB-PP synthase; IspH, HMB-PP reductase; IDI, isopentenyl diphosphate isomerase; GPPS, geranyl pyrophosphate synthase. FS, farnesene synthase; BS, bisabolene synthase; EIZS, epi-isozizaene synthase; pentS, pentalenene synthase; MrTPS2, α-isocomene synthase; PS, pinene synthase; LS, limonene synthase; GS geraniol synthase

overexpression of the MEP pathway or the MVA pathway alone, resulting in a production of 24.0 g/L isoprene with a yield of 0.267 g/g of glucose (Yang et al. 2016). Farnesene production was significantly increased using two cyanobacterial parental strains, whose MEP pathways were optimized by overexpressing DXS, IDI, and ispA coding genes to provide abundant FPP for farnesene synthesis (Lee et al. 2017). Meadows et al. (2016) improved farnesene production, reaching concentrations at commercial levels of up to 130 g/L. The authors added several heterologous reactions to the central carbon metabolism in *S. cerevisiae,* which increased the acetyl-CoA supply, decreased ATP requirements, and decreased carbon loss to CO_2. The production of α-farnesene was increased by modifying the cytosolic acetyl-CoA supply and the MVA biosynthesis module and weakening the FPP branch module in *P. pastoris,* which resulted in a concentration of up to 17.92 g/L, equivalent to a yield of 0.078 g/g glucose (Jian-Zhong et al. 2023). Overexpression of α-bisabolene has also been achieved in *Yarrowia lipolytica* by using the yeast's ability to accumulate lipids and applying a golden gate DNA assembly toolbox to HMG reductase and α-bisabolene synthase (El-Sharoud et al. 2023). The authors reported that the transformed *Y. lipolytica* strain produced 816 mg/L of α-bisabolene, which increased to 1243 mg/L by varying the culture conditions. Other promising jet-fuel sesquiterpenes, such as epi-isozizaene, pentalenene, and α-isocomene, have been produced separately through the endogenous MEP pathway and heterologous MVA pathway for efficient FPP supply using *E. coli* and *S. cerevisiae.* The concentrations of these compounds ranged between 77.5 mg/L (α-isocomene) and 780.3 mg/L (pentalenene) (Liu et al. 2018). In another study, the limonene biosynthesis pathway was engineered in the yeast *Rhodosporidium toruloides* by overexpressing the limonene synthase (LS), neryl pyrophosphate synthase (NPPS)/geranyl pyrophosphate synthase, and native hydroxy-methyl-glutaryl-CoA reductase (HMGR) (Liu et al. 2021). In addition, to improve the limonene titer, the acetoacetyl-CoA thiolase/HMGR (EfMvaE) and mevalonate synthase (EfMvaS) from *Enterococcus faecalis*, the mevalonate kinase from *Methanosarcina mazei* (MmMK), and the chimeric enzyme NPPS-LS were introduced in the carotenogenesis-deficient strain, resulting in concentrations of up to 393.5 mg/L. In another study, limonene titer was improved (2.23 g/L) by engineering the pyruvate dehydrogenase (PDH) bypass to directly push the acetyl-CoA flux into the MVA pathway and deleting target genes outside the MVA pathway in limonene production (Zhang et al. 2021). Improvements in pinene yields have also been reported. For example, Liu-Jing et al. (2021) reported that α-pinene production was improved using engineered *Y. lipolytica* by overexpressing α-pinene synthase, obtaining a production of up to 36.1 mg/L. The production of monoterpene geraniol has been improved in engineered strains of *S. cerevisiae.* A geraniol production of 1.69 g/L has been achieved by enhancing the MVA pathway flux, screening different sources of geraniol synthases (GESs) and GPP synthases (GPPSs), designing fusion proteins of GES and GPPS, dynamic control of *ERG20* expression, and deletion of *OYE2* (encoding an NADPH oxidoreductase) or *ATF1* (encoding an alcohol acetyltransferase), which are involved in the endogenous conversion of geraniol to other terpenoids (Zhao et al. 2016, 2017).

2.3 Fatty Acid Pathway

Oleochemical renewable materials, such as used oil and fat, have the potential to be transformed into a hydrocarbon component (Basumatary et al. 2023). The use of these feedstocks is through triglycerides, which are their main component and consist of fatty acid chains. These compounds have low proportions of oxygen and, when deoxygenated, can generate different chain hydrocarbons (Zhu et al. 2023). Alkanes and alkenes are two important hydrocarbon classes used for biofuels that can be obtained from fatty acids. Alkanes and alkenes have relatively high freezing points and viscosities compared to their branched-chain isomers, limiting their operability at low temperatures, which is an essential characteristic for diesel (Bruder et al. 2019). They are also the most important components of natural gas, gasoline, and aviation kerosene. Moreover, alkanes/alkenes are considered superior fuels because they have better properties than ethanol and diesel, since their energy density ranges between 42–47 MJ/kg, whereas it is lower than 43 MJ/kg in ethanol and diesel (Gujjala et al. 2018). Alkanes/alkenes that contain more than four carbons are present as liquids, and thus most of them can be produced using regular bioreactors. Generally, C3–C12 alkanes and alkenes are referred to as short-chain alkanes/alkenes, while C13–C17 ones are called long-chain alkanes/alkenes (Sun et al. 2015). Alkanes and alkenes from microbial biosynthesis are chemical products of high value because they can be used for a wide range of industrial purposes, as well as for drop-in fuels (Liu et al. 2015; Jiang et al. 2018; Karatzos and van Dyk 2016).

Some microorganisms are capable of naturally biosynthesizing alkanes and alkenes, but their production level is extremely low (Lu et al. 2022). The most notable is the pair formed by fatty acyl-ACP/CoA reductase (FAR) and aldehyde decarbonylase, discovered in the cyanobacteria *Synechococcus elongatus*, which converts fatty aldehydes to alkanes expressed in *Escherichia coli* (Zargar et al. 2017; Bruder et al. 2019) (Fig. 2.3). However, there are other pathways and enzymes involved in the biosynthesis of alkanes and alkenes from natural products, and the introduction of biosynthetic pathways in heterologous microbial hosts has allowed the production of different alkane and alkene structures (Table 2.1). The biosynthesis of fatty acids through metabolic pathways to obtain alkanes/alkenes for the production of drop-in fuels is described below.

2.3.1 Fatty Aldehyde Decarbonylation

Aldehyde decarbonylases (also called aldehyde-deformylating oxygenases) are carbon–carbon lyases that catalyze the decarbonylation of aldehydes to alkanes; in fact, the substrate of these enzymes results from the reduction of activated fatty acids to fatty aldehydes (Kang and Nielsen 2017). Given that the decarbonylation reaction produces formate as a by-product, aldehyde decarbonylases were redesignated as fatty aldehyde-deformylating oxygenases (ADOs) (Lu et al. 2022). Aldehyde-deformylating oxygenases have been discovered using different organisms, including plants, insects, and cyanobacteria. Aldehyde-deformylating oxygenases naturally convert long-chain fatty aldehydes into alkanes through the generation of formate and water as by-products (Bruder et al. 2019). They are also non-heme diiron ferritin-like enzymes that are structurally and mechanistically

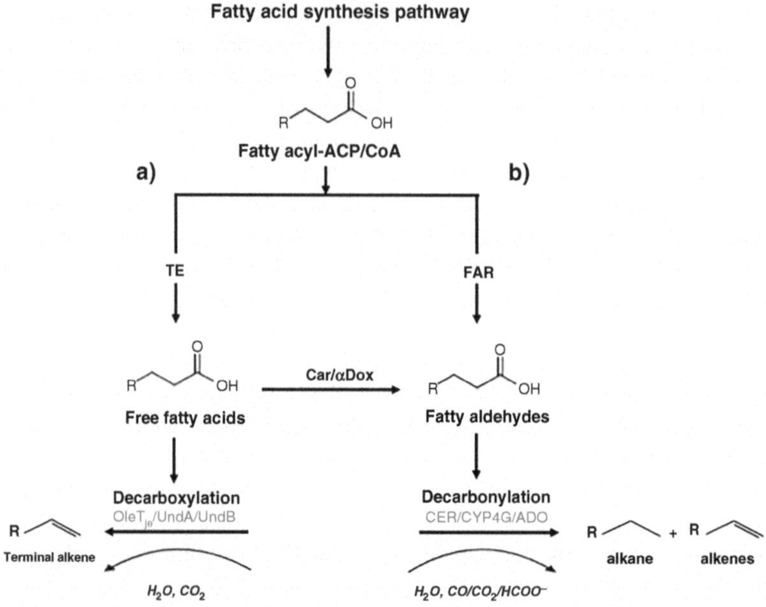

Fig. 2.3 Schematic representation of the metabolic pathways for the synthesis of alkanes/alkenes. (**a**) alkene production by decarboxylation enzymes: OleT$_{je}$, UndA, and UndB; (**b**) conversion of fatty aldehydes to alkanes/alkenes by FAR enzymes: CER (plants), CYP4G (insects), and ADO (cyanobacteria); Te, heterologous fatty acyl-ACP/CoA thioesterase; Car, carboxylic acid reductase; αDox, fatty acid α-dioxygenase

similar to ribonucleotide reductases and methane monooxygenases. The reaction requires molecular oxygen and reducing equivalents.

The pathway consists of a fatty acyl-ACP/CoA and an ADO. The NADPH-dependent reductase generates C_n fatty aldehydes from the reduction of fatty acyl-ACPs/CoAs. Subsequently, the C_n fatty aldehydes are transformed into C_{n-1} alkanes or alkenes by the ADO (Lu et al. 2022) (Fig. 2.3). Current studies indicate that fatty acyl-ACPs/CoAs in yeasts are insufficient for the overproduction of hydrocarbons, especially in *S. cerevisiae*, where fatty acyl-ACPs are only available in the mitochondria. However, yeasts can be engineered to accumulate free fatty acids (FFA) at a higher level, and the endogenous FFA pool can be used for the biosynthesis of hydrocarbons through alternative pathways. In addition to ADOs, two classes of eukaryote aldehyde decarbonylases have been discovered in plants and insects. In *Arabidopsis thaliana*, aldehyde decarbonylase Cer1 (also known as ECERIFERUM1) has been proposed to interact with Cer3 and cytochrome b5 Cytb5-B, catalyzing the decarbonylation of fatty aldehydes. In addition, Cer1, Cer3, and Cytb5-B form a multienzyme complex that converts very-long-chain fatty acyl-CoAs (C26–C32) to hydrocarbons (Santner et al. 2023).

Table 2.1 Microorganisms and enzymes are involved in the production of alkene/alkane derivatives of fatty acids

Enzyme	Microorganism	Precursor	Product	Chain length	Titer	Reference
$OleT_{JE}$	*Y. lipolytica*	FFA	Alkene	C12, C14, C16, C18, C20	0.9 mg/L	(Yang et al. 2018)
$OleT_{JE}$	*Y. lipolytica*	FFA	Alkene	C15, C17	600 mg/L	(Yang et al. 2019)
$OleT_{JE}$	*E. coli*	FFA	Alkene	C11, C13, C15, C17	97 mg/L	(Y. Liu et al. 2014)
$OleT_{JE}$	*S. cerevisiae*	FFA	Alkene	C11–C19	3.7 mg/L	(Chen et al. 2015)
UndA	*S. cerevisiae*	FFA	Alkene	C8, C10, C12	3 mg/L	(Z. Zhu et al. 2017)
UndB	*E. coli*	FFA	Alkene	C5–C17	10 mg/L	(Rui et al. 2014)
UndA	*E. coli*	FFA	Alkene	C9–C13	55 mg/L	(Rui et al. 2015)
ADO	*E. coli*	FAR	Alkane	C15, C17	24 mg/L	(Song et al. 2016)
ADO	*E. coli*	FAR	Alkane	C13	580 mg/L	(Choi & Lee 2013)
ADO	*S. cerevisiae*	FFA	Alkane/alkene	C15, C17	0.8 mg/L	(Zhou et al. 2016)
ADO	*E. coli*	FAR	Alkane/alkene	C13–C17	40 mg/L	(Schirmer et al. 2010)
ADO	*E. coli*	FAR	Alkane	C11, C13	4 mg/L	(Yan et al. 2016)

2.3.2 Decarboxylation of Fatty Acids

Decarboxylation of fatty acids provides a sustainable pathway for the production of alkane products (Zargar et al. 2017). Furthermore, the fatty acid decarboxylation pathway directly converts fatty acids to terminal alkenes through decarboxylase enzymes (Monteiro et al. 2022). Three different types of enzymes are involved in the direct enzymatic conversion of fatty acids to terminal alkenes: $OleT_{JE}$, UndA, and UndB. $OleT_{JE}$ is a cytochrome P450 enzyme belonging to the CYP152 family and was discovered from *Jeotgalicoccus* sp. ATCC 8456 (Kang and Nielsen 2017). Cytochrome P450 enzymes are able to catalyze a variety of reactions (not only physiological) and are thus considered among the most versatile biocatalysts in nature (Monteiro et al. 2022). The metabolic process is based on the fact that alkenes can be synthesized by cytochrome P450 $OleT_{JE}$ through the direct decarboxylation of fatty acids, with the emission of CO_2. The activity of $OleT_{JE}$ depends on the supply of heme as a cofactor and hydrogen peroxide as the only electron and oxygen donor (Lu et al. 2022). Additionally, this was the first described direct enzymatic decarboxylation of FAs, specifically long-chain C16–C20 FAs, to yield C1-shortened alkenes (Santner et al. 2023). Different studies have shown the capacity of using these enzymes, for example, Zhou et al. (2016) overexpressed $OleT_{JE}$ in an FFA-overproducing strain of *S. cerevisiae,* which resulted in an alkene titer of 0.12 mg/L.

Alkanes can also be catalyzed by the non-heme Fe (II)-dependent oxidase UndA and the enzyme UndB (Herman and Zhang 2016). These enzymes were identified for the first time in *Pseudomonas*, where they convert medium-chain fatty acids to C1-shortened terminal alkenes through oxidative decarboxylation (Santner et al. 2023). UndA represents a group of non-heme oxidases/decarboxylases that specifically convert medium-chain fatty acids (C10–C14) into C1-shortened alkenes. UndA activity in alkene biosynthesis was confirmed both by mutagenesis in the native *Pseudomonas* producer and functional reconstitution in the heterologous host *E. coli* (Herman and Zhang 2016). In addition, other biochemical and structural analyses have shown that UndA activates molecular oxygen to convert fatty acids to alkenes with CO_2 as a by-product (oxidative decarboxylation). Different studies have shown that the fatty acid substrate is activated via the abstraction of β-hydrogen, similar to the mechanism employed by $OleT_{JE}$. The enzyme UndB was originally classified as a fatty acid desaturase based on sequence homology, but it was also found to be an aldehyde decarboxylase (Sharma and Yazdani 2021; Herman and Zhang 2016). Compared to UndA, UndB shows a wide substrate range of C6 to C16 but prefers C10–C14 fatty acids like UndA. UndB has also shown a higher terminal olefin production efficiency compared to $OleT_{JE}$ and UndA (Sharma and Yazdani 2021). Rui et al. (2015) demonstrated that UndB was the most efficient catalyst among the three enzymes for producing 1-undecene in *E. coli*.

Alkane and alkene production by biosynthesis pathways for commercial purposes, fuels, and fatty acid-derived chemical products is still under study. However, advances in microbial and metabolic engineering have made progress. Furthermore, oleaginous yeasts are gaining much attention as potential hosts for the production of fatty acid-derived products (Gujjala et al. 2018). In addition, multienzymatic cascade reactions can be an efficient strategy for the conversion of triacylglycerol-based feedstocks to renewable drop-in fuels with the desired properties. However, these systems should be mainly designed to produce efficient and robust biocatalysts and then used in industrial processes (Santner et al. 2023).

2.4 α-Keto Acid Pathway

Drop-in biofuels are produced through different metabolic pathways, which include the Ehrlich pathway, also known as the α-keto acid pathway. It is a very important biochemical pathway because it synthesizes short-chain alcohols such as isobutanol, n-butanol, propanol, isopropanol, isoamyl alcohol, 3-methyl-1-butanol and 2-methyl-1-butanol, pentanol, hexanol, 1-octanol, and fatty alcohols. Figure 2.4 shows a general schematic representation of the α-keto acid pathway, illustrating that alcohol synthesis is possible through the decarboxylation and reduction of short-chain α-keto acids, which are intermediates of the amino acid synthesis pathway

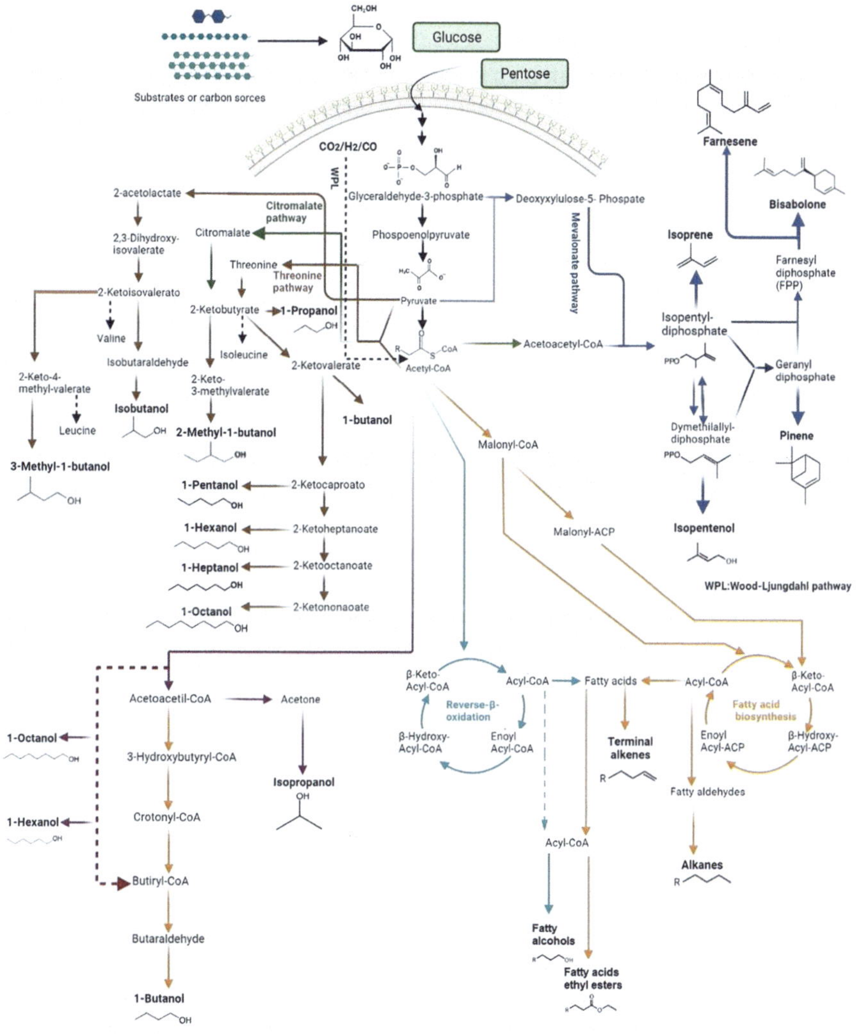

Fig. 2.4 Schematic representation of the metabolic pathways for the synthesis of biofuels: α-Keto acid and Claisen pathways and their relationship with other metabolic pathways for drop-in fuel production (Reproduced from Papoutsakis 2015 with modifications)

(Maurya et al. 2023; Jiang et al. 2018). In general, in the α-keto acid pathway, the carbon chain extends via α-keto acid elongation enzymes (αKAE) and, with the intervention of the α-keto acid decarboxylase, a mixture of alcohols is produced. However, given that the carbon chain extension mechanism is inefficient, there is a redox imbalance for medium-chain products, which limits the maximum efficiency of the metabolic pathway. In addition, isopentenol can be produced through the isoprenoid pathway (Fig. 2.4) but is limited to forming carbon chain lengths in multiples of five because it uses isopentenyl diphosphate as a building block, and it is limited by the inherent redox imbalance of the pathway (Sheppard et al. 2014).

Some higher alcohols (with high carbon content), such as n-Butanol, isobutanol, 1-Pentanol, isopentanol, and 1-Octanol, have generated substantial expectations, since they may be possibly used as fuels for diesel engines due to their high carbon content and their physicochemical characteristics that allow them to be directly mixed with diesel, particularly 1-octanol, since it is considered a potential diesel and aviation fuel substitute because it has similar characteristics, generating thus significant economic and environmental expectations (Atmanli and Yilmaz 2018; Henritzi et al. 2018). One of the main advantages of this type of alcohols is that they exhibit certain superior characteristics compared to ethanol (the most widely studied biofuel), among which are a higher energy density, ease of refining, and better compatibility with existing gasoline engines and technologies (Zhang et al. 2019).

These biofuels can be produced by fermentation, as can be observed in Fig. 2.4, where microbial synthesis occurs through the α-keto acid pathway. One main characteristic of this type of alcohols is that they can be produced from renewable sources since microorganisms can use them as a carbon source, avoiding thus the use of feedstocks intended for human or animal consumption (Santillan-Jimenez and Crocker 2012). However, their production through the α-keto acid pathway has been a challenge for many years because wild microorganisms, such as yeasts, are not able to produce the quantities of α-keto acids necessary for the production of alcohols or, in contrast, alcohol yields are not adequate for industrial production, and thus genetic engineering has emerged as a suitable alternative for the development of microorganisms with non-native genes that allow for increased production yields (Papoutsakis 2015).

Much attention has been given to l-butanol in recent years due to its potential as a gasoline substitute. It has been produced on a large scale using *Saccharomyces cerevisiae* through the optimization of the threonine biosynthetic pathway by genetic engineering, introducing the citramalate pathway for the synthesis of α-ketovalerate (Fig. 2.4), which is the precursor of butanol synthesis (Maurya et al. 2023). However, 1-butanol production is also carried out naturally by *Clostridium* species through acetone-butanol-ethanol (ABE) fermentation via a series of steps of reduction of CoA-linked intermediates. Nevertheless, given that the yields are not adequate for industrial production, they have been improved by replacing inefficient enzymes, developing co-culture systems, inducer-free promoters, flow analyses, and using metabolic engineering with *Escherichia coli*, as well as exploring the modification of metabolic pathways of amino acid synthesis, also known as the α-keto acid pathway (Wen and Shen 2016).

The production of butanol or its isoforms by the α-keto acid pathway depends on an adequate production of CoA and the reductive power (NADH) required for the synthesis of the precursor, in this case butyraldehyde (Wen and Shen 2016; Papoutsakis 2015), and depending on the microorganism used and the carbon metabolic flux, it could be directed toward 1-Octanol and/or 1-hexanol production, as observed in Fig. 4.4. It is important to mention that alcohol production through this metabolic pathway reduces the possibility of protein synthesis because amino acid precursors are used in the synthesis of alcohol (Wen and Shen 2016). There are studies where other strategies have been applied, for example, inducing the production of 3-methyl-butanol (Fig. 2.4) by adding 2-Ketoisovalerate to the culture medium, which is the precursor required for alcohol synthesis, forcing thus a metabolic flux of carbon toward its production. Amino acids such as valine and leucine have also been added to fermentations with *Saccharomyces cerevisiae*, strategically directing the flux of carbon toward the synthesis of alcohol by the α-keto acid pathway, thereby achieving a 3.3-fold increase in the yield of 3-methyl-1-butanol (Ramli et al. 2018). Genetic engineering using *Saccharomyces cerevisiae* has also allowed the overexpression of enzymes in mitochondria, increasing isobutanol yields by up to 260%, and the overexpression of enzymes in the cytoplasm increases yields by 10% (Avalos et al. 2013). There are some microorganisms that have been genetically modified to increase butanol yields, such as *Clostridium acetobutylicum*, which is capable of producing a mixture of solvents such as isopropanol-butanol-ethanol; the first two in particular by the α-keto acid pathway (Sheppard et al. 2014). The metabolic pathways of *Clostridium* generally start with acetyl-CoA for the synthesis of butyryl-CoA and hexanoyl-CoA, which are then reduced to the corresponding alcohols (1-butanol, 1-hexanol, 1-octanol, and isopropanol), as can be observed in Fig. 4.4 (Papoutsakis 2015). There has been substantial interest in 1-octanol in recent years, which can be synthesized from acetyl-CoA toward the synthesis of acetoacetyl-CoA, and from the threonine pathway toward the synthesis of 2-ketovalerate and 2-ketononanoate (Papoutsakis 2015).

Another biofuel that, like butanol, can be produced by fermentation is l-pentanol, and due to its physicochemical characteristics, and particularly its semi-polarity, it can be used mixed with diesel (in any proportion) and aviation fuel, demonstrating thus more favorable characteristics than other long-chain alcohols (Henritzi et al. 2018; Akhtar et al. 2015). In addition, various reports indicate that, when used in a mixture in percentages higher than 30% v/v, combustion efficiency increases and contaminant emissions decrease, particularly nitrogen oxide (NOx) (Atmanli and Yilmaz 2018; Atmanli and Yilmaz 2021). These compounds, as well as other higher alcohols (mentioned above), can be produced through the α-keto acid pathway, since it is amenable to genetic modifications for yield improvement, taking as a starting point the increased synthesis of amino acid precursors (Machado et al. 2012).

In general, the importance of the synthesis of higher alcohols or high carbon alcohols (C4–C8) by the α-keto acid pathway lies in the possibility of increasing their yields using genetic and protein engineering based on the vast knowledge available on metabolic pathways. The known information about metabolic pathways of amino acid synthesis could be applied to different types of agro-industrial waste that can be saccharified and used for the synthesis of drop-in fuels, reducing thus production costs (Lozada et al. 2020).

2.5 The Non-decarboxylative Claisen Condensation Pathway

The current environmental damage caused by the indiscriminate use of petroleum-based fuels has led to the search for alternatives that allow the mitigation of this damage. One of these alternatives is the production of biofuels or green fuels with the objective of a sustainable future where greenhouse gases are minimized. In this sense, biofuels such as alkanes and alkenes are an option because they can be produced from renewable sources, are considered a sustainable energy supply, and are also very similar to gasoline in their chemical composition (Chen et al. 2022; Herman and Zhang 2016). Furthermore, medium- to long-chain fatty alkanes and alkenes are currently considered ideal fuels since their chemical composition is similar to that of petroleum-based fuels and can also be used for the synthesis of many materials (Jaroensuk et al. 2020; Liu et al. 2014). Medium-chain alkanes (C5–C17) can be used as gasoline and diesel, as well as solvents for different industrial applications (Jaroensuk et al. 2020). The main advantage of this type of biofuels, compared to first-generation biofuels like ethanol, is that their characteristics make them compatible with the existing transport (drop-in fuel) infrastructure (technology) (Keasling and Chou 2008). However, even though the properties of fatty alkanes are very similar to those of conventional diesel, the higher degree of unsaturation in alkenes may cause the formation of radicals during combustion (Peralta-Yahya et al. 2012).

This type of biofuels is generally traditionally produced using lipid biomass by chemical decarboxylation at high temperatures, and thus the non-decarboxylative Claisen condensation becomes a suitable biochemical option, since it uses microorganisms or enzymes for this purpose and is considered a green or environmentally friendly technology (Xie et al. 2019). The advantage of microbial biosynthesis is that it is scalable and cost-effective, and it is considered one of the most promising strategies for biofuel production (Liu et al. 2014). Different reports indicate that *Escherichia coli, Saccharomyces cerevisiae,* and *Synechococcus* sp. have the capacity to produce alkanes and alkenes and have been studied at a genetic level to improve production yields (Kang and Nielsen 2017). *Vibrio furnissi* (halotolerant bacterium) has been reported to produce a range of intra- and extracellular hydrocarbons similar to kerosene (C15–C24). There are also reports indicating that yeasts can synthesize hydrocarbons in the range of C10–C34 (Kumar et al. 2017). This information holds significant relevance since, with the current advances in genetic and metabolic engineering, hydrocarbon production can be improved or genes can be expressed in microorganisms that can produce biofuels faster. Synthetic biology and metabolic engineering techniques have allowed the development of heterologous metabolic pathways in well-known microbial hosts such as *Escherichia coli* and *Saccharomyces cerevisiae* to produce hydrocarbons from unutilized biomass, or better known as agro-industrial waste, which is composed of metabolizable sugars. These microorganisms have been used for many years for industrial-scale production, they can be modified to tolerate toxic biofuels, and it is even possible to expand the range of biofuels that they can synthesize (Zhang et al. 2011).

The non-decarboxylative Claisen condensation metabolic pathway is an orthogonal anabolic pathway for the synthesis of functionalized small molecules that uses a primer and an extender molecule that operate in an iterative manner, and it was discovered in the metabolism of *Clostridium kluyveri* (Cheong et al. 2016; Jiang et al. 2018). In this pathway, two acetyl-CoA molecules condense to acetoacetyl-CoA, which is reduced to butyrate by β-reduction reactions, the carbon chain grows on CoA, and acetyl-CoA or other short-chain acyl-CoAs are used as primers (Jiang et al. 2018; Sarria et al. 2017) (Fig. 2.4). This pathway can be used to synthesize hydrocarbons such as alkanes (saturated) and alkenes (unsaturated) derived from fatty aldehydes using fatty acids as precursors (Jiang et al. 2018). The non-decarboxylative Claisen condensation is characteristic of an initial reaction by a thiolase enzyme (Wackett 2011). Studies at a genetic level have shown that four genes known as oleABCD are responsible for non-decarboxylative thiolytic condensation. These genes are distributed in various phyla of bacteria and can be located either in tandem or separately in genomes with oleCD, always clustered or fused in oleBC. OleA is responsible for the non-decarboxylative Claisen condensation of two long-chain fatty acyl-CoAs to form a β-ketoacid and release two CoAs to OleD, which is an NADPH-dependent reductase that stereospecifically reduces the β-ketoacid to a β-hydroxyl carboxylic acid, which is then converted to a β-lactone by OleC. Finally, the α/β-hydrolase OleB catalyzes the decarboxylation of cis-β-lactones to produce cis-alkenes. Collectively, the four-step pathway transforms two acyl-CoAs (C10–C15 + C10–C15) into a cis-alkene (C19–C29), which is within the carbon range of diesel, heavy fuel oil, or wax (Liu and Li 2020).

The importance of the non-decarboxylative Claisen condensation pathway lies in the possibility of increasing the yields of the final products, such as alkanes and alkenes, which are important in the drop-in fuel industry (Kang and Nielsen 2017). There is a wide diversity of microorganisms with the capacity of producing biofuels such as alkanes and alkenes from fatty acids through the non-decarboxylative Claisen condensation pathway, and even though this type of molecules can be obtained from plants using the same mechanism, production with microorganisms is more common because it can be facilitated by metabolic engineering and has been particularly enhanced by advances in genetic engineering using *Escherichia coli* as a model microorganism (Wackett 2011).

2.6 Polyketide Pathway

Polyketides are secondary metabolites of bacteria, fungi, plants, and animals that are synthesized by multienzyme complexes (polyketide synthases) (Yuzawa et al. 2017). Polyketide synthases (PKSs) are one of the biosynthetic factories to produce polyketides with different structures and biological activities. These enzymes have been historically studied and engineered to produce unnatural polyketides for drug

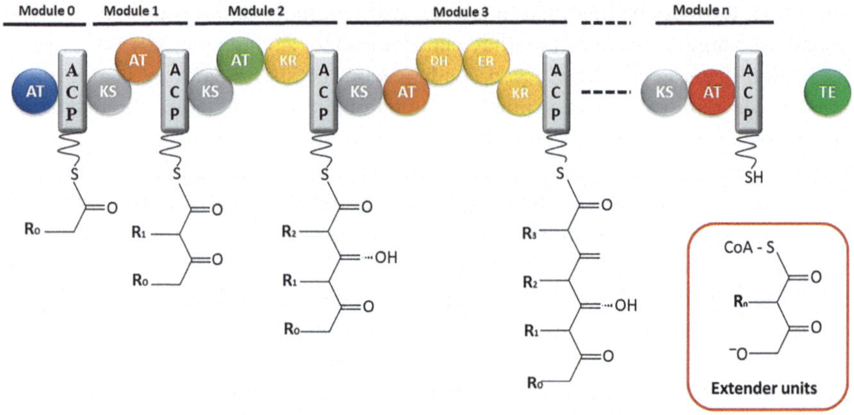

Fig. 2.5 Representation of the biosynthetic pathways of the type I PKS process. PKS domains are colored according to their function: Acyl Carrier Protein (ACP) (gray); Acyltransferase (AT) (blue, red, orange, and green); Ketosynthase (KS) (gray); Ketoreductase (KR), Dehydratase (DH) and Enoylreductase (ER) (yellow), and thioesterases (TE) (green)

discovery and have also been recently explored for synthesizing biofuels and industrial chemical products due to their versatility (Cai and Zhang 2018).

Polyketide synthases are classified into three types, I, II, and III, according to their organization and catalytic mechanisms (Junne and Kabisch 2016). Modular type 1 PKSs have a hierarchical organization where the whole enzyme complex is composed of various sub-units, each sub-unit is composed of various modules, and one module is composed of various domains (that join together diverse acetyl-CoA-building monomers) (Cai and Zhang 2018). Type II or III PKSs are composed of discrete enzymes used iteratively to generate aromatics. Polyketide synthases, in an iterative modular manner, involve three steps: initiation, elongation, and functionalization (Zargar et al. 2018). Some PKS domains are essential for the basic activity of the enzyme, such as the Ketosynthase (KS), Acyltransferase (AT), and Acyl Carrier Protein (ACP) domains (Hwang et al. 2020). They also have an additional thioesterase (TE) domain in the last elongation module, which allows the release of the molecule that is being synthesized. They can frequently exhibit auxiliary domains with Ketoreductase (KR), Dehydratase (DH), and Enoylreductase (ER) activities that appear on certain instances as a function of the type of additional modifications required for the synthesis of a particular compound (Fig. 2.5) (Kornfuehrer and Eustáquio 2019; Neves et al. 2022).

Polyketide synthases are efficient for producing bio-based gasoline substitutes, mainly for producing a bio-based acid or alcohol from an engineered PKS that, when chemically reduced (through hydrogenation), results in a molecule identical in structure to a petroleum-based compound such as drop-in fuels (Yang et al. 2021; Dongsoo et al. 2021). Moreover, compared to other metabolic pathways, such as the fatty acid, isoprenoid, and shikimate pathways, biosynthesis by modular type I PKSs is particularly important due to its versatility and shows the potential to synthesize compounds using a unique chemical space not offered by other biological pathways. However, PKS engineering for the production of biofuels and chemical products is still at the research stage, where the titers for any engineered systems in microorganisms are around 100 mg/L (Cai and Zhang 2018).

2.7 Conclusions

Drop-in biofuels are considered a sustainable alternative to replace fossil fuels. Currently, the commercial production of drop-in biofuels is only carried out through oleochemical processes, and thus other methodologies and strategies are being investigated in order to increase their production on a large scale. A viable alternative to increase drop-in biofuel production is based on the use of organisms and the exploration of their metabolic pathways to generate a diversity of compounds that can function as biofuels. This chapter presented recent information on the five principal biochemical pathways for the production of drop-in biofuels. Some routes have undergone more extensive exploration, such as the isoprenoid pathway, while studies are incipient regarding others, such as the polyketide pathway. To date, the productivities and yields of drop-in biofuels through natural biochemical pathways are low, prompting the investigation of different strategies in order to achieve production at an industrial level. Strategies have included metabolic and genetic engineering, upregulating pathway enzymes, deleting or downregulating competing pathways, engineering transcriptional factors, enhancing cell viability, synthetic biology, and regulating cell populations. The manipulation of biochemical pathways has allowed for increased concentrations of some drop-in biofuel compounds such as farnesene, where concentrations of up to 130 g/L have been obtained. However, the challenge remains significant in achieving the transition to the commercial use of drop-in biofuels.

References

Ajjawi I, Verruto J, Aqui M, Soriaga LB, Coppersmith J, Kwok K, Peach L, Orchard E, Kalb R, Xu WD et al (2017) Lipid production in *Nannochloropsis gaditana* is doubled by decreasing expression of a single transcriptional regulator. Nat Biotechnol 35:647–652. https://doi.org/10.1038/nbt.3865

Akhtar MK, Dandapani H, Thiel K, Jones PR (2015) Microbial production of 1-octanol: a naturally excreted biofuel with diesel-like properties. Metab Eng Commun 2:1–5. https://doi.org/10.1016/j.meteno.2014.11.001

Atmanli A, Yilmaz N (2018) A comparative analysis of n-butanol/diesel and 1-pentanol/diesel blends in a compression ignition engine. Fuel 234:161–169. https://doi.org/10.1016/j.fuel.2018.07.015

Atmanli A, Yilmaz N (2021) Comparative assessment of different diesel engines fueled with 1-pentanol and diesel blends. Environ Prog Sustain Energy 40(5):e13663. https://doi.org/10.1002/ep.13663

Avalos JL, Fink GR, Stephanopoulos G (2013) Compartmentalization of metabolic pathways in yeast mitochondria improves the production of branched-chain alcohols. Nat Biotechnol 31(4): 335–341. https://doi.org/10.1038/nbt.2509

Basumatary B, Brahma S, Nath B, Basumatary SF, Das B, Basumatary S (2023) Post-harvest waste to value-added materials: Musa Champa plant as renewable and highly effective base catalyst for Jatropha curcas oil-based biodiesel production. Bioresour Technol Rep 21:101338. https://doi.org/10.1016/j.biteb.2023.101338

Bouvier F, Rahier A, Camara B (2005) Biogenesis, molecular regulation and function of plant isoprenoids. Prog Lipid Res 44(6):357–429. https://doi.org/10.1016/j.plipres.2005.09.003

Bridgwater AV (2012) Review of fast pyrolysis of biomass and product upgrading. Biomass Bioenergy 38:68–94. https://doi.org/10.1016/j.biombioe.2011.01.048

Bruder S, Moldenhauer EJ, Lemke RD, Ledesma-Amaro R, Kabisch J (2019) Drop-in biofuel production using fatty acid photodecarboxylase from Chlorella variabilis in the oleaginous yeast *Yarrowia lipolytica*. Biotechnol Biofuels 12:1–13. https://doi.org/10.1186/s13068-019-1542-4

Cai W, Zhang W (2018) Engineering modular polyketide synthases for production of biofuels and industrial chemicals. Curr Opin Biotechnol 50:32–38. https://doi.org/10.1016/j.copbio.2017.08.017

Chen B, Lee DY, Chang MW (2015) Combinatorial metabolic engineering of *Saccharomyces cerevisiae* for terminal alkene production. Metab Eng 31:53–61. https://doi.org/10.1016/j.ymben.2015.06.009

Chen BS, Zeng YY, Liu L, Chen L, Duan P, Luque R, Zhang W (2022) Advances in catalytic decarboxylation of bioderived fatty acids to diesel-range alkanes. Renew Sust Energ Rev 158: 112178. https://doi.org/10.1016/j.rser.2022.112178

Cheong S, Clomburg JM, Gonzalez R (2016) Energy-and carbon-efficient synthesis of functionalized small molecules in bacteria using non-decarboxylative Claisen condensation reactions. Nat Biotechnol 34(5):556–561. https://doi.org/10.1038/nbt.3505

Choi YJ, Lee SY (2013) Microbial production of short-chain alkanes. Nature 502:571–574. https://doi.org/10.1038/nature12536

Dahman Y, Syed K, Begum S, Roy P, Mohtasebi B (2019) 14 – biofuels: their characteristics and analysis. In: Verma D (ed) Woodhead publishing series in composites science and engineering, biomass, biopolymer-based materials, and bioenergy. Woodhead Publishing, pp 277–325

Dickschat JS (2016) Bacterial terpene cyclases. Nat Prod Rep 33:87–110. https://doi.org/10.1039/c5np00102a

Dongsoo Y, Hengrui Z, Sang YL (2021) Production of Diversified Polyketides by Metabolic Engineering. Biochemistry 60(46):3424–3426. https://doi.org/10.1021/acs.biochem.1c00290

El-Sharoud WM, Zalma SA, Rios-Solis L, Ledesma-Amaro R (2023) Over-expression of α-bisabolene by metabolic engineering of *Yarrowia lipolytica* employing a golden gate DNA assembly toolbox. Biotechnol Notes 4:14–19. https://doi.org/10.1016/j.biotno.2022.12.005

Englund E, Shabestary K, Hudson EP, Lindberg P (2018) Systematic overexpression study to find target enzymes enhancing production of terpenes in *Synechocystis* PCC 6803, using isoprene as a model compound. Metab Eng 49:164–177. https://doi.org/10.1016/j.ymben.2018.07.004

Eroglu E, Melis A (2010) Extracellular terpenoid hydrocarbon extraction and quantitation from the green microalgae *Botryococcus braunii var. Showa*. Bioresour Technol 101:2359–2366. https://doi.org/10.1016/j.biortech.2009.11.043

Gao X, Gao F, Liu D, Zhang H, Nie X, Yang C (2016) Engineering the methylerythritol phosphate pathway in cyanobacteria for photosynthetic isoprene production from CO_2. Energy Environ Sci 9(4):1400–1411. https://doi.org/10.1039/C5EE03102H

Gujjala LKS, Kumar S, Rajak RC, Banerjee R (2018) Bioalkanes and bioalkenes: an ecofriendly and alternate fuel in bioenergy research. Biorefin 4:287–304. https://doi.org/10.1007/978-3-319-67678-4_13

Henritzi S, Fischer M, Grininger M, Oreb M, Boles E (2018) An engineered fatty acid synthase combined with a carboxylic acid reductase enables de novo production of 1-octanol in *Saccharomyces cerevisiae*. Biotechnol Biofuels 11:1–12. https://doi.org/10.1186/s13068-018-1149-1

Herman NA, Zhang W (2016) Enzymes for fatty acid-based hydrocarbon biosynthesis. Curr Opin Chem Biol 35:22–28. https://doi.org/10.1016/j.cbpa.2016.08.009

Hwang S, Lee N, Cho S, Palsson B, Cho BK (2020) Repurposing modular polyketide synthases and non-ribosomal peptide synthetases for novel chemical biosynthesis. Front Mol Biosci 7:1–27. https://doi.org/10.3389/fmolb.2020.00087

IEA (2019) Renewables 2019. Paris. See https://www.iea.org/reports/renewables-2019

Jaroensuk J, Intasian P, Wattanasuepsin W, Akeratchatapan N, Kesornpun C, Kittipanukul N, Chaiyen P (2020) Enzymatic reactions and pathway engineering for the production of renewable hydrocarbons. J Biotechnol 309:1–19. https://doi.org/10.1016/j.jbiotec.2019.12.010

Jeswani HK, Chilvers A, Azapagic A (2020) Environmental sustainability of biofuels: a review. Proc R Soc A 476:1–37. https://doi.org/10.1098/rspa.2020.0351

Jiang W, Gu P, Zhang F (2018) Steps towards 'drop-in' biofuels: focusing on metabolic pathways. Curr Opin Biotechnol 53:26–32. https://doi.org/10.1016/j.copbio.2017.10.010

Jian-Zhong X, Sheng-Ling C, Si-Qi N (2023) Systematic modulating carbon metabolism to improve α-farnesene production in *Pichia pastoris*. Fuel Process Technol 247:1–10. https://doi.org/10.1016/j.fuproc.2023.107757

Jung J, Lim JH, Kim SY, Im DK, Seok JY, Lee SJV, Oh MK, Jung GY (2016) Precise precursor rebalancing for isoprenoids production by fine control of gapA expression in *Escherichia coli*. Metab Eng 38:401–408. https://doi.org/10.1016/j.ymben.2016.10.003

Junne S, Kabisch J (2016) Fueling the future with biomass: processes and pathways for a sustainable supply of hydrocarbon fuels and biogas. Eng Life Sci 17:14–26. https://doi.org/10.1002/elsc.201600112

Kang MK, Nielsen J (2017) Biobased production of alkanes and alkenes through metabolic engineering of microorganisms. J Ind Microbiol Biotechnol 44:613–622. https://doi.org/10.1007/s10295-016-1814-y

Karatzos S, van Dyk S (2016) Drop-in biofuel production via conventional (lipid/fatty acid) and advanced (biomass) routes. Part I Biofuels, Bioprod Biorefin 8:743. https://doi.org/10.1002/BBB

Karatzos S, McMillan J, Saddler J, Donohoo PE, Weiss MA et al (2016) The potential and challenges of drop-in biofuels. IEA Bioenergy Task 39. (2014). http://task39.org. Accessed 16 Nov 2016

Karatzos S, van Dyk JS, McMillan JD, Saddler J (2017) Drop-in biofuel production via conventional (lipid/fatty acid) and advanced (biomass) routes. Part I Biofuels Bioprod Bioref 11:344–362. https://doi.org/10.1002/bbb.1746

Keasling JD, Chou H (2008) Metabolic engineering delivers next-generation biofuels. Nat Biotechnol 26(3):298–299. https://doi.org/10.1038/nbt0308-298

Khusnutdinov IS, Goncharova IN, Safiulina AG, Safina DN (2023) Synthesis of potential fuel additives via interaction of isoprene with various classes of oxygen-containing compounds. Egypt J Pet 32(1):9–14. https://doi.org/10.1016/j.ejpe.2022.11.002

Kornfuehrer T, Eustáquio AS (2019) Diversification of polyketide structures: via synthase engineering. Med Chem Comm 10:1256–1272. https://doi.org/10.1039/c9md00141g

Kumar M, Morya R, Gnansounou E, Larroche C, Thakur IS (2017) Characterization of carbon dioxide concentrating chemolithotrophic bacterium *Serratia* sp. ISTD04 for production of biodiesel. Bioresour Technol 243:893–897. https://doi.org/10.1016/j.biortech.2017.07.067

Lee HJ, Lee J, Lee SM, Um Y, Kim Y, Sim SJ, Choi JI, Woo HM (2017) Direct conversion of CO_2 to α-farnesene using metabolically engineered *Synechococcus elongatus* PCC 7942. J Agric Food Chem 65(48):10424–10428. https://doi.org/10.1021/acs.jafc.7b03625

Lim JH, Jung GYA (2017) Simple method to control glycolytic flux for the design of an optimal cell factory. Biotechnol Biofuels 10:160. https://doi.org/10.1186/s13068-017-0847-4

Liu K, Li S (2020) Biosynthesis of fatty acid-derived hydrocarbons: perspectives on enzymology and enzyme engineering. Curr Opin Biotechnol 62:7–14. https://doi.org/10.1016/j.copbio.2019.07.005

Liu Y, Wang C, Yan J, Zhang W, Guan W, Lu X, Li S (2014) Hydrogen peroxide-independent production of α-alkenes by OleT JE P450 fatty acid decarboxylase. Biotechnol Biofuels 7:1–12. https://doi.org/10.1186/1754-6834-7-28

Liu Q, Wu K, Cheng Y, Lu L, Xiao E, Zhang Y, Deng Z, Liu T (2015) Engineering an iterative polyketide pathway in *Escherichia coli* results in single-form alkene and alkane overproduction. Metab Eng 28:82–90. https://doi.org/10.1016/j.ymben.2014.12.004

Liu CL, Tian T, Alonso-Gutierrez J et al (2018) Renewable production of high density jet fuel precursor sesquiterpenes from *Escherichia coli*. Biotechnol Biofuels 11(285):1–15. https://doi.org/10.1186/s13068-018-1272-z

Liu S, Zhang M, Ren Y et al (2021) Engineering *Rhodosporidium toruloides* for limonene production. Biotechnol Biofuels 14(243):1–11. https://doi.org/10.1186/s13068-021-02094-7

Liu Y, Wang Z, Cui Z, Qi Q, Hou J (2022) Progress and perspectives for microbial production of farnesene. Bioresour Technol 347:1–12. https://doi.org/10.1016/j.biortech.2022.126682

Liu-Jing W, Zhong YT, Ming-Yue N, Shun-Cheng L, Qiang H (2021) Biosynthesis of α-Pinene by genetically engineered *Yarrowia lipolytica* from low-cost renewable feedstocks. J Agric Food Chem 69(1):275–285. https://doi.org/10.1021/acs.jafc.0c06504

Loreto F, Velikova V (2001) Isoprene produced by leaves protects the photosynthetic apparatus against ozone damage, quenches ozone products, and reduces lipid peroxidation of cellular membranes. Plant Physiol 127:1781. https://doi.org/10.1104/pp.010497

Lozada NJH, Simmons TR, Xu K, Jindra MA, Pfleger BF (2020) Production of 1-octanol in *Escherichia coli* by a high flux thioesterase route. Metab Eng 61:352–359. https://doi.org/10.1016/j.ymben.2020.07.004. Get rights and content

Lu R, Shi TQ, Lin L, Ledesma-Amaro R, Ji XJ, Huang H (2022) Advances in metabolic engineering of yeasts for the production of fatty acid-derived hydrocarbon fuels. Green Chem Eng 3:289–303. https://doi.org/10.1016/j.gce.2022.07.008

Machado HB, Dekishima Y, Luo H, Lan EI, Liao JC (2012) A selection platform for carbon chain elongation using the CoA-dependent pathway to produce linear higher alcohols. Metab Eng 14(5):504–511. https://doi.org/10.1016/j.ymben.2012.07.002

Mahapatra S, Kumar D, Singh B, Sachan PK (2021) Biofuels and their sources of production: a review on cleaner sustainable alternative against conventional fuel, in the framework of the food and energy nexus. Energy Nexus 4:1–12. https://doi.org/10.1016/j.nexus.2021.100036

Maurya R, Gohil N, Nixon S, Kumar N, Noronha SB, Dhali D, Singh V (2023) Rewiring of metabolic pathways in yeasts for sustainable production of biofuels. Bioresour Technol 372: 128668. https://doi.org/10.1016/j.biortech.2023.128668

Meadows A, Hawkins K, Tsegaye Y et al (2016) Rewriting yeast central carbon metabolism for industrial isoprenoid production. Nature 537:694–697. https://doi.org/10.1038/nature19769

Monteiro RRC, da Silva SSO, Cavalcante CL, de Luna FMT, Bolivar JM, Vieira RS, Fernandez-Lafuente R (2022) Biosynthesis of alkanes/alkenes from fatty acids or derivatives (triacylglycerols or fatty aldehydes). Biotechnol Adv 61:108045. https://doi.org/10.1016/j.biotechadv.2022.108045

Neves RPP, Ferreira P, Medina FE, Paiva P, Sousa JPM, Viegas MF, Fernandes PA, Ramos MJ (2022) Engineering of PKS Megaenzymes—a promising way to biosynthesize high-value active molecules. Top Catal 65:544–562. https://doi.org/10.1007/s11244-021-01490-5

Papoutsakis ET (2015) Reassessing the progress in the production of advanced biofuels in the current competitive environment and beyond: what are the successes and where progress eludes us and why. Ind Eng Chem Res 54(42):10170–10182. https://doi.org/10.1021/acs.iecr.5b01695

Peñuelas J, Llusia J, Asensio D, Munné-Bosch S (2005) Linking isoprene with plant thermotolerance, antioxidants and monoterpene emissions. Plant Cell Environ 28:278–286. https://doi.org/10.1111/j.1365-3040.2004.01250.x

Peralta-Yahya PP, Zhang F, Del Cardayre SB, Keasling JD (2012) Microbial engineering for the production of advanced biofuels. Nature 488(7411):320–328. https://doi.org/10.1038/nature11478

Ramli NA, Rahman RA, Illias RM (2018) Enhancement of Isobutanol and 3-Methyl-1-butanol production yields in *Saccharomyces cerevisiae* without genetic modification. JEST 1(2–2). https://doi.org/10.11113/jest.v1n2-2.26

Renewable Energy Policy Network for the 21st Century (REN21). 2019 Renewables 2019—Global status report. See https://www.ren21.net/wp-content/uploads/2019/05/gsr_2019_full_report_en.pdf

Rodriguez GM, Atsumi S (2014) Toward aldehyde and alkane production by removing aldehyde reductase activity in *Escherichia coli*. Metab Eng 25:227–237. https://doi.org/10.1016/j.ymben.2014.07.012

Rohmer M (1999) The discovery of a mevalonate-independent pathway for isoprenoid biosynthesis in bacteria, algae and higher plants. Nat Prod Rep 16(5):565–574. https://doi.org/10.1039/A709175C

Rui Z, Li X, Zhu X, Liu J, Domigan B, Barr I, Cate JHD, Zhang W (2014) Microbial biosynthesis of medium-chain 1-alkenes by a nonheme iron oxidase. Proc Natl Acad Sci USA 111:18237–18242. https://doi.org/10.1073/pnas.1419701112

Rui Z, Harris NC, Zhu X, Huang W, Zhang W (2015) Discovery of a family of desaturase-like enzymes for 1-alkene biosynthesis. ACS Catal 5:7091–7094. https://doi.org/10.1021/acscatal.5b01842

Sacchettini JC, Poulter CD (1997) Creating isoprenoid diversity. Science 277:1788–1789. https://doi.org/10.1126/science.277.5333.1788

Sanadze G (1956) Emission of gaseous organic substances from plants. Repertuar Akademiia Nauk Gruzinskoi SSR 17:429–433

Santillan-Jimenez E, Crocker M (2012) Catalytic deoxygenation of fatty acids and their derivatives to hydrocarbon fuels via decarboxylation/decarbonylation. J Chem Technol Biotechnol 87(8):1041–1050. https://doi.org/10.1002/jctb.3775

Santner P, Chanquia SN, Petrovai N, Benfeldt FV, Kara S, Eser BE (2023) Biocatalytic conversion of fatty acids into drop-in biofuels: towards sustainable energy sources. EFB Bioeconomy 3:100049. https://doi.org/10.1016/j.bioeco.2023.100049

Sarria S, Kruyer NS, Peralta-Yahya P (2017) Microbial synthesis of medium-chain chemicals from renewables. Nat Biotechnol 35(12):1158–1166. https://doi.org/10.1038/nbt.4022

Schirmer A, Rude MA, Li X, Popova E, Del Cardayre SB (2010) Microbial biosynthesis of alkanes. Science 5991:559–562. https://doi.org/10.1126/science.1187936

Seemann M, Campos N, Rodriguez-Concepción M, Ibañez E, Duvold T et al (2002) Isoprenoid biosynthesis in *Escherichia coli* via the methylerythritol phosphate pathway: enzymatic conversion of methylerythritol cyclodiphosphate into a phosphorylated derivative of (E)-2-methylbut-2-ene-1, 4-diol. Tetrahedron Lett 43:1413–1415. https://doi.org/10.1016/S0040-4039(02)00003-5

Sharma A, Yazdani SS (2021) Microbial engineering to produce fatty alcohols and alkanes. J Ind Microbiol Biotechnol 48:kuab011. https://doi.org/10.1093/jimb/kuab011

Sheppard MJ, Kunjapur AM, Wenck SJ, Prather KL (2014) Retro-biosynthetic screening of a modular pathway design achieves selective route for microbial synthesis of 4-methyl-pentanol. Nat Commun 5(1):5031. https://doi.org/10.1038/ncomms6031

Song X, Yu H, Zhu K (2016) Improving alkane synthesis in *Escherichia coli* via metabolic engineering. Appl Microbiol Biotechnol 100:757–767. https://doi.org/10.1007/s00253-015-7026-y

Sun X, Shen X, Jain R, Lin Y, Wang J, Sun J, Wang J, Yan Y, Yuan Q (2015) Synthesis of chemicals by metabolic engineering of microbes. Chem Soc Rev 44:3760–3785. https://doi.org/10.1039/c5cs00159e

Terry G, Stokes N, Hewitt C, Mansfield T (1995) Exposure to isoprene promotes flowering in plants. J Exp Bot 46:1629. https://www.jstor.org/stable/23695250

Wackett LP (2011) Engineering microbes to produce biofuels. Curr Opin Biotechnol 22(3):388–393. https://doi.org/10.1016/j.copbio.2010.10.010

Walid ME, Samar AZ, Leonardo RS, Rodrigo LA (2023) Over-expression of α-bisabolene by metabolic engineering of Yarrowia lipolytica employing a golden gate DNA assembly toolbox. Biotechnology Notes 4:14–19. https://doi.org/10.1016/j.biotno.2022.12.005
Wen RC, Shen CR (2016) Self-regulated 1-butanol production in *Escherichia coli* based on the endogenous fermentative control. Biotechnol Biofuels 9(1):1–15. https://doi.org/10.1186/s13068-016-0680-1
Westfall PJ, Pitera DJ, Lenihan JR, Eng D, Woolard FX, Regentin R, Horning T, Tsuruta H, Melis DJ, Owens A, Fickes S, Diola D, Benjamin KR, Keasling JD, Leavell MD, McPhee DJ, Renninger NS, Newman JD, Paddon CJ (2012) Production of amorphadiene in yeast, and its conversion to dihydroartemisinic acid, precursor to the antimalarial agent artemisinin. PNAS 109(3):E111–E118. https://doi.org/10.1073/pnas.1110740109
Xie S, Jia C, Prakash A, Palafox MI, Pfaendtne J, Lin H (2019) Generic biphasic catalytic approach for producing renewable diesel from fatty acids and vegetable oils. ACS Catal 9(4):3753–3763. https://doi.org/10.1021/acscatal.9b00215
Yan H, Wang Z, Wang F, Tan T, Liu L (2016) Biosynthesis of chain-specific alkanes by metabolic engineering in *Escherichia coli*. Eng Life Sci 16:53–59. https://doi.org/10.1002/elsc.201500057
Yang J, Xian M, Su S, Zhao G, Nie Q et al (2012) Enhancing production of bio-isoprene using hybrid MVA pathway and isoprene synthase in *E. Coli*. PLoS One 7(4):e33509. https://doi.org/10.1371/journal.pone.0033509
Yang C, Gao X, Jiang Y, Sun B, Gao F, Yang S (2016) Synergy between methylerythritol phosphate pathway and mevalonate pathway for isoprene production in *Escherichia coli*. Metab Eng 37:79–91. https://doi.org/10.1016/j.ymben.2016.05.003
Yang K, Li F, Qiao Y, Zhou Q, Hu Z, He Y, Yan Y, Xu L, Madzak C, Yan J (2018) Design of a new multienzyme complex synthesis system based on *Yarrowia lipolytica* simultaneously secreted and surface displayed fusion proteins for sustainable production of fatty acid-derived hydrocarbons. ACS Sustain Chem Eng 6:17035–17043. https://doi.org/10.1021/acssuschemeng.8b04401
Yang K, QiaoY LF, Xu Y, Yan Y, Madzak C, Yan J (2019) Subcellular engineering of lipase dependent pathways directed towards lipid related organelles for highly effectively compartmentalized biosynthesis of triacylglycerol derived products in *Yarrowia lipolytica*. Metab Eng 55:231–238. https://doi.org/10.1016/j.ymben.2019.08.001
Yang D, Zhou H, Lee SY (2021) Production of diversified polyketides by metabolic engineering. Biochemistry 60:3424–3426. https://doi.org/10.1021/acs.biochem.1c00290
Yuan J, Ching CB (2014) Combinatorial engineering of mevalonate pathway for improved amorpha-4,11-diene production in budding yeast. Biotechnol Bioeng 111(3):608–617. https://doi.org/10.1002/bit.25123
Yuzawa S, Keasling JD, Katz L (2017) Bio-based production of fuels and industrial chemicals by repurposing antibiotic-producing type i modular polyketide synthases: opportunities and challenges. J Antibiot 70:378–385. https://doi.org/10.1038/ja.2016.136
Zargar A, Bailey CB, Haushalter RW, Eiben CB, Katz L, Keasling JD (2017) Leveraging microbial biosynthetic pathways for the generation of 'drop-in'biofuels. Curr Opin Biotechnol 45:156–163. https://doi.org/10.1016/j.copbio.2017.03.004
Zargar A, Barajas JF, Lal R, Keasling JD (2018) Polyketide synthases as a platform for chemical product design. AICHE J 64:4201–4207. https://doi.org/10.1002/aic.16351
Zhang F, Rodriguez S, Keasling JD (2011) Metabolic engineering of microbial pathways for advanced biofuels production. Curr Opin Biotechnol 22(6):775–783. https://doi.org/10.1016/j.copbio.2011.04.024
Zhang F, Ouellet M, Batth TS, Adams PD, Petzold CJ, Mukhopadhyay A, Keasling JD (2012) Enhancing fatty acid production by the expression of the regulatory transcription factor FadR. Metab Eng 14:653–660. https://doi.org/10.1016/j.ymben.2012.08.009
Zhang Y, Lane S, Chen JM, Hammer SK, Luttinger J, Yang L, Avalos JL (2019) Xylose utilization stimulates mitochondrial production of isobutanol and 2-methyl-1-butanol in *Saccharomyces cerevisiae*. Biotechnol Biofuels 12:1–15. https://doi.org/10.1186/s13068-019-1560-2

Zhang X, Liu X, Yonghui M, Lijuan Z, Jianjun Q, Guang-Rong Z (2021) Combinatorial engineering of *Saccharomyces cerevisiae* for improving limonene production. Biochem Eng J 176:1–10. https://doi.org/10.1016/j.bej.2021.108155

Zhao J, Bao X, Li C, Shen Y, Hou J (2016) Improving monoterpene geraniol production through geranyl diphosphate synthesis regulation in *Saccharomyces cerevisiae*. Appl Microbiol Biotechnol 100:4561–4571

Zhao J, Li C, Zhang Y et al (2017) Dynamic control of *ERG20* expression combined with minimized endogenous downstream metabolism contributes to the improvement of geraniol production in *Saccharomyces cerevisiae*. Microb Cell Fact 16:1–11. https://doi.org/10.1186/s12934-017-0641-9

Zhou YJ, Buijs NA, Zhu Z, Qin J, Siewers V, Nielsen J (2016) Production of fatty acid-derived oleochemicals and biofuels by synthetic yeast cell factories. Nat Commun 7:11709. https://doi.org/10.1038/ncomms11709

Zhu Z, Zhou YJ, Kang MK, Krivoruchko A, Buijs NA, Nielsen J (2017) Enabling the synthesis of medium chain alkanes and 1-alkenes in yeast. Metab Eng 44:81–88. https://doi.org/10.1016/j.ymben.2017.09.007

Zhu H, Si X, Wang Y, Zhu P, Pang X, Wang X, Fauconnier ML, Ju N, Zhang S, Lv J (2023) Fatty acid, triglyceride, and kinetic properties of milk fat fractions made by the combination of dry fractionation and short-path molecular distillation. J Dairy Sci 106(10):6655–6670. https://doi.org/10.3168/jds.2022-22970

Chapter 3
Renewable Residues as Feedstock for Drop-in Biofuel Production

Luis Fernando Flores, Carlos Saul Osorio-Gonzalez, Rahul Saini, and Satinder Kaur Brar

Abstract There is a growing interest in promoting the use of residues as a bioenergy feedstock to produce drop-in biofuels. This approach is considered a promising and sustainable solution to replace conventional fossil fuels. Renewable residues offer a renewable source of carbon that addresses energy, economic, and environmental requirements. This chapter examines various types of renewable residues, categorizing them based on their state and origin, and highlights their potential as valuable raw materials for drop-in biofuel production. Biochemical methods used to convert these renewable residues into drop-in biofuels, as well as their pretreatment techniques are analyzed. It also discusses the associated advantages of producing these advanced biofuels from renewable residues, such as decreasing dependence on fossil fuels within the context of reducing greenhouse gas emissions and promoting a circular and sustainable economy. However, the chapter also emphasizes the limitations and challenges that need to be addressed, including appropriate raw material selection, residue availability, and process optimization. Furthermore, it mentions leading companies involved in the production of drop-in biofuels using renewable residues, as they are driving innovation and developing advanced technologies in this field. In summary, this chapter offers a comprehensive overview of drop-in biofuel production from renewable residues. It underscores the significance of this approach as a sustainable and promising alternative for the fuel industry while also highlighting potential areas for future research.

Keywords Renewable residues · By-products · Sustainable feedstocks · Renewable feedstocks

---

L. F. Flores
Escuela de Producción, Tecnología y Medio Ambiente, Universidad Nacional de Río Negro, Villa Regina, Argentina

C. S. Osorio-Gonzalez · R. Saini · S. K. Brar (✉)
Department of Civil Engineering, Lassonde School of Engineering, North York, ON, Canada
e-mail: satinder.brar@lassonde.yorku.ca

S. K. Brar et al. (eds.), *The Microbiology of the Drop-in Biofuel Production*, Biofuel and Biorefinery Technologies 15, https://doi.org/10.1007/978-3-031-61637-2_3

3.1 Introduction

The increasing demand for renewable fuels is driven by the need to address the scarcity of fossil resources and address sustainability concerns. As a result, the interest in using residues as renewable feedstocks for producing drop-in biofuels is on the rise, aiming for sustainable solutions to energy and environmental challenges. In this regard, renewable residues have emerged as valuable sources of renewable carbon and energy (van Dyk et al. 2019) offering a promising and sustainable alternative to conventional fossil fuels (Fig. 3.1).

Renewable residues, classified either as wastes, resources out of place (Salihu et al. 2012) or by-products, encompass a wide range of organic and inorganic materials. These include agricultural residues such as stalks, leaves, straw, husks, and shells, animal manure, wood residues like sawdust and wood chips, remnants and leftovers from food waste, as well as wastewater and certain gaseous by-products (Karatzos et al. 2014; Whalen et al. 2017; Chia et al. 2022). These materials can be utilized as valuable resources for energy production, providing a renewable source of carbon, energy, and other valuable products. In this context, various processes like anaerobic digestion, composting, pyrolysis, fermentation, combustion, conversion, treatment, or valorization can be employed to produce drop-in biofuels as a renewable energy source (Cabrera-Jiménez et al. 2022).

This chapter explores diverse forms of renewable residues used in the synthesis of drop-in biofuels, classifying them based on their state and origin. It emphasizes the importance of appropriate feedstock selection, pretreatment methods, and biochemical conversion techniques to ensure the sustainability and viability of drop-in biofuel production from renewable residues.

Furthermore, specific characteristics and potentials for sustainable biofuel production, highlighting the associated benefits such as economic viability, environmental sustainability, and social implications are presented. It also compares drop-in biofuel production from renewable residues with other generations of biofuels, considering factors like potential impacts on food supply, environment, and

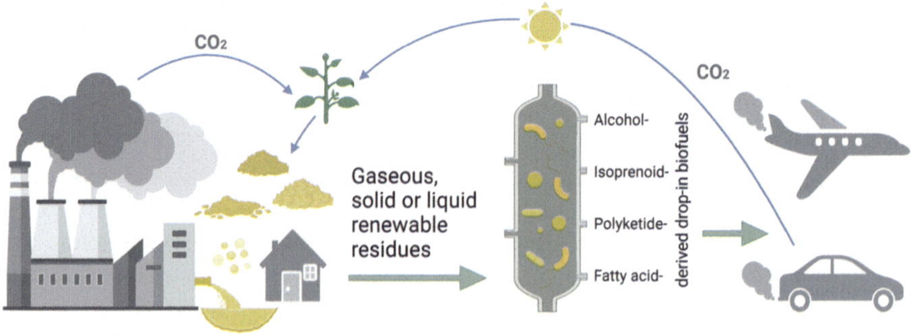

Fig. 3.1 Renewable residues conversion into drop-in biofuels to achieve life-cycle net zero emissions

economy. In addition to discussing the benefits, the chapter addresses the limitations and challenges that need to be overcome in order to optimize biochemical conversion processes and promote sustainable drop-in biofuel production from renewable residues, emphasizing the importance of strategies and technical considerations in addressing these challenges. Moreover, the chapter mentions leading companies that are actively involved in drop-in biofuel production using renewable residues, showcasing their contributions to innovation and technological advancements in this field.

3.2 Renewable Residues

A residue is considered renewable when it is derived from a renewable and sustainable source, and its production does not deplete or unsustainably harm natural resources or the environment. Renewable residues often come from organic or biological materials that can be regenerated or replenished within a relatively short period of time through natural processes and human intervention (Salihu et al. 2012; Chau et al. 2022; Kocaturk et al. 2023).

However, for a residue to be considered renewable, it generally needs to meet certain additional conditions that support its sustainability and ability to replenish over time. For drop-in biofuel production, a renewable residue can be analyzed according to the following conditions:

- **Availability and regenerative capacity:** The residue comes from a source that can be continuously and sustainably regenerated (Chavan et al. 2022; Gil 2022). This implies that the source of origin has a regeneration rate that is equal to or faster than the rate of consumption or waste generation. For instance, biomass waste, agricultural waste, food waste, municipal solid waste or sewage sludge are produced all over the world directly or indirectly by providing food, energy, or water to society (Usmani et al. 2021; Wang and Tester 2023).
- **Sustainable process or source:** If the residue is generated as a by-product or waste of a renewable process, such as the production of biogas from organic waste, it can be considered renewable. It is important that the production or generation of the residue does not harm the environment or deplete natural resources in an unsustainable manner. Therefore, renewable residues should originate from renewable processes that employ sustainable and environmentally friendly methods, without relying on finite resources or engaging in harmful practices. Factors to consider encompass resource utilization, energy consumption, avoidance of hazardous substances or chemicals, non-renewable waste generation, environmental impact, and compliance with environmental regulations and best practices. This involves adhering to waste management regulations, minimizing waste generation, and maximizing resource efficiency throughout the production process. Residues originating from renewable sources such as agricultural waste, forest waste, food waste, or organic matter are more likely to be considered renewable and a sustainable source of energy (Ouadi et al. 2019).

- **Low environmental impact:** The production of the residue for use in the production of drop-in biofuels should have a minimal environmental impact. This involves examining elements like water consumption, land utilization, potential effects on biodiversity, air quality, soil deterioration, water contamination, and excessive emissions of greenhouse gases.
- **Not competitive with food or feed production:** The production of residues should not compete with food or feed production, as this can have negative repercussions on the food chain and significantly impact food availability, food security, and agricultural resources. Within the framework of drop-in biofuel production, this condition opposes the cultivation of biofuel crops on a large scale, which can potentially harm the environment by exerting pressure on limited arable land thus competing with food crops, exacerbating deforestation, and leading to bush burning, resulting in additional emissions during the application of fertilizers or other chemicals that can contaminate not only terrestrial but also aquatic habitats (Kargbo et al. 2021; Awogbemi and Kallon 2022). Furthermore, this condition also includes avoiding the use of waste materials that could be utilized as food or feed and promoting the use of waste materials that are unsuitable for human or animal consumption.
- **Reuse or recyclability:** If the waste can be recycled or reused for other productive purposes, it is considered renewable. This promotes a circular economy approach, where waste materials are treated as valuable resources that can be used sustainably and efficiently in the production of drop-in biofuels.
- **Regulatory compliance:** The waste production and utilization must comply with relevant environmental regulations and standards related to waste management, renewable energy, and environmental protection. This is to prevent negative impacts on public health and the increase of pollution and environmental degradation.
- **Energy and economic efficiency:** The energy and economic efficiency of utilizing the waste as a renewable resource must be evaluated. Striving for a balance between the amount of energy required to transform the waste into biofuel and the energy obtained from it ensures that production is efficient and profitable. This involves considering the costs of collection, processing, and transformation, as well as the economic and energy benefits derived from its use in the drop-in biofuel production.

In conclusion, meeting these conditions is crucial to ensure that a waste can be genuinely renewable and utilized as a sustainable feedstock for drop-in biofuels production. However, it is important to recognize that determining the renewability of a waste can be complex and context-dependent. Different industries, processes, and waste streams may exhibit varying levels of renewability, making it essential to consider the specific characteristics and circumstances of each waste type or production process.

Evaluating the environmental impact and sustainability of the waste throughout its entire lifecycle, including production, processing, use, and disposal, is necessary to identify potential enhancements and ensure sustainability at all stages of the process. Consulting environmental experts, complying with relevant regulations, and incorporating sustainability principles are recommended to determine whether a waste can be classified as renewable. Thorough evaluation and verification of these conditions are crucial in guaranteeing the sustainability and viability of utilizing waste as renewable resources.

3.3 Types of Renewable Residues and Their Application in Drop-in Biofuel Production

Renewable residues from various sources can serve as an environmentally-responsible feedstock for drop-in biofuel production. The classification of these residues is diverse, taking into account factors such as their origin, material composition, and physical state (Peng et al. 2023). In this section, a categorization of renewable residues into solids, liquids, and gases, accompanied by their respective conversion methods to drop-in biofuels or its precursors, is shown in the diagram in Fig. 3.2.

It is important to emphasize that certain renewable residues can exist in multiple physical states, depending on their inherent composition and the specific processing techniques employed. For instance, biomass can undergo conversion into gaseous syngas, solid biochar, or liquid bio-oil, contingent upon the chosen technological pathway. An example is illustrated in Fig. 3.2, where biomass pyrolysis produces syngas as a by-product that acetogenic bacteria can then ferment to produce alcohol-derived drop-in biofuels.

To provide further insight, the following subsections address examples of drop-in biofuels derived from renewable residues, presented within each corresponding physical state category.

3.3.1 Gaseous Renewable Residues

Renewable gaseous residues encompass gases that are generated as by-products or emissions during the production process but are not intended to be part of the final product. Its characteristics may differ based on factors like the specific production process, raw materials used, and the treatment technologies applied. Consequently, these residues may contain impurities, contaminants, by-products, or other substances (Karatzos et al. 2017; Martinez-Villarreal et al. 2022).

Various renewable gaseous residues can be identified, including biogas waste, landfill gas, and syngas. Biogas waste is created by decomposing organic waste, landfill gas is generated from the degradation of solid waste, and synthesis gas or syngas is obtained via biomass gasification processes (Czyrnek-Delêtre et al. 2016; Kargbo et al. 2021).

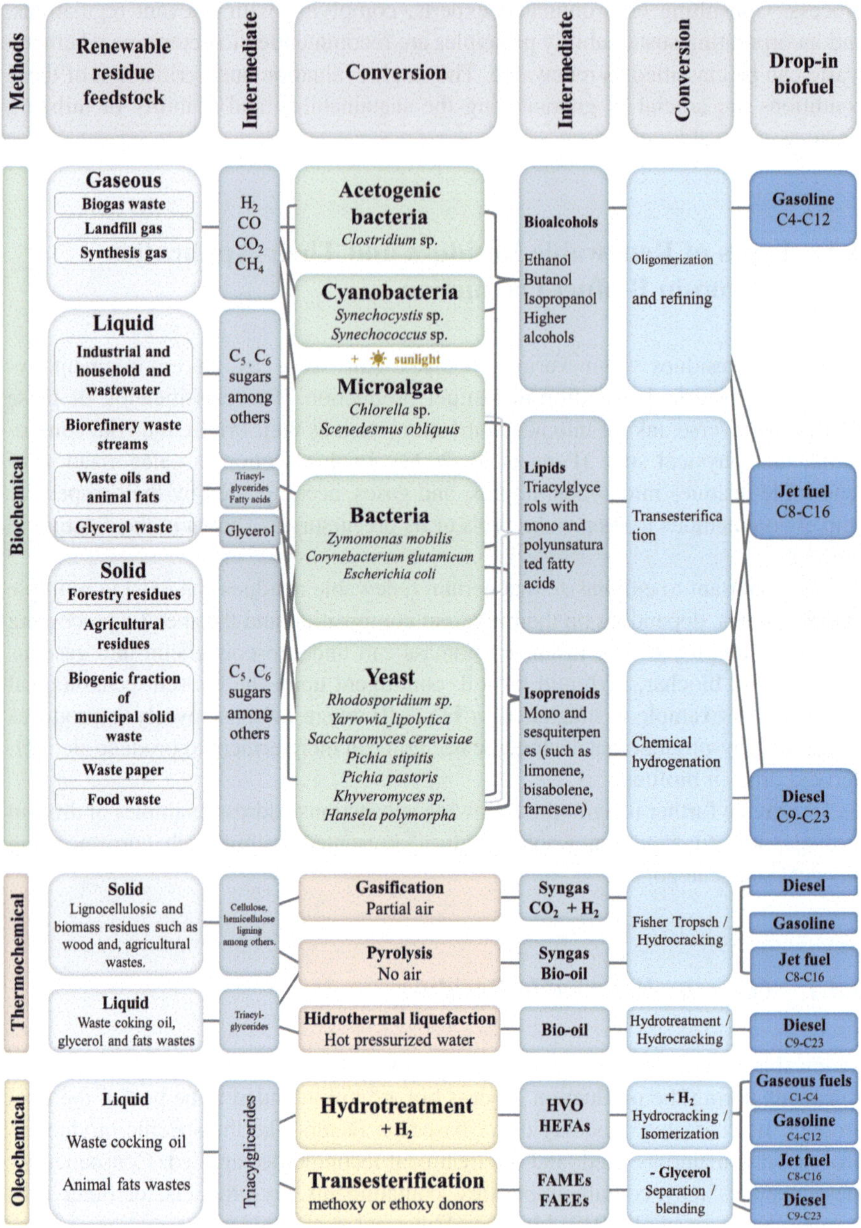

Fig. 3.2 Biochemical, thermochemical, and oleochemical conversion of renewable residues into drop-in biofuels and its precursors. Note that the residues could be potentially bio-converted by various engineered microorganisms according to the progress of metabolic engineering

Nevertheless, the classification of renewable gaseous waste can vary depending on the specific context and regional regulations. In this regard, it is crucial to consider the utilization of non-biological raw materials, such as CO_2 waste from industrial plant emissions, as an environmentally friendly option for sustainable biofuel production (Doliente et al. 2020). For example in this context, specific types of microalgae have demonstrated the ability to absorb CO_2 from combustion gases (Khan et al. 2023), offering environmental benefits and potential economic advantages, including reduced production costs associated with biofuels (Collotta et al. 2018). Consequently, in this chapter, combustion gases will be considered as renewable residues due to their potential to provide carbon as a source for the synthesis of drop-in biofuels, as long as the processes for generating these wastes become more sustainable and renewable.

- **Biogas, syngas, landfill, and biomethane waste:** It consists mainly of methane and carbon dioxide, with small amounts of other gases like nitrogen, oxygen, and volatile organic compounds. Biomethane is a purified form of biogas with a higher methane content. Its residues may contain residual impurities such as carbon dioxide, water vapor, and trace contaminants (Karatzos et al. 2017).

  In addition, in the effort to decrease the amount of biomass or organic waste generated in urban areas or industries, biorefineries could employ different gasification processes to transform organic waste into syngas, which can then be converted into valuable products using bacterial fermentation. Acetogenic bacteria are particularly useful in this context, as they not only capture CO_2 and CO generated as by-products of human activities but also serve as a biological substitute for converting synthesis gas into valuable products during the gasification process in a chemical refinery. However, there is a challenge in the efficiency of syngas fermentation because it is currently low and requires improvement in order to compete with chemical catalysis (García and Galán 2022).

- **Impurities or contaminants:** Renewable gases such as biomethane or biogas may require purification or upgrading processes, which can generate waste in the form of impurities or contaminants that are removed during the purification process, such as carbon dioxide, trace gases, or other pollutants. Furthermore, mixtures of CO_2, CO, and H_2 that are frequently present in certain industrial gas emissions, such as those originating from the steel industry, should also be taken into consideration (Cuellar-Bermudez et al. 2015; Schubert 2020). This is supported by the fact that studies on the fermentation of biogenic gases at laboratory, pilot, and demonstration scales have successfully produced a range of biofuels using algae, bacteria, or yeasts as biocatalysts, whether native or genetically modified, in mixed or pure aerobic and anaerobic cultures. Via anaerobic fermentation through acetogenic bacteria, gases can be valorized to produce ethanol and higher alcohols, among other biofuels. In addition, certain types of microalgae have the ability to directly utilize combustion gases containing high levels of CO_2 ($\sim$26% v/v) generated by various sectors including steel, cement, thermal energy, or incineration industries (Oh et al. 2018). These allow for the valorization, to the extent possible, of industrial biogenic gas emissions through their bioconversion into valuable added-value products (Rene et al. 2022).

3.3.1.1 Renewable Gaseous Residues Used in Drop-in Biofuel Production

The utilization of renewable gaseous waste for its conversion into drop-in biofuels can be achieved mainly through the utilization of microalgae or acetogenic bacteria.

3.3.1.1.1 Microalgae-Mediated Bioconversion of Gaseous Residues into Drop-in Fuels

The emission of CO_2 into the atmosphere has become a global concern due to its annual increase and its recognition as the primary cause of intensifying the greenhouse effect. Nevertheless, microalgae can be employed to sustainably mitigate these emissions (Duarte et al. 2016). While this presents itself as a highly promising solution, this subsection will not only mention its applications for harnessing renewable residue in the production of drop-in biofuels but also address challenges such as the required economic investment.

With the ability to transform inorganic substances (i.e., water, CO_2, PO_4^{3-}, and NO_3^-) into complex organic molecules (such as carbohydrates, proteins, and lipids) through photosynthesis, microalgae are suitable microorganisms for harnessing the potential of gas residues (Zhu et al. 2020). In this context, microalgae have the capability to physiochemically assimilate CO_2 and remediate nutrients from combustion industrial gases (Behera et al. 2022). Extensive research has described algae-based technologies for the treatment of flue gases, degradation of pollutants, and production of biofuels, utilizing CO_2 from gaseous waste in their metabolic pathways (Duarte et al. 2016).

Generally, microalgae have the capability to capture CO_2 from three sources: the atmosphere, emissions from power plants and industrial processes, and soluble carbonates. Atmospheric sequestration of CO_2 is likely the most basic method for carbon absorption, relying on mass transfer from the air to microalgae, however, the potential performance from the atmosphere is limited by the low concentration of CO_2 in the air. Therefore, CO_2 capture from industrial combustion gas emissions could achieve a better recovery due to the higher concentration. Studies show that certain species of microalgae can be cultivated with 10–15% CO_2, a concentration range typically present in the flue gases of power plants (Duarte et al. 2016). Furthermore, studies conducted to date have shown that elevated levels of CO_2 enhance microalgae growth, directly contributing to the reduction of other pollutants (Zieliński et al. 2023). In this respect, these microorganisms can be employed in industries associated with CO_2 emissions during energy generation, enabling the redirection of combustion gas into bioreactors to stimulate microalgae growth.

Consequently, the co-location of a microalgae-to-biofuel conversion plant with a CO_2-emitting industry has been suggested as a potentially efficient solution, both economically and environmentally (Collotta et al. 2018). However, several challenges are associated with this approach:

Firstly, although flue gases serve as a viable CO_2 source for algal photosynthesis, the elevated levels of CO_2, NO_x, and SO_x present in these gases can be detrimental to algae, potentially causing toxicity and acidification of the culture medium. As a result, this may lead to a decrease in both biomass and lipid productivities (Moon et al. 2022). Also, the high temperatures of the gaseous waste can negatively impact the efficiency of microalgae photosynthesis. Additionally, utilizing combustion gases presents obstacles such as low pressure and the need for energy input to supply them to the system (Cuellar-Bermudez et al. 2015). There is also the potential risk of introducing contaminants like dust or heavy metals into the system, which requires microorganisms capable of tolerating such conditions (Cuellar-Bermudez et al. 2015).

Therefore, the implementation of large-scale applications for producing drop-in biofuels from gas waste using microalgae has been limited by techno-economic challenges. To determine the commercial viability of a microalgae biorefinery for drop-in biofuel production, essential considerations should include the use of Life-Cycle Assessment and Techno-Economic Analysis approaches. These analytical tools will enable the evaluation of alternative processing methods for the long-term manufacturing of end products (Thanigaivel et al. 2022).

Alternatively, the carbon sequestration from flue gas and wastewater for algae cultivation has been proposed that could positively affect environmental performance and reduce costs associated with drop-in biofuel production. The direct use of these wastes as feedstock reduces the costs of supplying water and nutrients needed for algae to grow effectively (Zieliński et al. 2023). Addressing these barriers and achieving widespread implementation will require further advancements (Collotta et al. 2018). On the other hand, carbonaceous residues and biomass can undergo gasification to generate syngas with elevated concentrations of H_2 and CO. The feedstocks for gasification include wood residues, municipal or industrial waste, petroleum, natural gas, and chemical waste, coal, lignin, and tires. At the end, the produced syngas can undergo fermentation with the conversion of CO and H_2 into alcohols and organic acids (Phillips et al. 2017). It is claimed that syngas fermentation offers several advantages compared to alternative processes, such as sugar fermentation and thermo-catalytic conversion of syngas. For instance, this strategy can employ lignin in conjunction with the carbohydrate fractions of biomass, unlike sugar fermentation. Compared to thermo-catalytic conversion of syngas, it is considered more economically viable at smaller scales due to lower capital costs and less sensitivity to impurities (van Dyk et al. 2019).

However, a key bottleneck in syngas fermentation appears to be the limited solubility in aqueous solutions of H_2 and CO gases. Before microorganisms can assimilate them, these gases need to be initially absorbed by the aqueous fermentation broth. Due to the relatively low gas-liquid mass transfer and its low solubility in water, the rates of syngas fermentation often stem from physical mass transfer constraints rather than metabolic capacities (van Dyk et al. 2019). For instance, in 2012, a synthetic biogas containing CO_2 and $CH_4$was subjected to photosynthetic fixation by algae and subsequently treated by a methane-oxidizing culture of *Methylocystis parvus*. This groundbreaking achievement marked the first complete microbial conversion of both greenhouse gases into biomass without the need for external O_2 supply. Building upon these findings, it could be an innovative biotechnological method for the conversion of diverse biogas forms into valuable bioproducts (van der Ha et al. 2012). As a result, the implementation of methanotrophic-microalgae co-cultures was proposed as a viable and promising strategy for concurrently mitigating greenhouse gas emissions (Ruiz-Ruiz et al. 2020). Similarly, a mixotrophic cultivation of microalgae was suggested for converting biogas into microalgae-based biodiesel, wherein a co-culture of methanotrophs and *Scenedesmus obliquus* (microalgae) was established. The methanotrophs converted CH_4 into an organic intermediate (along with CO_2) to support the mixotrophic growth of microalgae. This co-culture system, fueled by biogas, presented a great opportunity to produce microalgae-based biodiesel utilizing biogas as a substrate (Xiong et al. 2023).

3.3.1.1.2 Bacteria-Mediated Bioconversion of Gaseous Residues into Drop-in Fuels

Alcohols like n-hexanol and n-butanol offer increased energy density compared to ethanol. Moreover, these higher alcohols are compatible with existing fuel infrastructure and can be converted into aviation fuels and diesel. In 2014, the great potential syngas fermentation to produce higher alcohols was achieved by mixing cultures between the CP15 strain of *Alkalibaculum bacchi* and *Clostridium propionicum*, a producer of propionic acid, because several organisms convert syngas into ethanol, but only a limited number has the ability to produce higher alcohols. The mixed culture resulted into n-butanol and n-propanol besides ethanol through syngas fermentation (Liu et al. 2014). Currently, over hundred anaerobic acetogens bacteria have been isolated, which are capable of synthesizing from CO or CO_2 plus H_2, the intermediate acetyl-CoA. Acetogens can grow using syngas and have been utilized to generate ethanol, acetate, butyrate or 2,3-butanediol, among other key products. Moreover, genetic modifications can enable the production of other compounds like acetone or butanol from syngas. *Clostridium*, for instance, produces not only butanol but also ethanol, acetone, acetic acid, and butyric acid as liquid products, along with H_2 and CO_2 as gaseous by-products (Nanda et al. 2017).

In this context, various industrially valuable acetogenic bacteria have been previously engineered through synthetic biology techniques, including *Clostridium ljungdahlii*, *Clostridium autoethanogenum*, *Acetobacterium woodii*, or *Moorella thermoacetica* (García and Galán 2022) to be improved to make better use of these gaseous renewable wastes. It is also important to note that these scientific endeavors have faced challenges such as the limited gas solubility in water-based solutions and the gas-converting bacteria genetic complexity. Although, LanzaTech, a leading company in the industry, has successfully implemented the bioconversion of syngas to ethanol at a commercial scale using the acetogen *Clostridium autoethanogenum* and industrial waste gases. The company has achieved successful genetic engineering of the species to produce farnesene using a waste gas containing CO (Walls and Rios-Solis 2020). Additionally, researchers have engineered cyanobacteria strains like *Synechococcus* and *Synechocystis* spp. to transform CO_2 into isobutanol through photosynthesis, achieving significant titers of up to 1 g/L (Keasling et al. 2021). Isoprenoids, on the other hand, have been derived from CO_2, synthesis gas, and CH_4. Although these studies primarily serve as proof-of-concept, featuring relatively low titers and yields, they offer potential for integrating renewable gas substrates into other organic complex molecules. These molecules, in turn, hold promise as potential blendstocks for aviation fuel and diesel (Keasling et al. 2021). These were some examples of organisms capable of capturing and recycling carbon from gaseous residues to produce advanced drop-in biofuels, which would otherwise be spread in the atmosphere contributing to environmental pollution.

3.3.2 Liquid Renewable Residues

Liquid renewable residues refer to by-products and organic or biomass-based materials derived from renewable sources that exist in a liquid form. These residues are commonly generated as a result of diverse agricultural, industrial, or municipal activities. Instead of being disposed of, these residues can be effectively utilized to produce energy or valuable products through various processes (Singh et al. 2022; Saba et al. 2023).

Examples of such liquid renewable residues include:

- **Residues derived from biorefineries:** Encompass a wide range of feedstocks, including liquids rich in lignin and waste streams, which are produced during the process of bio-refining. For example, residues like stillage or vinasse generated during bioethanol production can be classified as liquid renewable residues (Pina et al. 2015). Additionally, residues obtained from the anaerobic digestion of organic waste during the production of biogas, known as digestate, are liquid renewable residues that contain a high concentration of nutrients and organic matter (Vural Gursel et al. 2022). Glycerol derived from biodiesel production is another example of a liquid renewable residue (Chilakamarry et al. 2021). This category also includes residues from the cultivation of microorganisms, such as algae, for various applications. After the desired components have been extracted, the remaining liquid can be utilized as a renewable residue.
- **Vegetable oils and animal fats residues:** Liquid residues generated during the production of vegetable oils, as well as used cooking oil from households, restaurants, and food processing industries, can be considered as liquid renewable residues. These residues can be collected and processed into biodiesel or other renewable bioproducts (Walls and Rios-Solis 2020).

Due to its abundance, cost-effectiveness, and environmental benefits cooking oil has been identified as a suitable feedstock for the production of drop-in biofuels (Roy et al. 2014). Similarly, animal fats obtained from livestock or poultry processing industries can serve as liquid renewable residues and can be converted into drop-in biofuels (Malode et al. 2021).

- **Industrial effluents:** It is characterized by their organic and nutrient-rich composition and has been extensively studied as potential raw materials for biofuel production. Studies have shown that wastewater from olive mills, sugar confectioneries, and cheese whey holds promise as a medium for biofuel production, facilitating the production of single-cell oil by oleaginous yeasts (Singh et al. 2022). Furthermore, contaminated wastewater provides a favorable environment for the growth of microalgae, which allows for the removal of environmental contaminants while providing a valuable raw material for biofuel production (Kargbo et al. 2021).

The examples provided above represent just a few instances of liquid renewable residues used for bioenergy production through drop-in biofuel production.

3.3.2.1 Renewable Liquid Residues Used in Drop-in Biofuel Production

Various researchers have explored the production of biobutanol using cost-effective raw materials such as starch, algae biomass, and even gaseous waste, as previously demonstrated. For instance, an innovative approach to enhancing biobutanol production during fermentation involves increasing the carbohydrate content in algae biomass through the utilization of cheese whey. Then, this biomass was fermented by *C. acetobutylicum* achieving a successfully biobutanol production (Kallarakkal et al. 2021). Regarding liquid residues from biorefineries, glycerol, a major by-product of biodiesel manufacturing, holds promise as a cost-effective feedstock. Researchers successfully converted glycerol into farnesane, a precursor to approved jet fuel, using a modified *E. coli* strain (Walls and Rios-Solis 2020). While impressive yields of 8.7 g/L were obtained with pure glycerol, subsequent experiments using crude glycerol yielded lower titers. This decline was attributed to inhibitory compounds in the medium, underscoring the necessity for pretreatment to optimize the productivity when utilizing crude glycerol (Walls and Rios-Solis 2020). Additionally, it has been demonstrated that glycerol is the preferred source of carbon of an engineered *E. coli* strain to produce limonene. This strain attained a remarkable titer of final limonene around 3.6 g/L in a fed-batch process under glycerol-limited conditions. Furthermore, *E. coli* underwent genetic modification to produce the bicyclic monoterpene sabinene (Walls and Rios-Solis 2020).

On the other hand, lipid by-products such as waste cooking oil have been well studied in the production of drop-in biofuels, due to among the different types of biomass utilized to extract hydrocarbon-rich bio-oil, triglycerides derived from vegetable oil are widely acknowledged as excellent raw materials. Used cooking oil, for example, serves as a valuable raw material, effectively reducing the cost of biodiesel production since waste cooking oil accounts for around 70–95% of the total production cost (Xiong et al. 2023). In this regard, utilizing kitchen waste as a raw material for biofuel preparation offers additional advantages compared to other biomass sources. Firstly, kitchen waste is rich in organic content, providing sufficient material and energy for biofuel conversion without the need for additional substances. Secondly, the low refractory substance content in kitchen waste alleviates difficulties encountered during pretreatment (Xiong et al. 2023). Finally, the study of microbial processes to produce isoprenoid products from waste cooking oil is still in its early stages. However, the low cost and widespread availability of waste cooking oil make it a suitable feedstock for advanced biojet fuels production (Walls and Rios-Solis 2020). Table 3.1 offers a comprehensive overview with additional examples including the successful valorization of waste cooking oil into drop-in biofuels and its precursors.

3.3.3 Solid Renewable Residues

Renewable solid residues refer to solid state organic by-products generated from various renewable sources such as agriculture, forestry, and industrial production, among others. These renewable residues encompass solid or compressed materials that release their stored energy through combustion, incineration, or bioconversion via biochemical treatments (Taptich et al. 2018; Akolgo et al. 2021; Chau et al. 2022). In order to reduce reliance on fossil fuels and mitigate their adverse environmental impacts, researchers have extensively explored the potential of forestry, agricultural, and also industrial solid residues such as coconut husk, rice husk, sugarcane bagasse, wheat straw, and sawdust among others, as viable alternative energy sources (Akolgo et al. 2021).

Unlike liquid or gaseous waste, solid waste presents challenges in terms of storage, handling, and transportation. Therefore, there is a strong motivation to convert solid waste into liquid or gaseous fuels (Rahimi et al. 2022). Therefore, listed below are the main groups of solid renewable waste that can be bio-converted to drop-in biofuels:

- **Municipal Solid Waste:** Among these renewable wastes are solid wastes generated by several institutions, households, and commercial establishments. It is considered a significant by-product of urban lifestyles and is growing at a faster pace than urbanization itself. Organic municipal solid waste and sewage sludge flakes are potential easily accessible feedstock to be directly transformed into biofuels or drop-in biofuel precursors (Okoro et al. 2022).
- **Lignocellulosic residues:** These renewable residues are abundant in terrestrial plant biomass and consist of lignin and polymeric carbohydrates, making them attractive raw materials to bioenergy production. They offer a promising alternative to fossil fuels in achieving future sustainable development goals (Zhu et al. 2020). The main component of these lignocellulosic residues is lignin, a

Table 3.1 Examples of renewable wastes and microorganisms involved in their biochemical conversion into drop-in biofuels or its precursors

Renewable residue	Microorganism	Drop-in Biofuel	Reference
Alcohol-derived fuels			
Synthesis gas	*Alkalibaculum bacchi*-CP15 and *Clostridium propionicum* mixed culture	Ethanol n-propanol n-butanol	(Liu et al. 2014)
Industrial waste gases (synthesis gas and steel mill off-gas)	*Clostridium autoethanogenum*	Ethanol	(Heffernan et al. 2020)
Rice straw sugarcane bagasse	*Clostridium acetobutylicum* ATCC 824	n-butanol	(Tsai et al. 2020)
Waste papers	*Clostridium acetobutylicum* ATCC824	n-butanol	(Mondal et al. 2022)
Cheese whey water	*Clostridium acetobutylicum*	n-butanol	(Kallarakkal et al. 2021)
Food waste	*Clostridium saccharoperbutylacetonicum deltptabuk*	n-butanol	(Jin et al. 2020)
Coffee silverskin	*Clostridium acetobutylicum* DSM 792	n-butanol	(Niglio et al. 2019)
Municipal solid waste	*Clostridium acetobutylicum* and *Enterobacter aerogenes*	2,3-butanediol n-butanol	(Ebrahimian et al. 2022b)
Sugarcane trash	Engineered *Pichia pastoris*	Isobutanol	(Bumrungtham et al. 2022)
Isoprenoid-derived fuels			
Peanut hull	Engineered *Escherichia coli*	Isoprene	(Wang et al. 2017)
Wastewater biosolids	Engineered *Methanosarcina acetivorans*	Isoprene	(Carr et al. 2021)
Waste cooking oil	Engineered *Yarrowia lipolytica*	Limonene	(Pang et al. 2019)
Waste cooking oil	Engineered *Yarrowia lipolytica*	Limonene	(Li et al. 2022)
Waste cooking oil	Engineered *Yarrowia lipolytica*	β-Farnesene	(Liu et al. 2022)
Waste cooking oil	Engineered *Yarrowia lipolytica*	α-Bisabolene β-Bisabolene γ-Bisabolene	(Zhao et al. 2021)
Corn Stover	Engineered *Rhodosporidium toruloides*	α-Bisabolene 1,8-cineole	(Kirby et al. 2021)
Polyketide-derived fuels			
Corn stover	Engineered *Streptomyces albus*	Short-chain ketones (C5–C7)	(Yuzawa et al. 2018)
Municipal solid waste	Engineered *Escherichia coli*	Medium-chain ketones (C11–C17)	(Yan et al. 2019)
Fatty acid-derived fuels			

(continued)

Table 3.1 (continued)

Renewable residue	Microorganism	Drop-in Biofuel	Reference
Real textile wastewater	*Chlorella vulgaris*	Palmitic acid, Linoleic acid, linolenic acid	(Javed et al. 2022)
Discarded vegetables	*Cryptococcus curvatus*	Oleic acid Palmitic acid	(Gallego-García et al. 2023)
Livestock wastewater	*Desmodesmus* sp. EJ8–10	Lauric acid Myristic acid Myristoleic acid Palmitic acid stearic acid Oleic acid Linoleic acid	(Li et al. 2021)
Waste cooking oil	*Rhodococcus* sp. YHY01	Palmitoleic acid Palmitic acid Myristic acid	(Bhatia et al. 2021)
Sugarcane bagasse	*Rhodococcus opacus* DSM 43205	Palmitic acid Stearic acid Oleic acid α-Linolenic acid	(Mahmood and Singh 2023)
Waste office paper	*Rhodococcus opacus* PD630	Oleic acid Palmitic acid Palmitoleic acid Stearic acid Pentadecanoic acid Heptadecanoic acid	(Nair and Sivakumar 2023)
Rice straw	Evolved *Rhodosporidium toruloides*	Palmitic acid Stearic acid Oleic acid Linoleic acid Linolenic acid	(Park et al. 2023)
Apple pomace	*Rhodosporidium toruloides* DSM 4444	Palmitic acid Stearic acid Oleic acid Linoleic acid	(Tuhanioglu et al. 2022)
Crude glycerol and wood hydrolysate	*Rhodosporidium toruloides*-1588	Palmitic acid Stearic acid Oleic acid	(Saini et al. 2022)
Waste office paper	*Rhodosporidium toruloides*	Oleic acid Palmitic acid Stearic acid Myristic acid	(Nair and Sivakumar 2022)

polyphenolic compound, along with cellulose, representing one of the most abundant biopolymers found in nature. Thus, lignocellulosic residues refer to plant materials that are rich in cellulose, hemicellulose, and lignin, such as agricultural and forest residues.

Lignocellulosic residues also encompass forest residues generated from activities like logging, bark removal, sawdust, and wood chipping. These residues are abundant, cost-effective, and environmentally friendly resources that can be utilized for the production of biofuels, biomass heating, and electricity generation. Unlike first-generation bioenergy resources derived from edible food crops, forest residues do not compete with food production. Furthermore, their carbon balance is nearly neutral as the carbon dioxide (CO_2) emitted during combustion is offset by the CO_2 absorbed by trees during their growth (Da Costa et al. 2023).

The utilization of solid renewable residues to produce drop-in biofuels not only helps reduce waste generation but also contributes to the development of a circular economy and the transition toward a more sustainable and low-carbon future.

3.3.3.1 Renewable Solid Residues Used in Drop-in Biofuel Production

The continuous cultivation of trees and crops leads to the ongoing generation of waste. Nevertheless, the unwieldiness and challenges associated with handling these residues impede a smooth transition from fossil fuels to more sustainable alternatives (Rahimi et al. 2022). Given the substantial waste produced by the agricultural industry and its adverse effects on the environment, it is advisable to convert this waste into biofuels. Consequently, the biotransformation of agricultural residues into drop-in biofuels represents a key strategy for sustainable waste management, resource recovery, and the maintenance of the environmental carbon balance (Awogbemi and Kallon 2022).

An advantageous characteristic of agricultural waste found on modern farms is its relatively uniform composition. This consistent raw material facilitates the stable operation of the anaerobic digestion process (Lee 2017). Combining oxygen treatment with subsequent fermentation appears to be a successful strategy for enhancing the transformation of lignin compounds into potential drop-in biofuels, although further research is necessary (Zhu et al. 2020). There are the two sustainable pathways to convert agricultural waste into drop-in biofuels, biochemical and thermochemical based processes. The first one integrates biological and chemical processes and is commonly used to convert agricultural waste with high content of moisture, polysaccharides such as cellulose or hemicellulose, as well as a carbon/nitrogen ratio exceeding 30%, into biofuels like bioethanol, biomethane, biobutanol, and biodiesel. On the flip side, the thermochemical pathway is most suitable for agricultural waste characterized by C/N ratio below 30% with a low moisture content and high lignin content (Awogbemi and Kallon 2022). For instance, coffee silver skin is a significant by-product generated in substantial quantities during the roasting process of green coffee beans. By pretreatment and hydrolysis, hydrolysate fermentation was possible to produce butanol (Niglio et al. 2019). Alternatively, another study explored the production of biobutanol using potato peel waste and dilute orange peel extract as a growth medium for *Clostridium acetobutylicum* MTCC 11274. The findings revealed that the prepared dilute orange peel extract effectively stimulated the growth of this strain, facilitating the fermentation process to produce acetone, butanol, and ethanol (Kamboj and Ms 2021).

Fermentation of food residues also holds a significant potential for energy recovery. Considering its attributes, numerous organic food residues are suitable to produce energy-rich biofuels like butanol, which stands out as an economically viable approach for their utilization (Saba et al. 2023). Food wastes are also possible feedstocks to acetone-butanol-ethanol production through both batch and continuous fermentation of immobilized cells. In this process, the initial step involved liquefying and saccharifying the food waste to obtain fermentable sugars. Subsequently, these hydrolysates are introduced into the fermentation systems to produce acetone, butanol, and ethanol. This significant development holds promise for advancing the utilization of organic residues to produce advanced biofuels (Jin et al. 2020). In this way, by-products from processing industries, tomato pomace and its hydrolysate, both served as interesting feedstocks for acetone-butanol-ethanol-isopropanol biorefineries through *Clostridium* fermentation processes (Hijosa-Valsero et al. 2019).

However, the considerable variability in the composition of food waste has posed challenges to the consistency and predictability of biobutanol production, hindering the establishment of a robust industrial fermentation process (Jin et al. 2020). So far, the economic viability of cellulosic biofuels has been impeded by slow and incomplete saccharification, as well as the high costs associated with enzymes. While approaches like the utilization of thermotolerant microorganisms to enhance production or the simultaneous saccharification and fermentation, further improvements are necessary (Hoffman et al. 2021). For instance, a one-step conversion approach of sweet potato processing residues to butanol without pretreatment of liquefaction and saccharification was also demonstrated. This feasibility to utilize sweet potato processing waste as feedstock for butanol fermentation provides an innovative approach to mitigate environmental pollution while recycling these biomass residues (Jin et al. 2022). This demonstrates that it is essential to employ sustainable and technologically advanced methods of valorization to safeguard the environment and foster the growth of the bioeconomy (Saba et al. 2023).

Regarding municipal solid waste biodegradable fraction, it was studied as a zero-cost substrate to produce gasoline. Energy recovery was accomplished through the co-cultivation of *Clostridium acetobutylicum* and *Enterobacter aerogenes*, leveraging their complementary characteristics in the coproduction of hydrogen, 2,3-butanediol, and acetone-butanol-ethanol for biofuel production. The findings revealed that each kilogram of raw municipal solid waste biodegradable fraction yielded 311.8 mL of gasoline equivalent (equivalent to 10 MJ of energy) when considering the bioenergy perspective (Ebrahimian et al. 2022b). Furthermore, the municipal solid waste organic fraction has been assessed by harnessing the advantages of co-cultivating *Clostridium acetobutylicum* and *Mucor indicus*. From a bioenergy perspective, the co-cultivation approach yielded an impressive energy output of 7.2 MJ, surpassing the energy generated through the separate cultivation of *M. indicus* (4.4 MJ) and *C. acetobutylicum* (6.2 MJ). This notable difference emphasizes the significant biotransformation potential of fungal-bacterial co-cultivation of the organic fraction of municipal solid waste into bioenergy (Ebrahimian et al. 2022a).

Additional advances discussed in this section regarding renewable residues as feedstock to produce drop-in biofuels and their precursors are summarized in Table 3.1. In this table, advanced biofuels are categorized based on their derivation from alcohols, isoprenoids, polyketides, or fatty acids. It is important to note that some of these studies primarily concentrated on the biological generation of microbial fuels while neglecting the accompanying challenges of practical application. Although the implementation of synthetic and systems biology has shown advancements in biofuel production, there remains a significant disparity between demand and supply. As a result, there is a growing recognition of the necessity to explore alternative approaches, such as fuel design, to create fuel molecules that are both environmentally friendly and possess better desired physicochemical properties for the existing infrastructure (Okoro et al. 2022). Cases of successful companies that have overcome these challenges in large-scale advanced biofuel production are discussed in a later section.

3.4 Benefits of Using Renewable Residues in Drop-in Biofuel Production

Utilizing wastes as feedstocks for drop-in biofuel production has been identified as an effective strategy to promote sustainability, enhance resource value, and maximize ecological and socioeconomic benefits (Scarlat et al. 2019). Some of these benefits are shown in Fig. 3.3.

3.4.1 Food Security

The initial advancements in the field of biofuels faced various challenges, including the debate surrounding the competition between food production and fuel generation (Amer et al. 2020). In this regard, utilizing renewable residues as feedstock for drop-in biofuel production eliminates the need for biofuel crops and additional agricultural land, thus helping to mitigate deforestation and land use changes associated with agricultural expansion (Malode et al. 2021; García and Galán 2022; Yazdanparast et al. 2022).

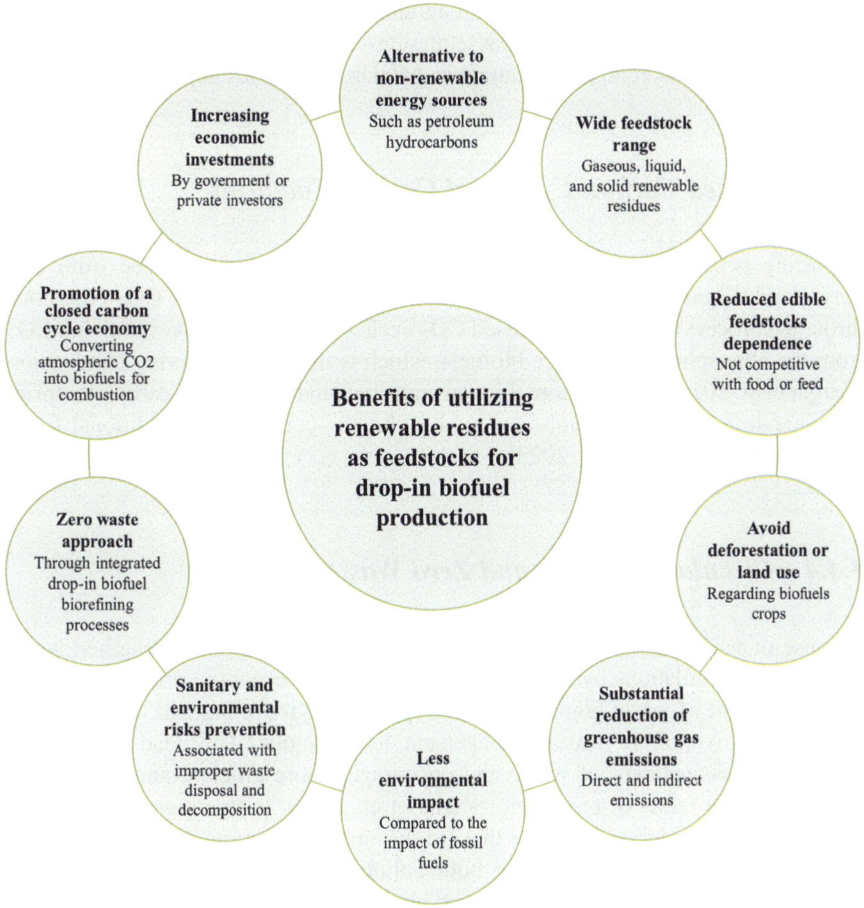

Fig. 3.3 Associated benefits of producing renewable residues-derived drop-in biofuels

3.4.2 Environmental Impact and Greenhouse Gas Emissions

The accelerated pace of urban development and industrial activities, coupled with the increasing human population, has associated an extensive waste generation and ecosystem degradation. Improper dumping or unregulated disposal of waste can have severe repercussions on the ecological environment while also depleting scarce land resources (Peng et al. 2023). In this respect, biofuels derived directly from waste offer a viable and sustainable solution to promote a renewable energy source with low environmental impact (Scarlat et al. 2019). Also benefits in air quality are achieved by reducing harmful atmospheric pollutant emissions, such as nitrogen oxides (NO_x) and particulate matter, compared to fossil fuels, contributing to a substantial reduction in greenhouse gas emissions (Malode et al. 2021; García and Galán 2022; Yazdanparast et al. 2022). In addition, drop-in biofuels produced from renewable sources such as biomass and agricultural residues can play a crucial role in reducing indirect greenhouse gas emissions related to deforestation and land conversion (Malode et al. 2021; García and Galán 2022; Yazdanparast et al. 2022).

3.4.3 Closed Carbon Cycle and Carbon Footprint

Replacing petrochemically-derived fuels with renewable fuels sourced from biomass, landfill gas, or atmospheric CO_2 can lead to a reduction in CO_2 emissions through a process known as the closed CO_2 cycle. This cycle involves utilizing CO_2 from the atmosphere to produce biofuels, which is then burned during combustion (Zargar et al. 2017). Furthermore this can contribute to reducing the carbon footprint by generating lower CO_2 emissions and pollutants compared to traditional fossil fuels (Barragán-Ocaña et al. 2023).

3.4.4 Circular Economy and Zero Waste

The use of waste as feedstock in the production of biofuels can strengthen waste management regulations by aligning them with circular economy principles, reducing the amount of waste going to landfills, promoting recycling and recovery, and fostering innovation in waste management technologies. This leads to a more sustainable management of waste and encourages more efficient and responsible development in terms of resources (Scarlat et al. 2019). In this regard, large-scale drop-in biofuel production systems that are sustainable over time imply the adoption of strategies focused on mitigating both polluting waste and the general carbon footprint from production to use (Amer et al. 2020). This is done within the goal to actively contribute to a circular economy by promoting the bioconversion of these materials (Iragavarapu et al. 2023).

3.4.5 Socioeconomic Benefits

The widespread use of fossil fuels as the primary anthropogenic source of energy has proven to be unsustainable, given its negative impact on climate change, human health, land use, and resource depletion (Cabrera-Jiménez et al. 2022; Awogbemi and Kallon 2022). The utilization of renewable residue for biofuel production can have a significant impact on improving public health and reducing the risks associated with improper waste management. In this context, drop-in biofuels derived from renewable residues represent a promising solution to achieve not only greater environmental but also social sustainability, reducing the impacts associated with waste dumping and the carbon footprint. This promotes sustainable economic development, a cleaner, more resilient, and equitable future (Scarlat et al. 2019). The benefits include diversifying the fuel supply, reducing dependence on fossil fuels, and boosting the local economy through job creation in the bioenergy sector. This promotes the local economy and creates employment opportunities in this sector (Cabrera-Jiménez et al. 2022). Moreover, drop-in biofuels derived from renewable residues can help promote international development and alleviate poverty in underdeveloped economies by providing an affordable and sustainable energy source. It can improve the quality of life and foster long-term sustainable development in these regions (Cuellar-Bermudez et al. 2015).

3.4.6 Growing Financial Support

The significant initial costs of drop-in biofuels in terms of capital and operation required to carry out production on an industrial scale were some of the initial challenges in this industry (Amer et al. 2020). Various government initiatives are presently supporting numerous projects aimed at expanding biofuel production, covering capital costs, and offering startup loans. Billions of dollars in investments have been deployed globally by public and foreign investors to contribute funds for the production of clean fuels (Saini et al. 2021). This support has primarily been driven by environmental concerns associated with fossil fuel use, price volatility, and the global rise in energy consumption, encouraging researchers and fuel producers to develop affordable and environmentally friendly energy sources (Rahimi et al. 2022; Awogbemi and Kallon 2022; Ardila-Suárez et al. 2023).

Finally, while government policies can play a critical role by providing economic incentives for drop-in biofuel production or by limiting certain taxes associated with the production or sale of biofuels, optimizing the processes will result in better economic-productive outcomes (Cabrera-Jiménez et al. 2022). In the end, the economic benefits associated will include increased job creation, additional sources of income for producers and local communities, as well as reduced energy costs (Barragán-Ocaña et al. 2023).

3.5 Selecting the Best Renewable Feedstock

The selection of optimal renewable residues for producing drop-in biofuels through microbial fermentation is a complex process that entails the evaluation of multiple factors. For instance, determining factors include:

- **Environmental effects.** Evaluating the environmental impact of acquiring renewable residues and their utilization in drop-in biofuel production, as opposed to alternative applications, is crucial. On the one hand, it is important to select residues that do not directly compete with food production to avoid potential conflicts and minimize the impact on land use and water resources. From an environmental standpoint, residues that would otherwise be discarded, such as agricultural or industrial waste, are preferred options. On the other hand, drop-in biofuels can be produced using various sources and processes, each of which has different environmental impacts and achieves different engine performance outcomes. Thus, identifying the most suitable biofuels necessitates considering process sustainability (Cabrera-Jiménez et al. 2022).

- **Availability and quantity of renewable residues.** It is crucial to assess the long-term availability of residues in the specific region or geographical area where drop-in biofuel production will occur, ensuring a scalable and sustainable production process. If production takes place elsewhere, the residues must be harvested, transported to processing sites, and then delivered to end-users. Thus, it is essential to identify the transportation implications in order to describe the environmental profile of the entire drop-in biofuel supply chain (Halog and Awuah 2013) and evaluate its economic feasibility (Cabrera-Jiménez et al. 2022).
- **Feedstock composition and its conversion technology.** It is important to analyze the chemical composition of the renewable residues, such as the concentration of fermentable compounds like sugars, starch, lignocellulose, or other organic compounds. Fermentation yields are generally sensitive to the varying composition of the feedstock due to the specialized nature of the involved microorganisms. The selected renewable residues should be amenable to efficient conversion through microbial fermentation, while avoiding residues with high concentrations of inhibitors to microbial growth. Residues with a high concentration of fermentable compounds are more suitable for biofuel production, although this may depend on the specific microorganism employed and its ability to decompose and ferment the residue (Hoffman et al. 2021).
- **Economic viability.** To be a viable alternative to fossil fuels, the resulting drop-in biofuel must demonstrate superior environmental advantages compared to the fossil fuel it replaces, prove economically competitive, and have the capability to be produced in significant amounts to make a substantial impact (Hill et al. 2006). Consequently, it is critical to consider the economic viability of drop-in biofuel production from renewable residues as a crucial factor. This entails considering costs associated with acquisition, pretreatment, residue fermentation, and product purification (Cabrera-Jiménez et al. 2022). Additionally, evaluating the potential added value of other possible products that can be derived from the use of these residues is necessary.
- **Regulations.** Finally, it is important to take into account the legal and regulatory frameworks that govern drop-in biofuel production in the specific region. Compliance with all applicable norms and regulations is of utmost importance. In conclusion, the selection of optimal renewable residues for drop-in biofuel production through microbial fermentation involves considering factors such as chemical composition, availability, environmental impact, conversion potential, economic viability, and legal and regulatory aspects. Adopting an integrated

approach that encompasses these factors will contribute to the selection of the most suitable residues and the successful development of sustainable biofuel production (Kolosz et al. 2020; Selvaggi and Valenti 2021).

3.6 Companies Currently Using Renewable Residues for Biofuel Production

Traditionally, commercial production of drop-in biofuels has been centered around the oleochemical platform, where vegetable oils have been the primary choice of feedstock due to their abundant availability, high energy density, and ease of transportation (van Dyk et al. 2019). However, in the past few years, there has been a growing interest in the exploration of biological solutions for fuel production, despite concerns about their commercial viability when compared to cheaper crude oil and synthetic chemical technologies.

Several emerging companies have developed innovative technologies to tackle challenges related to process efficiency, profitability, and greenhouse gas mitigation in drop-in biofuel production. These companies utilize synthetic biology techniques to design customized microbial-based cell factories to produce drop-in biofuels from renewable feedstocks. Achieving this has involved the use of genetically engineering tools to modify existing biological pathways, optimizing bioprocess productivity (Amer et al. 2020). Presented below are some of the most important companies in the field of drop-in biofuel production through biochemical and chemical conversion of renewable residues:

3.6.1 Renmatix

Renmatix, based in the USA, utilizes a patented process called Plantrose® to convert plant residues such as corn stover and wood chips into bio-based intermediate products, including biofuels. This company has employed and further enhanced this technology through collaborations with other firms. In addition, studies on microbial based biofuel production have also been included to replace the plant-based oils. For example, they declared a Joint Development Agreement to integrate Plantrose with two technologies from Gevo, evaluating the commercial feasibility of producing renewable aviation fuel (Karatzos et al. 2017; van Dyk et al. 2019). More recently, Renmatix collaborated with Amyris and Total on a project involving bioprocessing using *S. cerevisiae* strains overproducing farnesene using lignocellulosic feedstocks to reduce production costs to $2 per liter. Although this required facilitating xylose catabolism and resistance to growth inhibitors through metabolic engineering of the yeast, the projected minimum production cost was $4–5 per liter, leading to the premature termination of the project (Walls and Rios-Solis 2020).

3.6.2 Neste Corporation

Neste Corporation is a Finnish company specialized in the production of advanced biofuels, such as renewable diesel and renewable aviation fuel, using renewable residues such as animal fats and used cooking oil by chemical methods. In 2015, Neste Oil produced 46,000 barrels of hydrotreated vegetable oil biofuels per day at their major facilities in Rotterdam, Finland, and Singapore, primarily using palm and other vegetable oils (Karatzos et al. 2014). Additionally, Neste has successfully utilized used cooking oil and tallow at their large-scale renewable diesel facilities (van Dyk et al. 2019).

In 2020, Neste introduced a semi-biological process, known as the Neste process, which involved the synthetic catalytic conversion of glycerol biodiesel to produce propane. However, this process was not considered truly sustainable or renewable (Amer et al. 2020). Currently, commercial production of BioLPG, or bio-propane, by Neste relies on the extraction and processing of a 5% bio-propane by-product from the hydrotreatment of waste and vegetable oils for the production of renewable diesel (Currie et al. 2022). Aiming for an annual production capacity of 30 to 40 thousand tons of fuel, Neste has invested a total of EUR 60 million to construct a BioLPG facility in Rotterdam for its production and storage (Iragavarapu et al. 2023). It is essential to recognize that an increase in demand could potentially lead to higher feedstock costs. The availability of sustainable and efficient feedstocks is identified as the main challenge in the industry, and Neste suggests that improved agricultural practices, such as winter cropping, silvopasture, and cultivation on degraded or abandoned land, could provide additional feedstock volumes (Karatzos et al. 2014). As Neste expands, it also plans to explore other renewable residue sources, including waste plastic, municipal waste, and algae, as potential feedstocks for transportation fuels and other chemical productions, with support from universities, research institutes, and Business Finland (Chia et al. 2022).

3.6.3 Fulcrum BioEnergy

Fulcrum BioEnergy, a pioneering company in the USA, converts municipal solid waste into renewable aviation fuel and diesel through their patented thermochemical process known as "Bioconversion Technology" (Doliente et al. 2020). Fulcrum BioEnergy is planning to produce Fischer-Tropsch (FT) fuels in proposed facilities in the USA, utilizing municipal solid waste (MSW) as the feedstock (van Dyk et al. 2019). Fulcrum BioEnergy has been successful in commercially producing jet fuel through the FT process, processing around 30,000 tons per year of MSW (Doliente et al. 2020). However, it should be noted that the gasification/FT synthesis route faces challenges such as high capital costs, the cost-effective production of high-quality syngas, and the achievement of an economy of scale (van Dyk et al. 2019).

3.6.4 Velocys

Velocys, a British company, specializes in the production of drop-in biofuel from waste materials such as woody biomass, forest residues, and agricultural waste. Their focus is on utilizing the biomass gasification pathway for producing these drop-in biofuels. As part of their initiatives, Velocys has been concentrating on gas-to-liquids conversion through Fischer-Tropsch technology on a small scale at their ENVIA Energy plant. (van Dyk et al. 2019).

3.6.5 Gevo, Inc.

Gevo, Inc. has been operating a biorefinery in Texas since 2011, where they produce renewable isobutanol for commercial jet fuel using Alcohol-to-Jet processes (Rodriguez et al. 2023). Their production technology, known as Gevo Integrated fermentation technology, involves the fermentation of corn waste biomass using an engineered yeast strain that yields high amounts of isobutanol (Rodriguez et al. 2023). They have successfully developed and implemented this genetically engineered yeast strains capable of utilizing corn as a substrate for sustainable aviation fuel production (Saini et al. 2021). The bioprocess is integrated with a product recovery system that continuously removes the formed isobutanol (Rodriguez et al. 2023) which is further converted into jet fuel (Schubert 2020). Among the companies involved in renewable butanol synthesis, Gevo, along with Butamax, is a leading player in biobutanol production (Emmanuel et al. 2023). Because of its technical maturity and similarity to ethanol fermentation processes, these companies have opted for the lignocellulosic biomass fermentation pathway (Ro et al. 2023).

3.6.6 LanzaTech

Specializing in the production of low-carbon fuels, LanzaTech utilizes both industrial emissions and renewable resources (Doliente et al. 2020). They employ a gas fermentation process to convert waste gases, such as carbon dioxide and monoxide, into ethanol and other biofuels like acetone and isopropanol (Karatzos et al. 2017; Rodriguez et al. 2023). The LanzaTech method involves the utilization of industrial waste gases containing H_2, CO, and CO_2, like flue gas by-products from steel industry (Doliente et al. 2020). By recycling the carbon in these waste gases, LanzaTech prevents its emission into the atmosphere. The process is more feasible in this way than using or producing syngas or biogas from different feedstocks. Gasified biomass can also supplement these gases, and municipal waste can be utilized as well. Microbiological species are also employed to ferment the gases and produce alcohols (Doliente et al. 2020). LanzaTech is among the restricted number of companies that currently produce on an industrial scale bio-based 2,3-butanediol. Different microorganisms employed in their processes produce varying degrees of 2,3-butanediol stereoisomers, typically resulting in a mixture of two stereoisomers (Koller 2023). Likewise, LanzaTech, in collaboration with Virgin Atlantic Airways, has plans to produce over 13.5 million liters of bio-aviation fuel mixed with conventional jet fuel and diesel through the development of a new facility. The main goal is to use waste streams from municipal and industrial sources as raw materials (Doliente et al. 2020), increasing the range of renewable wastes to be transformed into drop-in biofuels.

Finally, other companies involved in biobutanol production include BUTALACO GmbH (Switzerland), Butyl Fuel (USA), GS Caltex (South Korea), TetraVitae Bioscience (USA), Green Biologics (UK), Metabolic Explorer (France), Energy Quest (Malaysia), Cathy Industrial Biotech (China), and SaBuCo (Saudi Arabia) (Tsai et al. 2020).

3.7 Future Aspects

The biochemical production of drop-in biofuels from renewable residue holds great potential for the future. As we move toward a more sustainable society, significant advancements in this field are likely to be seen, with a focus on ensuring responsible utilization of generated waste and minimizing environmental impact (Ho Kugele and Sarkar 2023).

To improve the large-scale production of drop-in biofuels from renewable residue using microorganisms, it is crucial to keep process costs low, demanding continuous advancements in conversion technologies. Enhanced pretreatment techniques are necessary to improve the breakdown and biotransformation of renewable residue into drop-in biofuel precursors. The development of more efficient enzymes and microorganisms for the conversion of renewable residue into drop-in biofuels through metabolic pathway optimization is an important area of research (Jiang et al. 2018). Overcoming bioprocess obstacles such as microbial product inhibition, which negatively affects performance, efficiency, and operability, is essential (Azimov et al. 2021). Ongoing advancements in microbial hosts for biofuel production are imperative to harness waste products, extending beyond agriculture to municipal sources, possibly encompassing plastics (Keasling et al. 2021). To reduce the costs, biofuel production in non-sterile environments will be required, necessitating specially engineered microorganisms capable of withstanding harsh conditions that few others can endure (Keasling et al. 2021). In response to these challenges, it is imperative to apply metabolic engineering, a crucial approach that enhances yields, titers, productivity, and robustness during large-scale production of microbial drop-in fuels (Jiang et al. 2018).

Additionally, interdisciplinary research is needed to analyze costs, assess sensitivities, conduct energy paired with exergy analyses and perform techno-economic and rigorously life-cycle evaluations. These efforts will enhance our comprehension of intricate processes kinetics involved in the breakdown and transformation of waste into drop-in biofuels. Employing cutting-edge technologies such as machine learning, or artificial intelligence, and optimization tools is necessary to enhance process parameters, designing and optimizing reactors, managing costs, labor and time requirements. These advancements will also enhance conversion efficiency and product quality (Awogbemi and Kallon 2022). Moreover, the development of predictive mathematical models for resource assessment and real-time prevention of productivity loss during large-scale production is essential (Behera et al. 2022).

Increased financial, infrastructure, and investments are crucial to promote the bioconversion of residues into renewable drop-in biofuels. In the future, governments should encourage farmers to benefit from converting their waste into useful bioproducts instead of burning residues that can further degrade the environment and reduce soil fertility. Ongoing research will update the current information providing essential knowledge to investors, farmers, biofuel refineries, and governmental authorities at various levels concerning the energy valorization of agricultural residues to meet the global demand for renewable fuels (Awogbemi and Kallon 2022). Small and medium-sized enterprises looking to invest in the production of drop-in biofuels should be granted incentives and tax exemptions. The availability of consistent and sustainable agricultural residues will ensure the cost-effective and accessible raw materials needed for their conversion into biofuels (Awogbemi and Kallon 2022). However, the range of waste used as feedstock must be diversified to maintain logistics, production, and supply chains so that products are readily available to consumers. This should be accompanied by consumer awareness campaigns highlighting the advantages of drop-in biofuels over their synthetic counterparts, reducing the market window for successful commercialization. In this way, energy security will be further enhanced by reducing dependence on conventional fuels, thereby helping to mitigate risks associated with fossil fuel price volatility, supply disruptions, and geopolitical tensions (Behera et al. 2022).

3.8 Conclusion

In nutshell, the production of drop-in biofuels from renewable residues presents a range of benefits, opportunities, and challenges in the pursuit of sustainable energy solutions. It represents a promising pathway toward a cleaner and more sustainable energy matrix. Utilizing these waste materials as valuable feedstock not only diminishes reliance on fossil fuels and addresses greenhouse gas emissions but also generates economic and social benefits, facilitates employment generation, and fosters local economic development, thereby promoting a circular and sustainable economy. With ongoing advancements in research and industry, significant progress in drop-in biofuel production from renewable residues is anticipated, driving a more sustainable and environmentally-responsible energy transition.

Acknowledgments The authors are thankful for the funding and support from Emerging Leaders in the Americas Program (ELAP) and James and Joanne Love Chair in Environmental Engineering at York University, Canada.

References

Akolgo GA, Awafo EA, Essandoh EO, Owusu PA, Uba F, Adu-Poku KA (2021) Assessment of the potential of charred briquettes of sawdust, rice and coconut husks: using water boiling and user acceptability tests. Sci Afr 12:e00789. https://doi.org/10.1016/j.sciaf.2021.e00789

Amer M, Toogood H, Scrutton NS (2020) Engineering nature for gaseous hydrocarbon production. Microb Cell Factories 19(1):209. https://doi.org/10.1186/s12934-020-01470-6

Ardila-Suárez C, Villegas JP, de Barros Neto EL, Ghislain T, Lavoie J-M (2023) Bee green: renewable hydrocarbon fuels from honeycomb residues. Fuel 340:127319. https://doi.org/10.1016/j.fuel.2022.127319

Awogbemi O, Kallon DVV (2022) Valorization of agricultural wastes for biofuel applications. Heliyon 8(10):e11117. https://doi.org/10.1016/j.heliyon.2022.e11117

Azimov U, Okoro V, Hernandez HH (2021) Recent progress and trends in the development of microbial biofuels from solid waste—a review. Energies 14(19):6011. https://doi.org/10.3390/en14196011

Barragán-Ocaña A, Merritt H, Sánchez-Estrada OE, Méndez-Becerril JL, del Pilar L-BM (2023) Biorefinery and sustainability for the production of biofuels and value-added products: a trends analysis based on network and patent analysis. PLoS One 18(1):e0279659. https://doi.org/10.1371/journal.pone.0279659

Behera B, Selvam SM, Paramasivan B (2022) Research trends and market opportunities of microalgal biorefinery technologies from circular bioeconomy perspectives. Bioresour Technol 351:127038. https://doi.org/10.1016/j.biortech.2022.127038

Bhatia SK, Gurav R, Choi Y-K, Lee H-J, Kim SH, Suh MJ, Cho JY, Ham S, Lee SH, Choi K-Y, Yang Y-H (2021) Rhodococcus sp. YHY01 a microbial cell factory for the valorization of waste cooking oil into lipids a feedstock for biodiesel production. Fuel 301:121070. https://doi.org/10.1016/j.fuel.2021.121070

Bumrungtham P, Promdonkoy P, Prabmark K, Bunterngsook B, Boonyapakron K, Tanapongpipat S, Champreda V, Runguphan W (2022) Engineered production of isobutanol from sugarcane trash hydrolysates in Pichia pastoris. J Fungi 8(8):767. https://doi.org/10.3390/jof8080767

Cabrera-Jiménez R, Mateo-Sanz JM, Gavaldà J, Jiménez L, Pozo C (2022) Comparing biofuels through the lens of sustainability: a data envelopment analysis approach. Appl Energy 307:118201. https://doi.org/10.1016/j.apenergy.2021.118201

Carr S, Aldridge J, Buan NR (2021) Isoprene production from municipal wastewater biosolids by engineered archaeon Methanosarcina acetivorans. Appl Sci 11(8):3342. https://doi.org/10.3390/app11083342

Chau KY, Moslehpour M, Tu Y-T, Tai NT, Tien NH, Huy PQ (2022) Exploring the impact of green energy and consumption on the sustainability of natural resources: empirical evidence from G7 countries. Renew Energy 196:1241–1249. https://doi.org/10.1016/j.renene.2022.07.085

Chavan S, Yadav B, Atmakuri A, Tyagi RD, Wong JWC, Drogui P (2022) Bioconversion of organic wastes into value-added products: a review. Bioresour Technol 344:126398. https://doi.org/10.1016/j.biortech.2021.126398

Chia SR, SBHJM N, Chew KW, HSH M, Shamsuddin AH, Show PL (2022) Algae as potential feedstock for various bioenergy production. Chemosphere 287:131944. https://doi.org/10.1016/j.chemosphere.2021.131944

Chilakamarry CR, Sakinah AMM, Zularisam AW, Pandey A (2021) Glycerol waste to value added products and its potential applications. Syst Microbiol and Biomanuf 1(4):378–396. https://doi.org/10.1007/s43393-021-00036-w

Collotta M, Champagne P, Mabee W, Tomasoni G (2018) Wastewater and waste CO2 for sustainable biofuels from microalgae. Algal Res 29:12–21. https://doi.org/10.1016/j.algal.2017.11.013

Cuellar-Bermudez SP, Garcia-Perez JS, Rittmann BE, Parra-Saldivar R (2015) Photosynthetic bioenergy utilizing CO2: an approach on flue gases utilization for third generation biofuels. J Clean Prod 98:53–65. https://doi.org/10.1016/j.jclepro.2014.03.034

Currie F, Twigg MS, Huddleson N, Simons KE, Marchant R, Banat IM (2022) Biogenic propane production by a marine Photobacterium strain isolated from the Western English Channel. Front Microbiol 13:1000247. https://doi.org/10.3389/fmicb.2022.1000247

Czyrnek-Delêtre MM, Ahern EP, Murphy JD (2016) Is small-scale upgrading of landfill gas to biomethane for use as a cellulosic transport biofuel economically viable?: upgrading of landfill gas to biomethane and using as a transport biofuel requires a minimum subsidy of €0.55/mn3 to allow economic viability. Biofuels Bioprod Biorefin 10(2):139–149. https://doi.org/10.1002/bbb.1627

Da Costa TP, Murphy F, Roldan R, Mediboyina MK, Chen W, Sweeney J, Capareda S, Holden NM (2023) Technical and environmental assessment of forestry residues valorisation via fast pyrolysis in Ireland. Biomass Bioenergy 173:106766. https://doi.org/10.1016/j.biombioe.2023.106766

Doliente SS, Narayan A, Tapia JFD, Samsatli NJ, Zhao Y, Samsatli S (2020) Bio-aviation fuel: a comprehensive review and analysis of the supply chain components. Front Energy Res 8:110. https://doi.org/10.3389/fenrg.2020.00110

Duarte JH, Fanka LS, Costa JAV (2016) Utilization of simulated flue gas containing CO2, SO2, NO and ash for Chlorella fusca cultivation. Bioresour Technol 214:159–165. https://doi.org/10.1016/j.biortech.2016.04.078

Ebrahimian F, Denayer JFM, Karimi K (2022a) Efficient coproduction of butanol, ethanol, and biohydrogen from municipal solid waste through a cocultivated biorefinery. Energy Convers Manag 255:115303. https://doi.org/10.1016/j.enconman.2022.115303

Ebrahimian F, Karimi K, Angelidaki I (2022b) Coproduction of hydrogen, butanol, butanediol, ethanol, and biogas from the organic fraction of municipal solid waste using bacterial cocultivation followed by anaerobic digestion. Renew Energy 194:552–560. https://doi.org/10.1016/j.renene.2022.05.067

Emmanuel MA, Bender SG, Bilodeau C, Carceller JM, DeHovitz JS, Fu H, Liu Y, Nicholls BT, Ouyang Y, Page CG, Qiao T, Raps FC, Sorigué DR, Sun S-Z, Turek-Herman J, Ye Y, Rivas-Souchet A, Cao J, Hyster TK (2023) Photobiocatalytic strategies for organic synthesis. Chem Rev 123(9):5459–5520. https://doi.org/10.1021/acs.chemrev.2c00767

Gallego-García M, Moreno AD, González A, Negro MJ (2023) Efficient use of discarded vegetal residues as cost-effective feedstocks for microbial oil production. Biotechnol Biofuels 16(1):21. https://doi.org/10.1186/s13068-023-02268-5

García JL, Galán B (2022) Integrating greenhouse gas capture and C1 biotechnology: a key challenge for circular economy. Microb Biotechnol 15(1):228–239. https://doi.org/10.1111/1751-7915.13991

Gil A (2022) Challenges on waste-to-energy for the valorization of industrial wastes: electricity, heat and cold, bioliquids and biofuels. Environ Nanotechnol Monit Manag 17:100615. https://doi.org/10.1016/j.enmm.2021.100615

Halog A, Awuah N (2013) Environmental assessment of a forest derived "drop-in" biofuel. In: Fang Z (ed) Biofuels – economy, environment and sustainability. InTech

Heffernan JK, Valgepea K, De Souza Pinto Lemgruber R, Casini I, Plan M, Tappel R, Simpson SD, Köpke M, Nielsen LK, Marcellin E (2020) Enhancing CO2-valorization using clostridium autoethanogenum for sustainable fuel and chemicals production. Front Bioeng Biotechnol 8:204. https://doi.org/10.3389/fbioe.2020.00204

Hijosa-Valsero M, Garita-Cambronero J, Paniagua-García AI, Díez-Antolínez R (2019) Tomato waste from processing industries as a feedstock for biofuel production. Bioenergy Res 12(4): 1000–1011. https://doi.org/10.1007/s12155-019-10016-7

Hill J, Nelson E, Tilman D, Polasky S, Tiffany D (2006) Environmental, economic, and energetic costs and benefits of biodiesel and ethanol biofuels. Proc Natl Acad Sci USA 103(30): 11206–11210. https://doi.org/10.1073/pnas.0604600103

Hoffman SM, Alvarez M, Alfassi G, Rein DM, Garcia-Echauri S, Cohen Y, Avalos JL (2021) Cellulosic biofuel production using emulsified simultaneous saccharification and fermentation (eSSF) with conventional and thermotolerant yeasts. Biotechnol Biofuels 14(1):157. https://doi.org/10.1186/s13068-021-02008-7

Iragavarapu GP, Imam SS, Sarkar O, Mohan SV, Chang Y-C, Reddy MV, Kim S-H, Amradi NK (2023) Bioprocessing of waste for renewable chemicals and fuels to promote bioeconomy. Energies 16(9):3873. https://doi.org/10.3390/en16093873

Javed F, Rehman F, Khan AU, Fazal T, Hafeez A, Rashid N (2022) Real textile industrial wastewater treatment and biodiesel production using microalgae. Biomass Bioenergy 165: 106559. https://doi.org/10.1016/j.biombioe.2022.106559

Jiang W, Gu P, Zhang F (2018) Steps towards 'drop-in' biofuels: focusing on metabolic pathways. Curr Opin Biotechnol 53:26–32. https://doi.org/10.1016/j.copbio.2017.10.010

Jin Q, An Z, Damle A, Poe N, Wu J, Wang H, Wang Z, Huang H (2020) High acetone-butanol-ethanol production from food waste by recombinant *clostridium saccharoperbutylacetonicum* in batch and continuous immobilized-cell fermentation. ACS Sustain Chem Eng 8(26): 9822–9832. https://doi.org/10.1021/acssuschemeng.0c02529

Jin Y, Ding F, Wang J, Yi Z, Gao Y, Yang L, Fang Y, Du A, Zhao H (2022) One-step conversion of sweet potato waste to butanol via fermentation by Clostridium acetobutylicum. Biomass Conv Bioref:1–12. https://doi.org/10.1007/s13399-022-03314-2

Kallarakkal KP, Muthukumar K, Alagarsamy A, Pugazhendhi A, Naina Mohamed S (2021) Enhancement of biobutanol production using mixotrophic culture of Oscillatoria sp. in cheese whey water. Fuel 284:119008. https://doi.org/10.1016/j.fuel.2020.119008

Kamboj R, Ms D (2021) Orange peel extract enhanced sugar recovery and butanol production from potato peel by *Clostridium acetobutylicum*. Int J Green Energy 18(10):987–997. https://doi.org/10.1080/15435075.2021.1890083

Karatzos S, McMillan JD, Saddler JN (2014) The potential and challenges of drop-in biofuels: a report by IEA bioenergy task 39. University of British Columbia, Vancouver

Karatzos S, van Dyk JS, McMillan JD, Saddler J (2017) Drop-in biofuel production via conventional (lipid/fatty acid) and advanced (biomass) routes. Part I Biofuels, Bioprod Bioref 11(2): 344–362. https://doi.org/10.1002/bbb.1746

Kargbo H, Harris JS, Phan AN (2021) "Drop-in" fuel production from biomass: critical review on techno-economic feasibility and sustainability. Renew Sust Energ Rev 135:110168. https://doi.org/10.1016/j.rser.2020.110168

Keasling J, Garcia Martin H, Lee TS, Mukhopadhyay A, Singer SW, Sundstrom E (2021) Microbial production of advanced biofuels. Nat Rev Microbiol 19(11):701–715. https://doi.org/10.1038/s41579-021-00577-w

Khan S, Das P, Abdul Quadir M, Thaher MI, Mahata C, Sayadi S, Al-Jabri H (2023) Microalgal feedstock for biofuel production: recent advances, challenges, and future perspective. Fermentation 9(3):281. https://doi.org/10.3390/fermentation9030281

Kirby J, Geiselman GM, Yaegashi J, Kim J, Zhuang X, Tran-Gyamfi MB, Prahl J-P, Sundstrom ER, Gao Y, Munoz N, Burnum-Johnson KE, Benites VT, Baidoo EEK, Fuhrmann A, Seibel K, Webb-Robertson B-JM, Zucker J, Nicora CD, Tanjore D, Magnuson JK, Skerker JM, Gladden JM (2021) Further engineering of R. Toruloides for the production of terpenes from lignocellulosic biomass. Biotechnol Biofuels 14(1):101. https://doi.org/10.1186/s13068-021-01950-w

Kocaturk E, Salan T, Ozcelik O, Alma MH, Candan Z (2023) Recent advances in lignin-based biofuel production. Energies 16(8):3382. https://doi.org/10.3390/en16083382

Koller M (2023) Biotechnological approaches to generate biogenic solvents and energy carriers from renewable resources. EuroBiotech J 7(2):96–120. https://doi.org/10.2478/ebtj-2023-0007

Kolosz BW, Luo Y, Xu B, Maroto-Valer MM, Andresen JM (2020) Life cycle environmental analysis of 'drop in' alternative aviation fuels: *a review*. Sustain Energ Fuels 4(7):3229–3263. https://doi.org/10.1039/C9SE00788A

Kugele ASH, Sarkar B (2023) Reducing carbon emissions of a multi-stage smart production for biofuel towards sustainable development. Alexandria Engineering Journal 70(1):93–113. https://doi.org/10.1016/j.aej.2023.01.003

Lee SY (ed) (2017) Consequences of microbial interactions with hydrocarbons, oils, and lipids: production of fuels and chemicals. Springer International Publishing, Cham

Li G, Zhang J, Li H, Hu R, Yao X, Liu Y, Zhou Y, Lyu T (2021) Towards high-quality biodiesel production from microalgae using original and anaerobically-digested livestock wastewater. Chemosphere 273:128578. https://doi.org/10.1016/j.chemosphere.2020.128578

Li S, Rong L, Wang S, Liu S, Lu Z, Miao L, Zhao B, Zhang C, Xiao D, Pushpanathan K, Wong A, Yu A (2022) Enhanced limonene production by metabolically engineered Yarrowia lipolytica from cheap carbon sources. Chem Eng Sci 249:117342. https://doi.org/10.1016/j.ces.2021.117342

Liu K, Atiyeh HK, Stevenson BS, Tanner RS, Wilkins MR, Huhnke RL (2014) Mixed culture syngas fermentation and conversion of carboxylic acids into alcohols. Bioresour Technol 152:337–346. https://doi.org/10.1016/j.biortech.2013.11.015

Liu Y, Zhang J, Li Q, Wang Z, Cui Z, Su T, Lu X, Qi Q, Hou J (2022) Engineering Yarrowia lipolytica for the sustainable production of β-farnesene from waste oil feedstock. Biotechnol Biofuels 15(1):101. https://doi.org/10.1186/s13068-022-02201-2

Mahmood Z, Singh L (2023) Rhodococcus opacus high-cell-density batch cultivation with a bagasse hydrolysate for possible triacylglycerol synthesis. Biomed Biotechnol Res J 7(2):209. https://doi.org/10.4103/bbrj.bbrj_55_23

Malode SJ, Prabhu KK, Mascarenhas RJ, Shetti NP, Aminabhavi TM (2021) Recent advances and viability in biofuel production. Energ Convers Manag: X 10:100070. https://doi.org/10.1016/j.ecmx.2020.100070

Martinez-Villarreal S, Breitenstein A, Nimmegeers P, Perez Saura P, Hai B, Asomaning J, Eslami AA, Billen P, Van Passel S, Bressler DC, Debecker DP, Remacle C, Richel A (2022) Drop-in biofuels production from microalgae to hydrocarbons: microalgal cultivation and harvesting, conversion pathways, economics and prospects for aviation. Biomass Bioenergy 165:106555. https://doi.org/10.1016/j.biombioe.2022.106555

Mondal S, Biswal D, Pal K, Rakshit S, Kumar Halder S, Mandavgane SA, Bera D, Hossain M, Chandra Mondal K (2022) Biodeinking of waste papers using combinatorial fungal enzymes and subsequent production of butanol from effluent. Bioresour Technol 353:127078. https://doi.org/10.1016/j.biortech.2022.127078

Moon M, Park W-K, Lee SY, Hwang K-R, Lee S, Kim M-S, Kim B, Oh Y-K, Lee J-S (2022) Utilization of whole microalgal biomass for advanced biofuel and biorefinery applications. Renew Sust Energ Rev 160:112269. https://doi.org/10.1016/j.rser.2022.112269

Nair AS, Sivakumar N (2022) Enhanced production of biodiesel by Rhodosporidium toruloides using waste office paper hydrolysate as feedstock: optimization and characterization. Fuel 327:125174. https://doi.org/10.1016/j.fuel.2022.125174

Nair AS, Sivakumar N (2023) Biodiesel production by oleaginous bacteria Rhodococcus opacus PD630 using waste paper hydrolysate. Biomass Conv Bioref 13(17):15827–15836. https://doi.org/10.1007/s13399-021-02135-z

Nanda S, Golemi-Kotra D, McDermott JC, Dalai AK, Gökalp I, Kozinski JA (2017) Fermentative production of butanol: perspectives on synthetic biology. New Biotechnol 37:210–221. https://doi.org/10.1016/j.nbt.2017.02.006

Niglio S, Procentese A, Russo ME, Sannia G, Marzocchella A (2019) Investigation of enzymatic hydrolysis of coffee silverskin aimed at the production of butanol and succinic acid by

fermentative processes. Bioenergy Res 12(2):312–324. https://doi.org/10.1007/s12155-019-09969-6
Oh Y-K, Hwang K-R, Kim C, Kim JR, Lee J-S (2018) Recent developments and key barriers to advanced biofuels: a short review. Bioresour Technol 257:320–333. https://doi.org/10.1016/j.biortech.2018.02.089
Okoro V, Azimov U, Munoz J (2022) Recent advances in production of bioenergy carrying molecules, microbial fuels, and fuel design – a review. Fuel 316:123330. https://doi.org/10.1016/j.fuel.2022.123330
Ouadi M, Bashir MA, Speranza LG, Jahangiri H, Hornung A (2019) Food and Market Waste-A Pathway to Sustainable Fuels and Waste Valorization. Energy Fuels 33(10):9843–9850. https://doi.org/10.1021/acs.energyfuels.9b01650.
Pang Y, Zhao Y, Li S, Zhao Y, Li J, Hu Z, Zhang C, Xiao D, Yu A (2019) Engineering the oleaginous yeast Yarrowia lipolytica to produce limonene from waste cooking oil. Biotechnol Biofuels 12(1):241. https://doi.org/10.1186/s13068-019-1580-y
Park GW, Shin S, Kim SJ, Lee J-S, Moon M, Min K (2023) Rice straw-derived lipid production by HMF/furfural-tolerant oleaginous yeast generated by adaptive laboratory evolution. Bioresour Technol 367:128220. https://doi.org/10.1016/j.biortech.2022.128220
Peng X, Jiang Y, Chen Z, Osman AI, Farghali M, Rooney DW, Yap P-S (2023) Recycling municipal, agricultural and industrial waste into energy, fertilizers, food and construction materials, and economic feasibility: a review. Environ Chem Lett 21(2):765–801. https://doi.org/10.1007/s10311-022-01551-5
Phillips J, Huhnke R, Atiyeh H (2017) Syngas fermentation: a microbial conversion process of gaseous substrates to various products. Fermentation 3(2):28. https://doi.org/10.3390/fermentation3020028
Pina JC, Bono JAM, AKD O, RDS R, Amorim DO (2015) Organic residues on rooting and yield of sugarcane in Typic Quartzipsamments soil. Rev Bras Eng Agríc Ambient 19(7):650–655. https://doi.org/10.1590/1807-1929/agriambi.v19n7p650-655
Rahimi Z, Anand A, Gautam S (2022) An overview on thermochemical conversion and potential evaluation of biofuels derived from agricultural wastes. Energy Nexus 7:100125. https://doi.org/10.1016/j.nexus.2022.100125
Rene ER, Veiga MC, Kennes C (2022) Recent trends in biogenic gas, waste and wastewater fermentation. Fermentation 8(8):347. https://doi.org/10.3390/fermentation8080347
Ro JW, Zhang Y, Kendall A (2023) Developing guidelines for waste designation of biofuel feedstocks in carbon footprints and life cycle assessment. Sustain Prod Consump 37:320–330. https://doi.org/10.1016/j.spc.2023.03.009
Rodriguez K, Pedroso M, Harris A, Garg S, Hine D, Köpke M, Schenk G, Marcellin E (2023) Gas fermentation for microbial sustainable aviation fuel production. Microbiol Aust 44(1):31–35. https://doi.org/10.1071/MA23008
Roy S, Biswas JK, Kumar S (2014) Nutrient removal from waste water by macrophytes—an eco-friendly approach to waste water treatment and management. Energ Environ Res 4(2):55. https://doi.org/10.5539/eer.v4n2p55
Ruiz-Ruiz P, Gómez-Borraz TL, Revah S, Morales M (2020) Methanotroph-microalgae co-culture for greenhouse gas mitigation: effect of initial biomass ratio and methane concentration. Chemosphere 259:127418. https://doi.org/10.1016/j.chemosphere.2020.127418
Saba B, Bharathidasan AK, Ezeji TC, Cornish K (2023) Characterization and potential valorization of industrial food processing wastes. Sci Total Environ 868:161550. https://doi.org/10.1016/j.scitotenv.2023.161550
Saini R, Osorio-Gonzalez CS, Brar SK, Kwong R (2021) A critical insight into the development, regulation and future prospects of biofuels in Canada. Bioengineered 12(2):9847–9859. https://doi.org/10.1080/21655979.2021.1996017
Saini R, Osorio-Gonzalez CS, Hegde K, Kaur Brar S, Vezina P (2022) A co-fermentation strategy with wood hydrolysate and crude glycerol to enhance the lipid accumulation in Rhodosporidium

toruloides-1588. Bioresour Technol 364:127821. https://doi.org/10.1016/j.biortech.2022.127821

Salihu A, Alam MZ, Abdul Karim MI, Salleh HM (2012) Lipase production: An insight in the utilization of renewable agricultural residues. Resour Conserv Recycl 58:36–44. https://doi.org/10.1016/j.resconrec.2011.10.007

Scarlat N, Fahl F, Dallemand J-F (2019) Status and opportunities for energy recovery from municipal solid waste in Europe. Waste Biomass Valor 10(9):2425–2444. https://doi.org/10.1007/s12649-018-0297-7

Schubert T (2020) Production routes of advanced renewable C1 to C4 alcohols as biofuel components—a review. Biofuels Bioprod Biorefin 14(4):845–878. https://doi.org/10.1002/bbb.2109

Selvaggi R, Valenti F (2021) Assessment of fruit and vegetable residues suitable for renewable energy production: GIS-based model for developing new frontiers within the context of circular economy. ASI 4(1):10. https://doi.org/10.3390/asi4010010

Singh S, Pandey D, Saravanabhupathy S, Daverey A, Dutta K, Arunachalam K (2022) Liquid wastes as a renewable feedstock for yeast biodiesel production: opportunities and challenges. Environ Res 207:112100. https://doi.org/10.1016/j.envres.2021.112100

Taptich MN, Scown CD, Piscopo K, Horvath A (2018) Drop-in biofuels offer strategies for meeting California's 2030 climate mandate. Environ Res Lett 13(9):094018. https://doi.org/10.1088/1748-9326/aadcb2

Thanigaivel S, Vickram S, Dey N, Gulothungan G, Subbaiya R, Govarthanan M, Karmegam N, Kim W (2022) The urge of algal biomass-based fuels for environmental sustainability against a steady tide of biofuel conflict analysis: is third-generation algal biorefinery a boon? Fuel 317:123494. https://doi.org/10.1016/j.fuel.2022.123494

Tsai T-Y, Lo Y-C, Dong C-D, Nagarajan D, Chang J-S, Lee D-J (2020) Biobutanol production from lignocellulosic biomass using immobilized Clostridium acetobutylicum. Appl Energy 277:115531. https://doi.org/10.1016/j.apenergy.2020.115531

Tuhanioglu A, Hamamci H, Alpas H, Cekmecelioglu D (2022) Valorization of apple pomace via single cell oil production using oleaginous yeast Rhodosporidium toruloides. Waste Biomass Valor 14(3):765–779. https://doi.org/10.1007/s12649-022-01884-w

Usmani Z, Sharma M, Awasthi AK, Sivakumar N, Lukk T, Pecoraro L, Thakur VK, Roberts D, Newbold J, Gupta VK (2021) Bioprocessing of waste biomass for sustainable product development and minimizing environmental impact. Bioresour Technol 322:124548. https://doi.org/10.1016/j.biortech.2020.124548

van der Ha D, Nachtergaele L, Kerckhof F-M, Rameiyanti D, Bossier P, Verstraete W, Boon N (2012) Conversion of biogas to bioproducts by algae and methane oxidizing bacteria. Environ Sci Technol 46(24):13425–13431. https://doi.org/10.1021/es303929s

van Dyk S, Su J, Mcmillan JD, Saddler J (2019) Potential synergies of drop-in biofuel production with further co-processing at oil refineries. Biofuels Bioprod Biorefin 13(3):760–775. https://doi.org/10.1002/bbb.1974

Vural Gursel I, Elbersen B, Meesters KPH, van Leeuwen M (2022) Defining circular economy principles for biobased products. Sustain For 14(19):12780. https://doi.org/10.3390/su141912780

Walls LE, Rios-Solis L (2020) Sustainable production of microbial isoprenoid derived advanced biojet fuels using different generation feedstocks: a review. Front Bioeng Biotechnol 8:599560. https://doi.org/10.3389/fbioe.2020.599560

Wang K, Tester JW (2023) Sustainable management of unavoidable biomass wastes. Green Energy Resour 1(1):100005. https://doi.org/10.1016/j.gerr.2023.100005

Wang S, Wang Z, Wang Y, Nie Q, Yi X, Ge W, Yang J, Xian M (2017) Production of isoprene, one of the high-density fuel precursors, from peanut hull using the high-efficient lignin-removal pretreatment method. Biotechnol Biofuels 10(1):297. https://doi.org/10.1186/s13068-017-0988-5

Whalen J, Xu CC, Shen F, Kumar A, Eklund M, Yan J (2017) Sustainable biofuel production from forestry, agricultural and waste biomass feedstocks. Appl Energy 198:281–283. https://doi.org/10.1016/j.apenergy.2017.05.079

Xiong X, Zhang W, Ha X, Li N, Chen S, Xing H, Yang J (2023) The preparation processes and influencing factors of biofuel production from kitchen waste. Fermentation 9(3):247. https://doi.org/10.3390/fermentation9030247

Yan J, Liang L, He Q, Li C, Xu F, Sun J, Goh E, Konda NVSNM, Beller HR, Simmons BA, Pray TR, Thompson VS, Singh S, Sun N (2019) Methyl ketones from municipal solid waste blends by one-pot ionic-liquid pretreatment, saccharification, and fermentation. ChemSusChem 12(18):4313–4322. https://doi.org/10.1002/cssc.201901084

Yazdanparast R, Jolai F, Pishvaee MS, Keramati A (2022) A resilient drop-in biofuel supply chain integrated with existing petroleum infrastructure: toward more sustainable transport fuel solutions. Renew Energy 184:799–819. https://doi.org/10.1016/j.renene.2021.11.081

Yuzawa S, Mirsiaghi M, Jocic R, Fujii T, Masson F, Benites VT, Baidoo EEK, Sundstrom E, Tanjore D, Pray TR, George A, Davis RW, Gladden JM, Simmons BA, Katz L, Keasling JD (2018) Short-chain ketone production by engineered polyketide synthases in Streptomyces albus. Nat Commun 9(1):4569. https://doi.org/10.1038/s41467-018-07040-0

Zargar A, Bailey CB, Haushalter RW, Eiben CB, Katz L, Keasling JD (2017) Leveraging microbial biosynthetic pathways for the generation of 'drop-in' biofuels. Curr Opin Biotechnol 45:156–163. https://doi.org/10.1016/j.copbio.2017.03.004

Zhao Y, Zhu K, Li J, Zhao Y, Li S, Zhang C, Xiao D, Yu A (2021) High-efficiency production of bisabolene from waste cooking oil by metabolically engineered *Yarrowia lipolytica*. Microb Biotechnol 14(6):2497–2513. https://doi.org/10.1111/1751-7915.13768

Zhu P, Abdelaziz OY, Hulteberg CP, Riisager A (2020) New synthetic approaches to biofuels from lignocellulosic biomass. Curr Opin Green Sustain Chem 21:16–21. https://doi.org/10.1016/j.cogsc.2019.08.005

Zieliński M, Dębowski M, Kazimierowicz J, Świca I (2023) Microalgal carbon dioxide (CO2) capture and utilization from the European Union perspective. Energies 16(3):1446. https://doi.org/10.3390/en16031446

Chapter 4
Microalgal-Based Drop-in Biofuel Production

Bridget Signe Price and Frank William Rowley Chaplen

Abstract Microalgae are photosynthetic microorganisms that show potential for biofuel production due to their ability to accumulate large quantities of neutral lipids. From a climate perspective, microalgae are valued as a biofuel feedstock because they use the greenhouse gas carbon dioxide as a carbon source. The production of microalgal-based drop-in biofuel requires four main steps: cultivation, harvesting and dewatering, extraction, and conversion. Over the years, the processes used to produce microalgal biofuels have improved, taking advantage of the unique characteristics of microalgae, but there is much improvement. The primary challenge in the implementation of industrial microalgal biofuel production systems is the cost of production, which cannot be solved simply.

Keywords Algae · Photosynthesis · Photobioreactor · Carbon sequestration · Drop-in biofuels

4.1 Introduction

Microalgae encompass up to 1,000,000 species spanning the phylogenetic tree (Correa et al. 2017; Rumin et al. 2020). Historically, microalgae have been grouped into two classes: bacterial microalgae (cyanobacteria) and eukaryotic microalgae, which includes green algae, red algae, and diatoms (Tibocha-Bonilla et al. 2018). These unicellular organisms can occupy a variety of habitats, including terrestrial environments like deserts and snowbanks, but they are predominantly found in aquatic systems like lakes and oceans (Terlova et al. 2021). Morphologically, microalgae are also diverse. Microalgae like diatoms possess siliceous frustule shells

B. S. Price
School of Chemical, Biological, and Environmental Engineering, Oregon State University, Corvallis, OR, USA

F. W. R. Chaplen (✉)
Department of Biological and Ecological Engineering, Oregon State University, Corvallis, OR, USA
e-mail: frank.chaplen@oregonstate.edu

S. K. Brar et al. (eds.), *The Microbiology of the Drop-in Biofuel Production*, Biofuel and Biorefinery Technologies 15, https://doi.org/10.1007/978-3-031-61637-2_4

while the green algae *Scenedesmus* sp. exist in chain-like colonies. Regardless of their differences, microalgae share, by definition, one characteristic: the capacity to photosynthesize. The ability to fix carbon dioxide and produce oxygen via photosynthesis makes microalgae an incredibly important primary producer, with over 50% of photosynthesis carried out by microalgae globally (Moroney and Ynalvez 2009).

Research into the commercial production of microalgae only began in the 1950s, a movement spurred by the need for alternative sources of foods and biochemicals in the aftermath of World War II (Borowitzka 2013). In addition to the nutritional value of microalgae as a food, they were, and still are, a desirable feedstock for the production of biofuels for three main reasons. First, microalgae cultivation can occur anywhere with access to water and sunlight, which makes it possible to produce microalgal-based biofuel on non-arable land. Second, microalgae can acclimatize to different environmental conditions. This makes it possible to cultivate algae in wastewater, meaning that microalgae can be simultaneously used as a bioremediation tool and a fuel feedstock. Third, microalgae are capable of achieving higher lipid yields on less land than oil-bearing agricultural crops (Zhang et al. 2022b). Over the last 30 years, interest in microalgal-based biofuels as a climate-friendly alternative to fossil fuels, conventional first-generation biofuels, and other microbe-based biofuels has increased because of their ability to remove carbon dioxide (CO_2) from the atmosphere during photosynthesis.

Microalgae can be used to produce multiple types of biofuels. One course of microalgal-based biofuel production is through the direct synthesis of fuels by microalgae. An example of this is the production of biohydrogen by microalgae through photosynthesis in oxygen and sulfur deprived conditions (Dębowski et al. 2021). More commonly, microalgal-based biofuels like bioethanol and biodiesel are produced through the conversion of storage compounds like carbohydrates and lipids. Compared to bioethanol, biodiesel produced from lipids is the ideal fuel candidate for a microalgal-based fuel (Fig. 4.1; Lakatos et al. 2019).

4.2 Cultivation

Microalgae are capable of accumulating high levels of lipids, which makes them an attractive source of microbial lipids. Unfortunately, microalgae have relatively low lipid and biomass productivities, posing a challenge in the industrialization of microalgal-based biofuels because the cost of microalgae production is high. To work around this shortcoming, the goal of cultivation systems is to increase the growth rate of microalgae to compensate for their low productivities while keeping the costs of cultivation low (Tan et al. 2020). Three factors can be considered in the cultivation of microalgae for biofuel production: strain selection, the conditions required for growth, and the design of the cultivation system.

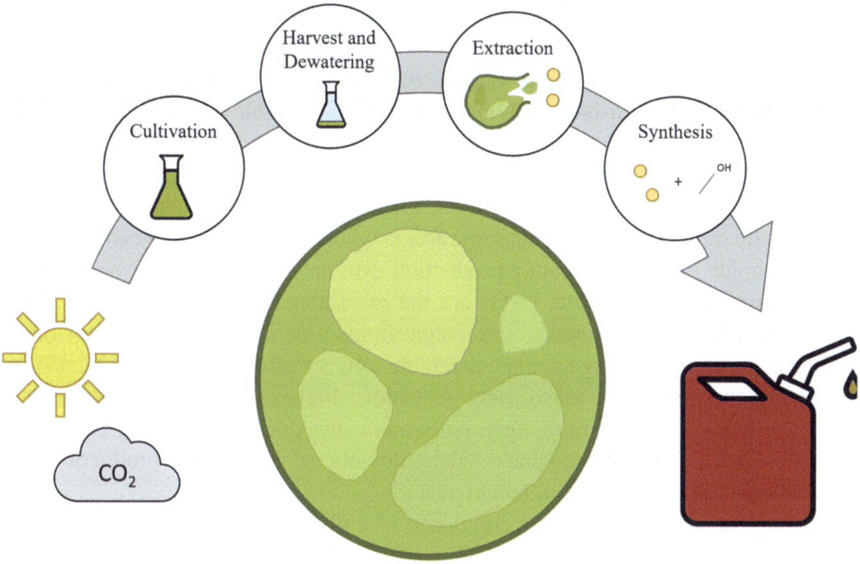

Fig. 4.1 Overview of microalgal drop-in biofuel production

4.2.1 Strain Selection

Microalgae strain selection is an important factor in the overall success of microalgal-based drop-in fuel production. Not all microalgae are suited for biofuel production, as shown through the great variation of biomass and lipid productivities between species (Shokravi et al. 2019). Ideally, the microalga strain selected for biofuel production can achieve both high lipid and biomass productivities, which would help make biofuel production more economical. For this reason, oleaginous microalgae, or algae that can accumulate more than 20% of their dry biomass in the form of lipids, are the most promising candidates for biofuel production (Tibocha-Bonilla et al. 2018). Approximately 300 strains of microalgae are considered oleaginous, but microalgae from the genera *Chlorella, Desmodesmus, Nannochloropsis, Tetraselmis, Scenedesmus, Phaeodactylum,* and *Dunaliella* are some of the most commonly researched for biofuel production (Yoshida et al. 2012; Shokravi et al. 2019). Regardless of the lipid accumulation potential, strains must have, at a minimum, high biomass productivities, which can be achieved through the optimization of cultivation conditions or strain improvement (Venteris et al. 2014). Specifically, strain improvement can be a useful tool in making strains more resilient to stress, which can improve their performance.

4.2.1.1 Lipid Metabolism

Microalgae synthesize two major classes of lipids: membrane and storage lipids. Membrane lipids, which include phosphoglycerides, betaine ether lipids, and glycosyl glycerides, play an important role in the structure and functionality of microalgal cells (Dean and Lodhi 2018; Li-Beisson et al. 2019). Meanwhile, microalgae store lipids in the form of non-polar glycolipids, of which triacylglycerol (TAG) is the most abundant. Triacylglycerol is what makes microalgae an attractive source of lipids for drop-in biofuel production. Not only can microalgae accumulate large quantities of TAGs, but TAGs are the most abundant storage lipid in other organisms, which is what makes microalgal lipids a perfect replacement for other renewable biofuel feedstocks like vegetable oils (Yoshida et al. 2012). While optimizing the cultivation conditions to achieve highest TAG accumulation has been a primary focus of microalgae cultivation, the potential of other classes of microalgal lipids, such as terpenes and hydrocarbons, for biofuel production has been increasingly recognized in recent years.

Triacylglycerol is glycerolipid composed of a glycerol backbone esterified with three fatty acids. In microalgae, TAGs are synthesized through two pathways (Li-Beisson et al. 2019). The first pathway is called the Kennedy pathway, which is the de novo synthesis of TAGs using acyl-CoA (Fig. 4.2). Fatty acids are produced in the chloroplast and are exported to the endoplasmic reticulum where they are sequentially added to a glycerol-3-phosphate molecule to form a TAG (Yoshida et al. 2012; Fattore et al. 2022). The second pathway is acyl-CoA independent and mediated by phospholipid:diacylglycerol acyltransferase (PDAT). In this pathway, PDAT catalyzes the transfer of a acyl group from membrane phospholipids to diacylglycerol, forming a TAG (Muñoz et al. 2021). This pathway becomes favorable during exponential growth and during periods of nitrogen starvation, where it becomes more energetically efficient in the short term, but negatively affects membrane homeostasis over long periods of time (Fattore et al. 2022). After TAG is synthesized, it is assembled into lipid droplets, which store TAG within a phospholipid monolayer. In addition to functioning as a storage vesicle, lipid droplets play a role in energy supply, membrane structure, and signal transduction, emphasizing the overall importance of lipid metabolism on cell functionality (Zienkiewicz and Zienkiewicz 2020).

Microalgae tend to accumulate triacylglycerol when subjected to stressful or unfavorable environmental conditions. Nutrient limitation and unfavorable culture conditions, such as high salt concentration and light intensity, are the primary environmental stressors that lead to lipid accumulation (Chen and Wang 2021). Nitrogen starvation is a well-studied mechanism for lipid accumulation in microalgae. When microalgae are subjected to nitrogen depletion, PDAT is overexpressed, drawing carbon away from photosynthetic pathways that contribute to biomass production. The flux of carbon is directed instead toward the synthesis of fatty acids and TAG production is mediated by membrane lipids (Tibocha-Bonilla et al. 2018; Fattore et al. 2022). For example, Rios et al. (2015) found that

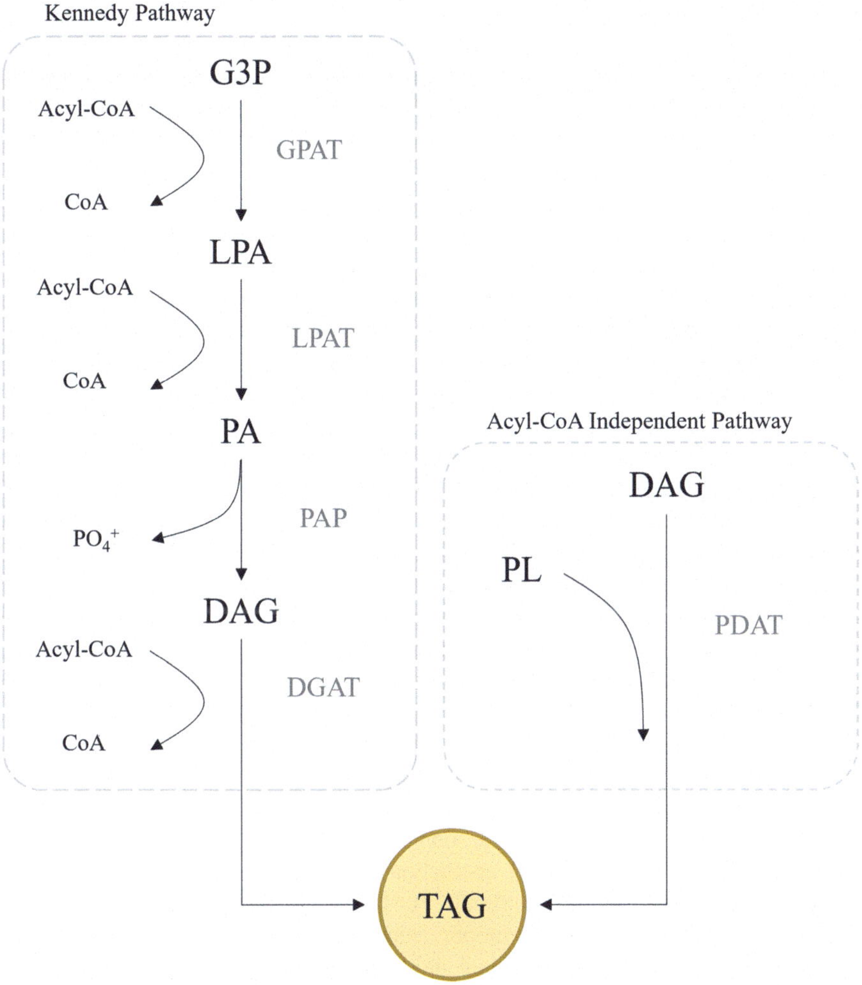

Fig. 4.2 Microalgal TAG synthesis pathways. G3P, glycerol-3-phosphate; GPAT, glycerol 3-phosphate acyltransferase; LPA, lysophosphatidic acid; LPAT, lysophosphatidic acid acyltransferase; PA, phosphatidic acid; PAP, phosphatidic acid phosphatase; DAG, diacylglycerol; DGAT, diacylglycerol acyltransferase; PL, (membrane) phospholipid; PDAT, phospholipid: diacylglycerol acyl transferase. Adapted from Muñoz et al. (2021)

Desmodesmus sp. cultured in the absence of nitrogen achieved a lipid accumulation 12% higher compared to those cultured in 750 mg/L of $NaNO_3$, but a final biomass concentration more than 50% lower. Similarly, iron starvation results in lower photosynthetic efficiency, triggering the overexpression of the protein that catalyzes the addition of a fatty acid to diacylglycerol, so carbon is directed toward lipid storage rather than growth (Devadasu and Subramanyam 2021).

4.2.1.2 Strain Improvement

Both biomass and lipid yields can be increased for more economic drop-in biofuels production through the better design and optimization of cultivation systems. However, some challenges associated with microalgae cultivation might be better addressed through the utilization of genetic modification or metabolic engineering to improve strain performance. These include biomass productivity, lipid yield, oxygen tolerance, photosynthetic efficiency, pH tolerance, harvestability, CO_2 assimilation, temperature tolerance, and cell wall stability or susceptibility to rupture for lipid recovery (Venteris et al. 2014; Ahmad et al. 2023). Common methods used to modify microalgae are mutagenesis, adaptive laboratory evolution, genetic engineering, and systems biology. The mathematical analysis of metabolism and the impact of genetic changes on cell metabolism could be an important adjunct to purely genetic methods. However, a major challenge with utilizing quantitative approaches for the improvement of microalgal strains is the lack of metabolic models, especially genome scale models. Presently, genome scale models exist for a limited number of microalgal species including *Chlamydomonas, Chlorella, Nannochloropsis, Synechocystis, Monoraphidium, Osteococcus, Tisochrysis,* and *Phaeodactylum* (Tibocha-Bonilla et al. 2018).

The key to any successful genetic modification is genetically modified strain stability. Species have evolved with functions optimized for specific ecological niche stability. The stable introduction of non-native genetic elements without the presence of antibiotic resistance is a significant barrier. In addition, there are significant concerns with the large-scale production of genetically modified organisms in open ponds. Genetic methods that might be used for strain improvement to improve the economics of algal lipid production for drop-in biofuel applications include methods required for the infiltration of DNA-based genetic elements into the algal strains (transformation) and construction of the genetic elements uses species-specific genetic tools (Khoo et al. 2023). Transformation is mediated through approaches such as electroporation, microbial mediation (*Agrobacterium tumefaciens*; extremely complex), and physical methods, including bead beating and particle bombardment. The absence of genetic tools can be a significant barrier to the genetic engineering strain improvements for many algal strains. However, there have been significant developments in generic genetic tools over the past two decades. Khoo et al. (2023) provide multiple examples of gene engineering applied to algal strains for improved lipid profile and production. Maximum potential for enhanced neutral lipid (TAG) production for biodiesel production was observed for modifications in the Kennedy pathway using both overexpression (over production) of existing genes and heterologous (non-native gene) expression (production for a given organism). Enhancements of 50–100% or more in TAG by weight% relative to biomass were not uncommon.

4.2.2 *Parameters of Growth*

4.2.2.1 Illumination

Light is a deterministic factor in the growth, functional state, and reproduction of microalgae, making it the most important variable in the cultivation of microalgae (Maltsev et al. 2021). In general, microalgae cultivation systems are optimized so that light is only variable limiting photosynthetic productivity. Since different photobioreactor systems have different sources of light, the method of controlling the illumination can look like manipulating an artificial light source to the system needs or altering a system to maximize natural light. Regardless of the light source, the main considerations for illumination in the maximization microalgal production are the lighting intensity, duration of lighting, and the spectral composition of light waves.

4.2.2.1.1 Spectral Composition of Light

Photosynthetically active radiation (PAR) is the range of light wavelengths usable for photosynthesis, which is around 400–700 nm (Chowdury et al. 2020). The exact spectrum that microalgal cells can absorb is dictated by the pigment composition of their light-harvesting antennae. The two primary groups of pigments that participate in light harvesting in microalgae are chlorophylls and carotenoids. These pigments are capable of absorbing photons from different parts of the light spectrum. Chlorophylls, most commonly chl a, chl b, and chl c, contribute to the absorption of red light (600–680 nm) and blue light (450–475 nm). Carotenoids, including carotenes and xanthophylls, absorb mostly green and blue light (420–470 nm) (Lehmuskero et al. 2018; Masojídek et al. 2021; Ma et al. 2022).

The composition and concentration of photosynthetic pigments differ between microalgal species, so the spectral composition can have an impact on the growth of microalgae, especially between species. Typically, microalgae grow best in white light, which is a combination of many wavelengths of light. An example of this trend can be seen in the case where *Isochrysis galbana* grown in mixed red and blue LED lighting achieved a specific growth rate of 0.042 h^{-1}, which dropped to 0.021 and 0.024 h^{-1} when *I. galbana* was grown in monochromatic blue and red LED light, respectively (Ra et al. 2018). In some cases, microalgal species do grow better in monochromatic light such as *Chlorella vulgaris* which obtained growth rates of 0.09 d^{-1} in red light but only 0.084 d^{-1} in white light (Esteves et al. 2020). While microalgae typically have a set variety of pigments, they can use the process of photo-acclimation to adjust their pigment content, composition, and concentration to be better suited to the light conditions. Photo-acclimation also refers to other macromolecular changes such as those in cell morphology, respiration, and photosynthetic enzyme production that microalgae undergo in response to differing light conditions (Straka and Rittmann 2018).

4.2.2.1.2 Light Intensity

The intensity of light in a microalgae culture can be described by the Beer-Lambert Law, where the intensity of light is the highest at the illuminated surface and then decreases exponentially with distance from the illuminated surface. In other words, as light travels through a culture, photons are absorbed by the cells closest to the illuminated surface. This attenuation of light through a microalgae culture can create three regions of light availability: limitation, saturation, and inhibition. Photosynthesis in microalgae increases until it reaches a maximum rate. The light intensity that this maximum occurs at is called the saturation point (Chowdury et al. 2020). Light intensities greater than what leads to the maximum rate cause inhibition of microalgal growth via photoinhibition, while lower intensities result in light-limited microalgae. The attenuation of light increases with increasing cell concentration, meaning that it becomes more difficult to maintain a region of light saturation because microalgae farther from the light source are shaded by other microalgae, becoming light limited. In cases like these, increasing the light intensity is not a solution, as the growth of microalgae closest to the light source can become inhibited. The range of light intensity in which microalgae are limited, saturated, and inhibited is species dependent and the proper light intensity is critical for optimal growth. For example, when the light intensity of a *Scenedesmus obliquus* was increased from 50 to 150 μE, the biomass production almost doubled, and the fatty acid content increased by 2% (Nzayisenga et al. 2020).

4.2.2.1.3 Photoinhibition

During photosynthesis, the photosynthetic electron transport chain is initiated when photosystem II (PSII) is excited by light, allowing water to be oxidized to oxygen (Tang and Blankenship 2013). Not all of the light absorbed by pigments is used to drive photosynthesis, some is dissipated as heat or re-emitted as fluorescence. The process of dissipating excessive light energy as heat are mechanisms designed to avoid photodamage is called non-photochemical quenching (NPQ) (Nikolaou et al. 2015). Photoinhibition occurs when the absorbed light intensity exceeds the threshold that can be managed by NPQ mechanisms. Excess light energy causes excited chlorophylls to form triplet states and transfer energy to oxygen to produce highly reactive single oxygen species ($^1O_2$). The reactive singlet oxygen causes irreversible reaction to the D1 protein of PSII leading to a decrease in photosynthetic productivity (Levin et al. 2023). Microalgal cells do have an additional defense mechanism against photoinhibition where they can remove and replace the damaged D1 protein with a new copy, restoring the function of PSII (Vinyard et al. 2013). Prolonged exposures to high light intensities can lead to the point where the rate of repair cannot keep up with PSII inactivation, leading to cell death.

4.2.2.1.4 Duration and Frequency of Lighting

A photoperiod is the length of time within a 24 h time frame that light is available. In nature this period is dictated by the rising and setting of the sun, but in microalgae cultivation where light can come from an artificial source, the photoperiod can be controlled. While artificial lighting makes it possible to supply a culture with light for 24 h, some microalgae require time to grow in the absence of light to synthesize metabolic compounds necessary for growth from the compounds produced via light reactions. For instance, *Nannochloropsis* sp. showed a decrease in cell density, lipid content, and specific growth rate when the photoperiod was extended 18:06 to 24:0 h light/dark at a light intensity of 100 μmol m^{-2} s^{-1} (Wahidin et al. 2013). The photoperiod can also result in variations in the cellular content of proteins, carbohydrates, and lipids, which can allow the light/dark regime to be selected with the production of a certain compound in mind. For example, *Ankistrodesmus falcatus* achieved the highest carbohydrate content of 7.7% when grown on a 06:18 h light/dark regime, which decreased to 5.92% under 24:0 h light/dark cycle, and the highest lipid production of 35.13% on a 12:12 h light/dark regime (George et al. 2014).

4.2.2.2 Carbon Supply

As photoautotrophic microorganisms, microalgae do not require an input of organic carbon to grow. The most common form of carbon supplied to microalgae is gaseous carbon dioxide. However, the supply of carbon dioxide is limited by both low water solubility and low CO_2 gas-liquid mass transfer efficiency which leads to two carbon supply issues (Kim et al. 2021). First, the fraction of carbon dioxide present in the atmosphere is too low to support optimal microalgae growth, meaning that an air stream enriched with CO_2 is typically necessary for cultivation. Second, up to 90% of the gaseous carbon dioxide supplied to a culture is lost to the atmosphere. This presents a challenge when microalgae are being utilized in carbon capture systems because it undermines the advantages of the system (Kim et al. 2013). The simplest solution to this issue is using bicarbonate as an inorganic carbon source, which microalgae can assimilate by using two different mechanisms involving carbonic anhydrase (Pancha et al. 2015). The parallel utilization of bicarbonate and carbon dioxide as a carbon source is shown to lead to higher photosynthetic productivity and improved growth as a result (White et al. 2013). For example, *Scenedesmus* sp. CCNM 1077 grown with CO_2 in addition 0.6 g/L sodium bicarbonate achieved a biomass production 23% higher than when it was grown in absence of bicarbonate (Pancha et al. 2015).

The solubility of CO_2 is dictated by the thermodynamic equilibrium of the dissociation of carbonic acid. This means that the form of inorganic carbon present in a culture is dictated by the pH and vice versa (Fig. 4.3). For example, in seawater, which has a pH of around 8.1, the primary form of inorganic carbon is bicarbonate, while in a more acidic culture more CO_2 and carbonic acid expected to be present

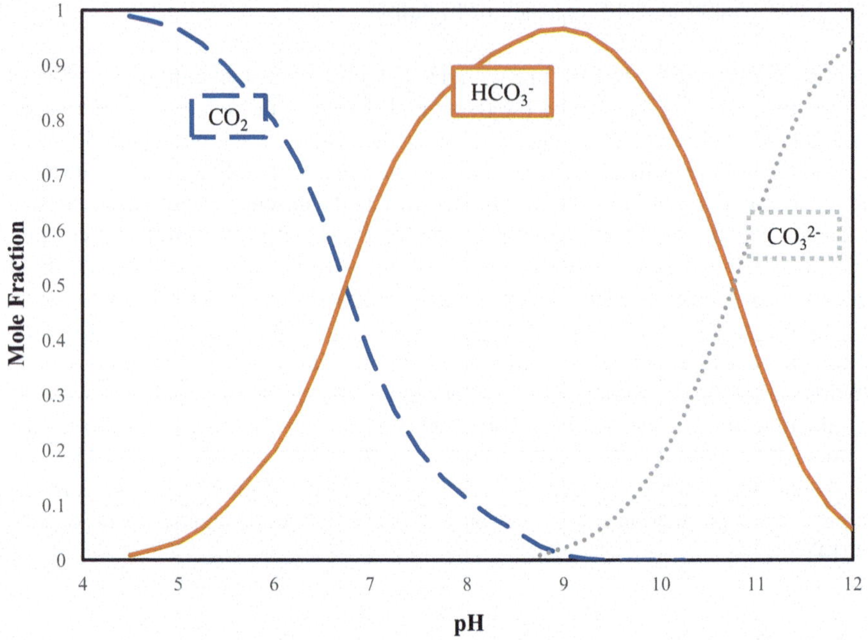

Fig. 4.3 pH impacts of inorganic carbon in an aqueous mixture

(White et al. 2013). The relationship between CO_2 and pH makes it possible to use CO_2 supply as pH control in microalgae cultivation, which is one way to prevent its limitation. As microalgae grow and consume CO_2 the pH increases, so an input in CO_2 mains the pH remains at an optimal set point while also ensuring that adequate CO_2 is available for growth.

4.2.2.2.1 Mixotrophic Carbon Supply

Microalgae are unique microorganisms because many species can use more than one trophic mode. In a few species, such as *Thraustochytrium* sp., *Chlorella* sp., *Scenedesmus* sp., and *Neochloris oleoabundans,* microalgae can act as heterotrophs in the absence of light (Castillo et al. 2021). While cultivation in heterotrophic conditions removes the limitation that light has on microalgae growth, this trophic mode is not necessarily better than cultivation in the presence of light. Studies have found that photoautotrophic conditions are preferable over heterotrophic conditions, *Chlorella vulgaris* achieved lipid productions 136% higher in photoautotrophic compared to heterotrophic conditions (Zhang et al. 2014). Heterotrophy is not the only unique trophic state that microalgae can have, some species are capable of using multiple trophic modes. This ability is called mixotrophy, which means that photoautotrophy and heterotrophy are used simultaneously, enabling microalgae to assimilate both organic and inorganic carbon from the environment. Not only does mixotrophic cultivation avoid the carbon supply challenges associated with photo-autotrophic cultivation, but it widens the capacity in which microalgae can be used to remove unwanted carbon from the environment. When grown in polluted river water, *Chlorella* sp. SL7A reduced the biological demand 85% and achieved a lipid accumulation of around 29.5%, showing that mixotrophy is a tool to meet lipid production goals while helping the environment (Ummalyma and Singh 2022).

4.2.2.3 Oxygen Accumulation

In photobioreactor systems, the oxygen produced via photosynthesis can accumulate to levels detrimental to microalgae growth. The overall extent of oxygen inhibition is species specific. For example, in *Chlorella vulgaris* oxygen concentrations over 30 g/m^3 lead to a 30% decrease in biomass productivity, whereas in *Nannochloropsis* sp. an oxygen concentration of 20.45 g/m^3 leads to a 37.5% decrease in specific growth rate (Raso et al. 2012; Kazbar et al. 2019). The processes that occur at high oxygen concentrations that negatively affect microalgae growth are the Mehler reaction, photoinhibition, and photorespiration. While the Mehler reaction and photoinhibition are driven by a mismanagement of light reaction products, photorespiration is a direct result of high oxygen (Kliphuis et al. 2011; Sforza et al. 2020).

4.2.2.3.1 Photorespiration

Photorespiration, or "respiration in light," is a metabolic pathway that exists in all organisms that perform oxygenic photosynthesis (Eisenhut et al. 2019). During this process, ribulose-1,5-bisphosphate is oxygenated by Rubisco rather than carboxylated, producing toxic phosphoglycolate and glycolate molecules. Phosphoglycolate decreases the regeneration of ribulose-1,5-bisphosphate and glycolate decreases the rate of photosynthesis in chloroplasts, causing the microalgal cell to work harder, inhibiting growth (Zhang et al. 2019). High concentrations of oxygen and the presence of light increase the possibility of microalgae substituting oxygen for carbon dioxide during photosynthesis, triggering this energy-intensive process that has been shown to cause biomass losses of up to 66% (Toro-Huertas et al. 2019). Maintaining high CO_2 concentrations can help prevent photorespiration from occurring.

4.2.2.4 Nutrient Supply

In addition to carbon, nitrogen and phosphorus are the most important nutrients for microalgae growth. Nitrogen is essential for the formation of proteins, nucleic acids, vitamins, and photosynthetic pigments. Only cyanobacteria are capable of nitrogen fixation, so nitrogen must be provided to microalgae such as ammonium, nitrate, nitrite, or in an organic form (Chowdury et al. 2020). While microalgae assimilate both nitrate and ammonium via active transport, many microalgae prefer ammonium as a nitrogen source because it requires less energy to consume. For example, when *Scenedesmus dimorphus* grown with ammonium hydroxide as a nitrogen source achieved a growth rate of 1.35 d^{-1} compared to 0.83 d^{-1} when grown on the same molar nitrogen concentration of sodium nitrate (Sun et al. 2022). Compared to nitrogen, microalgae need much less phosphorus, with the ratio of nitrogen needed compared to phosphorus being approximately 16:1. Even so, phosphorus is essential for growth and the synthesis of biomolecules like phospholipids, adenosine triphosphate, and deoxyribonucleic acids. The preferred form of phosphorus is orthophosphate, which is acquired by microalgae via active transport (Vu et al. 2021).

4.2.2.4.1 Micronutrients

In microalgae, micronutrients such as iron, copper, manganese, cobalt, and molybdenum play essential roles in facilitating metabolic processes. For example, iron aids in photosynthesis, where it is a cofactor in the transfer of electrons from photosystem II to photosystem I and assists in light-harvesting. A deficiency in iron leads to reduced carbon fixation, resulting in decreased growth (Rajabi Islami and Assareh 2019; Rana and Prajapati 2021). Under iron starvation conditions, the biomass production of *Chlamydomonas reinhardtii* was half that of the control which did contain iron (Devadasu and Subramanyam 2021). Meanwhile, molybdenum is a constituent of several metalloflavin enzymes and is specifically required for the synthesis of the molybdenum-molybdopterin cofactor which is used for the reduction of nitrate to ammonium (Healey et al. 2023).

The primary challenge in micronutrient supply is ensuring that the micronutrients are bioavailable. Iron is the micronutrient of greatest concern. Microalgal growth rate depends more on iron concentration compared to other micronutrients and it is difficult to maintain the bioavailability of iron (Dengg et al. 2022). For microalgae to assimilate iron from the environment, it must be in its water-soluble ferrous ion form. Unfortunately, in aquatic systems, the ferric ion form of iron is more thermodynamically stable than the ferrous ion, so the ferrous ion oxidizes to ferric ions, which is not bioavailable to microalgae (Rana and Prajapati 2021). To prevent ferrous iron from oxidizing, there are a few strategies. Most commonly is the utilization of chemical chelating agents to bind with iron to form water-soluble complexes that microalgae can assimilate. It is also possible to co-culture microalgae with bacteria that produce siderophores. This type of co-cultivation has been shown to make it possible for microalgae to accumulate iron easier than when grown axenically, such as when *Chlorella variabilis* ATCC 12198 was co-cultured with siderophore producing *Idiomarina loihiensis* RS14 in both iron sufficient and deficient conditions (Rajapitamahuni et al. 2019). Meanwhile, the form in which micronutrients are bioavailable can differ between microalgal species, meaning that the way micronutrients are supplied needs to be species dependent. For example, some microalgae, such as *Thalassiosira pseudonana,* lack the ability to synthesize cobalamin (Vitamin B12) from cobalt and require exogenous cobalamin (Croft et al. 2006; Facey et al. 2022).

4.3 Cultivation System Designs

The cultivation of microalgae occurs in a specialized bioreactor system, called a photobioreactor (PBR), that includes a light source to facilitate photosynthesis. Photobioreactors can be classified into two categories, open and closed systems. The ideal microalgae cultivation system should possess six key characteristics, including: an adequate source of light, effective transfer of material across the liquid–gas interface, a simple operating procedure, low contamination rate, cheap construction and production costs, and a high land efficiency (Tan et al. 2020). At a minimum, a PBR requires light, an input of carbon, the management of oxygen accumulation, a nutrient supply, and temperature and pH controls (Chowdury et al. 2020). Open and closed photobioreactors incorporate all these parameters into their design, but the method of optimization differs between systems.

4.3.1 Open Systems

Open systems are the simplest way to cultivate microalgae. As implied by their name, open systems are photobioreactors that are open to the atmosphere. Examples of open systems include natural water bodies like lakes and ponds, as well as man-made circular and raceway ponds, with raceway ponds being the most common type of open system photobioreactor (Tan et al. 2020).

4.3.1.1 Configuration

The design of a raceway pond consists of a loop, or many interconnected loops, with two straight zones and two turning zones. The most expensive, but important, part of a raceway pond is the paddlewheel. The paddlewheel is responsible for recirculating the culture, which prevents water stagnation and promotes aeration. This mixing is especially important for biomass productivity in a raceway ponds, where depths can range from 20 to 40 cm, because it prevents microalgae from residing in a region of light limitation for too long (Legrand et al. 2021; Inostroza et al. 2023). Unfortunately, paddlewheels are often not a sufficient method of introducing CO_2 into a culture and the mass transfer through contact with the atmosphere is also not sufficient to support optimal microalgae growth. For this reason, raceway ponds benefit from an additional input of CO_2 in order to achieve greater biomass productivities. One inexpensive carbon source for raceway ponds is flue gas from combustion plants, which in addition to promoting microalgal growth can reduce the environmental impact of these plants (Zhu et al. 2014).

Given the simple construction of a raceway pond, they are an affordable option for industrial microalgae cultivation. Raceway ponds are also easy to operate and maintain, have a low energy demand, and are scalable. Despite this, raceway pond systems are limited by low biomass production for several reasons. First, raceway ponds are susceptible to variations in the weather. Especially in terms of light availability, raceway ponds should be in regions that are consistently sunny year-round. In the United States, this confines the location raceway ponds to southern states like California, Texas, and Florida (Venteris et al. 2014). Additionally, unexpected drops in temperature can result in a drop in biomass productivity and even culture collapse, while rainy weather can result in pond overflow. Second, raceway ponds can be contaminated by unwanted species. While some contaminants do not negatively impact microalgal growth or biofuel production, others, like zooplankton, insects, and flagellate protozoan can be fatal predators of microalgae (Wang et al. 2018). Finally, the open nature of raceway ponds leads to water loss due to evaporation, increasing production costs.

4.3.1.2 Wastewater Treatment

Microalgae can be cultured in wastewater, presenting an opportunity to reduce the cost of biofuel production while preventing the release of excess nutrients and contaminants into the environment. In fact, man-made open cultivation systems were originally developed to treat wastewater. Wastewater typically contains all the nutrients required for microalgal growth, including organic carbon sources, meaning that wastewater can be used as a low to zero cost nutrient source that avoids CO_2 limitation (Pushpakumari Kudahettige et al. 2018). The inclusion of microalgae in wastewater treatment provides a platform that supports other micro-organisms beneficial to wastewater treatment. Through photosynthesis, microalgae produce oxygen which can be used by aerobic bacteria to degrade organic contaminants, while the CO_2 produced by the bacteria via cellular respiration can be consumed by the microalgae. This type of symbiotic relationship enhances nitrogen and phosphorus removal while still leading to high biomass yields. For example, in a mixed culture of microalgae and activated sludge, the final total nitrogen, phosphorus, and COD removal were 86%, 79%, and 99%, respectively, and a final biomass concentration of 1.12 g/L was achieved (Nguyen et al. 2020).

4.3.2 Closed Systems

Closed photobioreactors are bioreactor systems that are specially designed to culture photoautotrophic microorganisms like microalgae. The closed nature of closed PBRs does not allow direct exchange of materials between the culture and environment, allowing for highly controlled growth conditions that can produce contamination-free microalgae cultures.

4.3.2.1 Light Source

Aside from the utilization of tunable artificial light sources, closed photobioreactors manage light through the design of their shape. Closed PBRs come in a large variety of shapes with the most common being flat plate, tubular, and annular, all of which seek to optimize light in a different way (Fig. 4.4). The simplest type of closed PBR is the flat plate photobioreactor. The design has a large surface area to volume ratio, which exposes much of the culture to the light source, and a short light path length, reducing the formation of a growth limiting region. These design characteristics make it possible for flat plate systems to be highly productive (Lee et al. 2014; Fasaei et al. 2018). Mixing via an airlift system allows for adequate CO_2 supply while also enabling an equal light exposure throughout the culture (Legrand et al. 2021). Meanwhile, tubular photobioreactors, which consist of long tubes filled with microalgae, also have a large surface area and short light path, allowing for high biomass production. The tubes can also be configured in different directions, such as horizontal or vertical, around light sources, or in a certain geometry that allows maximal light exposure (Tan et al. 2020). Unfortunately, the successful implementation of tubular PBRs is limited by poor mass transfer, which leads to the buildup of oxygen. Especially in the presence of light, this buildup of oxygen negatively affects microalgal growth by triggering photorespiration and photoinhibition (Raso et al. 2012). The final common closed PBR shape is the annular photobioreactor. The annular shape allows for lighting to occur within the core, in addition to outside, decreasing the probability of light limitation and increasing the surface to volume ratio of the reactor.

There are two primary challenges associated with the lighting of a closed photobioreactor system. The first challenge is associated with the cost of artificial

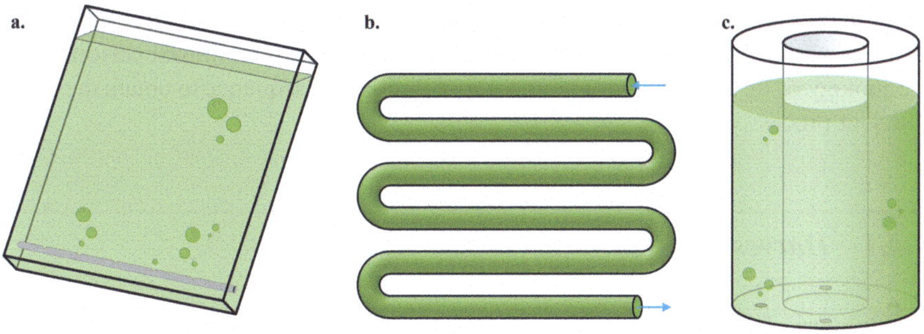

Fig. 4.4 Closed photobioreactor designs. The shape of a closed photobioreactor can be selected to make the most out of important parameters like light, CO_2 mass transfer, and temperature control. Common designs include (**a**) flat plate, (**b**) tubular, and (**c**) annular photobioreactors

lighting. When artificial lighting is used as a light source, it becomes the primary energy consumer in a microalgae cultivation system. An alternative to artificial light is the utilization of sunlight (Masojídek et al. 2021). While some closed PBR designs make the integration of natural light easy, such as flat plates, which can be angled toward the sun for maximal exposure, losses of productivity due to shorter range of daylight and weather inconsistencies make it undesirable alternative (Lee et al. 2014). Another challenge of closed PBR lighting is heat management. Overheating can be deadly to microalgae, necessitating the deployment of measures to minimize these detrimental effects. Methods of temperature control in closed PBR systems include heat exchange systems of passive-evaporative cooling systems (Nwoba et al. 2019). An example of a closed PBR system that makes special design features to make natural lighting more effective and reduce the deleterious effects of heats is the Christmas tree photobioreactor made by the German company GICON. As implied by its name, the Christmas tree PBR is shaped into a conical framework that resembles a Christmas tree. The angled geometry allows for sunlight exposure all throughout the day, with the angle of the cone sides correlating with the distribution of light over the entire reactor. To minimize temperature effects, the PBR includes an internal tube, which can be filled with water to regulate the temperature (Legrand et al. 2021).

4.4 Harvesting and Dewatering

Microalgae cultivation delivers a dilute algal solution ranging from 0.05–0.075% dry matter in open systems to 0.3–0.4% in closed systems, making it difficult to directly extract lipids from the microalgal biomass. Harvesting and dewatering exist to increase the total dry matter to 10–25% dry matter and a drying step can be added to achieve a dry product (Fasaei et al. 2018). While harvesting and dewatering are essential for downstream processes, they are energy intensive and costly, accounting for 20–30% of total production costs, because the small size (2–50 μ), negative surface charge, and low concentration of the microalgal cells make separation difficult (Elisabeth et al. 2021; Ray et al. 2021; Vu et al. 2021). Currently, harvesting and dewatering processes are either carried out in a single high impact performance technology or by combining multiple technologies into a sequence to obtain the most efficient process.

4.4.1 Harvesting

The primary objective of harvesting is to separate microalgal biomass from a bulk liquid suspension. Many methods of harvesting microalgae exist, such as sedimentation, filtration, flocculation, flotation, and centrifugation. However, flocculation is the most promising method for the harvesting of microalgae because the process is effective, low energy, and simple enough to be used in large-scale industrial settings (Fasaei et al. 2018; Matter et al. 2019). Flocculation aims to increase the "size" of the microalgae and neutralize the negative surface charge of microalgal cells through chemical, electrical, and biological means (Ray et al. 2021). This process results in the aggregation of microalgal cells and the formation of settleable flocs in the cultivation media, which aids in the recovery of biomass.

4.4.1.1 Chemical Flocculation

Chemical flocculation occurs due to charge neutralization and electrostatic bridging between suspended microalgal cells from applied chemical flocculants, including inorganic metal-salts and polymers. The efficiency of this flocculation method is dictated by the flocculants' solubility and electronegativity as well as the dosage and cultivation media characteristics (Matter et al. 2019). Chemical flocculation is the simplest and best understood method of flocculation in addition to being up to 95% effective in recovering microalgal biomass (Chatsungnoen and Chisti 2016). Despite these advantages, the chemical flocculants are difficult to remove from both the flocculated biomass and recovered cultivation media, which can negatively affect downstream processes and prevent the media from being re-used without additional processing. Additionally, the release of flocculants into the environment can be both an environmental and health concern.

4.4.1.2 Bio-Flocculation

Bio-flocculation refers to the process where flocculation is induced by other micro-organisms, such as auto-flocculating algae, bacteria, fungi, or yeast, or by bio-flocculants. In recent years, bio-flocculation has attached attention because it is a feasible method of harvesting microalgae with low energy inputs (Li et al. 2020). Bio-flocculation is also a promising flocculation method because it requires less pre-treatment, avoids chemical contamination, and allows the cultivation media to be easily recycled (Ray et al. 2021).

4.4.1.2.1 Autoflocculation

Autoflocculation is the spontaneous aggregation of microalgal cells. This phenomenon is highly species specific, and even only then is it observed in specific conditions, such as with a change in pH or aging (Matter et al. 2019; Li et al. 2021). From a large-scale microalga harvesting perspective, the encouragement of autoflocculation using a change in pH is viewed as the most practical because it can be controllable. While autoflocculation does occur because of natural fluxes in pH, an input to alter the pH cultivation media is typically required to truly control the autoflocculation of microalgae (Young et al. 2021). Both alkaline and acidic conditions can induce self-aggregation. Acidic conditions lead to the dissociation of carboxyl and amine groups in the algal cell wall, causing changes in to total microalgal surface charge. Meanwhile, alkaline conditions above pH 9 can cause either the secretion of extracellular polymers by the microalgal cells to change the surface charge or the neutralization of surface charges by positively charged ions present in the cultivation media (Matter et al. 2019). Typically calcium and magnesium salts are paired with an increase in pH to achieve alkaline conditions because alkaline conditions are the most favorable for autoflocculation (Beltrán-Rocha et al. 2021). Recent studies have shown that in alkaline conditions the autoflocculation efficiency of *Desmodesmus* sp.CHX1 could reach 94% and 80% for *Chlorella vulgaris*. The primary drawback to autoflocculation is that it can take a long time (>24 hrs) for flocculation to occur compared to other flocculation methods (Zhang et al. 2022a; Thu et al. 2023).

4.4.1.3 Electro-Flocculation

During electro-flocculation an electric current is used to induce floc formation. The electric current through the culture causes the charge neutralization of microalgal cells, which allows them to stick together. Electro-flocculation can achieve high biomass recovery in a short amount of time. For example, the electro-flocculation of *Scenedesmus acuminatus* at 20 V after 18.5 min achieved a flocculation efficiency of 90%, while a 98% flocculation efficiency was achieved in a consortia of *Chlorella* sp., *Spirulina* sp., *Scenedesmus* sp., and *Synechocystis* sp. at 20 V in 15 min (Bleeke et al. 2015; Nageshwari et al. 2022). An increase in voltage can improve the flocculation efficiency and lower the time it takes to reach the maximum flocculation efficiency (Phiri et al. 2021). However, higher voltages also cause the electrodes to degrade faster and consume more energy than lower voltages, which increases the operating costs of electro-flocculation. The degradation of the electrodes can release metal oxides into the cultivation media, which contaminates it and reduces its recyclability. Typically iron and aluminum are used to make electrodes, but magnesium has been proved to be an affordable and less toxic alternative (Bleeke et al. 2015).

4.4.2 Dewatering

The objective of dewatering is to increase the biomass concentration and lower the water content of microalgae. Flocculation is only effective for concentration up to 2–2.5% dry matter, which means that additional processes are necessary to reach the final concentration of 10–25% preferred for lipid extraction. Common methods of dewatering include filtration and centrifugation, but in general dewatering technologies only differ from harvesting methods in terms of the goal biomass concentration (Fasaei et al. 2018).

4.5 Extraction

Microalgal lipid extraction process includes cell drying, cell disruption, and lipid extraction. The principles governing microalgal lipid extraction are like those of other oleaginous microorganisms but include some distinctions that tailor process to unique characteristics of microalgae such as the presence of chlorophyll and a rigid cell wall. While drying microalgal biomass is not necessary for lipid extraction, the extraction of lipid from dry microalgal biomass is most efficient. In addition, using drying to reduce the water content of microalgal biomass to below 5% offers benefits beyond efficiency as it can extend the shelf-life of the biomass, reduce the weight for transport, and make more downstream processes possible (Fasaei et al. 2018). Several methods of drying microalgal biomass exist such as freeze drying, air drying, oven drying, microwave drying, spray drying, and solar drying. The most suitable drying method is freeze drying because it preserves biomass quality well, meaning that there is little degradation of the desired by-products. For example, when *Tetraselmis subcordiformis* was freeze dried the extractable lipid content and the amount of fatty acid methyl esters (FAME) were 4% and 393 ppm higher than the next best methods, respectively (Aljabri et al. 2023). However, from a cost and sustainability perspective, solar drying is the best method of drying microalgal biomass because it does not require high energy or capital input, which freeze drying does require. Unfortunately, the efficiency of sun drying is strongly influenced by geography and weather conditions, it is susceptible to contamination, and solar radiation can result in the degradation of valuable biomass by-products (Schmid et al. 2022).

Microalgae must undergo cell disruption prior to lipid extraction. Unlike higher plants, microalgae synthesize lipids within the same cell that they are synthesized, instead of possessing specialized organs for the production and storage of lipids, which simplifies the cell disruption process (Mulgund 2022). However, microalgae do possess a cell wall, which can be rigid and difficult to disrupt depending on the species. Microalgal cell walls typically possess compounds like glycoproteins and carbohydrates, but can contain algaenan, a non-hydrolyzable biopolymer which increases the rigidity and impenetrability of the cell wall (Machado et al. 2022). This issue of cell wall rigidity is pertinent to microalgal biofuel production because algaenan is a common cell wall constituent of species that are preferred for biofuel production such as *Chlorella* and *Scenedesmus* (Alhattab et al. 2019). To disrupt the cell wall, microalgal cells can undergo mechanical or non-mechanical disruption processing. Mechanical methods include bead milling, ultrasonication, microwaving, pasteurization, and homogenization, whereas non-mechanical methods involve the utilization of chemicals or enzymes (Tan et al. 2020; Zhang et al. 2022b).

After cell disruption, processing is required to separate the lipids from other cell by-products irrelevant to biofuel production. Conventional lipid extraction technology relies on organic solvents. Given that the microalgal lipids suitable for conversion to biofuel are neutral, the organic solvents are selected based on their polarity and solvent mixtures are tuned to be specific toward these lipids. The most recognized method of microalgal lipid extraction, the Bligh and Dyer solvent method, is a good example of this concept because it uses a mixture of chloroform and methanol to selectively extract neutral lipids (Venkata Subhash et al. 2022). While this method is effective in the extraction of lipids, capable of recovering over 95% of lipids from microalgal biomass, the solvents employed are toxic, volatile, poorly biodegradable, and cannot be recycled (Kanda et al. 2012). Alternative extraction process substitutes supercritical fluids, ionic liquids, deep eutectic solvents (DES), and switchable amine solvents for organic solvents (Lee et al. 2021).

4.5.1 Challenges and Solutions

As previously discussed, microalgal biomass is typically dried before lipid extraction. Drying biomass is an energy-intensive process, which increases the costs of microalgal biofuel production. Oftentimes this operational cost is justified because lipid extraction from wet biomass is hampered by limited contact between the negatively charged microalgal cells in the hydrophilic phase and hydrophobic extraction solvents (Ananthi et al. 2021). Regardless, the development of wet lipid extraction is a potential solution to this cost-related challenge because it eliminates the need to dry biomass.

4.5.1.1 Deep Eutectic Solvents

Deep eutectic solvents are composed of an organic salt and a hydrogen-bond donor that are capable of self-association. In the processing of wet biomass, this is useful because the hydrogen bonds can interact with the tough cellulose and hemicellulose of the microalgal cell well, modifying the cell walls in such a way that enhances lipid extraction (Lee et al. 2021). For example, the treatment of wet, undisrupted *Chlorella vulgaris* and *Chlorococcum* sp. GN38 with DES led to a 30% increase in FAME content from untreated biomass (Pan et al. 2017). Meanwhile, the utilization of DES composed of choline chloride and oxalic acid and an ethyl acetate and ethanol solvent to extract lipids from wet *Dunaliella salina* biomass enhanced recovery by 63.6% compared to biomass without DES treatment (Asevedo et al. 2023). Not only does this allow the drying step to be avoided, but it eliminates the need for cell disruption to occur before extraction, simplifying the overall process of lipid extraction and cutting costs. In addition to its potential for wet lipid extraction, DES shows promise for lipid extraction in general due to their low cost, recyclability, low toxicity, biodegradability, and are widely considered as a green extraction method because they can be composed of plant-based primary metabolites (Lee et al. 2021). These qualities make DES a potential solution to the challenges presented by the volatility and toxicity of conventional lipid extraction techniques.

4.6 Synthesis

Given the ability of microalgae to produce large quantities of neutral lipids, biodiesel production has been a dominating theme in microalgal-based drop-in biofuel research for many years. However, other compounds produced by microalgae can be used as a fuel after undergoing a conversion process or simply after extraction from the cell.

4.6.1 *Biodiesel*

The conversion of microalgal lipids to biodiesel occurs though a transesterification process. Typically, TAGs are reacted with methanol in the presence of a catalyst to produce fatty acid methyl esters and a glycerol by-product. Common catalysts for transesterification are sodium hydroxide or potassium hydroxide, but enzymes like lipase can be used as well. Biodiesel production using lipases is an economically feasible and sustainable catalysis option that offers benefits like high-purity yield, reduction of saponification, and recycling (Yoshida et al. 2012; Hossain et al. 2020). After transesterification, the biodiesel is separated from the crude glycerol and dried.

To reduce the number of processes required for microalgal-based drop-in biofuel production, lipid conversion can be carried out in situ, or be integrated into the extraction step. The combination of extraction and conversion are one-step methods, where the lipid extraction and lipid conversion occur simultaneously in a reactor, which offers cost benefits. One-step methods are as effective, if not more effective, compared to sequential extraction and conversion steps, as seen by the 30% increase in FAME content of *Chlorella vulgaris* from a two-step process (Pan et al. 2017). Other examples of in situ conversion include the integration of the Bligh and Dyer method with a HCl catalyzed transesterification, which was capable of achieving 90 wt% lipid conversion to FAME (Kim et al. 2015).

4.6.2 *Terpenes*

Terpenes are a class of hydrocarbons that naturally occur in microalgae, where they play essential functions in microalgal metabolism. The thermochemical and thermophysical properties of certain terpenes, including α-pinene, β-pinene, limonene, farnesene, and bisabolene, make terpenes ideal drop-in biofuel candidates, especially in the context of high energy content fuels like jet fuels (Walls and Rios-Solis 2020). Unlike microalgal-based biodiesel, terpenes can be directly used as a biofuel, eliminating the need for a conversion step. This increases the attractiveness of terpenes because conversion processes like transesterification are energy intensive and costly (Vidyashankar et al. 2015). A challenge in using microalgal terpenes for biofuel production is that microalgae accumulate terpenes in small quantities and great variety. An exception to this fact is the colonial microalgae *Botryococcus braunii* which can accumulate up to 70% of its biomass in the form of terpenes. Unfortunately, *B. braunii* is difficult to culture, making it an impractical species for scale-up. For this reason, the main technique used to increase the production of terpenes is strain improvement. For example, the introduction and overexpression of patchoulol synthase into the genome of *Chlamydomonas reinhardtii* made it possible to produce up to 117 μgg L^{-1} d^{-1} of sesquiterpenoid when grown on CO_2 (Lauersen et al. 2016). Meanwhile, the modification of *Chlorella saccharophila* simply by using ultraviolet radiation yielded two strains that were capable of achieving terpene ratios approximately 5 to 9.5% higher than the wild-type strain (Teco-Bravo et al. 2021).

4.7 Perspectives

Microalgal-based drop-in biofuels are an appealing fuel source due to the positive environmental impact microalgae can have. Aside from removing carbon dioxide from the atmosphere via photosynthesis, microalgae can be cultured in wastewaters with little modification to the system, presenting an opportunity to help the environment even more. Despite these potential benefits, the production of fuels from microalgae is prohibitively expensive, preventing wide-scale accessibility. Moving forward, the development of new technologies are necessary in all aspects of microalgal-based biofuel production. When compared to other oleaginous microorganisms, the microalgae cultivation introduces the parameter of light, which is difficult and expensive to optimize. Ideally, innovation can occur to minimize the variability of natural light to reduce the energy demands of cultivation system lighting. Another area where cultivation falls short is in the maximum of microalgal lipids. Currently, the best understood mechanisms for lipid accumulation involve subjecting microalgae to stressors, which negatively impact biomass production, negating the benefits of the accumulation. Strain improvement needs to occur to help microalgae to be able to accumulate lipids and grow optimally simultaneously. In the context of promising technologies for microalgae harvesting, flocculation is a good method to cut the costs of harvesting, but the ideal microalgal-based drop-in fuel production system would eliminate the need for extensive harvesting and dewatering by utilizing wet extraction and in situ transesterification.

4.8 Conclusions

The microalgae based drop-in biofuel production process consists of four major steps. First is cultivation, where microalgae are grown in either open or closed photobioreactors in conditions that optimize lipid production. After cultivation, microalgae are harvested from the cultivation media. A common low energy harvesting method is flocculation, where microalgae are encouraged to form flocs. Typically, a dewatering step is included in the harvesting step to aid in transportation and downstream processing. Next, molecules of interest are extracted from microalgae using a cell disruption and separation process. Once the lipids are extracted from the microalgae, a conversion step occurs to convert the lipids into a biofuel, such as biodiesel. In the context of some types of biofuel compounds, like terpenes, no conversion step is required before utilization as a fuel.

References

Ahmad A, Banat F, Alsafar H, Hasan SW (2023) Recent breakthroughs in integrated biomolecular and biotechnological approaches for enhanced lipid and carotenoid production from microalgae. Phytochem Rev 22(4):993–1013. https://doi.org/10.1007/s11101-022-09804-5

Alhattab M, Kermanshahi-Pour A, Brooks MS-L (2019) Microalgae disruption techniques for product recovery: influence of cell wall composition. J Appl Phycol 31(1):61–88. https://doi.org/10.1007/s10811-018-1560-9

Aljabri H, Cherif M, Siddiqui SA, Bounnit T, Saadaoui I (2023) Evidence of the drying technique's impact on the biomass quality of Tetraselmis subcordiformis (Chlorophyceae). Biotechnol Biofuels Bioprod 16(1):85. https://doi.org/10.1186/s13068-023-02335-x

Ananthi V, Balaji P, Sindhu R, Kim S-H, Pugazhendhi A, Arun A (2021) A critical review on different harvesting techniques for algal based biodiesel production. Sci Total Environ 780: 146467. https://doi.org/10.1016/j.scitotenv.2021.146467

Asevedo EA, das Chagas BME, de Oliveira Júnior SD, dos Santos ES (2023) Recovery of lipids and carotenoids from Dunaliella salina microalgae using deep eutectic solvents. Algal Res 69: 102940. https://doi.org/10.1016/j.algal.2022.102940

Beltrán-Rocha JC, Guajardo-Barbosa C, Rodríguez-Fuentes H, Reyna-Martínez GR, Osornio-Berthet L, García-Martínez M, Quintal ID-B, López-Chuken UJ, Beltrán-Rocha JC, Guajardo-Barbosa C, Rodríguez-Fuentes H, Reyna-Martínez GR, Osornio-Berthet L, García-Martínez M, Quintal ID-B, López-Chuken UJ (2021) Some implications of natural increase of pH in microalgae cultivation and harvest by autoflocculation. Lat Am J Aquat Res 49(5): 836–842. https://doi.org/10.3856/vol49-issue5-fulltext-2691

Bleeke F, Quante G, Winckelmann D, Klöck G (2015) Effect of voltage and electrode material on electroflocculation of Scenedesmus acuminatus. Bioresour Bioprocess 2(1):36. https://doi.org/10.1186/s40643-015-0064-6

Borowitzka M (2013) Energy from microalgae: a short history. In: Algae for biofuels and energy, pp 1–15

Castillo T, Ramos D, García-Beltrán T, Brito-Bazan M, Galindo E (2021) Mixotrophic cultivation of microalgae: an alternative to produce high-value metabolites. Biochem Eng J 176:108183. https://doi.org/10.1016/j.bej.2021.108183

Chatsungnoen T, Chisti Y (2016) Harvesting microalgae by flocculation–sedimentation. Algal Res 13:271–283. https://doi.org/10.1016/j.algal.2015.12.009

Chen H, Wang Q (2021) Regulatory mechanisms of lipid biosynthesis in microalgae. Biol Rev 96(5):2373–2391. https://doi.org/10.1111/brv.12759

Chowdury KH, Nahar N, Deb UK (2020) The growth factors involved in microalgae cultivation for biofuel production: a review. Comput Water Energ Environ Eng 9(4):185–215. https://doi.org/10.4236/cweee.2020.94012

Correa I, Drews P, Botelho S, de Souza MS, Tavano VM (2017) Deep learning for microalgae classification. In: 2017 16th IEEE international conference on machine learning and applications (ICMLA), pp 20–25
Croft MT, Warren MJ, Smith AG (2006) Algae need their vitamins. Eukaryot Cell 5(8):1175–1183. https://doi.org/10.1128/EC.00097-06
Dean JM, Lodhi IJ (2018) Structural and functional roles of ether lipids. dbzyxb 9(2):196–206. https://doi.org/10.1007/s13238-017-0423-5
Dębowski M, Dudek M, Zieliński M, Nowicka A, Kazimierowicz J (2021) Microalgal hydrogen production in relation to other biomass-based technologies—a review. Energies 14(19):6025. https://doi.org/10.3390/en14196025
Dengg M, Stirling CH, Reid MR, Verburg P, Armstrong E, Kelly LT, Wood SA (2022) Growth at the limits: comparing trace metal limitation of a freshwater cyanobacterium (Dolichospermum lemmermannii) and a freshwater diatom (Fragilaria crotonensis). Sci Rep 12(1):467. https://doi.org/10.1038/s41598-021-04533-9
Devadasu E, Subramanyam R (2021) Enhanced lipid production in Chlamydomonas reinhardtii caused by severe iron deficiency. Front Plant Sci 12:615577
Eisenhut M, Roell M-S, Weber APM (2019) Mechanistic understanding of photorespiration paves the way to a new green revolution. New Phytol 223(4):1762–1769. https://doi.org/10.1111/nph.15872
Elisabeth B, Rayen F, Behnam T (2021) Microalgae culture quality indicators: a review. Crit Rev Biotechnol 41(4):457–473. https://doi.org/10.1080/07388551.2020.1854672
Esteves AF, Soares OSGP, Vilar VJP, Pires JCM, Gonçalves AL (2020) The effect of light wavelength on CO2 capture, biomass production and nutrient uptake by green microalgae: a step forward on process integration and optimisation. Energies 13(2):333. https://doi.org/10.3390/en13020333
Facey JA, King JJ, Apte SC, Mitrovic SM (2022) Assessing the importance of cobalt as a micronutrient for freshwater cyanobacteria. J Phycol 58(1):71–79. https://doi.org/10.1111/jpy.13216
Fasaei F, Bitter JH, Slegers PM, van Boxtel AJB (2018) Techno-economic evaluation of microalgae harvesting and dewatering systems. Algal Res 31:347–362. https://doi.org/10.1016/j.algal.2017.11.038
Fattore N, Bucci F, Bellan A, Bossi S, Maffei ME, Morosinotto T (2022) An increase in the membrane lipids recycling by PDAT overexpression stimulates the accumulation of triacylglycerol in Nannochloropsis gaditana. J Biotechnol 357:28–37. https://doi.org/10.1016/j.jbiotec.2022.07.007
George B, Pancha I, Desai C, Chokshi K, Paliwal C, Ghosh T, Mishra S (2014) Effects of different media composition, light intensity and photoperiod on morphology and physiology of freshwater microalgae Ankistrodesmus falcatus—a potential strain for bio-fuel production. Bioresour Technol 171:367–374. https://doi.org/10.1016/j.biortech.2014.08.086
Healey EM, Flood S, Bock PK, Fulweiler RW, York JK, Coyne KJ (2023) Effects of nitrate and ammonium on assimilation of nitric oxide by Heterosigma akashiwo. Sci Rep 13(1):621. https://doi.org/10.1038/s41598-023-27692-3
Hossain SM, Razzak S, Al Shater A, Moniruzzaman M, Hossain M (2020) Recent advances in enzymatic conversion of microalgal lipids into biodiesel. Energ Fuels 34(6):6735–6750. https://doi.org/10.1021/acs.energyfuels.0c01064
Inostroza C, Dávila J, Román S, Fernández-Sevilla JM, Acién FG (2023) Use of airfoils for enhancement of photosynthesis rate of microalgae in raceways. J Appl Phycol 35:2571–2581. https://doi.org/10.1007/s10811-023-02996-z
Kanda H, Li P, Ikehara T, Yasumoto-Hirose M (2012) Lipids extracted from several species of natural blue–green microalgae by dimethyl ether: extraction yield and properties. Fuel 95:88–92. https://doi.org/10.1016/j.fuel.2011.11.064

Kazbar A, Cogne G, Urbain B, Marec H, Le-Gouic B, Tallec J, Takache H, Ismail A, Pruvost J (2019) Effect of dissolved oxygen concentration on microalgal culture in photobioreactors. Algal Res 39:101432. https://doi.org/10.1016/j.algal.2019.101432

Khoo KS, Ahmad I, Chew KW, Iwamoto K, Bhatnagar A, Show PL (2023) Enhanced microalgal lipid production for biofuel using different strategies including genetic modification of microalgae: a review. Prog Energy Combust Sci 96:101071. https://doi.org/10.1016/j.pecs.2023.101071

Kim G, Choi W, Lee C-H, Lee K (2013) Enhancement of dissolved inorganic carbon and carbon fixation by green alga Scenedesmus sp. in the presence of alkanolamine CO2 absorbents. Biochem Eng J 78:18–23. https://doi.org/10.1016/j.bej.2013.02.010

Kim B, Im H, Lee JW (2015) In situ transesterification of highly wet microalgae using hydrochloric acid. Bioresour Technol 185:421–425. https://doi.org/10.1016/j.biortech.2015.02.092

Kim G-Y, Heo J, Kim K, Chung J, Han J-I (2021) Electrochemical pH control and carbon supply for microalgae cultivation. Chem Eng J 426:131796. https://doi.org/10.1016/j.cej.2021.131796

Kliphuis AMJ, Martens DE, Janssen M, Wijffels RH (2011) Effect of O2:CO2 ratio on the primary metabolism of Chlamydomonas reinhardtii. Biotechnol Bioeng 108(10):2390–2402. https://doi.org/10.1002/bit.23194

Lakatos GE, Ranglová K, Manoel JC, Grivalský T, Kopecký J, Masojídek J (2019) Bioethanol production from microalgae polysaccharides. Folia Microbiol 64(5):627–644. https://doi.org/10.1007/s12223-019-00732-0

Lauersen KJ, Baier T, Wichmann J, Wördenweber R, Mussgnug JH, Hübner W, Huser T, Kruse O (2016) Efficient phototrophic production of a high-value sesquiterpenoid from the eukaryotic microalga Chlamydomonas reinhardtii. Metab Eng 38:331–343. https://doi.org/10.1016/j.ymben.2016.07.013

Lee E, Pruvost J, He X, Munipalli R, Pilon L (2014) Design tool and guidelines for outdoor photobioreactors. Chem Eng Sci 106:18–29. https://doi.org/10.1016/j.ces.2013.11.014

Lee SY, Khoiroh I, Vo D-VN, Senthil Kumar P, Show PL (2021) Techniques of lipid extraction from microalgae for biofuel production: a review. Environ Chem Lett 19(1):231–251. https://doi.org/10.1007/s10311-020-01088-5

Legrand J, Artu A, Pruvost J (2021) A review on photobioreactor design and modelling for microalgae production. React Chem Eng 6(7):1134–1151. https://doi.org/10.1039/D0RE00450B

Lehmuskero A, Skogen Chauton M, Boström T (2018) Light and photosynthetic microalgae: a review of cellular- and molecular-scale optical processes. Prog Oceanogr 168:43–56. https://doi.org/10.1016/j.pocean.2018.09.002

Levin G, Yasmin M, Liveanu V, Burstein C, Hanna R, Kleifeld O, Simanowitz MC, Meir A, Tadmor Y, Hirschberg J, Adir N, Schuster G (2023) A desert chlorella sp. that thrives at extreme high-light intensities using a unique photoinhibition protection mechanism. Plant J 115(2): 510–528. https://doi.org/10.1111/tpj.16241

Li S, Hu T, Xu Y, Wang J, Chu R, Yin Z, Mo F, Zhu L (2020) A review on flocculation as an efficient method to harvest energy microalgae: mechanisms, performances, influencing factors and perspectives. Renew Sust Energ Rev 131:110005. https://doi.org/10.1016/j.rser.2020.110005

Li T, Hu J, Zhu L (2021) Self-flocculation as an efficient method to harvest microalgae: a mini-review. Water 13(18):2585. https://doi.org/10.3390/w13182585

Li-Beisson Y, Thelen JJ, Fedosejevs E, Harwood JL (2019) The lipid biochemistry of eukaryotic algae. Prog Lipid Res 74:31–68. https://doi.org/10.1016/j.plipres.2019.01.003

Ma S, Zeng W, Huang Y, Zhu X, Xia A, Zhu X, Liao Q (2022) Revealing the synergistic effects of cells, pigments, and light spectra on light transfer during microalgae growth: a comprehensive light attenuation model. Bioresour Technol 348:126777. https://doi.org/10.1016/j.biortech.2022.126777

Machado L, Carvalho G, Pereira RN (2022) Effects of innovative processing methods on microalgae cell wall: prospects towards digestibility of protein-rich biomass. Biomass 2(2): 80–102. https://doi.org/10.3390/biomass2020006

Maltsev Y, Maltseva K, Kulikovskiy M, Maltseva S (2021) Influence of light conditions on microalgae growth and content of lipids, carotenoids, and fatty acid composition. Biology 10(10):1060. https://doi.org/10.3390/biology10101060

Masojídek J, Ranglová K, Lakatos GE, Silva Benavides AM, Torzillo G (2021) Variables governing photosynthesis and growth in microalgae mass cultures. Processes 9(5):820. https://doi.org/10.3390/pr9050820

Matter IA, Bui VKH, Jung M, Seo JY, Kim Y-E, Lee Y-C, Oh Y-K (2019) Flocculation harvesting techniques for microalgae: a review. Appl Sci 9(15):3069. https://doi.org/10.3390/app9153069

Moroney JV, Ynalvez RA (2009) Algal photosynthesis. In: Encyclopedia of life sciences. John Wiley & Sons, Ltd

Mulgund A (2022) Increasing lipid accumulation in microalgae through environmental manipulation, metabolic and genetic engineering: a review in the energy NEXUS framework. Energ Nexus 5:100054. https://doi.org/10.1016/j.nexus.2022.100054

Muñoz CF, Südfeld C, Naduthodi MIS, Weusthuis RA, Barbosa MJ, Wijffels RH, D'Adamo S (2021) Genetic engineering of microalgae for enhanced lipid production. Biotechnol Adv 52: 107836. https://doi.org/10.1016/j.biotechadv.2021.107836

Nageshwari K, Chang SX, Balasubramanian P (2022) Integrated electrocoagulation-flotation of microalgae to produce Mg-laden microalgal biochar for seeding struvite crystallization. Sci Rep 12(1):11463. https://doi.org/10.1038/s41598-022-15527-6

Nguyen T-T-D, Nguyen T-T, An Binh Q, Bui X-T, Ngo HH, Vo HNP, Andrew Lin K-Y, Vo T-D-H, Guo W, Lin C, Breider F (2020) Co-culture of microalgae-activated sludge for wastewater treatment and biomass production: exploring their role under different inoculation ratios. Bioresour Technol 314:123754. https://doi.org/10.1016/j.biortech.2020.123754

Nikolaou A, Bernardi A, Meneghesso A, Bezzo F, Morosinotto T, Chachuat B (2015) A model of chlorophyll fluorescence in microalgae integrating photoproduction, photoinhibition and photoregulation. J Biotechnol 194:91–99. https://doi.org/10.1016/j.jbiotec.2014.12.001

Nwoba EG, Parlevliet DA, Laird DW, Alameh K, Moheimani NR (2019) Light management technologies for increasing algal photobioreactor efficiency. Algal Res 39:101433. https://doi.org/10.1016/j.algal.2019.101433

Nzayisenga JC, Farge X, Groll SL, Sellstedt A (2020) Effects of light intensity on growth and lipid production in microalgae grown in wastewater. Biotechnol Biofuels 13(1):4. https://doi.org/10.1186/s13068-019-1646-x

Pan Y, Alam MA, Wang Z, Huang D, Hu K, Chen H, Yuan Z (2017) One-step production of biodiesel from wet and unbroken microalgae biomass using deep eutectic solvent. Bioresour Technol 238:157–163. https://doi.org/10.1016/j.biortech.2017.04.038

Pancha I, Chokshi K, Ghosh T, Paliwal C, Maurya R, Mishra S (2015) Bicarbonate supplementation enhanced biofuel production potential as well as nutritional stress mitigation in the microalgae Scenedesmus sp. CCNM 1077. Bioresour Technol 193:315–323. https://doi.org/10.1016/j.biortech.2015.06.107

Phiri JT, Pak H, We J, Oh S (2021) Evaluation of Pb, Mg, Al, Zn, and Cu as electrode materials in the electrocoagulation of microalgae. Processes 9(10):1769. https://doi.org/10.3390/pr9101769

Pushpakumari Kudahettige N, Pickova J, Gentili FG (2018) Stressing algae for biofuel production: biomass and biochemical composition of Scenedesmus dimorphus and Selenastrum minutum grown in municipal untreated wastewater. Front Energ Res 6:132

Ra CH, Sirisuk P, Jung J-H, Jeong G-T, Kim S-K (2018) Effects of light-emitting diode (LED) with a mixture of wavelengths on the growth and lipid content of microalgae. Bioprocess Biosyst Eng 41(4):457–465. https://doi.org/10.1007/s00449-017-1880-1

Rajabi Islami H, Assareh R (2019) Effect of different iron concentrations on growth, lipid accumulation, and fatty acid profile for biodiesel production from Tetradesmus obliquus. J Appl Phycol 31(6):3421–3432. https://doi.org/10.1007/s10811-019-01843-4

Rajapitamahuni S, Bachani P, Sardar RK, Mishra S (2019) Co-cultivation of siderophore-producing bacteria Idiomarina loihiensis RS14 with Chlorella variabilis ATCC 12198, evaluation of microalgal growth, lipid, and protein content under iron starvation. J Appl Phycol 31(1):29–39. https://doi.org/10.1007/s10811-018-1591-2
Rana MS, Prajapati SK (2021) Resolving the dilemma of iron bioavailability to microalgae for commercial sustenance. Algal Res 59:102458. https://doi.org/10.1016/j.algal.2021.102458
Raso S, van Genugten B, Vermuë M, Wijffels RH (2012) Effect of oxygen concentration on the growth of Nannochloropsis sp. at low light intensity. J Appl Phycol 24(4):863–871. https://doi.org/10.1007/s10811-011-9706-z
Ray A, Banerjee S, Das D (2021) Microalgal bio-flocculation: present scenario and prospects for commercialization. Environ Sci Pollut Res 28(21):26294–26312. https://doi.org/10.1007/s11356-021-13437-0
Rios LF, Klein BC, Luz LF, Maciel Filho R, Wolf Maciel MR (2015) Nitrogen starvation for lipid accumulation in the microalga species Desmodesmus sp. Appl Biochem Biotechnol 175(1):469–476. https://doi.org/10.1007/s12010-014-1283-6
Rumin J, Nicolau E, de Oliveira G, Junior R, Fuentes-Grünewald C, Flynn KJ, Picot L (2020) A bibliometric analysis of microalgae research in the world, Europe, and the European Atlantic area. Mar Drugs 18(2):79. https://doi.org/10.3390/md18020079
Schmid B, Navalho S, Schulze PSC, Van De Walle S, Van Royen G, Schüler LM, Maia IB, Bastos CRV, Baune M-C, Januschewski E, Coelho A, Pereira H, Varela J, Navalho J, Cavaco Rodrigues AM (2022) Drying microalgae using an industrial solar dryer: a biomass quality assessment. Food Secur 11(13):1873. https://doi.org/10.3390/foods11131873
Sforza E, Pastore M, Franke SM, Barbera E (2020) Modeling the oxygen inhibition in microalgae: An experimental approach based on photorespirometry. New Biotechnol 59:26–32. https://doi.org/10.1016/j.nbt.2020.06.003
Shokravi ZS, Shokravi H, Aziz MA, Hooman S (2019) The fourth-generation biofuel: a systematic review on nearly two decades of research from 2008 to 2019. In: Fossil free fuels. CRC Press
Straka L, Rittmann BE (2018) Light-dependent kinetic model for microalgae experiencing photoacclimation, photodamage, and photodamage repair. Algal Res 31:232–238. https://doi.org/10.1016/j.algal.2018.02.022
Sun Z, Sun L, Liu Y (2022) The potential impact of replacing nitrate with ammonium hydroxide in microalgae production on the biomass productivity and CO2 utilization efficiency. Algal Res 67:102870. https://doi.org/10.1016/j.algal.2022.102870
Tan JS, Lee SY, Chew KW, Lam MK, Lim JW, Ho S-H, Show PL (2020) A review on microalgae cultivation and harvesting, and their biomass extraction processing using ionic liquids. Bioengineered 11(1):116–129. https://doi.org/10.1080/21655979.2020.1711626
Tang K-H, Blankenship RE (2013) Photosynthetic electron transport. In: Roberts GCK (ed) Encyclopedia of biophysics. Springer, Berlin, Heidelberg, pp 1868–1873
Teco-Bravo JI, Barahona-Pérez LF, Peraza-Echeverria S, Baas-Espínola FM, Reyes-Sosa CF, Herrera-Valencia VA (2021) Lipid profiles of acid-tolerant mutants of the green microalga Chlorella saccharophila reveal hydrocarbons and high-value lipids with potential industrial applications. Bioresour Technol Rep 13:100636. https://doi.org/10.1016/j.biteb.2021.100636
Terlova EF, Holzinger A, Lewis LA (2021) Terrestrial green algae Show higher tolerance to dehydration than do their aquatic sister-species. Microb Ecol 82(3):770–782. https://doi.org/10.1007/s00248-020-01679-3
Thu TTN, Phuong NTD, Luan TV, Anh LTV (2023) Comparison of biofloc formation from microalgal cultivation by auto- and bacteria-associated types of flocculation. Vietnam J Biotechnol 21(1):121–127. https://doi.org/10.15625/1811-4989/17053
Tibocha-Bonilla JD, Zuñiga C, Godoy-Silva RD, Zengler K (2018) Advances in metabolic modeling of oleaginous microalgae. Biotechnol Biofuels 11(1):241. https://doi.org/10.1186/s13068-018-1244-3

Toro-Huertas EI, Franco-Morgado M, de los Cobos Vasconcelos D, González-Sánchez A (2019) Photorespiration in an outdoor alkaline open-photobioreactor used for biogas upgrading. Sci Total Environ 667:613–621. https://doi.org/10.1016/j.scitotenv.2019.02.374
Ummalyma SB, Singh A (2022) Biomass production and phycoremediation of microalgae cultivated in polluted river water. Bioresour Technol 351:126948. https://doi.org/10.1016/j.biortech.2022.126948
Venkata Subhash G, Rajvanshi M, Raja Krishna Kumar G, Shankar Sagaram U, Prasad V, Govindachary S, Dasgupta S (2022) Challenges in microalgal biofuel production: a perspective on techno economic feasibility under biorefinery stratagem. Bioresour Technol 343:126155. https://doi.org/10.1016/j.biortech.2021.126155
Venteris ER, Wigmosta MS, Coleman AM, Skaggs RL (2014) Strain selection, biomass to biofuel conversion, and resource colocation have strong impacts on the economic performance of algae cultivation sites. Front Energy Res 2:37
Vidyashankar S, VenuGopal KS, Swarnalatha GV, Kavitha MD, Chauhan VS, Ravi R, Bansal AK, Singh R, Pande A, Ravishankar GA, Sarada R (2015) Characterization of fatty acids and hydrocarbons of chlorophycean microalgae towards their use as biofuel source. Biomass Bioenergy 77:75–91. https://doi.org/10.1016/j.biombioe.2015.03.001
Vinyard DJ, Ananyev GM, Charles Dismukes G (2013) Photosystem II: the reaction Center of Oxygenic Photosynthesis. Annu Rev Biochem 82(1):577–606. https://doi.org/10.1146/annurev-biochem-070511-100425
Vu HP, Nguyen LN, Emmerton B, Wang Q, Ralph PJ, Nghiem LD (2021) Factors governing microalgae harvesting efficiency by flocculation using cationic polymers. Bioresour Technol 340:125669. https://doi.org/10.1016/j.biortech.2021.125669
Wahidin S, Idris A, Shaleh SRM (2013) The influence of light intensity and photoperiod on the growth and lipid content of microalgae Nannochloropsis sp. Bioresour Technol 129:7–11. https://doi.org/10.1016/j.biortech.2012.11.032
Walls LE, Rios-Solis L (2020) Sustainable production of microbial isoprenoid derived advanced biojet fuels using different generation feedstocks: a review. Front Bioeng Biotechnol 8:599560
Wang Y, Gong Y, Dai L, Sommerfeld M, Zhang C, Hu Q (2018) Identification of harmful protozoa in outdoor cultivation of chlorella and the use of ultrasonication to control contamination. Algal Res 31:298–310. https://doi.org/10.1016/j.algal.2018.02.002
White DA, Pagarette A, Rooks P, Ali ST (2013) The effect of sodium bicarbonate supplementation on growth and biochemical composition of marine microalgae cultures. J Appl Phycol 25(1): 153–165. https://doi.org/10.1007/s10811-012-9849-6
Yoshida M, Tanabe Y, Yonezawa N, Watanabe MM (2012) Energy innovation potential of oleaginous microalgae. Biofuels 3(6):761–781. https://doi.org/10.4155/bfs.12.63
Young P, Phasey J, Wallis I, Vandamme D, Fallowfield H (2021) Autoflocculation of microalgae, via magnesium hydroxide precipitation, in a high rate algal pond treating municipal wastewater in the south Australian Riverland. Algal Res 59:102418. https://doi.org/10.1016/j.algal.2021.102418
Zhang K, Sun B, She X, Zhao F, Cao Y, Ren D, Lu J (2014) Lipid production and composition of fatty acids in Chlorella vulgaris cultured using different methods: photoautotrophic, heterotrophic, and pure and mixed conditions. Ann Microbiol 64(3):1239–1246. https://doi.org/10.1007/s13213-013-0766-y
Zhang C, Zhang L, Liu J (2019) The interrelation between photorespiration and astaxanthin accumulation in Haematococcus pluvialis using metabolomic analysis. Algal Res 41:101520. https://doi.org/10.1016/j.algal.2019.101520
Zhang B, Liu L, Lin X, Xu Z, Luo W, Luo L (2022a) Response surface methodology to optimize self-flocculation harvesting of microalgae Desmodesmus sp.CHX1. Environ Technol 43(17): 2647–2655. https://doi.org/10.1080/09593330.2021.1892831
Zhang S, Zhang L, Xu G, Li F, Li X (2022b) A review on biodiesel production from microalgae: influencing parameters and recent advanced technologies. Front Microbiol 13:970028. https://doi.org/10.3389/fmicb.2022.970028

Zhu B, Sun F, Yang M, Lu L, Yang G, Pan K (2014) Large-scale biodiesel production using flue gas from coal-fired power plants with Nannochloropsis microalgal biomass in open raceway ponds. Bioresour Technol 174:53–59. https://doi.org/10.1016/j.biortech.2014.09.116
Zienkiewicz K, Zienkiewicz A (2020) Degradation of lipid droplets in plants and algae—right time, many paths, one goal. Front Plant Sci 11:579019

Chapter 5
Yeast-Based Drop-in Biofuels

Malvika Sharma, Kerthika Devi Athiyappan, Makary Nasser, Chloe Tan, Balasubramanian Paramasivan, and Guneet Kaur

Abstract Sustainable drop-in biofuel production has the potential to alleviate concerns about energy supply and climate change by offering a desirable replacement for fossil fuels. Considering the diminishing supply of fossil fuels, recent developments in the production schemes and technologies of drop-in biofuels by microbes have garnered attention. The rapid advancement of synthetic biology and metabolic engineering has demonstrated that microbes can be engineered to produce drop-in biofuels from diverse renewable feedstocks. This chapter focuses on recent developments in metabolism-targeted techniques and the yeast-based synthesis of various types of drop-in biofuels from sustainable feedstocks. It also focuses on the bioprocess engineering systems being utilized to produce and boosting the yield of drop-in biofuels. Several pathways are being followed for production using metabolic engineering; however, the presence of harmful inhibitors in the feedstock may hinder the yeasts from developing to their full capability. In summary, the production of drop-in biofuels employing yeasts could eventually provide a platform for more economic and effective drop-in biofuel production.

Keywords Drop-in biofuels · Bioprocess and metabolic engineering · Yeasts · Feedstocks · In-situ product recovery

---

Malvika Sharma and Kerthika Devi Athiyappan contributed equally with all other contributors.

M. Sharma · M. Nasser · C. Tan · G. Kaur (✉)
School of Engineering, University of Guelph, Guelph, ON, Canada
e-mail: gkaur41@uoguelph.ca

K. D. Athiyappan · B. Paramasivan
National Institute of Technology Rourkela, Rourkela, India

S. K. Brar et al. (eds.), *The Microbiology of the Drop-in Biofuel Production*, Biofuel and Biorefinery Technologies 15, https://doi.org/10.1007/978-3-031-61637-2_5

5.1 Introduction

The prominent characteristics of microbial-derived drop-in biofuels are like those of conventional petroleum-based fuels, and they can be readily integrated into the current transportation infrastructure without blending or engine modification (Peralta-Yahya et al. 2011). Some examples of drop-in biofuels and the precursors of drop-in biofuels are bisabolene (Serhan et al. 2019), butanol (Swidah et al. 2018), decanol (Rutter and Rao 2016), fatty alcohols (Cordova et al. 2020), fatty acid ester (Gao et al. 2018), hexanal (Santiago-Gómez et al. 2009), isoamyl alcohol (Yuan et al. 2017), isobutanol (Wess et al. 2019), isopropanol (Tamakawa et al. 2013a, 2013b), octanol (Henritzi et al. 2018), and propanol (Nishimura et al. 2018). The well-established drop-in biofuels are butanol and isobutanol. Bacteria generally possess the tendency to produce butanol through their metabolism, especially *Clostridium* species which produce butanol in large amounts. After gaining knowledge of the mechanism of production of drop-in biofuels in bacteria, metabolic engineering strategies have been used in certain yeasts to produce drop-in biofuels in large quantities (Liu et al. 2021). *Saccharomyces cerevisiae* is the most abundantly used variety of yeast to produce drop-in biofuels. The yeast strain was known to produce the biofuels of the three major classes, namely—fatty acid ethyl esters, terpenoids, and higher alcohols. The alcohol (long-chain alcohol) family consists of butanol and isobutanol, while terpenoids can be further classified into monoterpenes and sesquiterpenes, and biodiesel falls under the family of fatty esters (Buijs et al. 2013). Other oleaginous yeast strains involved in the production of drop-in biofuels include *Rhodosporidium toruloides* in the production of terpene biofuels (Serhan et al. 2019; Yaegashi et al. 2017), *Yarrowia lipolytica* in the production of decanol (Rutter and Rao. 2016), fatty acid ethyl esters (Gao et al. 2018), and fatty alcohols (Cordova et al. 2020), *Pichia pastoris* in the production of fatty acid esters (Tao et al. 2015), *Lipomyces starkeyi* in the production of long-chain fatty alcohols (McNeil and Stuart. 2018), and *Candida utilis* in the production of isopropanol (Tamakawa et al. 2013a, 2013b).

This chapter focuses on drop-in biofuels derived from yeast, spanning from the initial inoculation for fuel production to the application of synthetic biology techniques in the fuel production process as shown in Fig. 5.1. The first section discusses different types of yeast strains involved in the production of drop-in biofuels along with the fuel properties of the yeast-derived drop-in biofuels. The metabolic flexibility of yeasts to produce drop-in biofuels has been discussed which includes the variety of feedstocks such as lipids/fatty acids, lignocellulosic feedstocks, food waste, etc. that yeasts can use for drop-in biofuels which makes them good producers. The chapter also discusses the bioprocess aspect in the production of drop-in biofuels, which includes the bioprocess engineering strategies for drop-in biofuel production using yeast systems. While this section discusses the different bioreactor setups such as batch, fed-batch, and continuous cultivations involved in the production of these drop-in biofuels using yeasts, it also sheds some light on the in-situ product recovery strategies. Bio-engineering approaches such as metabolic

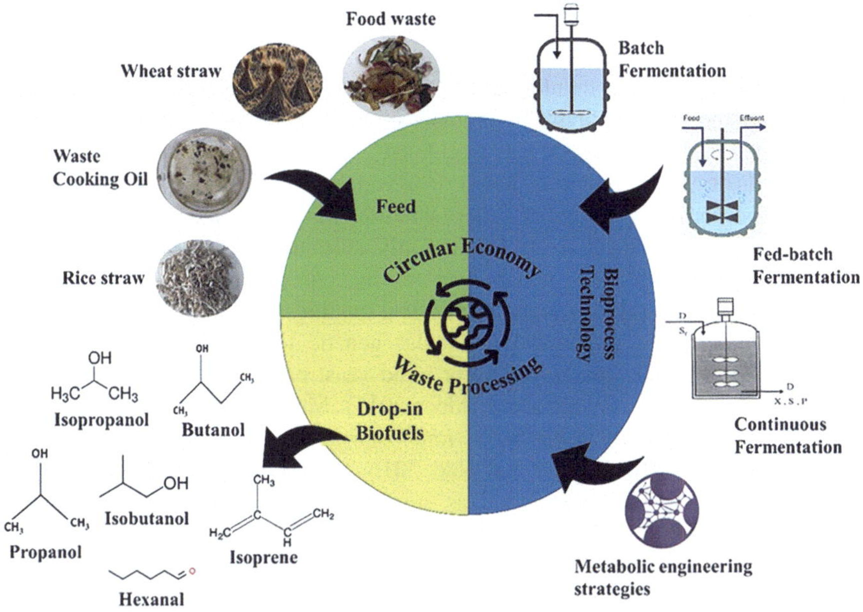

Fig. 5.1 Illustration of the concept of drop-in biofuel production using circular economy and bioprocess technology

engineering or synthetic biology approaches have also been discussed in the later part of the chapter. The metabolic engineering strategy in the drop-in biofuel production from the yeast has been divided into three subsections to focus on three different challenges and their solutions in fuel production which includes the metabolic engineering approaches to increase the yield, reduce the toxicity of lignocellulosic feedstocks to yeast and increased tolerance of yeast to drop-in biofuel. The final section focuses on the challenges and outlook for yeast-based drop-in biofuel production. Overall, the chapter provides perspectives on the drop-in biofuels derived from yeast and describes the future developmental goals of these fuels.

5.2 Yeast Strains for Production of Drop-in Fuels

Several strains of microorganisms have been used to produce a variety of drop-in biofuels. A detailed account of extensively investigated strains and the types of drop-in biofuels produced by these yeast strains is discussed in this section.

5.2.1 *Categories of Yeast Strains*

Various kinds of bacteria, moulds, and yeasts have been used out of which yeasts are regarded as the most desirable microbial cell producers for a range of biofuels (Nielsen et al. 2013). Yeasts are the "workhorse" microorganisms used in the current ethanol industry (Karatzos et al. 2014). It is widely recognized for its tolerance to alcohols and performs efficiently in fermentation when combined with lignocellulosic hydrolysates (Chen et al. 2011). Currently, substantial progress has been made towards retooling microbial metabolism for the generation of drop-in biofuels, which exhibit comparable or superior physical qualities to petroleum-derived fuels (Zhou et al. 2016a). Traditional molecular genetic techniques, combined with synthetic biology tools, are effective for rapid yeast prototyping and optimization of metabolic pathways (Wagner and Alper 2015). Multiple strains like *Saccharomyces cerevisiae*, *Yarrowia lipolytica*, *Rhodosporidium toruloides*, *Candida utilis* (Nishimura et al. 2018; Rutter and Rao 2016; Tamakawa et al. 2013a, 2013b; Yaegashi et al. 2017) are regarded as suitable microbial cell factories for the generation of drop-in biofuels (Liu et al. 2021). Table 5.1 represents the different yeast strains used to produce several drop-in biofuels. A detailed discussion is provided in the sections below.

5.2.1.1 Saccharomyces cerevisiae

Saccharomyces cerevisiae, commonly referred to as Baker's yeast, is an essential host for biotechnological applications like winemaking, bioethanol production, and brewing (Teo et al. 2015). This budding yeast has been regarded as an appealing microbial host for producing higher-chain alcohols (Yuan et al. 2017). It is characterized as being extremely robust, resilient to stress in fermentative processes, genetically flexible, and a phage infection-resistant organism that ferments carbohydrates to low pH values (de Jong et al. 2015; Schadeweg and Boles 2016). Currently, it is the industrial strain alcohol (ethanol) producer and has previously been altered to generate other heterologous metabolites (Steen et al. 2008). It has 16 chromosomes, and a genome size of 12 Mb which has been reduced to 8% by substituting 6 native chromosomes with synthetic counterparts containing vital modifications (Liu et al. 2021). It is an ideal host-microbe for engineering biofuel production since it has a fully sequenced genome and well-characterized metabolic pathway (Teo et al. 2015; Yuan et al. 2017). *S. cerevisiae* has been metabolically engineered for production of various drop-in biofuel: butanol (Swidah et al. 2018), propanol (Nishimura et al. 2018), isoamyl alcohol (Yuan et al. 2017), bisabolene (Peralta-Yahya et al. 2012), fatty acid esters (Gao et al. 2018), and fatty alcohols (Cordova et al. 2020). Butanol, propanol, and isoamyl alcohol are gasoline substitutes, bisabolene is jet fuel surrogate, and fatty acid esters and fatty alcohols are diesel substitutes.

Table 5.1 Various kinds of yeast strains producing drop-in biofuels

Drop-in biofuel	Yeast strain	Substrate	Substitute for fuel	Obtained yield	References
Propanol	*Saccharomyces cerevisiae*	Glucose	Gasoline	179 mg/L	Nishimura et al. 2018
Isopropanol	*Candida utilis*	Glucose		27 g/L	Tamakawa et al. 2013a, 2013b
Butanol	*Saccharomyces cerevisiae*	Glucose and Ketovalerate		2.4 g/L	Swidah et al. 2018
	Saccharomyces cerevisiae	Glucose		235 mg/L	Schadeweg and Boles 2016
	Saccharomyces cerevisiae	Glucose		835 mg/L	Shi et al. 2016
Isobutanol	*Saccharomyces cerevisiae*	Glucose		2.09 g/L	Wess et al. 2019
	Saccharomyces cerevisiae	Glucose		4.12 mg/g glucose	Chen et al. 2011
Isoamyl alcohol	*Saccharomyces cerevisiae*	Glucose and galactose		561 mg/L	Yuan et al. 2017
Octanol	*Saccharomyces cerevisiae*	Glucose	Diesel and jet fuels	49.5 mg/L	Henritzi et al. 2018
Decanol	*Yarrowia lipolytica*	Glucose		550 mg/L	Rutter and Rao 2016
Pentane	*Yarrowia lipolytica*	Glucose	Gasoline and jet fuels	4.98 mg/L	Blazeck et al. 2013
Hexanal	*Yarrowia lipolytica*	13-Hydroperoxides and dithiothreitols		600 mg/L	Santiago-Gómez et al. 2009
Bisabolene	*Rhodosporidium toruloides*	Lignocellulosic biomass	Jet fuel	680 mg/L	Yaegashi et al. 2017
	Saccharomyces cerevisiae	Glucose and galactose		994 mg/L	Peralta-Yahya et al. 2012
Fatty acid esters	*Yarrowia lipolytica*	Glucose and ethanol	Diesel	1.18 g/L	Gao et al. 2018
	Pichia pastoris	Glucose		169 mg/L	Tao et al. 2015
	Saccharomyces cerevisiae	Glucose		4.4 mg/L	de Jong et al. 2015
	Saccharomyces cerevisiae	Glucose		230 mg/L	Teo et al. 2015
Fatty alcohols	*Yarrowia lipolytica*	Glucose		5.8 g/L	Cordova et al. 2020
	Yarrowia lipolytica Lipomyces starkeyi	Glucose and xylose		167 mg/L, 770 mg/L	Wang et al. 2016a, 2016b, 2016c

(continued)

Table 5.1 (continued)

Drop-in biofuel	Yeast strain	Substrate	Substitute for fuel	Obtained yield	References
	Lipomyces starkeyi	Glucose		1.7 g/L	McNeil and Stuart 2018
	Saccharomyces cerevisiae	Glucose		330.2 mg/L (flask)	Feng et al. 2015
	Saccharomyces cerevisiae	Glucose		1.3 g/L	Sheng et al. 2016
	Rhodosporidium toruloides	Sucrose		<8 g/L	Fillet et al. 2015

5.2.1.2 Yarrowia lipolytica

Yarrowia lipolytica is a non-conventional oleaginous yeast that is "Generally Regarded as Safe" (GRAS), thus permitting use in industrial operations for products such as citric acid and polyunsaturated fatty acids (Cordova et al. 2020; Lu et al. 2021). Using oleaginous microbes for oleochemical synthesis may represent a novel approach to achieve commercially feasible output due to their inherent capacity to produce lipids (Wang et al. 2016a, 2016b, 2016c). According to current research, some genetically modified *Y. lipolytica* strains could even generate 73–100 g/L of lipids (Qiao et al. 2017; Xu et al. 2017) under optimal fermentation conditions (Gao et al. 2018). Furthermore, the yeast's ability to manufacture target molecules through metabolic engineering is made possible by the availability of high-quality annotated genome sequences, a growing set of genetic tools, and effective screening techniques (Cordova et al. 2020; Lu et al. 2021). For instance, 55 g/L lipid titre by engineered *Y. lipolytica* strain (Qiao et al. 2015) exhibited extensive metabolic flux to fatty acyl-CoA derivatives and tremendous potential for oleochemical synthesis (Wang et al. 2016a, 2016b, 2016c). Moreover, it is an incredible oleaginous workhorse due to its high intrinsic acetyl-Co-A flux and native lipid storage mechanism (Lu et al. 2021). It also can grow on a variety of inexpensive carbon sources and utilize a wide range of renewable materials as substrates (Gao et al. 2018). Hence, it is therefore attracting a great deal of attention for its potential to be an efficient host.

5.2.1.3 Rhodosporidium toruloides

Rhodosporidium toruloides is an oleaginous yeast belonging to the *Pucciniomycotina* subphylum in the Basidiomycota (Liu et al. 2013; Sampaio et al. 2003). It is a non-pathogenic, pink-coloured basidiomycetous fungus, that can accumulate lipids, like triglycerides, to more than 70% of its dry cell weight, making it a suitable host for the synthesis of oil for biodiesel and many other purposes from biomass (Liu et al. 2013a, 2013b; Zhu et al. 2012). Since it is a superior generator of β-carotenoid, it is also known as carotenogenic basidiomycete yeast and a producer of biotechnologically crucial enzymes, such as cephalosporin esterase and epoxide hydrolase (Koh et al. 2014; Eksemplaar et al. 1999; Politino et al. 1997). It is regarded as a unique strain of biotechnological importance to produce single cell oil and may find widespread uses in staple foods, biofuels, animal feed, surfactants, and industrial polymers (Zhu et al. 2012). Besides, this microorganism can metabolize several carbon sources including xylose, hexose, and pentose, suggesting prospective advantages over traditional lignocellulosic conversion hosts (Fillet et al. 2015; Yaegashi et al. 2017).

5.2.1.4 Candida Utilis

Candida utilis belongs to Crabtree—a negative group of yeasts which means that it can grow quickly under stringent aerobic conditions, and anaerobic fermentation with high cell density promotes efficient output (Tamakawa et al. 2013a, 2013b). It is an industrial potent microorganism, being utilized to produce various valuable compounds including L-lactic acid, carotenoids such as lycopene, α-amylase, glutathione, and RNA (Ichii et al. 1993; Ikushima et al. 2009; Li et al. 2004; Miura et al. 1998, 1999). Additionally, it has the maximum respiration activity among Crabtree-negative yeasts, which is essential for the effective generation of biomaterials (De Deken 1966). It can absorb nitrate, a naturally occurring mineral source of nitrogen that is available as potassium nitrate (Tomita et al. 2012) and can grow on xylose too, but does not ferment it to ethanol (Tamakawa et al. 2011). The dried *C. utilis* cells have been approved for use as a food additive by the Food and Drug Administration (FDA) (Boze et al. 1992). Tamakawa et al. (2013a, 2013b) produced isopropanol with engineered C. utilis which can used as a substitute for gasoline.

The use of the above strains for drop-in biofuels, both in native and/or engineered forms, using various bioprocessing strategies is discussed in the sections below.

5.2.2 *Drop-in Biofuels*

5.2.2.1 Alcohols

Propanol is a capable biofuel that can be utilized as a substitute for gasoline. It has not gained great attention as a biofuel because it only offers a slight increase in energy density over ethanol (Liu et al. 2021). Nishimura et al. (2018) produced 1-propanol by engineering the citramalate pathway which enhanced its production in S. cerevisiae. This was accomplished by introducing genes for citramalate synthase (CMS) from *Methanocaldococcus jannaschii,* citramalate hydrolyase (CMH), large subunit leuC, small subunit leuD, and overexpression of isopropyl malate dehydrogenase (IMDH) from *Clostridium beijerinckii* into the 2-Ketobutyrate (2-KB) pathway. Then, the threonine dehydratase (TODH) encoding gene was overexpressed followed by improvement of threonine biosynthesis from aspartate via the addition of *thrA*, *thrB,* and *thrC* from *Escherichia coli*. In order to restrict the conversion of threonine to glycine, *GLY1* was finally eradicated. In a high-density anaerobic fermentation process, the engineered strain produced 179 mg/L of 1-propanol while the control strain produced 12 mg/L of 1-propanol from glucose. Additionally, it has been observed that deletion of the acetolactate synthase gene (ILV2), which controls a competing metabolic pathway, causes yeast to produce 60 mg/L of 1-propanol (Eden et al. 2001).

Isopropanol can be utilized as an additive during the production of high-octane gasoline (Rassadin et al. 2006). It was produced using *Candida utilis* by Tamakawa et al. (2013a, 2013b) by expressing *cft*AB from *Clostridium acetobutylicum*, as well as adc and sadh from *Clostridium beijerinckii*. The resultant strain generated a massive quantity of ethanol alongside only 1.2 g/L of isopropanol. Additionally, the engineered *C. utilis* was overexpressed by the endogenous genes ACS2 and ERG10, and after 200 h of fed-batch growth in a bioreactor containing 200 g/L glucose, isopropanol production reached 27.2 g/L.

Butanol isomers like n-butanol or isobutanol are regarded as more suitable fuel alternatives than bioethanol (Schadeweg and Boles 2016). Since butanol has a longer chain length and is less hygroscopic than ethanol, it can be combined up to 85% with gasoline (Buijs et al. 2013) while ethanol can only be blended up to 10% due to regulatory restrictions and requirements for engine modification (Coyle 2010). Although still less than gasoline, butanol has a higher energy density than ethanol (Jin et al. 2011). Swidah et al. (2018) modified *S. cerevisiae* to implement the *Clostridia*-derived butanol synthesis pathway. First, yeast was overexpressed with the endogenous thiolase gene (*ERG10*). Then, the *C. beijerinckii* genes *hbd*, *bcd*, *crt*, and *adhe2* were expressed in the yeast. The engineered yeast generated approx. 300 mg/L of butanol. Steen et al. (2008) engineered various *S. cerevisiae* strains: ESY2, ESY3, ESY4, ESY5, ESY6, ESY7, and ESY11 out of which ESY7 produced a maximum concentration of n-butanol, i.e. 2.5 mg/L. It was overexpressed with HbCoA dehydrogenase (*Hbd*) and native thiolase (*ERG 10*) and utilized NADH, which cells should have copious amounts under fermentative conditions.

Isobutanol was produced by Chen et al. (2011) in *S. cerevisiae* by overexpressing biosynthetic genes *ILV2, ILV3,* and *ILV5* in valine metabolism during the anaerobic fermentation of glucose in the mineral medium. A rise in the yield from 0.16 to 0.97 mg/g glucose was observed. Additional overexpression of *BAT2*, which encodes the cytoplasmic branched-chain amino-acid aminotransferase, led to a two-fold increase in the yield. A schematic representation of the production of isobutanol and ethanol in *S. cerevisiae* has been demonstrated in Fig. 5.2. Under aerobic yeast extract/peptone/dextrose (YPD) complex medium the overexpression strain had an increment in the isobutanol yield from 3.86 mg/g glucose to 4.12 mg/g glucose while the reference strain had an increase from 0.28 mg/g glucose to 2.4 mg/g glucose. Wess et al. (2019) synthesized isobutanol through engineered *S. cerevisiae* which led to a 22-fold increase up to 0.22 g/L (5.28 mg/g glucose) in aerobic shake-flask cultures. Three endogenous enzymes acetolactate synthase (*Ilv2*), dihydroxy-acid dehydratase (*Ilv3*), and aceto-hydroxy acid reductoisomerase (*Ilv5*) of the valine synthesis pathway in the cytosol were overexpressed in

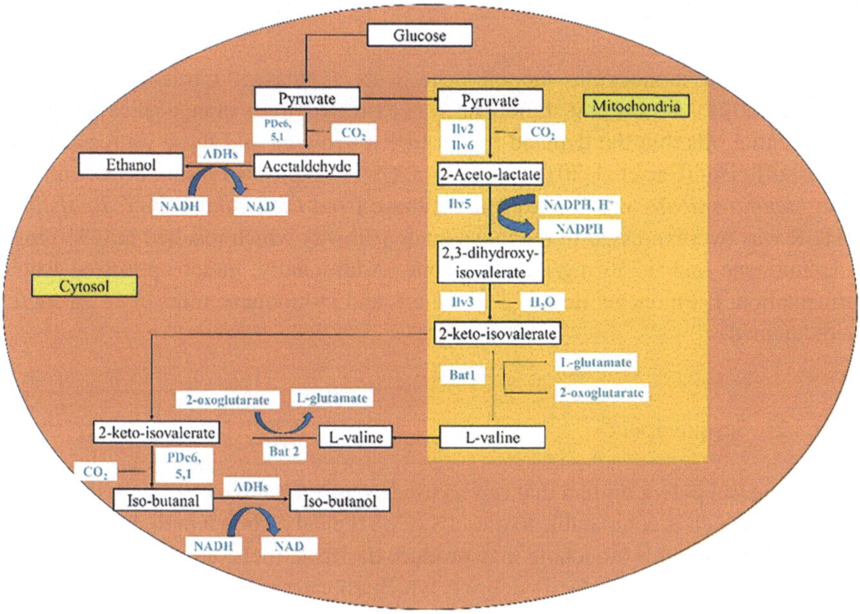

Fig. 5.2 The metabolic pathways from pyruvate to ethanol and isobutanol in *Saccharomyces cerevisiae*. The enzymes that catalyse the pathway from pyruvate to L-valine in the mitochondria are: acetolactate synthase (Ilv2 + Ilv6), dihydroxy-acid dehydratase (Ilv3), acetohydroxyacid reductoisomerase (Ilv5), and branched-chain amino-acid aminotransferase (Bat1). Ilv2 is the catalytic subunit of acetolactate synthase, and Ilv6 is the regulatory subunit. The enzymes that catalyse the pathway from L-valine to isobutanol in the cytosol are pyruvate decarboxylases (Pdc6, 5, 1) and alcohol dehydrogenases (ADHs). Pyruvate decarboxylases and alcohol dehydrogenases also catalyse the pathway from pyruvate to ethanol in S. cerevisiae. Adapted from Chen et al. (2011)

S. cerevisiae and disruption of endogenous *ILV2* blocked the first step of mitochondrial valine synthesis causing 22-fold increase in the isobutanol production.

5.2.2.2 Terpene-Based Advanced Biofuel

Terpenes are a diverse group of valuable chemicals having several applications in the chemical, flavour, fragrance, and pharmaceutical industries (Ma et al. 2018). Among these are monoterpenes and sesquiterpenes, which show potential as renewable substitutes for existing fuels because of their advantageous chain length, low clouding point, and high energy density (Peralta-Yahya et al. 2011). These were conventionally produced via chemical and plant synthesis, which requires extensive equipment and is unable to fulfil the expectations of the expanding fuel industry.

5.2.2.2.1 Monoterpenes

Limonene is a monocyclic monoterpene with a pleasant citrus-like fragrance (Lu et al. 2021). In the past, limonene was extracted from orange juice remnants and essential oils, but the demand for jet fuel is growing and these sources are no longer sufficient (Cao et al. 2016). Pang et al. (2019) expressed L-limonene synthase from *Mentha spicata* and D-limonene synthase from *Citrus limon* in *Y. lipolytica*. HMGR was overexpressed in the mevalonate pathway which resulted in 0.316 mg/L L-limonene and 0.256 mg/L D-limonene. Additionally, under optimized batch fermentation, L-limonene titres of 11.09 mg/L and D-limonene titres of 11.71 mg/L were attained.

5.2.2.2.2 Sesquiterpenes

Bisabolene is a potent biofuel that can be used as an alternative to jet fuel. It is a C_{15} compound which is comparable to diesel's (C_{16}) typical carbon length. Furthermore, the linear bisabolane's branching may produce desirable fuel cold characteristics. Its ring portion enhances the fuel's density, which enhances the fuel's energy density per unit volume (Peralta-Yahya et al. 2011). Zhao et al. (2021) synthesized α-, β-, γ-bisabolene by expressing three bisabolene synthases from *Y. lipolytica*. It resulted in production of 0.4 mg/L of α-bisabolene, 0.2 mg/L of β-bisabolene, and 0.05 mg/l of γ-bisabolene. The process was further followed by overexpression of hydroxymethylglutaryl-CoA reductase (HMGR). The titres reached 973.1 mg/L of α-bisabolene, 68.2 mg/L of β-bisabolene, and 20.2 mg/l of γ-bisabolene. Additionally, *R. toruloides* was incorporated with codon-optimized BIS which produced 680 mg/L bisabolene in a high-gravity fermentation of maize stover hydrolysate which was pre-treated with alkali and produced 261 mg/L of bisabolene when was pre-treated with choline α-ketoglutarate (a biocompatible ionic liquid) (Yaegashi et al. 2017).

5.2.2.3 Fatty Alcohols

Fatty alcohols are aliphatic alcohols with chain length ranging from C_8 to C_{32} with a polar hydrophilic hydroxyl moiety (McNeil and Stuart 2018; Wang et al. 2016a, 2016b, 2016c). Due to their amphiphilic nature, they have broad applications such as detergents, surfactants, pharmaceuticals, lubricants, and diesel-like fuels (Sheng et al. 2016). Presently, these are produced through petrochemical sources or from renewable sources like plant or animal fats, oils, and waxes (Sheng et al. 2016). Due to the emerging environmental concerns and competition with the food supply, these techniques have limitations (Liu et al. 2014). Fillet et al. (2015) generated 80% of fatty alcohols by expressing fatty acyl-CoA reductase from *Marinobactor aquaeolei* in *Rhodosporidium toruloides*. The titre ranged from 0.2 to 2 g/L in different fermentation media. A total of 8 g/L of C_{16}–C_{18} fatty alcohols were produced utilizing sucrose substrate.

Wang et al. (2016a, 2016b, 2016c) demonstrated the production of fatty alcohols by expressing the fatty acyl-CoA reductase gene from *Marinobactor aquaeolei* VT8 in *Lipomyces starkeyi* and *Y. lipolytica*. 167 mg/L and 770 mg/L of fatty alcohols by *Y. lipolytica and Lipomyces strakeyi* were produced, respectively. Hexadecanol (C16:0) and octadecanol (C18:0) were the predominant fatty alcohols and accounted for more than 85% of the total fatty alcohols produced. McNeil and Stuart (2018) expressed fatty acyl-Co A reductase and *mFAR1* from *Mus musculus* into *Lipomyces starkeyi.* Using batch cultivation, fatty alcohol output from glucose reached 1.7 g/L representing an output of 28 mg of fatty alcohols per gram of glucose.

5.2.2.4 Fatty Acid Esters

Fatty esters are produced from fatty acids and fatty alcohols via esterification (Liu et al. 2021). Fatty acid ethyl esters (FAEEs), for instance, are considered a potential sustainable fuel of the future due to comparable molecular characteristics with first-generation biodiesel (fatty acid methyl esters; FAME), crude oil (de Jong et al. 2015) and compatibility with the existing infrastructure (Gao et al. 2018). The generation of FAEEs by microbial conversion of sugars generated from plant biomass could lead to a more reliable, scalable, and efficient production process, in comparison with the existing chemical transesterification of vegetable oils and methanol (de Jong et al. 2012). Gao et al. (2018) engineered *Y. lipolytica* with wax ester (WS) genes that encode wax ester synthases (WSs). β oxidation was eliminated by deleting the *PEX10* gene and metabolic flux was redirected towards acetyl-CoA which led a production of 1.18 g/L FAEE in the shake flask. Teo et al. (2015) reported the generation of short- and branched-chain alkyl esters (FASBEs) by expressing two wax ester synthase genes (Maqu_0168 and ws2 from *Marinobactor*) into *S. cerevisiae.* A 230 mg/L of FASBEs including isoamyl, ethyl, active amyl esters and isobutyl were produced.

5.3 Metabolic Flexibility of Yeasts to Produce Drop-in Biofuels

It is essential to develop and support a bioeconomy centred on effective low-carbon emission technologies due to rising energy demands, environmental damage, and concerns about global warming brought on by using petroleum. The development of a bio-economic strategy to replace petroleum-derived fuels and chemicals with bio-based alternatives produced from sustainable carbon sources is necessary. This strategy also has the added benefit of boosting domestic economies through localized revenues and employment growth. Yeasts can produce drop-in biofuels from a variety of renewable resources, which are mentioned below.

5.3.1 Waste Cooking Oil

Currently domestic and industrial operations are resulting in surplus production of numerous low value by-products and/ or crude oil wastes. These may have detrimental consequences on human health as well as environment (Zhao et al. 2021). For instance, waste cooking oil (WCO) is mainly defined as vegetable oils that have been used to fry food at high temperatures along with other kitchen waste and greasy wastewater before being dumped directly into the sewer (Pang et al. 2019; Zhao et al. 2021). Currently, it has been observed that a significant amount of waste lipids is produced by households, fast food outlets, restaurants, and the food processing industry (Bhuiya et al. 2016). Following data showing that each individual contributes 4.1 kg of WCO annually (Maddikeri et al. 2012), it is projected that the amount of WCO produced globally each year is currently over 29 million tons (Lisboa et al. 2014). Consequently, disposal through prudent management is crucial (Kulkarni and Dalai 2006). WCO has conventionally been used in the manufacture of diesel engine fuel and domestic animal fodder (Domínguez et al. 2010). However, the presence of polar compounds, contaminants, high levels of free fatty acids, and water content in WCO may hinder the synthesis of biodiesel and reduce the final output. Furthermore, the European Union has sternly forbade the recycling of WCO from animal feedstock because it still includes undesirable elements that could be transferred to humans through the food chain (Lam et al. 2016). Recent research on the bioremediation of WCO has centred on utilizing WCO as raw and affordable feedstock for the production of value-added products, which would be more appealing from an economic and environmental conservation perspective for disposal of WCO (Katre et al. 2017, 2018).

The utilization of WCO as a fermentation media component to produce drop-in biofuel has been reported by various researchers (Fig. 5.3). The potential of WCO as a carbon source for the synthesis of limonene through engineered *Y. lipolytica* strain was investigated by Pang et al. (2019). There was a successful production of D-limonene and L-limonene by utilization of 70% WCO as a carbon source. A 20-fold increase in titre was observed from 0.124 mg/L D-limonene and 0.126 mg/L L-limonene (initial titres) to 2.514 ml/L D-limonene and 2.723 mg/L L-limonene. Wei et al. (2021) synthesized α-pinene, a vital biologically active monoterpene. There was a production of 19.6 mg/L of α-pinene by utilizing glucose as a single source of carbon which inclined to 33.8 mg/L on utilization of WCO. Zhao et al. (2021) produced α-bisabolene, β-bisabolene, and γ-bisabolene by feeding 1.18% (w/v) of WCO as a sole carbon source. The final engineered strains Po1g KαBS-AcrB, Po1g KβBS-AcrB, and Po1g KγBS-AcrB produced 157.8 mg/L of α-bisabolene, 20.9 mg/L of β-bisabolene, and 3.6 mg/L of γ-bisabolene, respectively. Overall, it was discovered that WCO is suitable for microbial growth because of its low cost, high carbon content, and lack of competition with current food plantations (Wei et al. 2021).

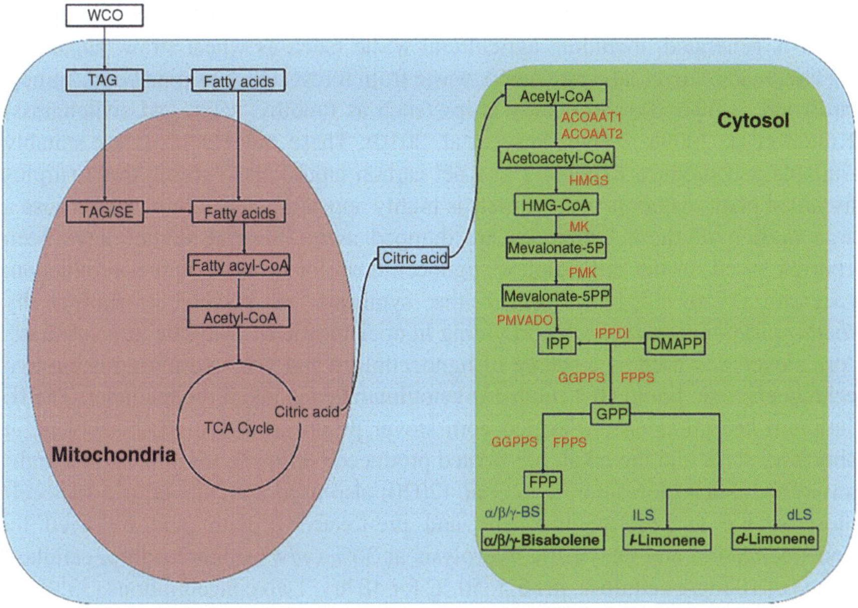

Fig. 5.3 Biosynthesis pathways for bisabolene and limonene production in the yeast *Y. lipolytica*. Yeast utilizes the MVA pathway to produce FPP for bisabolene and the mevalonate pathway to produce GPP for limonene from acetyl-CoA. To construct complete pathways, heterologous genes encoding α-bisabolene synthase (αBS), β-bisabolene synthase (βBS), γ-bisabolene synthase (γBS), d-limonene synthase (dLS), and l-limonene synthase (lLS) were introduced for bisabolene and limonene production, respectively. The endogenous MVA pathway enzymes are shown in red, and the heterologous enzymes in blue. Key enzyme abbreviations: DXPS—DXP synthase, DXPRI—DXP-reductoisomerase, MEPCT—MEP cytidylyltransferase, CDP-MEK—CDP ME kinase, MECDPS—MEDCP-synthase, HMBPPS—(E)-4-hydroxy-3-methylbut-2-enyl-diphosphate synthase, HMBPPR—HMBPP reductase, IPPDI—isopentenyl-diphosphate delta-isomerase, GPPS—geranyl-diphosphate synthase, FPPS—farnesyl diphosphate synthase, ACOAAT1—acetyl-CoA C-acetyltransferase 1, ACOAAT2—acetyl-CoA C-acetyltransferase 2, HMGS—hydroxymethylglutaryl-CoA synthase, HMGR—hydroxymethylglutaryl-CoA reductase, MK—mevalonate kinase, PMK—phosphomevalonate kinase, PMVADO—diphosphomevalonate decarboxylase, IPPDI—isopentenyl-diphosphate delta-isomerase, GGPPS—geranylgeranyl diphosphate synthase, type III, FPPS—farnesyl diphosphate synthase (Zhao et al. 2021 and Pang et al. 2019)

5.3.2 Carbohydrates

Lignocellulose comprises cellulose, hemicellulose, and lignin. Cellulose consists of glucose, and hemicellulose has pentose and hexose sugars, which are linked to the cellulose microfibrils. Lignin is a phenyl propane polymer that builds an intricate network that crosslinks cellulose and hemicellulose (Nanda et al. 2014). Lignocellulose is the most widespread sustainable carbon source currently available, cheap, and has been widely studied as a substrate for microbial production of bio-based fuels and chemicals (Yaegashi et al. 2017). It is a suitable alternative to the diminishing fossil fuels (Nanda et al. 2014). Every year, an ample of inedible plant matter is generated, including agricultural waste (such as wheat straw, sugarcane bagasse, maize stover, and rice husk), waste from forests (such as pinewood, spruce, hardwood, softwood), and energy crops (such as timothy, salix, and switchgrass) (Kumar et al. 2009a, 2009b; Sims et al. 2010). These raw resources are suitably available and produce extremely low net carbon emissions. Turning these surplus discarded plant materials into biofuels is highly appealing and significant because a large portion of these biomasses are dumped away. Several studies have been reported based on the utilization of lignocellulose for drop-in biofuel production. Yaegashi et al. (2017) reported the synthesis of bisabolene biofuel by *Rhodosporidium toruloides* by providing lignocellulosic hydrolysate as a substrate. Corn stover was used as a source of lignocellulose and was hydrolysed using two techniques, ionic liquid (IL) choline α-ketoglutarate and alkali pretreatment. The IL choline α-ketoglutarate pre-treated corn stover produced 261 mg/L bisabolene at laboratory scale and the alkali pre-treated produced 680 mg/L bisabolene in a high-gravity fed-batch bioreactor. Yao et al. (2020) also used corn stover as a lignocellulose source to produce limonene and pre-treated it with acid followed by biodetoxification and enzymatic hydrolysis at 30% (w/w) solids-loading, cellulase dosage of 10 mg/g cellulose, pH 5.5, 50 °C for 48 hrs. Two concentrations (33% and 50%, v/v) of hydrolysate were examined. With a 50% hydrolysate concentration, the levels of limonene increased and reached 20.57 mg/L (1.05 mg limonene/g of carbon). Wei et al. (2021) reported the synthesis of 36.1 mg/L of α-pinene by *Yarrowia lipolytica* in a 50% detoxified lignocellulosic hydrolysate as a sole carbon source for 72 hrs. These studies show sustainable and economical production of drop-in biofuels from renewable feedstocks.

The main sugar in hemicellulose is xylose, which is also the second-most prevalent sugar in lignocellulosic hydrolysates after glucose. Zhang et al. (2019) engineered *Saccharomyces cerevisiae* by deleting *BAT*1, *ALD*6, and *PHO*13, to develop a strain (YZy197) that produced 3.10 ± 0.18 g/L of isobutanol by consuming xylose as a carbon source. A 28-fold increase was witnessed which is the greatest isobutanol titre derived from xylose until now reported in yeast. Yao et al. (2020) engineered *Yarrowia lipolytica* and produced 9.00 mg/L of limonene in 72 hrs providing xylose as a sole carbon source whereas the control strain only produced 0.26 mg/L of limonene with glucose as a single source of carbon. A ten-fold increase in the concentration of limonene was observed.

5.3.3 Lipids/Fats

Production of microbial oil/lipid by yeasts can be utilized for biofuel production. The synthesized oil is converted to Fatty Acid Methyl Ester (FAME) or Fatty Acid Ethyl Ester (FAEE), whose properties are analysed for biofuel production. Carmona-Cabello et al. (2021) derived microbial oil from potato peels discarded by the hospitality sector. The waste was hydrolysed and was utilized by *Rhodosporidium toruloides* Y-27012 in a shake flask producing 32.9 g/L total dry weight and 36.4% (w/w) intracellular lipid content. The quantities increased in fed-batch bioreactor with 53.9 g/L of total dry weight and 26.7 g/L of lipid content. Tsakona et al. (2014) reported the production of microbial lipids from flour-rich waste (FRW) generated by bakery, confectionery, and wheat milling units. A total dry weight of 30.5 g/L with a microbial oil content of 40.4% (w/w) was achieved by *Lipomyces starkeyi* in a shake flask. There was an increment of 109.8 g/L in total dry weight and microbial oil content of 57.8% (w/w) observed in the fed-batch bioreactor. Angerbauer et al. (2008) investigated the production of lipids by utilization of sewage sludge. The sludge was pre-treated before lipid synthesis. The maximum lipid content was observed at a C/N concentration of 150. The lipid produced was around 6.4 g/L with a total dry weight of 9.5 g/L. Huang et al. (2009) pre-treated rice straw with sulphuric acid which resulted in a sulphuric acid treated rice straw hydrolysate (SARSH). The detoxified SARSH gave a lower lipid yield of 1.7 g/L when utilized by *Trichosporon fermentans* as a carbon source. SARSH was then detoxified with overliming, concentration, and adsorption by Amberlite XAD-4, which improved the fermentability by removing the inhibitors. The yield increased to 11.5 g/L on utilization of detoxified SARSH.

5.4 Bioprocess Engineering Strategies for Drop-in Biofuel Production Using Yeast Systems

5.4.1 Types of Bioprocess Engineering Strategies

With advancing research in bioprocess engineering strategies for yeast-based drop-in biofuel production, there has been an increase in awareness of alternatives to fossil fuels. This advances the potential to reduce greenhouse gas emissions. Bioprocess engineering has an important role in optimizing biofuel production from yeast. From synthesizing many literature reports, many strategies have been integrated into bioprocess engineering methods. These strategies are analysed in the production of drop-in biofuels using different yeast-based systems. In a batch process, nutrients and substrates are added to a bioreactor for the duration of contained nutrients and inhibitory by-products. Fed-batch cultivations are commonly used for assessing the sustainability of high cell densities and the reduction of accumulated inhibitory metabolites. Continuous cultivation is the process of fresh media and cells being added to the bioreactor while the culture broth is removed. In-situ product recovery aims to separate biofuel products from the culture broth during the bioprocess, further allowing for continuous production. The use of different bioprocess strategies is developed based on varying constraints (Table 5.2).

Table 5.2 Several bioprocess methods used for drop-in biofuel production and the production yields obtained

Bioprocess method	Feedstock	Strain	Product	Concentration (g/L)	Biofuel yield (g/g)	Productivity (g/L/h)	Reference
Batch fermentation	Glucose	ILV235_XCY561	Isobutanol	–	0.00097	–	Chen et al. 2011
	Glucose	Isoy8	Isobutanol	0.63027 1.62	0.01486	0.00656	Brat et al. 2012
	Glucose	BSW205	Isobutanol	1.61	0.016	–	Matsuda et al. 2013
	Glucose	BSW206	Isobutanol	0.9731	0.016	–	Matsuda et al. 2013
	Waste cooking oil	Po1g KαBS	α-Bisabolene	0.0682	–	0.00811	Zhao et al. 2021
	Waste cooking oil	Po1g KβBS	β-Bisabolene	0.0202	–	0.00056	Zhao et al. 2021
	Waste cooking oil	Po1g KγBS	γ-Bisabolene	35	–	0.00017	Zhao et al. 2021
	Crude glycerol	YL-ad9	Biodiesel	18.5	0.23	0.14	Qiao et al. 2015
	Crude glycerol A	AS2.1389	(triacylglycerol)	13.4	–	–	Xu et al. 2012
	Crude glycerol B	AS2.1389	Biodiesel (triacylglycerol) Biodiesel (triacylglycerol)		0.175	–	Xu et al. 2012
Fed-batch fermentation	Glucose	YL-ad9	Biodiesel	55	0.234	0.707	Qiao et al. 2015
	Glucose	XF3	(triacylglycerol)	1.11	0.03	0.042	Feng et al. 2015
	Sugarcane molasses feed	ATCC2040	1-Hexadecanol	39.2	–	0.25	Singh et al. 2020
		91	Fatty acid methyl	0.68	–	–	Yaegashi et al. 2017
	Hydrolysed lignocellulose	BIS_3	Ester (FAME)	10.4	0.0861	–	Zhou et al. 2016a
		YJZ47	Bisabolene	9.97		0.0906	Zhang et al. 2021
	Glucose	Δku70-AbWS	Free fatty acids				
	Glucose		Biodiesel (fatty acid Ethyl esters)				
Continuous fermentation	Food crops	–	Bioethanol	–	–	–	Jiang et al. 2018
	Glucose	LOCKR13	Biodiesel	14.68	0.242	0.1	Gientka et al. 2022
In-situ product Recovery	Glucose	NJ4	Isobutanol	33.90	–	–	Xin et al. 2018; Amiri 2020
	Glucose	–	Butanol	–	–	–	
	Microalgae	RUT-C30	Methane	80.6	–	0.24	Ambaye et al. 2021; Xue et al. 2016
	Activated carbon	–	Butanol	21.9	0.22	0.45	
	Sodium acetate	BA101	Butanol	33	0.33	1.5	Kujawska et al. 2015
	Spent mushroom	–	Biodiesel	30.21	0.3	0.31	Zhu et al. 2016
	Substrate	Pichia Pastoris	Isobutanol	20	–	–	Bumrungtham et al. 2022
	Sugarcane						

5.4.2 Batch Fermentation

Batch fermentation is often used in the brewing industry because of its high efficiency and control (Li et al. 2011). In batch recovery, a fixed volume of medium is introduced to the microbial culture (Bolmanis et al. 2023). The substrate and microorganisms are introduced to the system at time zero and are not removed, refilled, or changed until the fermentation process is complete. Batch fermentation is referred to as a "closed system" process because there are no additional alterations or interventions done to the contents within the bioreactor once the media and nutrients have been initially added (Burke et al. 2013). As the microorganisms multiply, they eventually consume all the nutrients present in the medium and create by-products (Bolmanis et al. 2023). The generation of these by-products and the depletion of the nutrients are what slows the growth of microorganisms and cause them to enter a stationary growth period (Bolmanis et al. 2023). After the bioprocess is complete, the biomass is then removed from the bioreactor and processed to produce the desired product. While this method may be quick and convenient, because of the eventual depletion of nutrients and the gradual increase of by-products, the biomass and product yields are limited. A yeast that is often used in biofuel production is *Saccharomyces cerevisiae. S. cerevisiae* is a yeast that is engineered to produce drop-in biofuels such as butanol, biodiesel, and fatty alcohols (Buijs et al. 2013). *S. cerevisiae* is commonly used for ethanol production because this microbe can ferment a wide range of sugars, tolerate a variety of pH values, and possess a high ethanol tolerance and productivity (Mohd Azhar et al. 2017). Due to these traits, *S. cerevisiae* is also used for the synthesis of other biofuels like isobutanol (Chen et al. 2011). A drawback to this yeast is that it cannot naturally ferment pentoses, only hexoses, which is why those that aim to ferment sugars such as xylose, require the engineering of *S. cerevisiae* to do so (Kumar et al. 2009a, 2009b).

Butanol can be created using yeast strains; however, current methods in butanol batch fermentation production are not economically feasible and generate relatively low yields (Xue et al. 2013). This can be attributed to high feedstock consumption, resulting from the production of by-products such as ethanol and acetone, in acetone-butanol-ethanol (ABE) fermentation (Xue et al. 2013). Furthermore, butanol titres are also limited by butanol's toxicity to host cells; but, if one were to dilute substrate concentration to resolve this issue, it causes a larger production of wastewater which makes this process energy intensive. So, for sustainable butanol production, one must modify strains to have greater butanol yield and tolerance, to facilitate a process that is economically competitive to methods that are currently in place (Xue et al. 2013). For butanol production with ABE fermentation, batch fermentation is often chosen in the industry, fed batch is used primarily in laboratories and continuous cultivation is used in large-scale synthesis (Xue et al. 2013).

Isobutanol is a biofuel that is of interest because it has a higher energy density and lower hygroscopicity and volatility compared to ethanol (Dürre 2007). Additionally, isobutanol is more versatile than ethanol, in that it can be used as jet fuel and used outside of the fuel industry (Connor and Liao 2009). The production of isobutanol by

Chen et al. (2011) was completed using *S. cerevisiae* and this study was the first reported isobutanol overproduction, using *S. cerevisiae,* overexpressing the first three steps of the valine biosynthesis pathway (Buijs et al. 2013; Chen et al. 2011). This yeast was chosen because of its robustness in industrial fermentations, relative tolerance to alcohols and its potential to ferment lignocellulosic materials for isobutanol production (Chen et al. 2011). Not to mention, *S. cerevisiae* also naturally produces isobutanol, in addition to other higher alcohols in small quantities, when it degrades amino acids (Hazelwood et al. 2008; Dickinson et al. 1998).

Another biofuel that is of interest due to the rise in environmental concerns and shortage of fossil fuels is biodiesel (Xu et al. 2012). Traditional diesel fuel is derived from animal fat or vegetable oils (Degfie et al. 2019). In comparison, biodiesel can be created from microbial lipids, as opposed to vegetable oils, since they have similar fatty acid compositions (Xu et al. 2012). Microbial lipids can be a sustainable alternative to vegetable oils because they do not require agricultural land, have a short life cycle, and do not compete with the food industry (Xu et al. 2012). Crude glycerol, the main by-product of biodiesel production, can be converted into lipids suitable for biodiesel production (Xu et al. 2012). To test the viability of using microbial lipids, the yeast *Rhodosporidium toruloides* was used in a batch fermentation process. *R. toruloides* was selected for testing industrial applications because it can store intracellular lipids up to a maximum of 60% of its cellular dry weight (Li et al. 2007). Furthermore, *R. toruloides* has been proven to be able to grow on a variety of substrates including sugars such as glucose and xylose, as well as on lignocellulosic and excess sludge hydrolysate (Li et al. 2007; Xu et al. 2011). In one such study, *R. toruloides* AS2.1389 was tested with two crude glycerol samples. One of the samples, referred to as "Glycerol A", contained 85% glycerol and small amounts of glyceride and fatty acid methyl ester (FAME) (Xu et al. 2012). In contrast, the other sample, "Glycerol B", contained 32.97% glycerol and small amounts of glyceride, FAME, and ash (Xu et al. 2012). When testing both samples, it was found that Glycerol A had a lipid concentration of 18.5 g/L and a maximum yield of 17.5 g/100 g of consumed glycerol (Xu et al. 2012). Additionally, Glycerol B had a lipid concentration of 13.4 g/L (Xu et al. 2012). Upon comparison against refined glycerol substrate, it was illustrated that both crude Glycerol A and B had higher biomass and lipid concentrations, demonstrating the advantages of using crude glycerol as a carbon source for *R. toruloides* (Xu et al. 2012). This study proved that *R. toruloides* has the potential for use in the lipid production of biodiesel, using crude glycerol. Furthermore, these experiments demonstrated that the use of crude glycerol as a feedstock led to higher biomass and lipid concentrations than both refined glycerol and glucose, showcasing the viability of sustainable feedstock for biodiesel creation (Xu et al. 2012).

Additionally, on the topic of biodiesel, bisabolene is the immediate precursor to bisabolane, a drop-in biofuel alternative to petroleum-derived diesel (Özaydın et al. 2013; Peralta-Yahya et al. 2011). Presently, bisabolene is produced through direct extraction from plant tissue; however, this process is limited by low yields, raw material availability and complex separation procedures (Lopresto et al. 2014; Zhao et al. 2021). Furthermore, the production of bisabolene is energy intensive and may cause issues to the environment. These factors create a demand for sustainable and renewable methods to produce bisabolene. From the choices under consideration, using microbial cell factories generated through synthetic biology and metabolic engineering to biosynthesize bisabolene shows promise, in overcoming the bottleneck that is preventing further utilization of bisabolene (Zhao et al. 2021). Zhao et al. (2021) explored this avenue using the oleaginous yeast, *Yarrowia lipolytica*. A key trait of *Y. lipolytica* is that it can use a wide range of substrates for growth and is an attractive host platform for terpenes because it possesses an endogenous cytosolic mevalonate (MVA) pathway that produces geranyl diphosphate (GPP), farnesyl diphosphate (FPP), and geranylgeranyl diphosphate (GGPP) (Zhao et al. 2021). These products are the direct substrates for biosynthesizing monoterpenes, sesquiterpenes, and diterpenes, respectively (Schmidt et al. 2010).

5.4.3 Fed-Batch Fermentation

Fed-batch fermentation accounts for some of the shortcomings seen in batch fermentation. In fed-batch fermentation, to prevent nutrient depletion from being a limiting factor in the cultivation process, this method adds additional nutrients and substrate to the bioreactor, intermittently or continuously, once the microorganisms have consumed the ones that were added during the onset of the process (Bolmanis et al. 2023; Poontawee and Limtong 2020). This re-supply of nutrients and substrate allows for an extension of the active growth phase (Bolmanis et al. 2023). In fermentation, it is important to keep in mind the amount of substrate that the cells can utilize. If too much substrate is added to the bioreactor, then the increased concentration of substrate may inhibit biomass growth. Similarly, the accumulation of some products may also impede biomass growth. Fed-batch fermentation allows the regulation of limiting factors that would otherwise restrain growth, thereby promoting product concentration and yield (Poontawee and Limtong 2020). This fermentation strategy leads to significant biomass accumulation, which is ideal for bioprocesses that aim to generate high product titres and densities. When compared to batch fermentation, a fed-batch produces a greater product titre, per fermentation.

A demonstration of fed-batch's advantage over batch fermentation was illustrated in a study that compared batch and fed-batch fermentation, using sugarcane molasses as a substrate, to produce lipids with five oleaginous yeast strains; however, only *Rhodosporidium toruloides* ATCC204091 and *Rhodotorula glutinis* ab1 were used for further fermentation studies and scaling of procedure (Singh et al. 2020). The oleaginous yeast strain *Rhodosporidium* was chosen for the study because it can store about 70% lipids with respect to their dry cell weight (Singh et al. 2020). In addition, the lipids that are produced by *Rhodosporidium* are mainly triacylglycerols, which are comparable to the conventional vegetable oils that are used in biodiesel production (MacDougall et al. 2011). Despite the benefits of using oleaginous yeast for biodiesel production, the major drawback to this method is the prohibitive cost of microbial oils, in comparison with vegetable oils (Singh et al.

2020). A potential solution to this was explored, when the researchers used carbon-rich media waste extract and sugarcane molasses as a supplement, to produce lipids under both batch and fed-batch fermentation conditions (Singh et al. 2018). This was an economically effective method because sugarcane molasses were inexpensive and locally available, in the area of their study. The batch fermentation lipid concentrations of the experiment using waste extract and sugarcane molasses were found to be 14.2 ± 1.72 g/L and 13.4 ± 0.98 g/L, for *R. toruloides* ATCC204091 and *R. glutinis ab1*, respectively (Singh et al. 2020). In contrast, under fed-batch conditions and analysing *R. toruloides* ATCC204091, it was witnessed that 39.2 ± 0.6 g/L of lipids were produced, with a productivity of 0.25 g/L/h. So, in comparing the results obtained under batch against fed-batch conditions, it is evident that fed-batch promotes greater productivities and titres since the fed-batch method demonstrated nearly a threefold increase in lipid production compared to batch fermentation.

Another study that used *R. toruloides* investigated the bioproduction of fatty acid ethyl esters (FAEEs). Generating FAEEs is of interest because the current method of industrial biodiesel production consumes plant oils or animal fats for feedstock, which poses a challenge since the use of these feedstocks contributes to environmental issues and also generates competition for land and materials needed for food production (Hill et al. 2006; Santori et al. 2012). Due to these concerns, the creation of FAEEs from microbial oils using renewable feedstock is desirable. *R. toruloides* was chosen because it can naturally ferment a wide assortment of plant-derived feedstocks such as cellobiose and xylose, giving it a broader industrial application (Zhang et al. 2016; Zhang et al. 2021). Furthermore, *R. toruloides* was selected over a yeast such as *Y. lipolytica,* because the former produces greater titres than the latter (Zhang et al. 2021). This study examined five wax synthases; however, only the AbWS from *A. baylyi* was successfully expressed and was able to produce FAEEs, making the recombinant Δku70-AbWS the only strain used for further studies (Zhang et al. 2021). FAEE yield was enhanced by modifying the strain through a point mutation, by substituting the glycine to isoleucine at position 355 (Zhang et al. 2021). This change resulted in a shift in substrate selectivity towards shorter-chain alcohols and led to impaired acyl-CoA-diacylglycerol acyltransferase (DGAT) activity (Röttig et al. 2015). Exogenous ethanol had to be added to the fermentation medium because *R. toruloides* cannot produce ethanol independently, under laboratory conditions (Zhang et al. 2021). It was important to optimize the ethanol concentration since its toxicity can inhibit cell growth and decrease FAEE yield. Fed-batch fermentation was used to observe engineered *R. toruloides'* capabilities for FAEE production, and it was found that 9.97 g/L of FAEE was maximally obtained, with a yield of 0.0861 g of FAEE/ g of glucose and a productivity of 0.0906 g/L/hr (Zhang et al. 2021). Additionally, while it was detected that FAEE was mainly secreted outside of the cell, which lowers the production costs associated with extraction, this method of producing FAEEs for use as biodiesel is not economically viable yet, due to the expenses associated with supplying the system with enough ethanol to sustain adequate yields.

Moreover, free fatty acids (FFA) can be produced from *S. cerevisiae* in fed-batch fermentation. Of which, alkanes and fatty alcohols can be obtained from these FFAs. Building upon previous discussions about *Saccharomyces cerevisiae*, this yeast is widely used for bioethanol production and is also one that displays robustness against harsh fermentation conditions (Caspeta et al. 2014; Mattanovich et al. 2014). Zhou et al. (2016b) focused their study on FFA pathways compared to other paths such as fatty acyl-CoA because FFAs can accumulate greater than 200-fold compared to fatty acyl-CoA and can also be biosynthesized to produce fatty alcohols and alkanes (Schjerling et al. 1996). They were able to generate FFAs by blocking fatty acid activation and degradation, optimizing, and utilizing an acetyl-coA pathway, overexpressing acetyl-coA carboxylase and expressing a more efficient fatty acid synthase (Zhou et al. 2016b). They were the first to create a plasmid-free FFA overproducing strain of *S. cerevisiae* through the integration of all pathway components into the genome, allowing for industrial applications. In fed-batch fermentation, a maximum of 10.4 g/L of FFAs were produced, resulting in up to 0.8 mg/L of alkane and 1.5 g/L of fatty alcohols extracted. This fatty alcohol titre was noted to be the highest obtained to date (Zhou et al. 2016b).

5.4.4 Continuous Fermentation

Technological advancements in pretreatment are heavily emphasized along with enzyme development and genetic engineering of yeast strains. This enables sustainable bioethanol production from lignocellulosic biomass playing a crucial role in optimizing biofuel production. Creating efficient utilization of yeast as a microorganism for biofuel production allows for enhanced titres and yields (Ambaye et al. 2021). Continuous cultivation allows the process to maintain a steady-state operation and higher productivity. The goal of high titres, yields, and productivity displays the success of bioprocess engineering in enhancing the performance of yeast-based drop-in biofuel production. Continuous cultivation is an effective bioprocess strategy that offers many advantages for drop-in biofuel production using yeast systems. In contrast to batch and fed-batch cultivation, continuous cultivation allows for a steady-state operation through a continuously added medium. This approach enables sustained growth and production, leading to higher overall productivity and potentially improved yields. A focused benefit of continuous cultivation is the elimination of growth inhibition from the accumulation of inhibitory by-products or depletion of essential nutrients. In comparison with batch cultivation, as the culture progresses and cells metabolize nutrients, the environment depletes the necessity for growth and product formation.

A study was conducted to achieve high-cell-density cultures of *Rhodotorula glutinis var. rubescens* LOCKR13 while maximizing de novo lipid productivity. As a medium for cultivation, deproteinated potato wastewater (DPW) was utilized. Glucose as a feedstock for DPW, researchers obtained a high yield of *Rhodotorula glutinis var. rubescens* LOCKR13 biomass, reaching a weight of 52 g (Gientka et al.

2022). Those who collaborated in this study established a two-stage fed-batch process at 20 °C. This resulted in the highest value of lipid accumulation, production, yield coefficients, and volumetric productivity for the strain. The lipid composition of the yeast-based biomass contained a large amount of oleic and palmitic acid. The lower incubation temperature increased the monounsaturated fatty acids (MUFAs). This further concluded that the manipulation of temperature during cultivation is associated with yeast oil when optimizing to a specific fatty acid profile. In this study, an analysis of microbial oil was conducted. The isotherms of the oxidation kinetics (PDSC) demonstrated that the lipids extracted from the biomass of the red yeast exhibited high oxidative stability. It is described that yeast oil contains desirable attributes for long-shelf-life food products. The high oxidative stability creates a strategy for biodiesel production. This research provides great insight into the development of sustainable and efficient strategies for biofuel production using microbial systems.

To further achieve high performance, continuous cultivation must be optimized. The flow rate of the fresh medium must be controlled to balance cell growth and product formation. A high dilution rate may promote cell growth but can result in lower product yields. A low rate may lead to higher product yields but lower productivity. From a case review, the authors conducted a study to analyse the advancements in butanol synthesis, exploring biotechnological methodologies involving microorganisms like *Clostridia*, and chemical configurations relying on petroleum-based processes. For drop-in biofuel production using yeast systems, continuous cultivations and in-situ product recovery strategies are crucial considerations. These strategies include the removal of biofuel from the fermentation broth allowing the fermentation process to continue. Extraction, membrane separation, and adsorption can be utilized for in-situ product recovery.

5.4.5 In-Situ Product Recovery

Bioprocess engineering strategies play an important role in the development of sustainable drop-in biofuels using yeast systems. Continuous cultivation along with in-situ product recovery has been encouraging the enhancement of biofuel production, efficiency, and yield. These innovative bioprocess engineering strategies create the potential for more optimized and sustainable drop-in biofuel production using yeast-based systems. The advancements show potential in increasing the development of renewable energy sources.

In-situ product recovery is an essential strategy that involves removing biofuels from the fermentation broth without interrupting the fermentation process. Fungal and bacterial membrane efflux pumps are used as a method of enhancing biofuel secretion in microbial cell factories. Through the interchanging of these pumps, the efficiency of biofuel secretion can be enhanced, leading to higher product titres and yields. Butanol is a key performance indicator as a fuel in many engines. Extensive research efforts have been established to replenish the acetone-butanol-ethanol

(ABE) fermentation process. In the context of the upstream phase, the design process and the optimization of the pretreatment methods are focused on maximizing the recovery of cellulose and hemicellulose (Amiri 2020). Through the optimization of carbohydrate resources, microbial fermentation can be improved creating higher yields of ABE. Different microbial systems and process integrations were explored to facilitate ABE production, with a particular emphasis on in-situ butanol removal. In-situ product recovery techniques enable the ability to continually remove butanol from the fermentation broth. This further reduces the product toxicity, enhancing productivity.

Within the context of acetone-butanol-ethanol (ABE) fermentation, the adsorption of n-butanol uses various adsorbents, such as activated carbon Norit ROW 0.8, zeolite CBV901, and polymeric resins Dowex Optipore L-493 and SD-2 (Xue et al. 2016). Adsorbents, such as activated carbon, offer selective adsorption of n-butanol, reducing the risk of its effects on yeast cells and leading to higher yields and productivity. The research study describes the performance of activated carbon in n-butanol production resulting in 21.9 g/L of serum-bottle fermentation with in-situ adsorption. With the use of a fermenter, the combination of free-cell and immobilized-cell fermentation with adsorption further results in a higher butanol production. The thermal desorption of the adsorbent presented the recovery of a highly concentrated butanol solution. These findings highlight the efficiency and compatibility of in-situ product recovery processes, with the use of carbon. This can be integrated into ABE fermentation processes in yeast systems. Through the adoption of these bioprocess engineering strategies, the production of drop-in biofuels using yeast systems can be further enhanced.

One study investigated the potential to convert spent mushroom substrate (SMS), a by-product from industrial mushroom production, into biobutanol using bioprocess engineering strategies and integrating an in-situ butanol removal strategy, while focusing on yeast systems. To enhance the availability of fermentable sugars from SMS, a modified organosolv pretreatment method was employed under mild conditions without the use of an acid catalyst. From this pretreatment, the lignin content of SMS had reduced, and the sugar yield improved. In using organosolv-treated SMS enzymatic broth in batch fermentation experiments, the addition of biodiesel extractant to the batch fermentation experiment improved butanol production compared to the control batch. Higher butanol yields were denoted through the integration of in-situ butanol removal using biodiesel as an extractant. The coherence of yeast systems, such as *S. cerevisiae*, with biodiesel extractant, was also emphasized in this study; however, there was no prominent toxicity to the yeast growth. By integrating strategies such as fed-batch fermentation with sugar feeding, the study attained a maximum butanol production of 30.21 g/L (Zhu et al. 2016). From this, butanol removal for drop-in biofuel production can be made sustainable and economically feasible tactics for biobutanol production from SMS. This research assists with the innovations of bioprocess engineering strategies for drop-in biofuel production by using yeast-based bioprocess engineering strategies, specifically yeast systems, to optimize biofuel production from renewable feedstocks.

A research study reviews the significant applications and considerations surrounding the implementation of in-situ product recovery (ISPR) technologies in bioprocesses. It emphasizes the favourable utilization of ISPR when dealing with inhibitory or unstable products, resulting in notable enhancements in overall product titre, productivity, yield, and reduced process flows. The study focused on laboratory-scale investigations through the successful implementation at an industrial scale improving the development of robust and reliable unit operations with clear demonstrations of long-term stability and reduced energy consumption (Van Hecke et al. 2014). The integration of ISPR with conventional downstream processing techniques in hybrid processes emerges as a promising approach to achieving optimal performance. The review highlighted the roles of continuous online monitoring, process modeling, and the synergistic integration of metabolic engineering to further enhance the effectiveness of ISPR technologies. With regard to the future development of metabolic engineering and ISPR, they are anticipated to unlock new and exciting possibilities to produce biochemicals. There is a careful consideration of the challenges related to toxicity as this is imperative.

Further research and development of potential methods of optimizing these strategies, along with the exploration of new feedstocks and downstream processing techniques, will continue to advance the field of biofuel production through yeast systems. The integration of these approaches creates the opportunity to expand the production of sustainable and efficient drop-in biofuels which aid the transition to a more sustainable energy future.

5.5 Production Enhancement of Drop-in Biofuels Using Metabolic Engineering Strategies

The titre enhancement has been reported for three major types of drop-in biofuels which includes short-chain, medium-chain, and long-chain drop-in biofuels using metabolic engineering strategies. This section discusses these strategies and the production enhancements achieved.

5.5.1 Metabolic Engineering for Short-Chain (C_2–C_6) Drop-in Biofuels

Propanol and isopropanol can be used as an additive or surrogates for high-octane gasoline (Nishimura et al. 2018; Rassadin et al. 2006). The metabolic engineering strategies for enhancing the production of propanol and isopropanol have been listed in Table 5.3.

Butanol is a well-known example of drop-in biofuel and the best alternative to gasoline substitutes (Schadeweg and Boles 2016). Butanol was produced in

Table 5.3 Metabolic engineering strategies used in different species of yeast for the enhancement of short-chain biofuels

Compounds	Yeast	Metabolic modifications	Titre	References
Propanol	*S. cerevisiae*	Overexpression of the following genes—KDC, ADH, cimA, leuC, leuD, tdcB, thrA, thrB, thrC in combination with knocking out of gly1 gene	179 mg/L	Nishimura et al. 2018
Isopropanol	*Candida utilis*	Overexpression of the following genes—cftAB, adc, sadh, ACS2, and ERG10	27 g/L	Tamakawa et al. 2013a, 2013b
Butanol	*S. cerevisiae*	Overexpression of the following genes—ERG10, hbd, crt, adhe2, bcd, ACS2, and ALD6 in combination with knocking out of adh1 gene	2.4 g/L	Swidah et al. 2018
		Overexpression of the following genes—ADH7, LEU5, LEU4s, ARO10, NFS1, LEU2, LEU1, CimA in combination with knocking out of the adh1 gene	835 mg/L	Shi et al. 2016
Isobutanol	*S. cerevisiae*	Overexpression of the following genes—KDC, ILV5, ILV3, ILV2, ADH along with knocking out of the following genes—ald6, adh1, gdp2, and gdp1	2.09 g/L	Wess et al. 2019
Isoamyl alcohol	*S. cerevisiae*	Overexpression of the following genes—ILV5, ILV3, ILV2, LEU1, LEU9, LEU2, OAC1, ADH7, and ARO10	561 mg/L	Yuan et al. 2017
Isoprene	*S. cerevisiae*	Overexpression of the following genes—ISPSM4, GAL4, ACS2, and ERG10 along with knocking out the following genes—Ty4, galL80, gal7, gal1, and GAL 10	3.7 g/L	Lv et al. 2016
Hexanal	*Y. lipolytica*	Overexpressing HPL gene	600 mg/L	Santiago-Gómez et al. 2009

S. cerevisiae after a few metabolic engineering modifications which includes overexpression of the endogenous thiolase gene (ERG10), expression of *hbd, crt, adhe2,* and *bcd* genes isolated from *Clostridium beijerinckii,* the yield obtained was about 300 mg/L (Swidah et al. 2018). An endogenous pathway was found in the *S. cerevisiae* that produced 835 mg/L n-butanol in a shake-flask culture (Shi et al. 2016). Overexpression of ALD6 and ACS2 genes in combination with the deletion of the ADH1 gene leads to an enhanced yield of butanol of about 2.4 g/L in *S. cerevisiae* (Swidah et al. 2018). Other short-chain drop-in biofuels are isobutanol, isoamyl alcohol, hexanal, and isoprene. Detailed metabolic engineering strategies involved in the enhancement of yield of these drop-in biofuels have been given in Table 5.3.

5.5.2 *Metabolic Engineering for Medium-Chain (C7–C14) Fuels*

1-octanol can be used as an intermediate to produce biodiesels and jet fuels (Li et al. 2018). A study by Henritzi et al. (2018) produced 49.5 mg/L 1-octanol using the *S. cerevisiae* strain by using metabolic engineering modification (Table 5.4) which was cultivated in a YPD medium where glucose was used as a carbon source. The study modified the chain length by mutating fatty acid synthase expression, resulting in the expression of phosphopantetheinyl transferase (SFP), carboxylic acid reductase (CAR), and aldehyde reductase (AHR) in the strain.

Table 5.4 Metabolic engineering strategies used in different species of yeast for the enhancement of medium-chain biofuels

Compounds	Yeast	Metabolic modifications	Yield/Titre	References
Octanol	*S. cerevisiae*	Overexpression of the following genes FAS2, FAS^{RK}, BsSfp, EcAhr, and MmCAR in combination with knocking out of the following genes faa2, fas2, and fas1	49.5 mg/L	Henritzi et al. 2018
Decanol	*Y. lipolytica*	Overexpression of AtFAR and CpFAT genes along with knock-in	550 mg/L	Rutter and Rao 2016
Limonene	*S. cerevisiae* *S. cerevisiae*	Overexpression of the following genes—ClLimS (H570Y), SlNPPS1, HMG2 (K6R), ERG20 (N127W) along with the introduction of ERG9 and ERG20 genes	166 mg/L	Ignea et al. 2019
		Overexpression of IDI1, SlNDPS1, tHMG1, and CltLS2 genes P_{ERG20}:: P_{HXT1} along with knocking out of gal80 gene	917.7 mg/L	Cheng et al. 2019a, 2019b
	Y. lipolytica *Y. lipolytica*	Overexpression of HMG1, tNDPS1, tLS, ERG12	23.56 mg/L	Cao et al. 2016
	Y. lipolytica *Y. lipolytica*	Overexpression of HMG1, tNDPS1, tLS, and ERG12 along with a change in carbon source as citrate and glucose	165.3 mg/L	Cheng et al. 2019a, 2019b
D-limonene		Overexpression of dLS and HMGR genes, along with knocking out the KU70 gene	11.70 mg/L	Pang et al. 2019
L-limonene		Overexpression of lLS and HMGR genes, along with knocking out the KU70 gene	11.08 mg/L	

1-decanol can also be used as a potential precursor for several applications which includes drop-in biofuel, lubricants, solvents, anti-foaming agents, and surfactants (Nanthagopal et al. 2019). But these are produced in a low titre within the yeast cell, so to enhance the production of 1-decanol, *Y. lipolytica* was metabolically engineered to express heterologous fatty-acyl-CoA reductase and deleted peroxisome assembly factor Pex10. The yeast produced 500 mg/L of 1-decanol in shake-flask cultivation (Rutter and Rao 2016).

Limonene, a significant monoterpene has been incorporated into aviation fuels and has undergone a trial flight experiment. A study involved genetic modification of *S. cerevisiae* to enhance the production of limonene up to 76 mg/L. This was achieved by upregulating the cascade of geranyl pyrophosphate (GPP), a precursor to monoterpene, through the downregulation of cytoplasmic farnesyl pyrophosphate synthase using an ergosterol-responsive transcription modulator circuit (Peng et al. 2018). Another pathway was developed using neryl diphosphate (NPP) as a precursor (alternative) to produce orthogonal monoterpenoids. The optimized orthogonal pathway comprising SlNPPS1 and ClLimS variant (H570Y) was further integrated with dynamic regulation of orthogonal pathways to facilitate the absorption of alternative substrates. The *S. cerevisiae* strain yielded a limonene production of 166 mg/L (Ignea et al. 2019). A slight modification in the orthogonal pathway was done by Cheng et al. (2019a, 2019b) in which the expression of limonene synthase (LS) and neryl diphosphate synthase 1 (NDPS1) converts IPP and DMAPP to limonene in *S. cerevisiae.* This modified pathway was combined with the overexpression IDI1 and tHMG1 and the downregulation of ERG20, producing a limonene yield of 917.7 mg/L.

5.5.3 *Metabolic Engineering for Long-Chain (More than C15) Fuels*

Fatty alcohols (FAs) are the best example of long-chain biofuels and are also known for their broad applications such as plasticizers, detergents, and emollients. In a study, a well-established system of *S. cerevisiae* was used to produce these FA (Table 5.5), the endogenous transcription factor expression was manipulated and a few genes (acetyl-CoA carboxylase (ACC), FAR, and ACL) were overexpressed, about 1.1 g/L hexadecanol was produced in the fed—*Rhodosporidium toruloides* fermentation (Feng et al. 2015). Another study by Guo et al. (2016) used the same yeast system with a few modifications, this research group introduced the fungal xylose utilization pathway and produced 1.2 g/L hexadecanol in the same fed-batch fermentation. The compartmentalization of the FA synthesis pathway in the peroxisome of *S. cerevisiae* resulted in the successful generation of 1.3 g/L of fatty alcohols, wherein hexadecanol constituted the predominant product proportion which exceeds 60% (Sheng et al. 2016). Metabolically engineering the FA elongation system along with the expression of FA reductase in the *S. cerevisiae* strain

Table 5.5 Metabolic engineering strategies used in different species of yeast for the enhancement of long-chain biofuels

Compounds	Yeast	Metabolic modifications	Yield/ Titre	References
Fatty alcohols	*S. cerevisiae* *S. cerevisiae* *S. cerevisiae* *S. cerevisiae*	Overexpression of ACC1, YlACL, and TαFAR gene along with knockout of ino and rpd3 genes	330.2 mg/ L in flask culture and 1.1 g/ L fed-batch cultivation	Feng et al. 2015
		Overexpression of TαFAR, ACC1, YlACL, CsXR, CtXDH, and PpXKS genes along with knocking out of ino1 and rpd3 genes	1.2 g/L	Guo et al. 2016
		Overexpression of TαFAR, PEX7, and ACC1 genes	1.3 g/L	Sheng et al. 2016
		Overexpression of CAR, ADH5, FaCoAR, and npgA along with gal80, adh6, hfd1, faa1, pox1, and faa4 genes	1.5 g/L	Zhou et al. 2016b
	Y. lipolytica *Y. lipolytica* *Y. lipolytica*	Overexpression of the MaFAR gene	167 mg/L	Wang et al. 2016a, 2016b, 2016c
		Overexpression of MaFAR and EcfadD genes	2.15 g/L	Xu et al. 2016
		Overexpression of DGA1 and MhFAR genes	5.8 g/L	Cordova et al. 2020
	Lipomyces starkeyi *Lipomyces starkeyi*	Overexpression of the MαFAR gene	770 mg/L	McNeil and Stuart. 2018
		Overexpression of the Mfar1 gene	1.7 g/L	Gao et al. 2018
	Rhodotorula toruloides	Overexpression of the MαFAR gene	A yield higher than 8 g/L	Fillet et al. 2015
Fatty acid esters	*Y. lipolytica*	Overexpression of WS and ACS2 genes along with deletion of pox10 gene	1.18 g/L	Gao et al. 2018
	S. cerevisiae	Overexpression of ADH7, ARO10, ILV3, ILV5, ILV2, and WS genes along with deletion of rpd3 and opi1 genes	230 mg/L	Teo et al. 2015
	Pichia pastoris	Overexpression of ADH2, ARO1,0 and WS genes	169 mg/L	Tao et al. 2015

(continued)

Table 5.5 (continued)

Compounds	Yeast	Metabolic modifications	Yield/ Titre	References
Fatty alka (e)nes	*S. cerevisiae* *S. cerevisiae*	Overexpression of PEX34, npgA, SeFD, SeFNR, SeADO, and MmCAR along with deletion of pox1, hfd1, pex31, pex32, and pex34 genes	3.55 mg/L	Zhou et al. 2016a
		Overexpression of camA, camB, PfundB, and FATP1 genes along with deletion of gal80, pox1, faa1, and faa4	35.3 mg/L	Zhou et al. 2018
	Y. lipolytica *Y. lipolytica*	Overexpression of pm MmCAR, PmADO, and BsSfp genes	23.3 mg/L	Xu et al. 2016
		Overexpression of TGL4, DGA2, CvFAP, and KlTGL3 genes along with deletion of *faa1* and *mfe1* genes	58.7 mg/L	Bruder et al. 2019
Farnesene	*S. cerevisiae* *S. cerevisiae*	Overexpression of FarnSyn, tHMG1, GDH2, ERG20 along with deletion of dpp1, lpp1, gdh1 genes	170 mg/L	Tippmann et al. 2016
		Overexpression of ADA, xPK, PTA, HMGr, PDR3, GAL4, ACS2, DzeutE, ERG20, tHMGR, ERG8, ERG9, ERG10, ERG12, ERG13, BjHMGS, CaTHL, ZmPDC, AaFS, IDI1, MVD1 genes along with deletion of ald4, ald6, rhr2, erg13, ygr250, adh1, bdh1, gal80, hxt3, dit1, ho, nat80, gal2, and adh5 genes	A yield greater than 130 g/L	Meadows et al. 2016
	Y. lipolytica	Overexpression of MdFS, IDI, ERG20, and Sc-tHMG genes	260 mg/L	Yang et al. 2016

(continued)

Table 5.5 (continued)

Compounds	Yeast	Metabolic modifications	Yield/ Titre	References
Bisabolene	*S. cerevisiae*	Overexpression of tHMG1, erg20, AgBIS, ups2–1, ERG2 genes	994 mg/L	Peralta-Yahya et al. 2011
	R. toruloides	Overexpression of the BIS gene	680 mg/L	Yaegashi et al. 2017

produced 83.5 g/L docosanol (Yu et al. 2017). With the deletion of genes such as POX1, FAA4, FAA1, HFD1, and ADH6 along with overexpression of FA synthesis genes, the *S. cerevisiae* produced 1.5 g/L fatty alcohol (Zhou et al. 2016a). The production of FAs can further be increased by 4.5-fold with the introduction of the human FATP1 transporter gene into the *S. cerevisiae* strain (Hu et al. 2018a, 2018b). Furthermore, the synthesis of FA mixtures (comprising odd and even chain alcohols) in *S. cerevisiae* was achieved through a synthetic pathway that employed cytosolic thioesterase, α-dioxygenase, and native ADHs, resulting in a yield of 20.3 mg/L (Jin et al. 2016). Another well-known yeast strain that is oleaginous in nature is *Yarrowia lipolytica* which has also been reported to produce FAs. The deletion of the DGA1/FAO1 gene along with the co-expression of the *Tyto alba* FAR gene produced 660.8 mg/L of hexadecanol (Wang et al. 2016a, 2016b, 2016c). Without any co-expression, just expressing the FAR gene from *Marinobactor aquaeolei* produced a yield of 167 mg/L. A study by Rigouin et al. (2018) overexpressed the EgDGAT1–1 and αfas-I1220Wgenes in *Y. lipolytica* with a defective β-oxidation produced 1.25 g/L C14 FAs. An increased yield of 2.15 g/L of the FA was produced in the *Y. lipolytica* strain when the co-expression of the FAR gene with fatty acyl-CoA synthetase (FadD) was done, which in turn activated the FFAs by the addition of coenzyme A (CoA) (Xu et al. 2016). The expression of MhFAR gene in *Y. lipolytica* produced a yield of about 5.8 g/L fatty alcohols. Another oleaginous yeast known as *Lipomyces starkeyi* produced a fatty alcohol yield of about 770 mg/L when the FAR gene was expressed within it (Cordova et al. 2020). A yield of 8 g/L FAs (specifically C16 and C18) was produced in *Rhodosporidium toruloides* when the FAR gene (codon-optimized) was introduced with the system (Fillet et al. 2015).

Fatty alka(e)nes are also recently developing into drop-in biofuels. Optimization of the enzymes involved in the alkane biosynthetic pathway; the yield of alkane was 0.8 mg/L in *S. cerevisiae* (Zhou et al. 2016b). The yield was further enhanced up to 3.55 mg/L when the pathways to produce peroxisomes (peroxisomal biogenesis) were targeted and engineered (Zhou et al. 2016a). Expression of CAR (from *Mycobacterium marinum*) gene which is an acyl carrier protein activation module, in combination with Sfp gene (isolated from *Bacillus subtilis*) and aldehyde deformylating oxygenase (ADO) (from *Prochlorococcus marinus*) increased the yield of alkane up to 23.3 mg/L in *Y. lipolytica* (Xu et al. 2016). Another study by Bruder et al. (2019) introduced a light-driven oxidase also known as FA photodecarboxylase (CvFAP) (from *Chlorella variabilis*) into *Y. lipolytica* which produced 58.7 mg/L of both alkane and alkene. A study by Li et al. (2020) introduced multiple copies of CvFAP gene copies into the genome of *Y. lipolytica* along with knocking out genes that were responsible for the alkane assimilation. This resulted in increased production of alkenes which was about 35 mg/mL. Once the exposure to light, C/N ratio, and substrate concentration (glucose) were optimized, the *Y. lipolytica* produced alkene of about 1.47 g/L in a fed-batch fermentation after a period of 17 days. The alkenes were produced in *S. cerevisiae* via fatty acid decarboxylation reaction by combinational engineering strategy and produced 3.7 mg/L of 1-alkenes (Chen et al. 2015). A yield of 35.3 mg/L 1-alkenes was produced from *S. cerevisiae* in which incorporation of a few genes was done which were desaturase-like enzyme UndB isolated from *Pseudomonas fluorescens*, long-chain fatty acid transporter protein 1 (FATP1) isolated from *Homo sapiens,* and ferredoxin-ferredoxin reductase (Fd/Fnr) isolated from *E. coli* together modified the fatty acid metabolism, wherein the increased level of alkene was observed (Zhou et al. 2018).

Farnesene is a building block of farnesane, an acyclic sesquiterpene that possesses high octane/cetane numbers, favourable energy content, and tends to be suitable for cold-weather applications. It is one of the simplest sesquiterpenes which serves as a potentially viable biofuel alternative to substitute jet fuel (George et al. 2015; Mohd Azhar et al. 2017). In addition, Meadows et al. (2016) reported that they achieved a farnesene production level of 130 g/L in an industrial bioreactor. This was accomplished by introducing four non-native metabolic enzymes, namely ADA, NADH-HMGr, xPK, and PTA, into an industrial-grade *S. cerevisiae* strain that is typically used for artemisinin production. In their study, Yang et al. (2016) utilized a codon-optimized α-farnesene synthase (α-FS) gene derived from an apple and expressed it in *Y. lipolytica.* The outcome of this experiment was the enhanced production of α-farnesene, which was about 2.75 mg/L. The α-FS was fused to the ERG20 gene and co-overexpressed with the tHMG1 gene isolated from *S. cerevisiae* and native IDI, resulting in α-farnesene titre of 57.08 mg/L. A scale-up experiment was conducted using a 1.5 L fermenter and the optimized *Y. lipolytica* culture, resulting in a production of 259.98 mg/L α-farnesene. Another study by Liu et al. (2020) achieved a yield of 34.5 mg/L α-farnesene when overexpression of a few genes including OptFSLERG20, Sc-tHMG1, and IDI is under the control of a hybrid promoter UAS4B-TEF in *Y. lipolytica.* Also, when two native genes—HMGR and ERG19 were overexpressed, the α-farnesene yield reached up to 90 mg/L. Using glycerol as a carbon source in a fed-batch cultivation experiment, the yield of α-farnesene reached 2.5 g/L. The highest yield of α-farnesene was about 25.55 g/L in the *Y. lipolytica* strain reported by Liu et al. (2019) after a period of 288 h in a fed-batch cultivation system using an α-farnesene overproducing strain modified with the integration of NHEJ-mediated modular strategy. This strategy involves the co-overexpression of Nacetyl-CoA acetyltransferase and FSERG20 from *E. coli,* NADH-dependent HMGR gene isolated from *Bordetella petrii,* ERG8, ERG 12, ERG19, and GPPS genes. Another study by Shi et al. (2021) produced 22.80 g/L of β-farnesene in a fed-batch cultivation system using the *Y. lipolytica* strain which was overexpressed for the AaBFS gene. The metabolic engineering

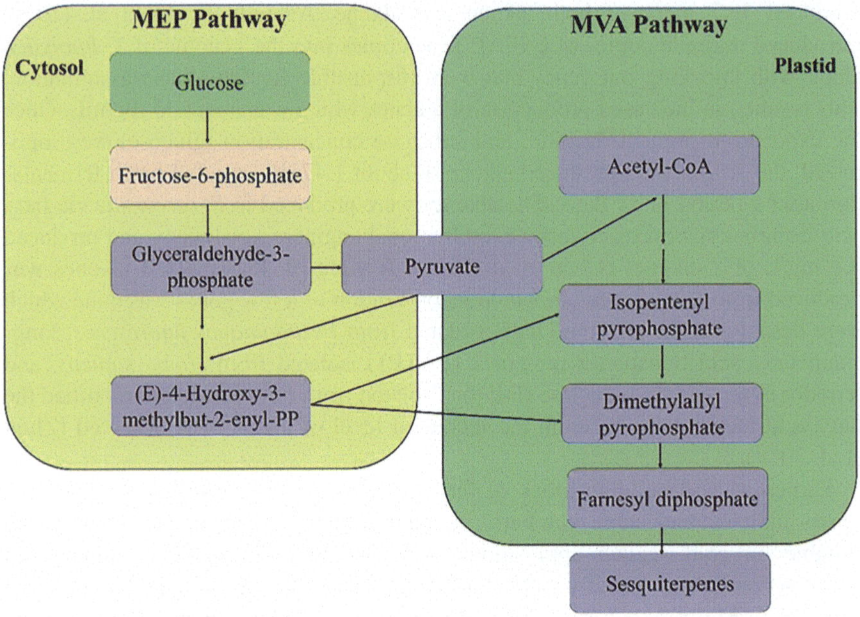

Fig. 5.4 Schematic representation of biosynthesis of sesquiterpenes from MEP and MVA pathways occurring in cytosol and plastid, respectively

strategy followed was reducing the auxotrophic requirement for leucine and augmenting the copy number of AaBFS, in conjunction with the interference of the β-carotene pathway and the elimination of DGA1 and DGA2.

Bisabolene is a precursor of the novel alternative to the existing D2 diesel fuel known as bisabolane. It belongs to the class of sesquiterpenes (terpene family), and two pathways are majorly involved in its production, which are the mevalonate (MVA) dependent pathway and the 2-C-methyl-D-erthrito-1,4-phosphate (MEP) pathway that occur in cytosol and plastid, respectively (Liu et al. 2022). Though the pathways occur in two different compartments, these are connected via intermediates and produce a variety of sesquiterpenes (Fig. 5.4). In general, bisabolene is produced in a very minimal amount in the yeast cells, but the yield of bisabolene can further be enhanced using different pretreatment and metabolic engineering strategies.

The yield of bisabolene produced by *Rhodosporidium toruloides* (a non-conventional yeast) was about 680 mg/L when the lignocellulosic biomass was alkaline pre-treated, and the culture was grown in a fed-batch (high-gravity) bioreactor (Yaegashi et al. 2017). Other metabolic engineering strategies include (1) Overproduction of the farnesyl diphosphate (FPP) which in turn converts into bisabolene, and using chemical hydrogenation the bisabolene was converted into bisabolane (with superior fuel properties) in *Saccharomyces cerevisiae* (Peralta-Yahya et al. 2011) (2) Replacing the Amorpha-4,11-diene synthase (ADS) gene

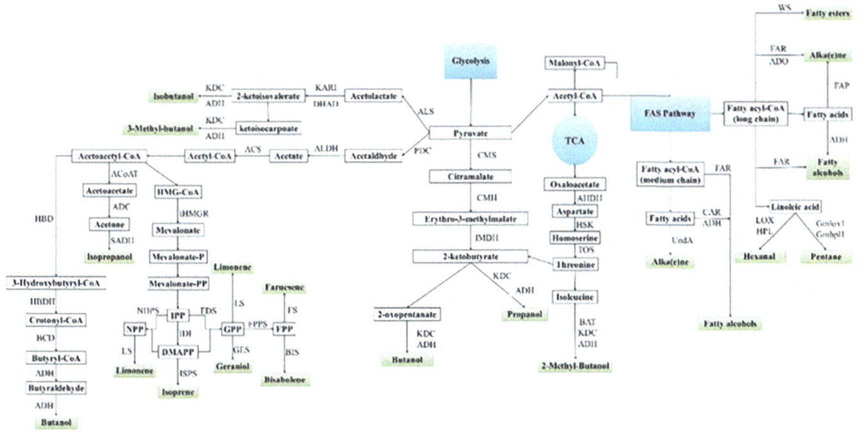

Fig. 5.5 Illustration of the production of drop-in biofuels and their precursors using metabolic engineering strategies in different pathways including glycolysis, TCA, and FAS in yeast

with the bisabolene synthase (BIS) gene in the *Saccharomyces cerevisiae* EPY224 which increased the bisabolene yield by 375 mg/L (Bröker et al. 2020). Another study by Ignea et al. (2012) altered the genome of *Saccharomyces cerevisiae* CEN. PK2.1C by tandem deletion of three genes which are UBC7, SSM4/DOA10, PHO86 and increased the yield by 11 folds. (3) A combination of deletion of the ROX1 gene and a few modifications in the mevalonate pathway increased the yield of bisabolene drastically up to 5.2 g/L (Özaydın et al. 2013). (4) A combination of overexpression of endogenous genes involved in the mevalonate pathway and introducing transporters that possess heterologous multidrug efflux which in turn increases the production of bisabolene in *Yarrowia lipolytica* (Zhao et al. 2021). The production of various drop-in biofuels with different metabolic engineering strategies have been mentioned in Fig. 5.5.

5.5.4 Improved Tolerance of Yeast Towards Toxicity of Inhibitors in Lignocellulosic Biomass

Increasing the yield of the drop-in biofuels within the yeast system has a few drawbacks which includes toxicity towards the yeast strain. There are only a few studies reported to enhance the tolerance of the yeast strain against the toxicity of the components in the lignocellulosic biomasses during the production of drop-in biofuels (Brennan et al. 2015). Phenolic aldehydes are reported to be potential inhibitors for yeast, Zhang and Bao (2022) experimented with four strains of oleaginous yeasts including *Rhodosporidium toruloides, Yarrowia lipolytica Trichosporon cutaneum, and Rhodotorula glutinis,* out of which *Trichosporon cutaneum* showed high tolerance towards the phenolic aldehyde present in the lignocellulosic biomass and produced 40.87 ± 1.85 g/L of lipid, suggesting that the strain of yeast can be used in the production of drop-in biofuel. A study by Brennan et al. (2015) reported an *S. cerevisiae* strain that was altered using a metabolic engineering strategy (truncation) for a protein named tTcb3p$^{1-989}$, the reported protein was tolerant to β-pinene, myrcene, and a bio-jet fuel blend AMJ-700 of the composition cymene 10%, 50% limonene, and 40% farnesene. Another study by Li et al. (2021) identified eight genes that were overexpressed when *Y. lipolytica* was exposed to external limonene, the genes identified were dLS, FTR, PHC, YALI0F19492p, WARS, MUC19, BAMA, and NDK, out of which overexpression of YALI0F19492p gene increased the limonene titre by 8-folds as well as increased the tolerance of the yeast towards limonene. The *S. cerevisiae* strain was metabolically engineered for two transcription factors, Pdr1p and Pdr3p which were site-mutated and found to have a high tolerance towards the alkane biofuels (Ling et al. 2015). An oleaginous yeast strain named *R. toruloides* has been found to possess a good tolerance level against most of the inhibitory compounds present in its substrate (Park et al. 2018). The *T. cutaneum* ACCC 20271 strains have been reported to have a high tolerance against the 4-hydroxybenzaldehyde (a substrate for lipid accumulation) (Hu et al. 2018a, 2018b) and an inhibitor degradation pathway has also been demonstrated by Wang et al. (2016a, 2016b, 2016c) to provide insight into the inhibitor degradation within the yeast system. The *S. cerevisiae* strain has been engineered for aldehyde reduction, increased extracellular potassium levels and pH to enhance the tolerance of the strain against a broad variety of inhibitors in the feedstock (Lam et al. 2021). The tolerance of *S. cerevisiae* against branched-chain alcohols has been achieved by the deletion of the GLN3 gene. The deletion of the gene not only increases the tolerance but also enhances the isobutanol production by 4.9 folds (Kuroda et al. 2019). Using a metabolic engineering strategy, the IPP pathway has been constructed in *S. cerevisiae* which not only increased the yield of isoprenol but also enhanced the tolerance of the yeast towards the high titre of the drop-in biofuel (Kim et al. 2021). Another study by Kirke (2019) has shown that the yeast *R. toruloides* which has been constructed with lipid droplet-associated GFP-tagged protein Ldp1-GFP has shown a greater tolerance to the lignocellulosic biomass (sugarcane bagasse) and furfural. A novel approach has been employed which was known as artificial simulation bio-evolution to improve the production as well as to enhance the tolerance of butanol production in *Clostridium acetobutylicum* T64 yeast. The approach is based on evolutionary dynamics and the natural selection of the best-yielding and tolerant strain (Liu et al. 2013a, 2013b). Mutations in the two elongation factors known as elf2 and elf2B have shown an improved tolerance and production of the medium-chain alcohols in the *S. cerevisiae* strain (Davis López et al. 2018). There have been only a few studies that have explored the tolerance of the yeast strain towards the toxicity of the lignocellulosic biomass so far, many insights into the tolerant mechanism must be studied in the future.

5.6 Conclusion

Significant interest has been expressed in using yeasts as effective production platforms for drop-in biofuels. Several strains have been utilized to produce drop-in biofuels including *Saccharomyces cerevisiae*, *Yarrowia lipolytica*, *Rhodosporidium toruloides*, and *Candida utilis*. Every strain has its own metabolic flexibility towards feedstock and multiple feeds like lignocellulosic, food waste, and lipids/fats are consumed to produce the drop-in biofuel. Bioprocessing using yeast systems and strategic methods plays a vital role in the yield and the kind of products generated. There are various methods for production of drop-in biofuels like Batch Fermentation, Fed- Batch Fermentation, and Continuous Cultivation. Most of the production currently employs batch and fed-batch cultivation, while fed-batch promotes a higher titre and productivity. Another emerging cultivation method is continuous cultivation. Continuous cultivation, compared to batch and fed-batch cultivation, enables a steady-state operation through a continually added medium. This strategy permits steady growth and production, which raises total productivity and may boost yields. The integration of continuous cultivation with in-situ product recovery enhances the possibility for more efficient and long-term drop-in biofuel generation using yeast-based systems.

Conventionally, yeasts were primarily employed for producing ethanol or to act as fatty crudes for biodiesel. Recent developments in metabolic engineering and synthetic biology have rendered the ability to easily design and build complex yeast systems. Yeasts can produce drop-in biofuels with various properties like high energy density and low recovery costs; however, genetically modifying them for desirable biofuel production along with improved characteristics is growing rapidly. In addition to this, the toxicity of inhibitors in the feedstock could prevent the yeasts from reaching their maximum potential. A limited study has been done in this field. An insight into the tolerant mechanism needs to be analysed in the future. Utilizing metabolic engineering strategies can enhance the production by the yeasts, which provides a potential substitute for other species like bacteria and viruses. Thus, yeast can also serve as a potential resource for drop-in biofuel production.

References

Ambaye TG, Vaccari M, Bonilla-Petriciolet A, Prasad S, van Hullebusch ED, Rtimi S (2021) Emerging technologies for biofuel production: A critical review on recent progress, challenges and perspectives. In. J Environ Manag 290:112627. https://doi.org/10.1016/j.jenvman.2021.112627

Amiri H (2020) Recent innovations for reviving the ABE fermentation for production of butanol as a drop-in liquid biofuel. Biofuel Res J 7(4):1256–1266. https://doi.org/10.18331/BRJ2020.7.4.4

Angerbauer C, Siebenhofer M, Mittelbach M, Guebitz GM (2008) Conversion of sewage sludge into lipids by Lipomyces starkeyi for biodiesel production. Bioresour Technol 99(8): 3051–3056. https://doi.org/10.1016/j.biortech.2007.06.045

Bhuiya MMK, Rasul MG, Khan MMK, Ashwath N, Azad AK (2016) Prospects of 2nd generation biodiesel as a sustainable fuel – part: 1 selection of feedstocks, oil extraction techniques and conversion technologies. In: *Renewable and sustainable energy reviews*, vol 55. Elsevier Ltd., pp 1109–1128. https://doi.org/10.1016/j.rser.2015.04.163

Blazeck J et al (2013) Heterologous production of pentane in the oleaginous yeast Yarrowia lipolytica. *J Biotechnol* 165:3–4. https://doi.org/10.1016/j.jbiotec.2013.04.003

Bolmanis E, Dubencovs K, Suleiko A, Vanags J (2023) Model predictive control—A stand out among competitors for fed-batch fermentation improvement. Fermentation 9(3):206. https://doi.org/10.3390/fermentation9030206

Boze H, Moutin G, Galzy P (1992) Production of food and fodder yeasts. Crit Rev Biotechnol 12(2):65–86

Brat D, Weber C, Lorenzen W, Bode HB, Boles E (2012) Cytosolic re-localization and optimization of valine synthesis and catabolism enables increased isobutanol production with the yeast Saccharomyces cerevisiae. Biotechnology for biofuels 5:1–16

Brennan TCR, Williams TC, Schulz BL, Palfreyman RW, Krömer JO, Nielsen LK (2015) Evolutionary engineering improves tolerance for replacement jet fuels in Saccharomyces cerevisiae. Appl Environ Microbiol 81(10):3316–3325. https://doi.org/10.1128/AEM.04144-14

Bröker JN, Müller B, Prüfer D, Gronover CS (2020) Combinatorial metabolic engineering in saccharomyces cerevisiae for the enhanced production of the fpp-derived sesquiterpene germacrene. Bioengineering 7(4):1–13. https://doi.org/10.3390/bioengineering7040135

Bruder S, Moldenhauer EJ, Lemke RD, Ledesma-Amaro R, Kabisch J (2019) Drop-in biofuel production using fatty acid photodecarboxylase from Chlorella variabilis in the oleaginous yeast Yarrowia lipolytica. Biotechnol Biofuels 12(1):1–13. https://doi.org/10.1186/s13068-019-1542-4

Buijs NA, Siewers V, Nielsen J (2013) Advanced biofuel production by the yeast saccharomyces cerevisiae. Curr Opin Chem Biol 17(3):480–488. https://doi.org/10.1016/j.cbpa.2013.03.036

Bumrungtham P, Promdonkoy P, Prabmark K, Bunterngsook B, Boonyapakron K, Tanapongpipat S, Champreda V, Runguphan W (2022) Engineered Production of Isobutanol from Sugarcane Trash Hydrolysates in Pichia pastoris. Journal of Fungi 8(8). https://doi.org/10.3390/jof8080767

Burke DG, Cotter PD, Ross RP, Hill C (2013) 14 – microbial production of bacteriocins for use in foods. In: McNeil B, Archer D, Giavasis I, Harvey L (eds) Microbial production of food ingredients, enzymes and nutraceuticals. Woodhead Publishing, pp 353–384. https://doi.org/10.1533/9780857093547.2.353

Cao X, Lv YB, Chen J, Imanaka T, Wei LJ, Hua Q (2016) Metabolic engineering of oleaginous yeast Yarrowia lipolytica for limonene overproduction. Biotechnol Biofuels 9(1):1–11. https://doi.org/10.1186/s13068-016-0626-7

Carmona-Cabello M, García IL, Papadaki A, Tsouko E, Koutinas A, Dorado MP (2021) Biodiesel production using microbial lipids derived from food waste discarded by catering services. Bioresour Technol 323:124597. https://doi.org/10.1016/j.biortech.2020.124597

Caspeta L, Chen Y, Ghiaci P, Feizi A, Buskov S, Hallström BM, Petranovic D, Nielsen J (2014) Altered sterol composition renders yeast thermotolerant. Science 346(6205):75–78. https://doi.org/10.1126/science.1258137

Chen X, Nielsen KF, Borodina I, Kielland-Brandt MC, Karhumaa K (2011) Increased isobutanol production in Saccharomyces cerevisiae by overexpression of genes in valine metabolism. http://www.biotechnologyforbiofuels.com/content/4/1/21

Chen B, Lee D-Y, Chang MW (2015) Combinatorial metabolic engineering of Saccharomyces cerevisiae for terminal alkene production. http://www.elsevier.com/open-access/userlicense/1.0/

Cheng S, Liu X, Jiang G, Wu J, Zhang JL, Lei D, Yuan YJ, Qiao J, Zhao GR (2019a) Orthogonal engineering of biosynthetic pathway for efficient production of limonene in Saccharomyces cerevisiae. ACS Synth Biol 8(5):968–975. https://doi.org/10.1021/acssynbio.9b00135

Cheng BQ, Wei LJ, Lv YB, Chen J, Hua Q (2019b) Elevating limonene production in oleaginous yeast Yarrowia lipolytica via genetic engineering of limonene biosynthesis pathway and optimization of medium composition. Biotechnol Bioprocess Eng 24(3):500–506. https://doi.org/10.1007/s12257-018-0497-9

Connor MR, Liao JC (2009) Microbial production of advanced transportation fuels in non-natural hosts. Curr Opin Biotechnol 20(3):307–315. https://doi.org/10.1016/j.copbio.2009.04.002

Cordova LT, Butler J, Alper HS (2020) Direct production of fatty alcohols from glucose using engineered strains of Yarrowia lipolytica. Metab Eng Commun 10:e00105. https://doi.org/10.1016/j.mec.2019.e00105

Coyle WT (2010) *Near-term challenges and implications for agriculture*. www.ers.usda.gov

Davis López SA, Griffith DA, Choi B, Cate JHD, Tullman-Ercek D (2018) Evolutionary engineering improves tolerance for medium-chain alcohols in Saccharomyces cerevisiae. Biotechnol Biofuels 11(1):1–15. https://doi.org/10.1186/s13068-018-1089-9

De Deken RH (1966) The Crabtree effect: A regulatory system in yeast. J gen Microbiol 44:149–156

de Jong B, Siewers V, Nielsen J (2012) Systems biology of yeast: enabling technology for development of cell factories for production of advanced biofuels. Curr Opin Biotechnol 23(4):624–630. https://doi.org/10.1016/j.copbio.2011.11.021

de Jong BW, Shi S, Valle-Rodríguez JO, Siewers V, Nielsen J (2015) Metabolic pathway engineering for fatty acid ethyl ester production in Saccharomyces cerevisiae using stable chromosomal integration. J Ind Microbiol Biotechnol 42(3):477–486. https://doi.org/10.1007/s10295-014-1540-2

Degfie TA, Mamo TT, Mekonnen YS (2019) Optimized biodiesel production from waste cooking oil (WCO) using calcium oxide (CaO) nano-catalyst. Sci Rep 9(1):18982. https://doi.org/10.1038/s41598-019-55403-4

Dickinson JR, Harrison SJ, Hewlins MJE (1998) An investigation of the metabolism of valine to isobutyl alcohol in Saccharomyces cerevisiae. J Biol Chem 273(40):25751–25756. https://doi.org/10.1074/jbc.273.40.25751

Domínguez A, Deive FJ, Angeles Sanromán M, Longo MA (2010) Biodegradation and utilization of waste cooking oil by Yarrowia lipolytica CECT 1240. Eur J Lipid Sci Technol 112(11):1200–1208. https://doi.org/10.1002/ejlt.201000049

Dürre P (2007) Biobutanol: an attractive biofuel. Biotechnol J 2(12):1525–1534. https://doi.org/10.1002/biot.200700168

Eden A, Van Nedervelde L, Drukker M, Benvenisty N, Debourg A (2001) Involvement of branched-chain amino acid aminotransferases in the production of fusel alcohols during fermentation in yeast. Appl Microbiol Biotechnol 55(3):296–300. https://doi.org/10.1007/s002530000506

Eksemplaar H, Onder M, Omstandighede G, Die V, Verwyder B, Nie W (1999) Universiteit Vrystaat

Feng X, Lian J, Zhao H (2015) Metabolic engineering of Saccharomyces cerevisiae to improve 1-hexadecanol production. Metab Eng 27:10–19. https://doi.org/10.1016/j.ymben.2014.10.001

Fillet S, Gibert J, Suárez B, Lara A, Ronchel C, Adrio JL (2015) Fatty alcohols production by oleaginous yeast. J Ind Microbiol Biotechnol 42(11):1463–1472. https://doi.org/10.1007/s10295-015-1674-x

Gao Q, Cao X, Huang YY, Yang JL, Chen J, Wei LJ, Hua Q (2018) Overproduction of fatty acid ethyl esters by the oleaginous yeast Yarrowia lipolytica through metabolic engineering and process optimization. ACS Synth Biol 7(5):1371–1380. https://doi.org/10.1021/acssynbio.7b00453

George KW, Alonso-Gutierrez J, Keasling JD, Lee TS (2015) Erratum to: isoprenoid drugs, biofuels, and chemicals—artemisinin, farnesene, and beyond [Adv Biochem Eng Biotechnol, 10.1007/10_2014_288]. In: *Advances in biochemical engineering/biotechnology*, vol 148. Springer Science and Business Media Deutschland GmbH, p 469. https://doi.org/10.1007/10_2015_310

Gientka I, Wirkowska-Wojdyła M, Ostrowska-Ligęza E, Janowicz M, Reczek L, Synowiec A, Błażejak S (2022) Enhancing red yeast biomass yield and lipid biosynthesis by using waste nitrogen source by glucose fed-batch at low temperature. Microorganisms 10(6):1253. https://doi.org/10.3390/microorganisms10061253

Guo W, Sheng J, Zhao H, Feng X (2016) Metabolic engineering of Saccharomyces cerevisiae to produce 1-hexadecanol from xylose. Microb Cell Factories 15(1):24. https://doi.org/10.1186/s12934-016-0423-9

Hazelwood LA, Daran JM, Van Maris AJ, Pronk JT, Dickinson JR (2008) The Ehrlich pathway for fusel alcohol production: a century of research on Saccharomyces cerevisiae metabolism. Appl Environ Microbiol 74(8):2259–2266. https://doi.org/10.1128/AEM.02625-07

Henritzi S, Fischer M, Grininger M, Oreb M, Boles E (2018) An engineered fatty acid synthase combined with a carboxylic acid reductase enables de novo production of 1-octanol in Saccharomyces cerevisiae. Biotechnol Biofuels 11(1):1–12. https://doi.org/10.1186/s13068-018-1149-1

Hill J, Nelson E, Tilman D, Polasky S, Tiffany D (2006) Environmental, economic, and energetic costs and benefits of biodiesel and ethanol biofuels. Proc Natl Acad Sci 103(30):11206–11210. https://doi.org/10.1073/pnas.0604600103

Hu M, Wang J, Gao Q, Bao J (2018a) Converting lignin derived phenolic aldehydes into microbial lipid by Trichosporon cutaneum. J Biotechnol 281:81–86. https://doi.org/10.1016/j.jbiotec.2018.06.341

Hu Y, Zhu Z, Nielsen J, Siewers V (2018b) Heterologous transporter expression for improved fatty alcohol secretion in yeast. Metab Eng 45:51–58. https://doi.org/10.1016/j.ymben.2017.11.008

Huang C, Zong M, Wu H, Liu QP (2009) Microbial oil production from rice straw hydrolysate by Trichosporon fermentans. Bioresour Technol 100(19):4535–4538. https://doi.org/10.1016/j.biortech.2009.04.022

Ichii T, Takehara S, Konno H, Ishida T, Sato H, Suzuki A, Yamazump K (1993) Development of a new commercial-scale airlift fermentor for rapid growth of yeast. *Journal of Fermentation and Bioengineering* 75(5):375–379

Ignea C, Trikka FA, Kourtzelis, Argiriou A, Kanellis AK, Kampranis SC, Makris AM (2012) Positive genetic interactors of HMG2 identify a new set of genetic perturbations for improving sesquiterpene production in Saccharomyces cerevisiae. http://www.microbialcellfactories.com/content/11/1/162

Ignea C, Raadam MH, Motawia MS, Makris AM, Vickers CE, Kampranis SC (2019) Orthogonal monoterpenoid biosynthesis in yeast constructed on an isomeric substrate. Nat Commun 10(1): 3799. https://doi.org/10.1038/s41467-019-11290-x

Ikushima S, Fujii T, Kobayashi O, Yoshida S, Yoshida A (2009) Genetic engineering of candida utilis yeast for efficient production of L-lactic acid. Biosci Biotechnol Biochem 73(8): 1818–1824. https://doi.org/10.1271/bbb.90186

Lu R, Cao L, Wang K, Ledesma-Amaro R, Ji XJ (2021) Engineering Yarrowia lipolytica to produce advanced biofuels: Current status and perspectives. Bioresour Technol 341:125877. https://doi.org/10.1016/j.biortech.2021.125877

Jiang W, Gu P, Zhang F (2018) Steps towards ‘drop-in’ biofuels: focusing on metabolic pathways. In Current Opinion in Biotechnology 53. https://doi.org/10.1016/j.copbio.2017.10.010

Jin C, Yao M, Liu H, Lee CFF, Ji J (2011) Progress in the production and application of n-butanol as a biofuel. Renew Sustain Energy Rev 15(8):4080–4106). Elsevier Ltd. https://doi.org/10.1016/j.rser.2011.06.001

Jin Z, Wong A, Foo JL, Ng J, Cao Y-X, Chang MW, Yuan Y-J (2016) Engineering Saccharomyces cerevisiae to produce odd chain-length fatty alcohols; engineering Saccharomyces cerevisiae to produce odd chain-length fatty alcohols. Biotechnol Bioeng 113:842–851. https://doi.org/10.1002/bit.25856/abstract

Karatzos S, McMillan JD, Saddler JN (2014) The potential and challenges of drop-in biofuels : a report by IEA Bioenergy Task 39

Katre G, Ajmera N, Zinjarde S, Ravikumar A (2017) Mutants of yarrowia lipolytica ncim 3589 grown on waste cooking oil as a biofactory for biodiesel production. Microb Cell Factories 16(1):1–14. https://doi.org/10.1186/s12934-017-0790-x

Katre G, Raskar S, Zinjarde S, Ravi Kumar V, Kulkarni BD, RaviKumar A (2018) Optimization of the in situ transesterification step for biodiesel production using biomass of Yarrowia lipolytica NCIM 3589 grown on waste cooking oil. Energy 142:944–952. https://doi.org/10.1016/j.energy.2017.10.082

Kim J, Baidoo EEK, Amer B, Mukhopadhyay A, Adams PD, Simmons BA, Lee TS (2021) Engineering Saccharomyces cerevisiae for isoprenol production. Metab Eng 64:154–166. https://doi.org/10.1016/j.ymben.2021.02.002

Kirke G (2019) Development of a genetic and molecular toolkit for the oleaginous red yeast Rhodotorula toruloides. In *Research (biological sciences)*

Koh CMJ, Liu Y, Moehninsi DM, Ji L (2014) Molecular characterization of KU70 and KU80 homologues and exploitation of a KU70-deficient mutant for improving gene deletion frequency in Rhodosporidium toruloides. BMC Microbiol 14:1–10

Kujawska A, Kujawski J, Bryjak M, Kujawski W (2015) ABE fermentation products recovery methods - A review. In Renewable and Sustainable Energy Reviews 48. https://doi.org/10.1016/j.rser.2015.04.028

Kulkarni MG, Dalai AK (2006) Waste cooking oil – an economical source for biodiesel: A review. Ind Eng Chem Res 45(9):2901–2913. https://doi.org/10.1021/ie0510526

Kumar P, Barrett DM, Delwiche MJ, Stroeve P (2009a) Methods for pretreatment of lignocellulosic biomass for efficient hydrolysis and biofuel production. Ind Eng Chem Res 48(8):3713–3729. https://doi.org/10.1021/ie801542g

Kumar A, Singh LK, Ghosh S (2009b) Bioconversion of lignocellulosic fraction of water-hyacinth (Eichhornia crassipes) hemicellulose acid hydrolysate to ethanol by Pichia stipitis. Bioresour Technol 100(13):3293–3297. https://doi.org/10.1016/j.biortech.2009.02.023

Kuroda K, Hammer SK, Watanabe Y, Montaño López J, Fink GR, Stephanopoulos G, Ueda M, Avalos JL (2019) Critical roles of the pentose phosphate pathway and GLN3 in Isobutanol-specific tolerance in yeast. Cell Syst 9(6):534–547.e5. https://doi.org/10.1016/j.cels.2019.10.006

Lam SS, Liew RK, Jusoh A, Chong CT, Ani FN, Chase HA (2016) Progress in waste oil to sustainable energy, with emphasis on pyrolysis techniques. In: *Renewable and sustainable energy reviews*, vol 53. Elsevier Ltd, pp 741–753. https://doi.org/10.1016/j.rser.2015.09.005

Lam FH, Turanlı-Yıldız B, Liu D, Resch MG, Fink GR, Stephanopoulos G (2021) Engineered yeast tolerance enables efficient production from toxified lignocellulosic feedstocks. Sci Adv 7: eabf7613. https://www.science.org

Li Y, Wei G, Chen J (2004) Glutathione: a review on biotechnological production. Appl Microbiol Biotechnol 66(3):233–242. https://doi.org/10.1007/s00253-004-1751-y

Li Y, Zhao ZK, Bai F (2007) High-density cultivation of oleaginous yeast Rhodosporidium toruloides Y4 in fed-batch culture. Enzym Microb Technol 41(3):312–317. https://doi.org/10.1016/j.enzmictec.2007.02.008

Li S-Y, Srivastava R, Suib SL, Li Y, Parnas RS (2011) Performance of batch, fed-batch, and continuous A–B–E fermentation with pH-control. Bioresour Technol 102(5):4241–4250. https://doi.org/10.1016/j.biortech.2010.12.078

Li G, Zhang C, Wang Z, Huang H, Peng H, Li X (2018) Fabrication of mesoporous Co3O4 oxides by acid treatment and their catalytic performances for toluene oxidation. Appl Catal A Gen 550: 67–76. https://doi.org/10.1016/j.apcata.2017.11.003

Li J, Ma Y, Liu N, Eser BE, Guo Z, Jensen PR, Stephanopoulos G (2020) Synthesis of high-titer alka(e)nes in Yarrowia lipolytica is enabled by a discovered mechanism. Nat Commun 11(1): 6198. https://doi.org/10.1038/s41467-020-19995-0

Li J, Zhu K, Miao L, Rong L, Zhao Y, Li S, Ma L, Li J, Zhang C, Xiao D, Foo JL, Yu A (2021) Simultaneous improvement of limonene production and tolerance in Yarrowia lipolytica through tolerance engineering and evolutionary engineering. ACS Synth Biol 10(4):884–896. https://doi.org/10.1021/acssynbio.1c00052

Ling H, Pratomo Juwono NK, Teo WS, Liu R, Leong SSJ, Chang MW (2015) Engineering transcription factors to improve tolerance against alkane biofuels in Saccharomyces cerevisiae. Biotechnol Biofuels 8(1):1–11. https://doi.org/10.1186/s13068-015-0411-z

Lisboa P, Rodrigues AR, Martín JL, Simões P, Barreiros S, Paiva A (2014) Economic analysis of a plant for biodiesel production from waste cooking oil via enzymatic transesterification using supercritical carbon dioxide. J Supercrit Fluids 85:31–40. https://doi.org/10.1016/j.supflu.2013.10.018

Liu XB, Gu QY, Yu XB (2013a) Repetitive domestication to enhance butanol tolerance and production in clostridium acetobutylicum through artificial simulation of bio-evolution. Bioresour Technol 130:638–643. https://doi.org/10.1016/j.biortech.2012.12.121

Liu Y, Koh CMJ, Sun L, Hlaing MM, Du M, Peng N, Ji L (2013b) Characterization of glyceraldehyde-3-phosphate dehydrogenase gene RtGPD1 and development of genetic transformation method by dominant selection in oleaginous yeast Rhodosporidium toruloides. Appl Microbiol Biotechnol 97(2):719–729. https://doi.org/10.1007/s00253-012-4223-9

Liu R, Zhu F, Lu L, Fu A, Lu J, Deng Z, Liu T (2014) Metabolic engineering of fatty acyl-ACP reductase-dependent pathway to improve fatty alcohol production in Escherichia coli. Metab Eng 22:10–21. https://doi.org/10.1016/j.ymben.2013.12.004

Liu Y, Jiang X, Cui Z, Wang Z, Qi Q, Hou J (2019) Engineering the oleaginous yeast Yarrowia lipolytica for production of α-farnesene. Biotechnol Biofuels 12(1):1–11. https://doi.org/10.1186/s13068-019-1636-z

Liu SC, Liu Z, Wei LJ, Hua Q (2020) Pathway engineering and medium optimization for α-farnesene biosynthesis in oleaginous yeast Yarrowia lipolytica. J Biotechnol 319:74–81. https://doi.org/10.1016/j.jbiotec.2020.06.005

Liu Z, Moradi H, Shi S, Darvishi F (2021) Yeasts as microbial cell factories for sustainable production of biofuels. In: *Renewable and sustainable energy reviews*, vol 143. Elsevier Ltd. https://doi.org/10.1016/j.rser.2021.110907

Liu CL, Xue K, Yang Y, Liu X, Li Y, Lee TS, Bai Z, Tan T (2022) Metabolic engineering strategies for sesquiterpene production in microorganism. Crit Rev Biotechnol 42(1):73–92). Taylor and Francis Ltd. https://doi.org/10.1080/07388551.2021.1924112

Lopresto CG, Petrillo F, Casazza AA, Aliakbarian B, Perego P, Calabrò V (2014) A non-conventional method to extract D-limonene from waste lemon peels and comparison with traditional Soxhlet extraction. Sep Purif Technol 137:13–20. https://doi.org/10.1016/j.seppur.2014.09.015

Lv X, Wang F, Zhou P, Ye L, Xie W, Xu H, Yu H (2016) Dual regulation of cytoplasmic and mitochondrial acetyl-CoA utilization for improved isoprene production in Saccharomyces cerevisiae. Nat Commun 7:12851. https://doi.org/10.1038/ncomms12851

Ma Y, Gao Z, Wang Q, Liu Y (2018) Biodiesels from microbial oils: opportunity and challenges. In: Bioresource technology, vol 263. Elsevier Ltd., pp 631–641. https://doi.org/10.1016/j.biortech.2018.05.028

MacDougall KM, McNichol J, McGinn PJ, O'Leary SJB, Melanson JE (2011) Triacylglycerol profiling of microalgae strains for biofuel feedstock by liquid chromatography-high-resolution mass spectrometry. Anal Bioanal Chem 401(8):2609–2616. https://doi.org/10.1007/s00216-011-5376-6

Maddikeri GL, Pandit AB, Gogate PR (2012) Intensification approaches for biodiesel synthesis from waste cooking oil: A review. Ind Eng Chem Res 51(45):14610–14628. https://doi.org/10.1021/ie301675j

Matsuda F, Ishii J, Kondo T, Ida K, Tezuka H, Kondo A (2013) Increased isobutanol production in Saccharomyces cerevisiae by eliminating competing pathways and resolving cofactor imbalance. Microb Cell Factories 12(1):119. https://doi.org/10.1186/1475-2859-12-119

Mattanovich D, Sauer M, Gasser B (2014) Yeast biotechnology: teaching the old dog new tricks. Microb Cell Factories 13(1):34. https://doi.org/10.1186/1475-2859-13-34

McNeil BA, Stuart DT (2018) Optimization of C16 and C18 fatty alcohol production by an engineered strain of Lipomyces starkeyi. J Ind Microbiol Biotechnol 45(1):1–14. https://doi.org/10.1007/s10295-017-1985-1

Meadows AL, Hawkins KM, Tsegaye Y, Antipov E, Kim Y, Raetz L, Dahl RH, Tai A, Mahatdejkul-Meadows T, Xu L, Zhao L, Dasika MS, Murarka A, Lenihan J, Eng D, Leng JS, Liu CL, Wenger JW, Jiang H, Tsong AE (2016) Rewriting yeast central carbon metabolism for industrial isoprenoid production. Nature 537(7622):694–697. https://doi.org/10.1038/nature19769

Miura Y, Kondo K, Shimada H, Saito T, Nakamura K, Misawa N (1998) Communication to the editor production of lycopene by THE food yeast Candida utilis that does not naturally synthesize carotenoid. Biotechnol Bioeng 58:306–308

Miura Y, Kettoku M, Kato M, Kobayashi K, Kondo K (1999) High level production of thermostable α-amylase from Sulfolobus solfataricus in high-cell density culture of the food yeast Candida utilis. J Molec Microbiol Biotechnol 1(1):129–134. www. caister.com/bacteria-plant

Mohd Azhar SH, Abdulla R, Jambo SA, Marbawi H, Gansau JA, Mohd Faik AA, Rodrigues KF (2017) Yeasts in sustainable bioethanol production: A review. Biochem Biophys Rep 10:52–61. https://doi.org/10.1016/j.bbrep.2017.03.003

Nanda S, Dalai AK, Kozinski JA (2014) Butanol and ethanol production from lignocellulosic feedstock: biomass pretreatment and bioconversion. Energ Sci Eng 2(3):138–148. https://doi.org/10.1002/ese3.41

Nanthagopal K, Ashok B, Saravanan B, Ramesh Pathy M, Sahil G, Ramesh A, Nurun Nabi M, Golam Rasul M (2019) Study on decanol and Calophyllum Inophyllum biodiesel as ternary blends in CI engine. Fuel 239:862–873. https://doi.org/10.1016/j.fuel.2018.11.037

Nielsen J, Larsson C, van Maris A, Pronk J (2013) Metabolic engineering of yeast for production of fuels and chemicals. Current Opinion in Biotechnology 24(3):398–404. https://doi.org/10.1016/j.copbio.2013.03.023

Nishimura Y, Matsui T, Ishii J, Kondo A (2018) Metabolic engineering of the 2-ketobutyrate biosynthetic pathway for 1-propanol production in Saccharomyces cerevisiae. Microb Cell Factories 17(1):1–11. https://doi.org/10.1186/s12934-018-0883-1

Özaydın B, Burd H, Lee TS, Keasling JD (2013) Carotenoid-based phenotypic screen of the yeast deletion collection reveals new genes with roles in isoprenoid production. Metab Eng 15:174–183. https://doi.org/10.1016/j.ymben.2012.07.010

Pang Y, Zhao Y, Li S, Zhao Y, Li J, Hu Z, Zhang C, Xiao D, Yu A (2019) Engineering the oleaginous yeast Yarrowia lipolytica to produce limonene from waste cooking oil. Biotechnol Biofuels 12(1):1–18. https://doi.org/10.1186/s13068-019-1580-y

Park YK, Nicaud JM, Ledesma-Amaro R (2018) The engineering potential of Rhodosporidium toruloides as a workhorse for biotechnological applications. Trends Biotechnol 36(3):304–317). Elsevier Ltd. https://doi.org/10.1016/j.tibtech.2017.10.013

Peng B, Nielsen LK, Kampranis SC, Vickers CE (2018) Engineered protein degradation of farnesyl pyrophosphate synthase is an effective regulatory mechanism to increase monoterpene production in Saccharomyces cerevisiae. Metab Eng 47:83–93. https://doi.org/10.1016/j.ymben.2018.02.005

Peralta-Yahya PP, Ouellet M, Chan R, Mukhopadhyay A, Keasling JD, Lee TS (2011) Identification and microbial production of a terpene-based advanced biofuel. Nat Commun 2(1):483. https://doi.org/10.1038/ncomms1494

Peralta-Yahya PP, Zhang F, Del Cardayre SB, Keasling JD (2012) Microbial engineering for the production of advanced biofuels. *Nature* 488(7411):320–328. https://doi.org/10.1038/nature11478

Politino M, Tonzi SM, Burnett WV, Romancik G, Usher JJ (1997) Purification and characterization of a cephalosporin esterase from Rhodosporidium toruloides. Appl Environ Microbiol 63(12):4807–4811

Poontawee R, Limtong S (2020) Feeding strategies of two-stage fed-batch cultivation processes for microbial lipid production from sugarcane top hydrolysate and crude glycerol by the oleaginous red yeast Rhodosporidiobolus fluvialis. Microorganisms 8(2):151. https://doi.org/10.3390/microorganisms8020151

Qiao K, Hussain S, Abidi I, Liu H, Zhang H, Chakraborty S, Watson N, Kumaran Ajikumar P, Stephanopoulos G (2015) Engineering lipid overproduction in the oleaginous yeast Yarrowia lipolytica. http://www.elsevier.com/open-access/userlicense/1.0/

Qiao K, Wasylenko TM, Zhou K, Xu P, Stephanopoulos G (2017) Lipid production in Yarrowia lipolytica is maximized by engineering cytosolic redox metabolism. Nat Biotechnol 35(2): 173–177. https://doi.org/10.1038/nbt.3763

Rassadin GVG, Shlygin OY, Likhterova NM, Slavin VN, Zharov AV (2006) Current problems. Motor fuels-the new quality level problems in production of high-octane, unleaded automotive. Chem Technol Fuels Oils 42(4)

Rigouin C, Croux C, Borsenberger V, Ben Khaled M, Chardot T, Marty A, Bordes F (2018) Increasing medium chain fatty acids production in Yarrowia lipolytica by metabolic engineering. Microb Cell Factories 17(1):142. https://doi.org/10.1186/s12934-018-0989-5

Röttig A, Zurek PJ, Steinbüchel A (2015) Assessment of bacterial acyltransferases for an efficient lipid production in metabolically engineered strains of E. Coli. Metab Eng 32:195–206. https://doi.org/10.1016/j.ymben.2015.09.016

Rutter CD, Rao CV (2016) Production of 1-decanol by metabolically engineered Yarrowia lipolytica. Metab Eng 38:139–147. https://doi.org/10.1016/j.ymben.2016.07.011

Sampaio JP, Gadanho M, Bauer R, Weiß M (2003) Taxonomic studies in the Microbotryomycetidae: Leucosporidium golubevii sp. nov., Leucosporidiella gen. Nov. and the new orders Leucosporidiales and Sporidiobolales. Mycol Prog 2(1):53–68. https://doi.org/10.1007/s11557-006-0044-5

Santiago-Gómez MP, Thanh HT, De Coninck J, Cachon R, Kermasha S, Belin JM, Gervais P, Husson F (2009) Modeling hexanal production in oxido-reducing conditions by the yeast Yarrowia lipolytica. Process Biochem 44(9):1013–1018. https://doi.org/10.1016/j.procbio.2009.04.028

Santori G, Di Nicola G, Moglie M, Polonara F (2012) A review analyzing the industrial biodiesel production practice starting from vegetable oil refining. Appl Energy 92:109–132. https://doi.org/10.1016/j.apenergy.2011.10.031

Schadeweg V, Boles E (2016) Increasing n-butanol production with Saccharomyces cerevisiae by optimizing acetyl-CoA synthesis, NADH levels and trans-2-enoyl-CoA reductase expression. Biotechnol Biofuels 9(1):1–11. https://doi.org/10.1186/s13068-016-0673-0

Schjerling CK, Hummel R, Hansen JK, Børsting C, Mikkelsen JM, Kristiansen K, Knudsen J (1996) Disruption of the gene encoding the acyl-CoA-binding protein ( ) perturbs acyl-CoA metabolism in. J Biol Chem 271(37):22514–22521. https://doi.org/10.1074/jbc.271.37.22514

Schmidt A, Wächtler B, Temp U, Krekling T, Séguin A, Gershenzon J (2010) A bifunctional geranyl and geranylgeranyl diphosphate synthase is involved in terpene oleoresin formation in Picea abies. Plant Physiol 152(2):639–655. https://doi.org/10.1104/pp.109.144691

Serhan M, Sprowls M, Jackemeyer D, Long M, Perez ID, Maret W, Tao N, Forzani E (2019) Total iron measurement in human serum with a smartphone. *AIChE Annual Meeting, Conference Proceedings, 2019-November*. https://doi.org/10.1039/x0xx00000x

Sheng J, Stevens J, Feng X (2016) Pathway compartmentalization in peroxisome of saccharomyces cerevisiae to produce versatile medium chain fatty alcohols. Sci Rep 6:26884. https://doi.org/10.1038/srep26884

Shi S, Si T, Liu Z, Zhang H, Ang EL, Zhao H (2016) Metabolic engineering of a synergistic pathway for n-butanol production in Saccharomyces cerevisiae. Sci Rep 6:25675. https://doi.org/10.1038/srep25675

Shi T, Li Y, Zhu L, Tong Y, Yang J, Fang Y, Wang M, Zhang J, Jiang Y, Yang S (2021) Engineering the oleaginous yeast Yarrowia lipolytica for β-farnesene overproduction. Biotechnol J 16(7):2100097. https://doi.org/10.1002/biot.202100097

Sims REH, Mabee W, Saddler JN, Taylor M (2010) An overview of second generation biofuel technologies. Bioresour Technol 101(6):1570–1580). Elsevier Ltd. https://doi.org/10.1016/j.biortech.2009.11.046

Singh G, Dhar YV, Asif MH, Misra P (2018) Exploring the functional significance of sterol glycosyltransferase enzymes. Prog Lipid Res 69:1–10. https://doi.org/10.1016/j.plipres.2017.11.001

Singh G, Sinha S, Kumar KK, Gaur NA, Bandyopadhyay KK, Paul D (2020) High density cultivation of oleaginous yeast isolates in 'mandi' waste for enhanced lipid production using sugarcane molasses as feed. Fuel 276:118073. https://doi.org/10.1016/j.fuel.2020.118073

Steen EJ, Chan R, Prasad N, Myers S, Petzold CJ, Redding A, Ouellet M, Keasling JD (2008) Metabolic engineering of Saccharomyces cerevisiae for the production of n-butanol. Microb Cell Factories 7:1–8. https://doi.org/10.1186/1475-2859-7-36

Swidah R, Ogunlabi O, Grant CM, Ashe MP (2018) N-butanol production in S. cerevisiae: co-ordinate use of endogenous and exogenous pathways. Appl Microbiol Biotechnol 102(22): 9857–9866. https://doi.org/10.1007/s00253-018-9305-x

Tamakawa H, Ikushima S, Yoshida S (2011) Ethanol production from xylose by a recombinant candida utilis strain expressing protein-engineered xylose reductase and xylitol dehydrogenase. Biosci Biotechnol Biochem 75(10):1994–2000. https://doi.org/10.1271/bbb.110426

Tamakawa H, Ikushima S, Yoshida S (2013a) Construction of a Candida utilis strain with ratio-optimized expression of xylose-metabolizing enzyme genes by cocktail multicopy integration method. J Biosci Bioeng 115(5):532–539. https://doi.org/10.1016/j.jbiosc.2012.12.001

Tamakawa H, Mita T, Yokoyama A, Ikushima S, Yoshida S (2013b) Metabolic engineering of Candida utilis for isopropanol production. Appl Microbiol Biotechnol 97(14):6231–6239. https://doi.org/10.1007/s00253-013-4964-0

Tao H, Guo D, Zhang Y, Deng Z, Liu T (2015) Metabolic engineering of microbes for branched-chain biodiesel production with low-temperature property. Biotechnol Biofuels 8(1):1–11. https://doi.org/10.1186/s13068-015-0270-7

Teo WS, Ling H, Yu AQ, Chang MW (2015) Metabolic engineering of Saccharomyces cerevisiae for production of fatty acid short- and branched-chain alkyl esters biodiesel. Biotechnol Biofuels 8(1):1–9. https://doi.org/10.1186/s13068-015-0361-5

Tippmann S, Scalcinati G, Siewers V, Nielsen J (2016) Production of farnesene and santalene by Saccharomyces cerevisiae using fed-batch cultivations with RQ-controlled feed; production of farnesene and santalene by Saccharomyces cerevisiae using fed-batch cultivations with RQ-controlled feed. Biotechnol Bioeng 113:72–81. https://doi.org/10.1002/bit.25683/abstract

Tomita Y, Ikeo K, Tamakawa H, Gojobori T, Ikushima S (2012) Genome and transcriptome analysis of the food-yeast candida utilis. PLoS One 7(5):e37226. https://doi.org/10.1371/journal.pone.0037226

Tsakona S, Kopsahelis N, Chatzifragkou A, Papanikolaou S, Kookos IK, Koutinas AA (2014) Formulation of fermentation media from flour-rich waste streams for microbial lipid production by Lipomyces starkeyi. J Biotechnol 189:36–45. https://doi.org/10.1016/j.jbiotec.2014.08.011

Van Hecke W, Kaur G, De Wever H (2014) Advances in in-situ product recovery (ISPR) in whole cell biotechnology during the last decade. Biotechnol Adv 32(7):1245–1255. https://doi.org/10.1016/j.biotechadv.2014.07.003

Wagner JM, Alper HS (2015) Synthetic biology and molecular genetics in non-conventional yeasts: current tools and future advances. http://www.elsevier.com/open-access/userlicense/1.0/

Wang J, Gao Q, Zhang H, Bao J (2016a) Inhibitor degradation and lipid accumulation potentials of oleaginous yeast Trichosporon cutaneum using lignocellulose feedstock. Bioresour Technol 218:892–901. https://doi.org/10.1016/j.biortech.2016.06.130

Wang W, Wei H, Knoshaug E, Van Wychen S, Xu Q, Himmel ME, Zhang M (2016b) Fatty alcohol production in Lipomyces starkeyi and Yarrowia lipolytica. Biotechnol Biofuels 9(1):1–12. https://doi.org/10.1186/s13068-016-0647-2

Wang G, Xiong X, Ghogare R, Wang P, Meng Y, Chen S (2016c) Exploring fatty alcohol-producing capability of Yarrowia lipolytica. Biotechnol Biofuels 9(1):1–10. https://doi.org/10.1186/s13068-016-0512-3

Wei LJ, Zhong YT, Nie MY, Liu SC, Hua Q (2021) Biosynthesis of α-Pinene by genetically engineered Yarrowia lipolytica from low-cost renewable feedstocks. J Agric Food Chem 69(1): 275–285. https://doi.org/10.1021/acs.jafc.0c06504

Wess J, Brinek M, Boles E (2019) Improving isobutanol production with the yeast Saccharomyces cerevisiae by successively blocking competing metabolic pathways as well as ethanol and glycerol formation. Biotechnol Biofuels 12(1):1–15. https://doi.org/10.1186/s13068-019-1486-8

Xin F, Liu J, He M, Wu B, Ni Y, Dong W, Zhang W, Hu G, Jiang M (2018) High biobutanol production integrated with in situ extraction in the presence of Tween 80 by Clostridium acetobutylicum. Process Biochem 67. https://doi.org/10.1016/j.procbio.2018.01.013

Xu J, Oura T, Liu D, Kajiwara S (2011) Heat-alkaline treatment of excess sludge and the potential use of hydrolysate as nitrogen source for microbial lipid production. Sheng wu gong cheng xue bao = Chin J Biotechnol 27(3):482–488. http://europepmc.org/abstract/MED/21650031

Xu J, Zhao X, Wang W, Du W, Liu D (2012) Microbial conversion of biodiesel byproduct glycerol to triacylglycerols by oleaginous yeast Rhodosporidium toruloides and the individual effect of some impurities on lipid production. Biochem Eng J 65:30–36. https://doi.org/10.1016/j.bej.2012.04.003

Xu P, Qiao K, Ahn WS, Stephanopoulos G (2016) Engineering Yarrowia lipolytica as a platform for synthesis of drop-in transportation fuels and oleochemicals. Proc Natl Acad Sci USA 113(39):10848–10853. https://doi.org/10.1073/pnas.1607295113

Xu P, Qiao K, Stephanopoulos G (2017) Engineering oxidative stress defense pathways to build a robust lipid production platform in Yarrowia lipolytica. Biotechnol Bioeng 114(7):1521–1530. https://doi.org/10.1002/bit.26285

Xue C, Zhao X-Q, Liu C-G, Chen L-J, Bai F-W (2013) Prospective and development of butanol as an advanced biofuel. Biotechnol Adv 31(8):1575–1584. https://doi.org/10.1016/j.biotechadv.2013.08.004

Xue C, Liu F, Xu M, Tang IC, Zhao J, Bai F, Yang ST (2016) Butanol production in acetone-butanol-ethanol fermentation with in situ product recovery by adsorption. Bioresour Technol 219:158–168. https://doi.org/10.1016/j.biortech.2016.07.111

Yaegashi J, Kirby J, Ito M, Sun J, Dutta T, Mirsiaghi M, Sundstrom ER, Rodriguez A, Baidoo E, Tanjore D, Pray T, Sale K, Singh S, Keasling JD, Simmons BA, Singer SW, Magnuson JK, Arkin AP, Skerker JM, Gladden JM (2017) Rhodosporidium toruloides: A new platform organism for conversion of lignocellulose into terpene biofuels and bioproducts. Biotechnol Biofuels 10(1):1–13. https://doi.org/10.1186/s13068-017-0927-5

Yang X, Nambou K, Wei L, Hua Q (2016) Heterologous production of α-farnesene in metabolically engineered strains of Yarrowia lipolytica. Bioresour Technol 216:1040–1048. https://doi.org/10.1016/j.biortech.2016.06.028

Yao F, Liu SC, Wang DN, Liu ZJ, Hua Q, Wei LJ (2020) Engineering oleaginous yeast Yarrowia lipolytica for enhanced limonene production from xylose and lignocellulosic hydrolysate. FEMS Yeast Res 20(6):foaa046. https://doi.org/10.1093/femsyr/foaa046

Yu T, Zhou YJ, Wenning L, Liu Q, Krivoruchko A, Siewers V, Nielsen J, David F (2017) Metabolic engineering of Saccharomyces cerevisiae for production of very long chain fatty acid-derived chemicals. Nat Commun 8:1–11. https://doi.org/10.1038/ncomms15587

Yuan J, Chen X, Mishra P, Ching CB (2017) Metabolically engineered Saccharomyces cerevisiae for enhanced isoamyl alcohol production. Appl Microbiol Biotechnol 101(1):465–474. https://doi.org/10.1007/s00253-016-7970-1

Zhang Y, Bao J (2022) Tolerance of Trichosporon cutaneum to lignin derived phenolic aldehydes facilitate the cell growth and cellulosic lipid accumulation. J Biotechnol 343:32–37. https://doi.org/10.1016/j.jbiotec.2021.09.009

Zhang S, Skerker JM, Rutter CD, Maurer MJ, Arkin AP, Rao CV (2016) Engineering Rhodosporidium toruloides for increased lipid production. Biotechnol Bioeng 113(5): 1056–1066. https://doi.org/10.1002/bit.25864

Zhang Y, Lane S, Chen JM, Hammer SK, Luttinger J, Yang L, Jin YS, Avalos JL (2019) Xylose utilization stimulates mitochondrial production of isobutanol and 2-methyl-1-butanol in Saccharomyces cerevisiae. Biotechnol Biofuels 12(1):1–15. https://doi.org/10.1186/s13068-019-1560-2

Zhang Y, Peng J, Zhao H, Shi S (2021) Engineering oleaginous yeast Rhodotorula toruloides for overproduction of fatty acid ethyl esters. Biotechnol Biofuels 14(1):115. https://doi.org/10.1186/s13068-021-01965-3

Zhao Y, Zhu K, Li J, Zhao Y, Li S, Zhang C, Xiao D, Yu A (2021) High-efficiency production of bisabolene from waste cooking oil by metabolically engineered Yarrowia lipolytica. Microb Biotechnol 14(6):2497–2513. https://doi.org/10.1111/1751-7915.13768

Zhou YJ, Buijs NA, Zhu Z, Gómez DO, Boonsombuti A, Siewers V, Nielsen J (2016a) Harnessing yeast peroxisomes for biosynthesis of fatty-acid-derived biofuels and chemicals with relieved side-pathway competition. J Am Chem Soc 138(47):15368–15377. https://doi.org/10.1021/jacs.6b07394

Zhou YJ, Buijs NA, Zhu Z, Qin J, Siewers V, Nielsen J (2016b) Production of fatty acid-derived oleochemicals and biofuels by synthetic yeast cell factories. Nat Commun 7(1):11709. https://doi.org/10.1038/ncomms11709

Zhou YJ, Hu Y, Zhu Z, Siewers V, Nielsen J (2018) Engineering 1-alkene biosynthesis and secretion by dynamic regulation in yeast. ACS Synth Biol 7(2):584–590. https://doi.org/10.1021/acssynbio.7b00338

Zhu Z, Zhang S, Liu H, Shen H, Lin X, Yang F, Zhou YJ, Jin G, Ye M, Zou H, Zhao ZK (2012) A multi-omic map of the lipid-producing yeast Rhodosporidium toruloides. Nat Commun 3:1112. https://doi.org/10.1038/ncomms2112

Zhu Y, Chang Y, Guan J, Shanguan G, Xin F (2016) Butanol production from organosolv treated spent mushroom substrate integrated with in situ biodiesel extraction. Renew Energy 96:656–661. https://doi.org/10.1016/j.renene.2016.04.048

Chapter 6
Bacterial Based Drop-In Biofuel Production

Mandana Lak, Ariel Marchlewicz, Seyed Ali Nojoumi, Mohadasseh Ramezani, and Sharareh Harirchi

Abstract Currently, renewable resources are promising feedstocks for advanced biofuels (drop-in biofuels) production that can be directly used for the transportation sector in the world to overcome climate change and air pollution and contribute to the conservation of biodiversity. These renewable feedstocks, including agriculture residues, oily wastes, animal manures, or forest residues, can be utilized by various types of microorganisms from microalgae to bacteria to produce biofuels. Presently, existing biofuels cannot be directly employed into the engines and require specific infrastructures and modified engines, and this issue multiplies the importance of using drop-in biofuels. These advanced biofuels which have well-defined chemical constituents can be easily used in the existing fuel infrastructures. Different processes are used for the production of advanced biofuels, while the biochemical approach seems to be more effective as pure molecular streams of functionalized

M. Lak
Department of Microbiology, Faculty of Biological Sciences, Alzahra University, Tehran, Iran

A. Marchlewicz
Faculty of Natural Science, Institute of Biology, Biotechnology and Environmental Protection, University of Silesia in Katowice, Katowice, Poland

Faculty of Mathematics and Science, Department of Biological and Environmental Science, University of Jyväskylä, Jyväskylä, Finland
e-mail: ariel.marchlewicz@us.edu.pl

S. A. Nojoumi
Microbiology Research Center, Pasteur Institute of Iran, Tehran, Iran

Department of Mycobacteriology and Pulmonary Research, Pasteur Institute of Iran, Tehran, Iran

M. Ramezani
Microorganisms Bank, Iranian Biological Resource Centre (IBRC), ACECR, Tehran, Iran

S. Harirchi (✉)
Swedish Centre for Resource Recovery, University of Borås, Borås, Sweden

Department of Biotechnology, Iranian Research Organization for Science and Technology, Tehran, Iran
e-mail: sh.harirchi@irost.ir

S. K. Brar et al. (eds.), *The Microbiology of the Drop-in Biofuel Production*, Biofuel and Biorefinery Technologies 15, https://doi.org/10.1007/978-3-031-61637-2_6

molecules can be manufactured. The heart of biochemical approach is microorganisms including yeasts, fungi, bacteria, and algae that hold the main answer for drop-in biofuels production. Among various microorganisms, bacteria are more likely to be well-matched for industrial production of advanced biofuels. In this chapter, after a short introduction to these biofuels, various feedstocks for bacterial utilization are explained, and latterly, technological frameworks and essential bacterial pathways for drop-in biofuels are discussed. Moreover, engineered bacterial pathways are included and in conclusion future perspectives of bacterial production approaches for drop-in biofuels are stated.

Keywords Metabolic pathways · Drop-in biofuels · Synthetic biology · Bacterial strains · Biomass

6.1 Introduction

The use of fossilized biomass for energy and chemicals since the Industrial Revolution has changed the natural carbon cycle, with significant effects now felt on a global scale Intergovernmental Panel on Climate Change (IPCC) report on climate change (Love 2022). A convenient source of concentrated energy, liquid transport fuels have well-established production guidelines, storage infrastructure, and distribution systems. A third of all energy use worldwide is accounted for by transportation, which uses around 65% of the world's total petroleum production (Solaymani 2019). Therefore, the decarbonization of the transportation sector is dependent on the use of transport fuels made from non-fossil resources (Shears 2019).

By replacing fossil fuels like coal, oil, and natural gas with fuels made from biomass, or by using direct carbon capture and nature-based carbon sequestration strategies (known as "Bioenergy with Carbon Capture Solutions (BioCCS or BECCS)") which are currently utilized to offset emissions in the form of carbon credits, decarbonization aims to lessen the environmental impact of the use of fossil fuels (Osman et al. 2021a). The market penetration of emission-free electric cars is accelerating in the transportation sector, but it is heavily dependent on accessible rapid charging infrastructures and C-free electrical generation capacity, both of which may be challenging to implement in non-urban or more remote settings, or in developing countries, which explains the ongoing (and currently increasing) demand for biofuels. Alcohols (1), biodiesel made from lipids (2), and biomass- or microbially produced combustible gases like biomethane and H_2 (3) that can be catalytically transformed into sustainable synthetic fuels are examples of current transport biofuels. Biofuels are combustible substances that can be solid, liquid, or gaseous and are sourced from or created by living things (Sims et al. 2008).

Currently available commercial biofuels include alcohols produced by microorganisms that are combined with gasoline, namely ethanol (C_2H_5OH) from *Saccharomyces cerevisiae* fermentation and n-butanol (C_4H_9OH) from *Clostridium* spp. or chemical conversion of ethanol as shown in Fig. 6.1. Furthermore, tri-acyl-glycerides (TAG) are catalytically converted into fatty acid mono-alkyl esters (crude

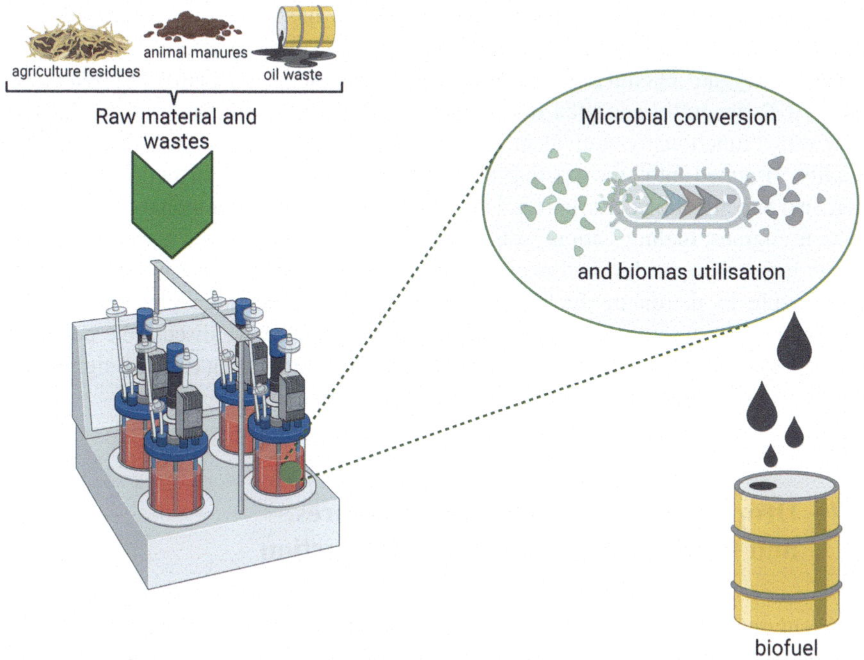

Fig. 6.1 Displays the microbial-based biofuel production using renewable substrates

biodiesel), glycerol, and methanol using alkaline catalysts like sodium hydroxide, potassium hydroxide, or sodium methoxide. Additionally, plant oils can be chemically hydro-treated to produce hydrogenated vegetable oils (often known as "renewable diesel fuels") (Wojcik et al. 2017), which are linear alkane hydrocarbons that are de-oxygenated and can be used in place of automobile gas oil that is generated from fossil fuels (Dimitriadis et al. 2018). The term "advanced biofuels" refers to a variety of alternative biomass or combustible molecules that are often generated from microbes. Researchers in academia frequently refer to biomass and fuel precursors obtained from heterotrophic microorganisms and microalgae other than those typically employed in commercial first- and second-generation (1G and 2G) biofuel production as being of the third generation (3G) (Dutta et al. 2014; Patel et al. 2017). Fourth-generation (4G) biofuels are those made from genetically modified microalgae, bacteria, and fungi (Malik et al. 2021).

Since the main part of global transportation and shipping armada still carries on fossil fuels, various harmful impacts, particularly air pollution, have been reported. Hence, the use of biofuels as transportation and vehicle fuels has the potential to have positive effects on the environment and financial markets, such reduced greenhouse gas (GHG) emissions, the creation of jobs, and increased energy security. The two most common commercially accessible biofuels are bioethanol and biodiesel. Based on the International Energy Agency (IEA) report, more than 110 billion liters bioethanol was produced in the world during 2022 while the global quantity of biodiesel was approximately 14 billion liters in 2022 (Agency 2023). These fuels cannot fully utilize the processing and distribution systems for petroleum because they are chemically and functionally distinct from petroleum-generated fuels. It is generally accepted that the use of biofuels would rise substantially if they were “functionally equivalent” to current petroleum-derived fuels and could be readily “dropped-into” the existing infrastructure (petroleum distribution and refining, fuel specifications, etc.). This is because infrastructure components, including vehicle engines, fueling stations, refineries, etc., are very expensive to change. Drop-in biofuels are liquid bio-hydrocarbons having characteristics that are functionally comparable to petroleum fuels and are completely compatible with the current petroleum infrastructure (Bioenergy 2019). In this chapter, we discuss the potential and features of drop-in biofuels produced by bacterial strains that can be a significant replacement of fossil fuels and previous biofuel generations.

6.2 Drop-In Biofuels: Definition, Features, and the Feedstocks for Their Production

Conventional biofuels, bioethanol, and biodiesel, which have separate chemical functional groups, may be precisely defined based on their chemical makeup alone. The qualities of drop-in biofuels, in contrast, are normally defined by the functional characteristics of the mixture, which includes the distillation profile, viscosity, acidity, etc., just like petroleum fuels. Drop-in biofuels typically consist of a mixture of several distinct types of hydrocarbons (Bohre et al. 2015). These kinds of biofuels are mostly plant-based but they can be increasingly combined with common existing fuels. The traditional biofuels such as methanol, ethanol, and even gaseous ones can be only added to the adjusted engines. Another drawback of conventional biofuels refers to this issue that they cannot fully utilize the infrastructure for petroleum processing and delivery because they differ chemically and functionally from fuels obtained from petroleum. These biofuels may only be used in relatively small mixes (ethanol maximum 15% and biodiesel maximum 10%), unless specialized engines are used (Gollakota et al. 2018). Moreover, conventional biofuels do not meet the density standards or other specifications needed for usage in industries like aviation and long-distance haulage. Currently, “food-based” feedstocks, including corn, sugarcane, and vegetable oils, are mostly used as feedstocks for conventional biofuels. Concerns about food security and environmental issues will continue to support the manufacture of biofuels from waste or more plentiful lignocellulosic resources that do not compete with food (Kararzos et al. 2017).

Although this is not always true, conventional biofuels’ ability to cut emissions (or carbon intensity) is typically lower than that of biofuels developed from lignocellulosic feedstocks. Even though drop-in biofuels are occasionally referred to as being oxygen-free, it should be remembered that crude oil can occasionally include up to 2 weight percent oxygen (rarely even more) (Speight 2014). It should be noted that drop-in biofuels may occasionally be subject to blend restrictions. For instance, hydro-treated renewable jet fuels often lack aromatics, which are necessary to preserve the seal integrity of the fuel systems in aircraft. As a result, ASTM (American Society for Testing and Materials) D7566 permits blends up to 50% to satisfy the overall fuel specification. Similar to other drop-in biofuels, other drop-in biofuels may not meet all of the requirements within a particular fuel standard and may thus need to be used in specified blends. It is also clear that drop-in biofuels, as opposed to bioethanol or biodiesel, are better positioned to prevent blend wall issues (although mix restrictions may still spread on) (Tong et al. 2014). Most nations apart from Brazil mix ethanol at levels that are below 10% by volume with the possible exception of flex-fuel vehicles (FFVs) and E85 vehicles, and they may be able to use the infrastructure already in place (the current stock of petroleum refineries, supply routes, and liquid fuel-powered combustion engines) more effectively (Van Der Kroft 2020). These points emphasize the significance of drop-in biofuels as the current infrastructures and engines do not need to be adjusted or little modification may be necessary. Some common drop-in biofuels include hydro-treated vegetable oil (HVO), fatty acid methyl esters (FAME), hydrothermal liquefaction (HTL) with full hydrodeoxygenation biofuels, fast-pyrolysis with full hydrodeoxygenation biofuels, and gasification with consequent Fischer–Tropsch (GFT) synthesis biofuels (Van Der Kroft and Pruyn 2021).

Since drop-in biofuels operate on the idea of a short carbon cycle, they have the potential to reduce GHG. The biomass utilized as a feedstock for the production of biofuels absorbs CO_2 from the air over their existence, in contrast to fossil goods. The same quantity of CO_2 is emitted into the atmosphere when the biofuel is burned (Bouman et al. 2017). There is no competitiveness for land usage in this regard when waste products are taken into consideration. Therefore, the overall emission is regarded as zero. However, because biomass is not distributed equally around the world, some areas may need to import or export significant amounts of feedstock. This means that the emissions from production and transportation must be taken into account when calculating the fuel's emissions. Therefore, the long-term viability of the biofuel is mostly determined by manufacturing and transportation. This results in a requirement for biofuel production and transportation. Therefore, production and transportation need to be optimized in order to lower costs and emissions (Initiative 2019).

The 2G biofuels that are produced from different feedstocks can be considered for drop-in biofuels. These feedstocks have high content of stored energy that can support the microbial and even bacterial growth. In Table 6.1, various types of feedstocks for biofuels production are shown. Interestingly, biomass-based feedstocks adsorbed and sequestrated air carbon dioxide (CO_2) during their lifespan, and whenever the produced biofuels are combusted, only the same amount of CO_2 is emitted into the atmosphere that contributes to the carbon neutral strategy. Furthermore, waste products as feedstock can be treated to reduce the large quantities and not to dispose in the environment.

Table 6.1 Different feedstocks used for biofuels production (Van Der Kroft and Pruyn 2021; Mańkowska et al. 2021)

Biomass feedstocks	Category
Palm oil	Vegetable oils
Sunflower oil	
Soybean oil	
Rapeseed oil	
Miscanthus	Energy crops and their residues
Poplar	
Willow	
Grain residues	Agricultural residues
Lignocellulosic residues of plants	
Municipal solid waste	Wastes
Animal manures	
Wastewater sludge	
Tallow (animal fat)	Waste oils
Used cooking oil	
Leftover from logging operations (stumps, branches, bark, or tree tops)	Forest residues
Early thinning of wood	
Final felling of wood (sawdust)	
By-products and co-products of industrial wood-processing operation (wood chips, sawmill slabs, sawdust, or bark)	

6.3 Technological Frameworks for Drop-In Biofuels Production from Feedstocks

Three basic technological frameworks, along with the hybrid category, can be used to manufacture drop-in biofuels (Fig. 6.2) (Ben-Iwo et al. 2016). Oleochemical methods built on lipid feedstocks can be improved to produce drop-in fuels that are different from FAME biodiesel, including hydro-treated essential fatty acids (HEFA) or HVOs. Gasification or pyrolysis processes that create intermediates that can be converted to drop-in biofuels (such as Fischer–Tropsch liquids and hydro-treated pyrolysis oils (HPOs)) are examples of thermochemical technologies based on lignocellulosic feedstocks. Hydrothermal liquefaction is another thermochemical technology, and when pyrolysis is specifically mentioned, it refers to all types of pyrolysis, including rapid and catalytic pyrolysis (De Jong et al. 2018).

In general, biochemical technologies generate single product hydrocarbons like farnesene, alcohols, or fatty acids (FAs) that can be modified further to drop-in biofuels. These sugars are frequently produced from conventional feedstocks like corn or sugarcane, or cellulosic sugars. These platforms are used in other hybrid technologies, such as the conversion of alcohol to jet (ATJ) and the fermentation of off-gases producing alcohols that can be upgraded (Karatzos et al. 2014). Drop-in biofuels are typically produced through the biochemical conversion of sugar, cellulose, or by enzymatic hydrolyzed starch or starch, followed by fermentation of lipids,

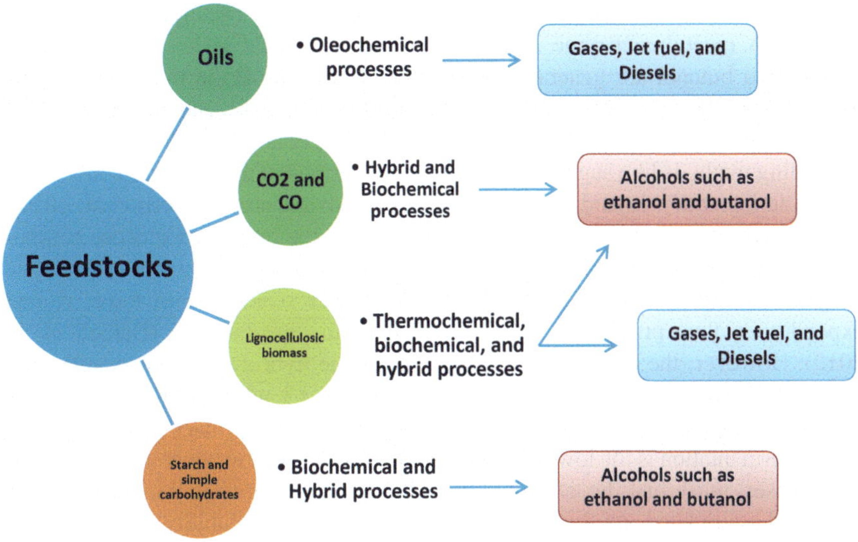

Fig. 6.2 Simplified technological frameworks for the manufacturing of drop-in biofuels

alcohols, or isoprenoids. These advanced biocatalytic processes change sugars to less oxygenated, more energy-dense molecules like longer-chain alcohols (butanol and butanediol), and higher molecular weight compounds like isoprenoids and FAs. This is similar to how glucose is fermented to ethanol in conventional (or so-called 1G ethanol production) biofuel production. The synthesis of drop-in biofuels and their intermediates has been proposed using a variety of biological routes, feedstocks, and microorganisms (Weber et al. 2010).

Although the majority of drop-in biofuel technologies fall under one of the oleochemical, thermochemical, or biochemical platforms previously mentioned, certain methods integrating two or more of the platforms are referred to as hybrid pathways to drop-in biofuels for this reason. The three main techniques in this group are ATJ, aqueous phase reforming, and syngas fermentation (Bioenergy 2019). Despite the fact that other microorganisms can use the hydrogen and carbon monoxide in syngas, this is covered in the section on hybrid technology that follows.

6.3.1 Syngas Fermentation

Thermo-catalytic conversion from syngas to liquid biofuels is possible. To manufacture potential drop-in biofuel intermediates, several fermentative microorganisms can also use syngas as their source of energy and carbon. Some autotrophic bacteria can utilize energy-rich molecules like CO and H_2 as well as single carbon compounds like CO and CO_2 as carbon sources. Syngas can be fermented by bacteria like *Clostridium ljungdahlii* and *Clostridium carboxidivorans* to produce intermediates like ethanol, butanol, and acetic acid. These bacteria convert CO to acetyl-CoA (acetyl-coenzyme A) via the Wood-Ljungdahl pathway, from which ethanol and butanol are generated (Sun et al. 2019). The H_2 in the syngas or the oxidation of CO to CO_2 by a carbon monoxide dehydrogenase enzyme provides the reducing power needed for the first process (Sánchez et al. 2019). In comparison to competing methods like sugar fermentation and thermo-catalytic syngas conversion, syngas fermentation is said to have a number of advantages (Ciliberti et al. 2020). For instance, this method can use lignin along with the biomass's glucose components, unlike sugar fermentation. There have been suggestions that thermo-catalytic syngas conversion is more cost-effective at smaller scales due to lower capital expenditures thus proving to be less sensitive to contaminants (Daniell et al. 2012). However, the volumetric productivity of the hybrid approach is still quite low, with current volumetric productivities around 0.3 g/L/h, similar to the biochemical routes to butanol and farnesene. It appears that the primary limiting step for syngas fermentation is the low solubility of CO and H_2 gases in aqueous solutions, as these gases must first be absorbed into the aqueous fermentation broth to be microbial assimilated. As the gas-liquid mass transfer and solubility of both molecular hydrogen and carbon monoxide in water are quite low, compared to more conventional sugar substrates (where gas-liquid mass transfer is not needed), syngas fermentation rates are typically constrained by physical mass transfer limitations rather than being limited by metabolic capacities. The syngas must be cooled down from a production temperature of around 700 °C or higher to approximately 50 °C or less before it can be fermented from an energy balance standpoint. This substantial cooling need could be troublesome because gasification systems frequently experience temperature swings that present difficult technical problems and generally have a negative impact on operational energy efficiency. Utilizing syngas fermentation technology, INEOS Bio and LanzaTech have constructed and are running sizable demonstration/small commercial operations. But up until now, ethanol rather than drop-in biofuels has been the dominant product (Kararzos et al. 2017).

6.4 Bacteria and Their Metabolic Pathways for Drop-In Biofuels Production

Cyanobacteria (a group of photosynthetic bacteria) are examples of microorganisms that can directly absorb CO_2 from the environment and transform it into ethanol or lipids. As an alternative, bacteria, yeasts, and heterotrophic algae can use sugars produced by the hydrolysis of cellulose and hemicellulose in cellulosic biomass, as well as sugars from sugarcane, sugar beet, starch, and other energy-storing polysaccharides (Bruder et al. 2019; Patel et al. 2017).

Despite the fact that other microorganisms can use the hydrogen and carbon monoxide in syngas; this is covered in the section on hybrid technology that follows. Although different intermediates can be used to produce either diesel or gasoline, the

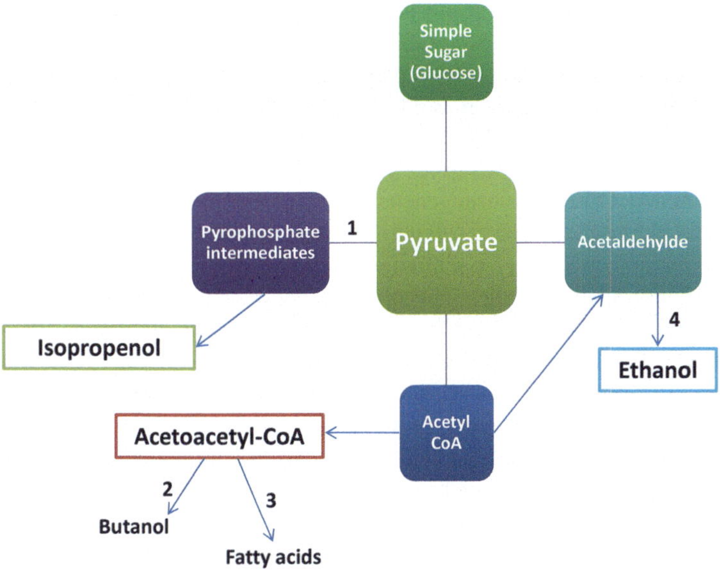

Fig. 6.3 Four primary metabolic routes for drop-in biofuels production by microorganisms. 1. Isoprenoid pathway; 2. Acetone-butanol-ethanol (ABE) pathway; 3. FA pathway; and 4. Ethanol pathway

production of gasoline is more suited to the higher oxygenated higher alcohols, whereas the production of diesel and jet fuel is more suited to the larger molecular weight, longer chain, more saturated lipids and isoprenoids. It is possible to choose or modify cyanobacteria, yeasts, and bacteria so that they create isoprenoids and lipids or higher alcohols, with the possibility because the metabolic pathways from one microbe may be heterologously produced in another (Lü et al. 2011; Howard et al. 2013).

The generated intermediate will likely need to undergo some type of hydro processing after biological production, despite the biosynthetic pathway employed, in order to produce a drop-in biofuel. As was previously mentioned, more hydrogen is needed to upgrade an intermediate to a fungible drop-in functional similar to diesel, jet fuel, or gasoline blend stock the more oxygenated and unsaturated the intermediate. One of these technologies' most alluring features from the standpoint of commercialization is the ability to utilize existing ethanol infrastructure by shifting the microorganisms employed in the fermentation procedure to a microbe-to-drop-in fuel or fuel intermediate method (Malik et al. 2021). For the process of transformation of glucose into the chemically reduced compounds that might make up drop-in biofuels, four primary metabolic routes have been proposed (Fig. 6.3).

The oxidation of glucose to two pyruvate molecules (glycolysis) is the starting point for each route. As the pyruvate is transformed into other intermediates, for example acetyl-CoA or acetaldehyde, the routes diverge from there. While the routes that lead to more saturated, longer-chain compounds like isoprenoids or FAs are aerobic, the mechanisms that result in butanol and alcohol are anaerobic (Jin et al. 2011). In regard to productivity and suitability for the production of drop-in biofuels, each pathway has a particular role to play. Using the specific pathway for palmitate biosynthesis as an example, it is clear that each elongation cycle requires the reducing power of two NADPH (nicotinamide adenine dinucleotide phosphate) molecules. A portion of the feedstock carbon must normally be aerobically respired into CO_2 and H_2O in order to regenerate the $NADP^+$ to NADPH (Jovanovic et al. 2010). The pentose phosphate pathway (PPP) is where NADPH is produced, and NADPH's reducing capacity essentially acts as a catalyst for hydrogen insertion. It has been demonstrated that the yields of metabolic processes involving the chemical reduction of carbohydrates can be increased (and the requirement for carbon-consuming NADPH synthesis decreased) by the alternate introduction of molecular hydrogen (Liu et al. 2021; Stein et al. 2022). The more elongation cycles and decreasing power needed depend on how long and saturated the FA chain is. This reducing power must come from either alternate hydrogen supplies or from the yield-detrimental oxidation of feedstock carbon (Kararzos et al. 2017).

The biological conversion performance and efficiency and the quantity of process inputs (such as nutrients, process energy, and equipment) needed to create and recover the target molecules from the production broth are the main factors that affect the commercial feasibility of the biochemical platform. Due to the high expense of sugar feedstocks, a biofuel (or biofuel intermediary) must be produced (Plugge 2017). To reach the kind of performance that can be attained by the "gold standard" of the existing sugar-to-ethanol fermentation technologies, production and recovery process stages must be extremely efficient and durable. The alternative drop-in biological methods, in contrast, produce substantially lower productivities than the traditional sugar-to-ethanol fermentation procedures and are more energy and carbon costly. For instance, production of butanol using *Clostridium* spp. can only accomplish 0.08–0.46 g/L/h, while production of ethanol using *S. cerevisiae* may frequently reach volumetric productivity of 2.00–3.00 g/L/h. In biological processes for manufacturing chemicals important to drop-in biofuels, there is still a large amount of technological risk and uncertainty, particularly when genetic engineering of the production microorganisms is involved (Stein et al. 2022; Liu et al. 2019). The lipids and solvent-type molecules that are employed as precursors for drop-in blend stocks are typically secondary metabolites that are more closely related to cell development, maintenance, and survival than to cellular energy metabolism. Unless the microorganism is stressed, they are usually only produced in modest volumes (Kararzos et al. 2017). Another issue is that, often, cell growth is sacrificed when higher quantities of these kinds of products are created. For instance, when algae are nitrogen-starved, they normally output more lipids but are unable to carry out the necessary protein synthesis for growth. Saturated substances, like as lipids, accumulate in dormant cells as a sort of energy storage to safeguard the organism against probable future energy restrictions from the perspective of microbial survival. Additionally, there is typically a delay between the start of the biological process and the beginning of the build-up of secondary metabolites like lipids. By contrast, yeasts (or bacteria) use a core metabolic route connected to main energy metabolism to produce ethanol (Bai et al. 2008). As a result, there is no delay in ethanol synthesis once the process is started, and fermentative ethanol production is closely related to cell development (Tse et al. 2021). The successful marketing of drop-in relevant goods, which rely on rapidly sophisticated biological processes, will be greatly influenced by the industrial resilience of various microbial strains and the ease of scaling up biological-based production processes (Kararzos et al. 2017).

It should be emphasized that the industrial difficulties for aerobic (isoprenoid and FA) and anaerobic (n-butanol and isobutanol) processes are very different. However, because they may be manufactured with high selectivity, biochemical-based platforms offer a clear advantage over thermochemical procedures in the manufacturing of renewable compounds (Osman et al. 2021b). Though achieving cost-competitive biofuel production may be difficult due to the reduced volumetric outputs and yields of these still-evolving microbial-derived industrial processes, their application as higher-value chemical feedstocks has shown to be profitable. It has been demonstrated that different creative approaches are needed in order to recover the products from the aqueous broth due to the smaller yields and various chemical characteristics as compared to ethanol (Adegboye et al. 2021).

Lower titers typically indicate the need for more production vessels (or volume) to produce a given quantity of product, while higher amounts of process energy may be necessary to recover a given quantity of product, as the final product titer is closely linked with process capital and operating costs. Because of this, it is doubtful that the majority of biochemically produced goods will be employed as drop-in biofuels before these higher-value chemical, cosmetic, pharmaceutical, etc., sectors are fully developed (Jiang et al. 2018). Even more valuable applications are likely when the chain length of these biochemically produced compounds increases (e.g., Amyris' farnesene, which commands greater pricing in the cosmetics/detergent sectors). It should be highlighted that creating oxidized products on biochemical platforms is more effective and appropriate than doing so with substantially reduced molecules. Highly oxidized compounds like lactic, acetic, and adipic acids are difficult to produce using non-biochemical platforms and are priced much higher than ethanol or biodiesel (Kararzos et al. 2017). Since sugar molecules naturally include large amounts of oxygen, converting them into oxidized intermediates rather than drop-in hydrocarbon fuels is probably more economically appealing. Due to anticipated rapid expansion in the markets for bio-based materials, market saturation difficulties for renewable chemical substances in the near future shouldn't be a concern (Kargbo et al. 2021).

Thoroughly, the biochemical approach to manufacturing drop-ins produces pure molecular streams of functionalized molecules, which now fetch greater chemical cost than as intermediates for drop-in biofuels. When this is compared to the oleochemical and thermochemical platforms, the main function of the metabolic pathways involved in the manufacture of drop-in biofuel is to biologically deoxygenate the sugar feedstock to produce more highly reduced or saturated products such as isoprenoids and Fas (Kararzos et al. 2017). However, compared to traditional sugar-to-ethanol fermentation processes, these biological processes are significantly less productive and relatively energy and carbon costly. As markets for bio-based chemicals keep growing, the need for biochemical drop-in products is anticipated to increase and stay considerable in the near- to mid-term.

6.5 Essential Bacterial Pathways Leading to "Drop-In" Biofuel Production

In recent years, there has been a rapid and remarkable advancement in the microbial synthesis of "drop-in" biofuels derived from renewable resources. This progress has been underpinned by the development of a plethora of metabolic engineering techniques, specifically tailored to elevate biofuel titers, enhance yields, improve productivities, and bolster the overall robustness of biofuel production systems. Moreover, various biosynthetic pathways have been ingeniously designed to produce biofuels with a diverse array of molecular architectures, as explored by Jiang et al. (2018). The evolution of the biofuels landscape has been nothing short of extraordinary, spanning a wide spectrum of scales, from small-scale laboratory setups to large industrial reactors. With the advent of 2G biofuels, including bioethanol and biodiesel, the biofuels sector has made substantial strides. A pivotal transformation has occurred in the feedstock options, transitioning from reliance on food crops such as corn and oilseeds to tapping into abundant and cost-effective non-food biomass sources, exemplified by corn stover and switchgrass (Frederix et al. 2016). Furthermore, the expansion of available feedstocks encompasses solid wastes from industries and sewage sludge (Gaeta-Bernardi and Parente 2016), as well as the harnessing of natural resources like syngas, methanol, and methane (Whitaker et al. 2017), and even the utilization of CO_2 as a feedstock (Oliver et al. 2016). These efforts are aimed at establishing the industrial-scale production of drop-in biofuels. In addition to the conventional model microorganisms like *Escherichia coli* and *S. cerevisiae*, there is now a diverse array of non-model organisms that serve as fermentation hosts. Each of these novel hosts exhibits unique capabilities, enabling them to efficiently consume specific feedstocks, adapt to particular fermentation conditions, and proficiently produce the desired biofuels. This diversification has been a significant focus, accompanied by intensive research to engineer the structures of bioproduced fuels, ensuring that their physiochemical and combustion properties align with existing engines and transportation infrastructures (Peralta-Yahya et al. 2012). The majority of these drop-in biofuels originate from five well-established biosynthetic pathways: the FA pathway, the isoprenoid pathway, the non-decarboxylative Claisen condensation process, the polyketide pathway, and the pathway of α-keto acids. What distinguishes these biofuels is their compatibility with a wide spectrum of engines, including spark-ignition, compression-ignition, and gas turbine engines. Remarkably, they often exhibit qualities that closely resemble those of petroleum-based fuels, differentiating them from the conventional bioethanol and biodiesel, as described by Zargar et al. (2017) and Cai and Zhang (2018).

6.5.1 α-Keto Acids Pathway

By decarboxylating and reducing cellular short-chain α-keto acids, essential intermediates in the synthesis of branched-chain amino acids by lactic acid bacteria, a combination of α-keto acid decarboxylase and an alcohol dehydrogenase has been harnessed to generate short-chain alcohols as part of the α-keto acid pathway. This pathway encompasses various alcohols, including isobutanol, 3-methyl-1-butanol, and 2-methyl-1-butanol. Furthermore, through a recursive process involving acetyl-CoA condensation and CO_2 release, the acyl chain length of α-keto acids can be further extended, making use of altered substrate specificity in LeuABCD enzymes. This method has been effectively implemented in multiple engineered *E. coli* strains to craft alcohols with medium-chain lengths, reaching up to C_8, or imbuing them with branched or aromatic characteristics, as demonstrated in the work of Jiang et al. (2018). Furthermore, it is noteworthy that cyanobacteria have been demonstrated as a successful utility in the production of isobutanol. In a notable example, within *Synechococcus elongatus* PCC 7942, an approach involving the expression of various enzymes, including AlsS from *Bacillus subtilis*, IlvC and IlvD from *E. coli*, Kivd from *Lactococcus lactis*, and Adh (YqhD) from *E. coli*, led to a significant increase in isobutanol synthesis, as demonstrated in the study by Atsumi et al. (2009). This pioneering work introduced a fascinating concept of harnessing the photosynthetic capabilities of cyanobacteria to play a pivotal role in the production of biofuels, showcasing the potential of these microorganisms in this innovative field. Moreover, the α-keto acid pathway offers the distinct advantage of precise chain length control, allowing for incremental carbon atom additions to the acyl chain. The somewhat reduced titer of medium-chain alcohols (C_6–C_8), in comparison to isobutanol, may be attributed to the chain elongation process, which comes at the cost of carbon efficiency and results in the accumulation of the reducing cofactor NADH, as detailed in the study by Jiang et al. (2018). Furthermore, α-keto acids or their corresponding alcohols have the potential to be used for the production of acetate-based esters, as explored by Rodriguez et al. (2014), as well as short-chain alkanes, as investigated by Zhang et al. (2016), serving a significant role in biofuel applications.

6.5.2 FA Pathway

In response to the limitations of oilseed-derived biodiesel, characterized by poor low-temperature performance and competition with food supplies, there has been a concerted effort to employ advanced technologies for the substantial production of microbial-based fuels derived from FAs. In vivo, the biosynthesis of FAs is primarily orchestrated by the distinct FA synthase (FAS) complexes, FASII, or multifunctional FASIE synthase, within eukaryotic cells and the majority of bacterial cells. This intricate process commences with the condensation of acetyl-CoA and malonyl-ACP, leading to the formation of a four-carbon acyl chain. Subsequently, this acyl chain is extended on acyl carrier proteins (ACPs) through a series of repetitive reactions, yielding straight long-chain acyl-ACPs spanning the range of C_{14} to C_{18}, predominantly intended for lipid synthesis. To cater to the specific requirements of biodiesel production, an approach involving the overexpression of a medium-chain-specific thioesterase has been successfully implemented in *E. coli*, resulting in the generation of medium-chain FAs, ranging from C_8 to C_{12}, as described by Torella et al. (2013). These medium-chain FAs, having favorable properties for biodiesel applications, can be further directed toward diverse biofuel products through tailored pathways, including alkanes, alcohols, and esters, as illustrated by Choi and Lee (2013), Youngquist et al. (2013), and Fan et al. (2013), respectively. In contrast to their branched-chain counterparts, biofuels derived from straight-chain fatty acids (SCFAs) exhibit comparably high freezing temperatures and viscosities, which pose constraints on their low-temperature operability—a critical parameter for applications in diesel and jet fuels. This limitation has prompted the development of strategies for engineering the production of branched-chain fatty acids (BCFA). The engineered *E. coli* BCFA pathway incorporates key components, including a branched-chain acyl-CoA dehydrogenase (BKD), a branched-chain-specific FabH, and a thioesterase, which change branched-chain acyl-CoAs to branched-chain acyl-ACPs and BCFAs, respectively. However, early iterations of modified strains displayed a modest production of BCFAs alongside a substantial co-production of SCFAs, primarily due to competition with the conventional FA biosynthetic pathway (Haushalter et al. 2014; Jiang et al. 2015). To enhance BCFA titers and percentages, several techniques were developed, such as dynamic modulation of the branched-chain specific FabH (Haushalter et al. 2014) and deletion of the *E. coli fabH* gene (Jiang et al. 2015). Interestingly, it was noted that overexpressing the rate-limiting enzyme complex BKD led to adverse effects, hampering the host cell's ability to lipoylate proteins and impeding cell growth, as reported by Bentley et al. (2016). The implementation of a synthetic protein lipoylation pathway, coupled with an increase in the BCFA titer to 276 mg/L, effectively mitigated this BKD-related toxicity. Further control over the branch structure of FAs was achieved by managing the cellular pool of each α-keto acid, as these α-keto acids ultimately contribute to the terminal groups of BCFAs. This approach has led to the distinct production of anteiso, even-chain-iso, and odd-chain-iso BCFAs, collectively accounting for 77%, 61%, and 77%, respectively, of the total free FAs (Bentley et al. 2016). This versatile capacity to engineer branch termini, regulate FA chain length and functional groups, and establish multiple pathways for the creation of drop-in biofuels significantly amplifies the array of compounds obtainable through the FA biosynthetic route. While naturally occurring FAS pathways have proven highly effective, offering FAs as desirable intermediates for biofuel production, challenges arise due to the necessity of FAs. Three key factors hinder the interaction of short- and medium-chain acyl-ACPs with desired heterologous enzymes in biofuel pathways: firstly, the incorporation of non-native FAs into the cell membrane may disrupt membrane fluidity and inhibit cell growth; secondly, the expression of native FA biosynthetic genes is often associated with growth, limiting the production of FA-derived fuels to specific growth phases; and thirdly, interactions among natural FAS enzymes can present complications, as explained by Jiang et al. (2018).

6.5.3 Non-decarboxylative Claisen Condensation Pathway

Clostridium kluyveri was the source of the first non-decarboxylative Claisen condensation process seen in nature (Yan and Dong 2018). Moreover, in the phyla *Verrucomicrobia*, *Planctomyces*, *Chloroflexi*, *Proteobacteria*, and *Actinobacteria*, *oleABCD* genes are responsible for the production of olefinic hydrocarbons and their expressed proteins have shown high similarity to non-decarboxylative Claisen condensation thiolases (Sukovich et al. 2010). Two molecules of acetyl-CoA are combined in this process to form acetoacetyl-CoA, which is then reduced via reduction events to butyrate. The non-decarboxylative Claisen condensation mechanism grows the carbon chain on CoA and utilizes acetyl-CoA or other short-chain acyl-CoAs as primers, compared to the FA biosynthetic pathway that grows the carbon chain on ACP. By reversing the oxidation route, Tarasava et al. (2022) indicated a synthetic non-decarboxylative Claisen condensation mechanism. An Acyl-CoA reductase and an alcohol dehydrogenase can stop the elongation cycle, resulting in straight-chain alcohols, or an acyl-CoA thioesterase can do so, resulting in carboxylic acids. Researchers then extended the primer from acetyl-CoA to six other functionalized primers and integrated three other extender units by switching thiolases and reduction enzymes. This platform has significantly increased the structural variety of molecules that are produced by biosynthesis, producing a number of novel bioproducts such as 2,3-dihydroxybutyric acid and 2-methylpentanoic acid. The non-decarboxylative Claisen condensation pathway has two key advantages over the FA and α-keto acid pathways: the capacity to integrate internal functional groups (including branch structures) into the final product and the capacity to orthogonally regulate the structures of both primers and extenders (Cheong et al. 2016). As more extenders are incorporated by new enzymes and new techniques are created for precisely regulating the number and sequence of elongation cycles, this route will become more adaptable.

6.5.4 Isoprenoid Pathway

Alternate fuel sources include terpenes and terpenoids produced by the isoprenoid pathway. C_5 isopentenyl pyrophosphate (IPP) and dimethylallyl pyrophosphate (DMAPP), which are made by either the mevalonate (MVA) or methylerythritol-4-phosphate (MEP) pathways, are used as building blocks in the isoprenoid pathway. Terpene synthases or polyisoprenoid diphosphate modifying enzymes such as hydrolases or esterases subsequently convert polyisoprenoid diphosphates into terpenes or terpenoids, lengthening the carbon chain as a result of the repetitive construction of IPP or DMAPP. By spreading the pathway at geranyl pyrophosphate (GPP) and farnesyl pyrophosphate (FPP), a number of terpenes and terpenoids with biofuel potential have been produced. These include monoterpenes (C_{10}), sesquiterpenes (C_{15}), and their derivatives. Pinene and limonene are two of them, and their hydrogenated by-products, farnesene and bisabolene, have qualities that are comparable to jet fuels and diesel fuels, respectively (Zargar et al. 2017). According to Chuck and Donnelly (2014), limonene has been identified as a highly effective alternative to aviation kerosene. It has the potential to be employed as a drop-in replacement without requiring any chemical modifications. Additionally, its lower viscosity offers advantages such as improved atomization and pumping. Novel compounds have also been biosynthesized and suggested as fuel options, such as linalool, 1,8-cineole, and geraniol (Mendez-Perez et al. 2017; Jiang et al. 2017).

Although products biosynthesized through the isoprenoid pathway can be naturally occurring, they are typically found in low amounts. Hence, the researchers are focused on the systems biology, metabolic engineering, and microbial cell factories to overproduce these terpene compounds as substitutes for current jet fuels. *E. coli* is a well-known microbial model that can be engineered to form precursors (IPP and DMAPP) for isoprenoid-derived biofuels (Walls and Rios-Solis 2020). For instance, Peralta-Yahya et al. (2011) successfully utilized an eight-gene MVA pathway to transform acetyl-CoA into FPP within *E. coli*. This was achieved by introducing a single plasmid, resulting in a sufficient supply of the essential sesquiterpene precursor. By co-expressing a second plasmid containing a bisabolene synthase gene, they were able to produce 388 mg/L of a potential jet fuel. Through a series of metabolic engineering techniques such as codon optimization, modularization, and promoter optimization, the researchers further enhanced bisabolene production to reach a titer of 912 mg/L. Moreover, Wu et al. (2019) studied the optimization of limonene production in *E. coli* using a modularization approach. The limonene pathway was divided into three modules. The upstream module expressed *mvaE* and *mvaS* genes, derived from *Enterococcus faecalis*, which facilitated the conversion of acetyl-CoA to MVA. The midstream module expressed four genes responsible for converting MVA to IPP/DMAPP, sourced from *S. cerevisiae*. Lastly, the downstream module contained the plant geranyl diphosphate synthase (GPPS) and limonene synthase genes. Through careful optimization of the upstream and midstream modules using ribosome binding site (RBS) and promoter engineering techniques, respectively, the researchers achieved an optimal strain that produced 182 mg/L of limonene.

In addition to conventional engineered *E. coli*, terpene biosynthesis can be performed in the engineered algal or cyanobacterial cells. Due to the fact that 4G biofuels primarily utilize CO_2 as the main carbon source, it is expected that their environmental impact will be significantly mitigated in comparison to biofuels produced using bacterial or yeast cells (Walls and Rios-Solis 2020). For example, *Synechocystis* sp., a cyanobacterial host, has been recently manipulated to produce bisabolene (Sebesta and Peebles 2020). The most noteworthy characteristic of the isoprenoid pathway is its ability to produce structurally diverse products because of the varied activity of terpene synthases. Numerous terpene synthases are currently being sequenced, and thousands of distinct terpene structures have already been discovered (Rabe et al. 2016; Reddy et al. 2020). The relationships among most terpene synthases and the terpene compounds they produce, however, are still unknown for the majority of them. The identification of novel biosynthetic pathways to improved biofuels could be facilitated by new techniques for high-throughput terpene synthase characterization.

The 4G of biofuels is becoming possible due to synthetic biology. In order to design and regulate genetic circuitry systems in microorganisms and create a biological fuel manufacturing platform as easy to use as configuring a computer, it applies engineering concepts, including modularization and compartmentalization, to biological systems (Andrianantoandro et al. 2006). Any non-essential cellular metabolic pathway which diverts the product by side reactions needs to be carefully evaluated and managed in order to boost output and lower costs of targeted bioenergy molecules of interest. To accomplish this goal, a collection of fundamental host microorganisms with related synthetic biology technologies are required to facilitate the rapid development, testing, and optimization of targeted metabolic pathways. Presently, there aren't many microbial synthetic biology prototyping systems that are tailored for the reliable and predictable design, development, and implementation of modified microorganisms. Pyruvate, acetyl-CoA, and phosphoenolpyruvate are the three fundamental pathway metabolites that are the foundation of microbially produced advanced biofuel design. Instead of relying solely on random evolutionary permutation selection, better assembly techniques can hasten the development of a specific metabolic route (Pfleger et al. 2006).

6.5.5 *Polyketide Pathways*

There are three different forms of polyketide synthases (PKSs), but Type I PKSs are particularly desirable as biofuels since they can undergo full reductive processing to produce saturated carbon skeletons. Simultaneously, the polyketide biosynthetic pathway is one of the most widely versatile pathways for producing structure-diverse hydrocarbons (Yuzawa et al. 2016). When designing a synthesis reaction using polyketide pathways, the process can be directed toward the production of 1-butene, 1-hexene, 1-pentanol, and 1-hexanol, which can be oligomerized to produce jet and diesel fuels (Liu et al. 2015b). The assembly line formation and rounds of elongation present an easily discernible biosynthetic rationale for the targeted production of particular hydrocarbons, making Type I PKSs an appealing engineering target. Ketosynthase (KS), acyl-transferase (AT), dehydratase (DH), methyltransferase (MT), enoyl reductase (ER), ketoreductase (KR), ACP, and thioesterase (TE) are the domains that make up a fully reducing module (Zargar et al. 2017).

PKSs are a potentially significant platform for producing biofuels with desired physical qualities because of their adaptability and ability to tune their products, as well as their reprogrammable domains. However, the fact that PKSs have been shown to be catalytically slower than FASs poses a possible problem when using them. For instance, the 6-deoxyerythronolide B synthase (DEBS) system takes around 2 min to complete the synthesis of palmitic acid, but FAS can do so in less than a second, at least in vitro (Smith and Tsai 2007). Although the exact mechanism underlying this variation in catalytic efficiency is unknown, expression in *E. coli*, which may not be the ideal environment for this specific class of enzymes to fold in, may cause measurements of PKS kinetics to be delayed (Garg et al. 2014).

Recent research conducted suggests that in vivo PKS kinetics in other hosts may resemble FAS dynamics more closely than previously believed. Menendez-Bravo et al. (2014) distracted the FFA (free fatty acid) pool to an iterative Type I PKS for the formation of fatty alcohols and esters in the first exploitation of PKSs for biofuel generation. The team generated 98 mg/L of multi-methyl branching esters in *E. coli* by producing an iterative *Mycobacterium* TB PKS that accepts methylmalonyl-CoA as a substrate. In another investigation, Liu et al. (2015b) expressed the *Streptomyces iterative* Type I PKS, SgcE, and its corresponding thioesterase, SgcE10, in *E. coli* using a two-plasmid system. In a fed-batch fermentation followed by chemical hydration, the group improved the ratio of SgcE expression to SgcE10 expression, and the study showed a yield of 140 mg/L of pentadecane. Modular Type I systems offer exceptional chances to precisely customize the chemical structure, even though these iterative PKS pathways are technically straightforward and served as a starting point for developing PKS pathways for the synthesis of biofuels. Our team recently built a chimeric Type I PKS with fully reductive processing domains into the first module of borrelidin PKS, enabling the bio-based manufacture of the common chemical, adipic acid, as a significant illustration of this adaptability (Hagen et al. 2016).

Because employing FAS is difficult, the high level of control offered by modular type I PKSs is particularly helpful for producing selectively branched biofuels, which are necessary to lower the freezing point of fuels. The inclusion of methylmalonyl-CoA as a substrate or the use of S-adenosyl methionine (SAM)-dependent C-methyltransferases is two mechanisms by which PKSs branch. Through AT domain exchanges, AT knockouts with complementation of a trans-AT domain, or AT-site guided mutagenesis, AT domain incorporation can be achieved artificially (Dunn and Khosla 2013). These techniques frequently reduced the AT's affinity for the substitute substrate, leading to slower kinetics than the AT domain's original affinity (Yuzawa et al. 2017).

Alternative methylation methods involving SAM-dependent C-methyltransferases have not received much research. The process of reaction of methyltransferases has been clarified by recent research, which shows that methylation can happen prior to or following Claisen condensation (Wagner et al. 2016) and that removed mono- and di-methyltransferases can operate in vitro (Stevens et al. 2016). With such fine product formation control, alternatives for biofuels could be created that are specifically tailored and have the appropriate combustion characteristics.

6.5.6 Strategies for Pathway Optimization

New approaches are being devised to optimize production pathways and boost product titers, yields, and productive capacities in addition to innovative biosynthetic pathways and biofuel prospects. The traditional idea behind metabolic engineering is to introduce metabolic flux into the supply chain. This phase can be facilitated via a variety of techniques, such as upregulating pathway enzymes (Jung et al. 2016), deleting or downregulating rival pathways, and modifying the overall metabolism by altering transcriptional factors (Zhang et al. 2012; Ajjawi et al. 2017). In the meantime, a growing variety of techniques have been created to regulate cell populations, improve cell viability, and strengthen host cells in order to build metabolic pathways (Lim and Jung 2017). The host's homeostasis for certain metabolites or cofactors is disrupted by metabolite overproduction, which frequently interferes with natural metabolism for scarce cellular resources (Ohtake et al. 2017). These result in decreased cell development (Ceroni et al. 2015). Therefore, the modified metabolic pathway must be balanced by removing route bottlenecks while preventing the build-up of excess proteins or intermediates in order to boost the system robustness and thereby to achieve high efficient production. A metabolic pathway can be stabilized statically using engineered promoters or RBSs (Guo et al. 2016), or dynamically using a metabolite biosensor to modify flow in response to a rate-limiting or hazardous metabolite by controlling its production and/or consumption (Liu et al. 2015a). Separating the manufacturing process from the cell growth phase is an alternate method to prevent negative impacts on cell growth. Several methods, such as stationary phase signaling (Fang et al. 2021), quorum sensing (Gupta et al. 2017), and nutrient depletion (Niehus et al. 2018), have been devised to encourage the change from the growth phase to the production phase. For instance, a GAL regulatory circuit was developed in *S. cerevisiae* based on the Crabtree effect to autosense glucose depletion and simultaneously trigger the generating pathway (Kang et al. 2017). Additionally, the build-up of particular compounds may result in a range of negative outcomes for the host cells, chiefly physiological alterations of cell membranes. The engineered hosts could exhibit improved tolerance to various toxic substances such as alcohols, carboxylic acids, and aromatic compounds.

The selection of both the microorganism and the specific technological conditions holds paramount significance in biofuel production. This significance stands for the fact that biofuels predominantly are based on hydrocarbons, which are integrally characterized by their high toxicity, leading to considerable interferences in cell growth and metabolic processes (Mukhopadhyay 2015). To illustrate, the oxidation product of limonene, limonene hydroperoxide, exerts a considerable inhibitory effect on the production of the target compounds within cells (Zargar et al. 2017). In this point of view, it becomes evident that the design of the biofuel production process should not only consider the toxic impacts of substrates and final biofuel products but also account for the influence of intermediate metabolites and the potential accumulation of dead-end metabolites. A notable example is the case of isoprenol, which has been observed to exhibit toxicity to *E. coli* when present at concentrations exceeding 2% (Mukhopadhyay 2015). Likewise, biofuels and their terpene-based precursors have demonstrated a high degree of toxicity toward the microorganisms employed in their synthesis (Niu et al. 2018; Mendez-Perez et al. 2017). It is important to note that the primary mechanism underlying the toxicity of hydrocarbon-based biofuels largely revolves around interactions, incorporation, and damage to biological membranes. These membrane-related factors constitute the central facets contributing to the toxic effects associated with biofuels derived from hydrocarbons. Also, unfavorable industrial conditions (such as low pH and high temperature) through modifying the structure of cell membrane FA (e.g., chain length, degree of saturation, and abundance of cyclic rings) (Tan et al. 2016) and the arrangement of phospholipid head groups should be considered as critical factors in processes (Tan et al. 2017). As well, the overall output of the entire culture is significantly impacted by the heterogeneity of the generating cells (Schmitz et al. 2017). A recently created method called PopQC combines metabolite production and cell growth in a positive way, allowing for the enrichment of high-producing variants from a heterologous culture and increasing overall product titers by many orders of magnitude (Xiao et al. 2016). The production yield can be increased by choosing non-genetic variants, which offers great promise for unique system-based development. Although the science of metabolic engineering has recently focused more on the synthesis of high-value products (pharmaceuticals, fine chemicals, etc.), significant progress has nonetheless been achieved in recent years toward better biofuels. We anticipate that in the near future, fresh approaches will be continuously developed to boost titers, yields, productivities, and resilience in large-scale production, making microbial production of "drop-in" fuels more and more competitive from an economic standpoint.

6.6 Engineering Bacterial Metabolisms for Advanced Biofuels

Although the present mitigation strategy for fossil fuels in the transportation sector relies on short-chain alcohols, biodiesel, and HVOs from 1G and 2G feedstocks, these additives damage gasoline quality at high concentrations (Gopalan et al. 2020). Therefore, it is preferable to directly substitute fossil-derived base fuel with straight and cyclic hydrocarbons of different C-chain lengths and saturation levels. Except for the few oleaginous microbes mentioned above, which have widespread metabolic capacity for hydrocarbon biosynthesis, these molecules are only produced in trace amounts (0.1% of dry mass) and serve physiological rather than energy storage functions, such as regulating membrane fluidity and permeability, cell–cell signaling, and defense against desiccation or environmental toxins (Ladygina et al. 2006). Therefore, the creation of direct fuel substitutes must be engineered in bacteria. Through genetic engineering, bacteria can be manipulated and modified to introduce non-native biochemical pathways, enhance the organism's general metabolic functioning, and select desirable end products. Genes involved in the biosynthesis pathway and/or other controlled genes and pathways can be effectively treated to enable organisms to produce more with improved selectivity and yield for particular items. Increased production yields can be achieved by further engineering these microorganisms to withstand unfavorable growing conditions, like hydrolysate inhibitor resistance, through the use of synthetic biology. Since these bacteria can be engineered to withstand the toxicity of biofuel products and can synthesize high-value compounds from a variety of carbon sources, they have found widespread application in the industrial chemical manufacturing process. In addition to facilitating the insertion of heterologous pathways for the purpose of improving production concentrations and yields, the broad metabolic capacity of them and the abundance of synthetic biology techniques available to them also allow for the selection of specific biofuels (Jovanovic et al. 2010). Without the discovery, characterization, and engineering of enzymes from various phyla, such as aldehyde deformylating oxygenases (ADO) from cyanobacteria (Mund et al. 2022), ECERIFERUMs (CER) from higher plants (Bourdenx et al. 2011), insect cytochrome P450s (CYP) (Qiu et al. 2012), fatty acid decarboxylases P450 $OleT_{JE}$ from *Jeotgalicoccus* (Belcher et al. 2014), UndA from *Pseudomonas* (Rui et al. 2014), and the FAP (Sorigué et al. 2017) completed the transformation of endogenous metabolic products into potential fuel molecules. For the creation of advanced bacterial biofuels in these studies, compounds generated from the fatty acid, terpenoid, and polyketide anabolic pathways are generally addressed. In Table 6.2, characterizations of some engineered bacteria for the production of advanced biofuels are shown. The expanding application of synthetic biology (Savage et al. 2008) in the design, engineering, and iterative development of process-tailored microorganisms expressing novel, proof-of-concept metabolic pathways for the production of prospective fuel molecules is largely responsible for these advancements.

Despite considerable applications of engineered bacteria, it might have unanticipated negative effects. Bacteria are self-replicating microorganisms, which makes it challenging to stop them from discharging more in order to stop an adverse consequence. Recombinant genes can spread from their host bacterium to other bacteria after being introduced into the environment, which is a typical complaint. The results show that inter-specific horizontal gene transfer has been observed for a wide variety of bacteria, genes, and environmental conditions. Therefore, unless further precautions are taken, it would be incorrect to believe that recombinant genes introduced into the environment won't spread to related bacteria. However, a number of techniques have been devised to reduce the spread of genes to other bacteria by horizontal transfer (Davison 2002).

Table 6.2 Characterizations of some engineered bacteria used for advance biofuels production

Engineered strain	Modified gene(s)	Engineering method	Biofuel type	Yield	References
E. coli BW25113	*metA*, *tdh*, *ilvB*, *ilvI*, and *adhE*	Host gene deletion	1-butanol and 1-propanol	2 g/L (1:1 ratio of 1-butanol and 1-propanol)	Shen and Liao (2008)
E. coli strain C41	P450 $OleT_{JE}$ from *Jeotgalicoccus* sp. ATCC 8456	Gene expression in the host	Alkene	–	Belcher et al. (2014)
Synechococcus sp. 7942	*thl*, *atoAD*, *adc*, and *adh*	Gene knockout	Isopropanol	0.02 g/L	Kusakabe et al. (2013)
Corynebacterium glutamicum	Pyruvate decarboxylase (*pdc*) and alcohol dehydrogenase (*adhB*) from *Zymomonas mobilis*	Gene expression	Ethanol	10.5 g/L	Inui et al. (2005)
Anabaena sp. 7120	*faS*	Gene knockout	Farnesene	0.3 mg/L	Halfmann et al. (2014)
L. lactis	Codon-optimized pyruvate decarboxylase (*pdc*) from *Z. mobilis*	Gene expression	Ethanol	–	Solem et al. (2013)
*B. subtilis*168	*bdhA* and *acoA*	Gene knockout	2,3-Butanediol	103.7 g/L	Fu et al. (2016)

6.6.1 Engineering Fatty Acid Advanced Derived Biofuels

FA synthesis is mechanistically constant between prokaryotes and eukaryotes, despite the fact that the enzymes involved are different. Through thioester connections between the carboxyl group of the intermediate metabolites and the Ser36 of the ACP, the intermediate metabolites are covalently linked to ACP during FA production. FA synthesis happens in two stages: C-chain cyclic elongation and initiation. Malonyl Co-A is transformed into malonyl-ACP, which is then combined with acetyl-CoA to create α-keto-butyryl-ACP, during the beginning of straight-chain FAs. This substance is next dehydrated to 2-enoyl-ACP, reduced to produce D3-hydroxybutyryl-ACP, and finally reduced to butyryl-ACP. The C-chain is created through condensation with malonyl-ACP followed by reduction-dehydration-reduction in the intermediate metabolites (Love 2022). TEs cleave the elongating fatty acyl-ACP from the acyl-ACP when it reaches a certain length, producing FFA.

It is possible to customize FFA to particular C-chain lengths with the overexpression of various heterologous TEs. The creation and degradation of cellular FFA pools are delicately balanced (−oxidation), with the first step being FFA conjugation with acyl-Co by acyl-CoA synthase (FadD) to form fatty acyl-CoAs, which are then converted to acetyl-CoA by degradative thiolases to produce metabolic energy. It's interesting to note that the C-efficiency of the entire process can be increased by reversing FFA degradation (a procedure known as "reverse oxidation"), in which the thiolases act as synthetic enzymes to produce long-chain FAs directly from acetyl-CoA rather than first requiring malonyl-CoA (Dellomonaco et al. 2011). Early metabolic transformations centered on designing organisms as catalysts for already existing biofuels, such as the production of biodiesel-like compounds in *E. coli* through conversion of free FAs to FAME by heterologous expression of a *Mycobacterium marinum* fatty acid O-methyl transferase, using endogenous SAM as the methyl donor and FFAs as the methyl donor. Additionally, FA metabolism has been modified to produce fuel molecules that can be substituted for petroleum distillates. The luxCED complex or CAR (carboxylic acid reductase) (Howard et al. 2013) can convert FFA to fatty aldehydes, which are further reduced by ADO (Schirmer et al. 2010) to alkanes and alkenes with different C-chain lengths. Both systems have advantages, with the CAR having a wider range of C-chain selectivity and the luxCED system permitting more precise customization of C-chain length (Winkler 2018).

As an alternative, fatty acyl-ACPs can be converted to alkanes by means of an NAD(P)H-dependent enzyme called fatty acyl-ACP reductase (AAR) (Liu et al. 2014). Aldehyde reductases (AHR) and acyl-CoA reductases (ACR) can convert fatty aldehydes and acyl-CoAs into fatty alcohols, respectively (Akhtar et al. 2015). Finally, as previously mentioned, a number of other enzymes have been discovered that catalyze the conversion of the by-products of lipid biosynthesis into appropriate, advanced biofuels. These enzymes include CER, CYP, $OleT_{JE}$ UndA, and FAP. Some bacteria, most notably *B. subtilis*, have the ability to produce branched-chain FAs, in which the initial acetyl Co-A is substituted by a different primer, such as isovaleryl-CoA, isobutyryl-CoA, or 2-methylbutyryl-CoA, which are all produced from the valine, leucine, and isoleucine metabolic pathways. The fuels can be customized and optimized by combining molecules with branched and straight chains. Although these proofs of concept are encouraging, it is both required and feasible to enhance the titers of biofuels or precursors produced in this way (Das et al. 2020) given the levels of world fuel demand. For instance, it has been demonstrated that TEs overexpression raises FFA titers (Tee et al. 2014), potentially by boosting the "pull" of the metabolic sink, and that deletion of FadD, the first enzyme in FA catabolism, raises cellular FFA concentration even more (Lennen and Pfleger 2012).

6.6.2 Terpenoid-Derived Biofuels

The mevalonate and the MEP pathways (Chang et al. 2013) are two alternate metabolisms that produce isoprenoids and terpenoids. In the first, mevalonic acid—the precursor of IPP and DMAPP—is produced by the condensation of three molecules of acetyl-CoA via acetoacetyl-CoA to 3-hydroxy-3-methylglutaryl-CoA (HMG-CoA). The basic 5-carbon isoprene building blocks IPP and DMAPP go through a head-to-tail condensation to produce the monoterpene (C_{10}) GPP, which is used to synthesize further terpenoid compounds (Love 2022). Two isoprene units can combine and produce cyclic or acyclic monoterpenes that are appropriate to be used for jet fuels (Walls and Rios-Solis 2020). FPP, the direct precursor of C_{15} sesquiterpenes, is produced by further condensations with C_5 isoprenes, while geranylgeranyldiphosphate (GGPP), the precursor of C_{20} diterpenes, is produced by an additional isoprene addition. By head-to-tail condensation, FPP and GPP can also combine to generate homodimers that contain C_{30} triterpenes and C_{40} tetraterpenes (Love 2022). IPP and DMAPP are produced by the glycolytic cycle in the MEP pathway. Because of the branches and rings that can be found in terpenoid hydrocarbon chains and high-octane numbers compared with primary alcohols, along with lower hygroscopicity and lower freezing point, these compounds are good candidates for building improved biofuels. *S. cerevisiae* and *E. coli* have been used to produce C_{15} bisabolane (Madavi et al. 2022; Peralta-Yahya et al. 2011) as well as limonene and oxidosqualene in *Y. lipolytica*, farnesane, a substitute for jet fuel, in *E. coli*, and the tricyclic sesquiterpenes epi-isozizaene and pentalenene in *S. cerevisiae* and isocomene in *E. coli*. However, recent advancements in artificial intelligence and machine-guided genomic mining hold promise to facilitate the quick identification and testing of potential biosynthetic pathways (Ramzi et al. 2020), thereby accelerating and streamlining the biosynthesis of novel biofuels. The vast number of potential terpenoids is a daunting bottleneck for further development.

6.7 Future Perspectives

The transportation industry faces a critical demand for innovative biofuels that can effectively serve as alternatives to conventional petroleum distillates. Initiating and actively pursuing the development of versatile, easily adaptable, and commercially viable bacterial biocatalysts becomes paramount. This endeavor aims not only to meet the immediate requirements of the transport sector but also to establish a foundation for ongoing advancements in biofuel technology, ensuring flexibility and adaptability in the face of evolving energy needs and environmental considerations. It is conceivable that drop-in biofuels derived from oleochemicals will play a pioneering role in laying the foundations for markets and establishing protocols for the practical application of these biofuels. A wealth of successful high-viscosity oil

(HVO)-based biofuel flight tests, evaluations of refinery processing, and the recent approval of oleochemical-derived jet fuel blend stocks by ASTM serve as compelling and illustrative instances that highlight the notable progress in this field. Despite the foundational role that oleochemical-based biofuels have played in establishing the groundwork for drop-in biofuels, there is anticipation that, in the mid-to-long-term outlook, thermochemical-based processes will constitute a significant proportion of the anticipated upsurge in drop-in biofuels production. These processes are strategically poised to spearhead the initial surge, particularly until saturation in chemical markets occurs. While the biochemical and hybrid-based procedures bring inherent value, they generally result in lesser quantities of more oxygenated drop-in biofuel intermediates.

Interestingly, these intermediate products often garner higher prices when positioned as chemicals in the rapidly expanding bio-based chemicals sector compared to their valuation as biofuels. Looking forward, there is a conceivable trajectory where future biorefineries will embrace biomass analogous to how petroleum refineries currently handle crude oil. This strategic approach involves the conversion of raw feedstock into a diverse array of fuels and chemical products within an intricately integrated facility. This forward-looking perspective underscores the potential evolution of biofuel production and signals a shift toward more efficient and comprehensive processing methods in line with the broader objectives of sustainability and industrial advancement. In contrast, more extensive thermochemical-based facilities will likely concentrate on transforming biomass feedstocks into dimensionally tailored drop-in biofuels and bioenergy products. In a slightly smaller-scale context, bio-chlorination facilities may engage in similar endeavors. While both paths to drop-in biofuels have witnessed substantial technical advancement in their development and enhancement, it is essential to emphasize the importance of supportive policies to expedite the growth of drop-in biofuels. These policies play a critical role in ensuring drop-in biofuels' sustained expansion and accelerating their eventual commercial success.

6.8 Conclusions

The transition toward achieving net-zero carbon emissions by 2050 poses a formidable challenge, especially when it comes to the decarbonization of the transport sector. This endeavor is further complicated by the backdrop of escalating energy demands, the proliferation of decentralized global energy systems, and the often-unpredictable adoption of emerging technologies. To address these challenges, innovative biofuels can be harnessed, crafted through diverse microbial metabolisms and utilizing biomass or alternative renewable substrates. Presently, the biofuels landscape predominantly relies on various microorganisms, spanning from microalgae to bacteria, for the production of commercial biofuels. However, the transport sector is in urgent need of cutting-edge biofuels with the potential to substitute for petroleum distillates. To bridge this gap, it is imperative to embark on the development of versatile, easily modifiable, and industrially viable bacterial biocatalysts. These biocatalysts should be equipped with the essential molecular tools required to cover the way for further innovation and optimization of engineered pathways. Complementing this, rigorous techno-economic and life-cycle evaluations of these novel opportunities are indispensable to facilitate the translation of advanced biofuels, often referred to as "drop-in biofuels," from laboratory-scale production to full-fledged commercial implementation. While it's true that drop-in biofuels, functionally on a par with petroleum-derived transportation fuels, are currently available in limited quantities on the market, there has been a notable surge in scientific progress and commercialization efforts in recent years. Much like conventional (1G) biofuels, derived from sugars and starch, laid the groundwork for the subsequent development and utilization of advanced (2G) biofuels, it's plausible that oleochemical-derived drop-in biofuels will initially serve as the vanguard in establishing markets and protocols for the application of these biofuels. The numerous successful HVO-based biofuel flight tests, refinery processing assessments, and the recent ASTM approval of oleochemical-derived jet fuel blend stocks serve as compelling examples of this progression. However, it's important to acknowledge that the availability, cost, and long-term sustainability of food-grade (vegetable oil) or animal oil/fat-based feedstocks may impose limitations on the continued advancement of the oleochemical platform. Similar to the experiences in the commercialization of cellulosic ethanol, where the adoption of more accessible and cost-effective non-food lignocellulosic feedstock often comes with increased technological challenges and elevated investment costs for feedstock preparation, it's an intricate balance of factors. These processes are designed to lead the initial surge, especially until chemical markets become saturated. The biochemical and hybrid-based procedures, while valuable, generally yield lower quantities of more oxygenated drop-in biofuel intermediates. These intermediates often command higher prices as chemicals in the rapidly expanding bio-based chemicals sector than they do as biofuels. While both paths to drop-in biofuels have witnessed substantial technical advancement in their development and enhancement, it is essential to emphasize the importance of supportive policies aimed at expediting the growth of drop-in biofuels. These policies play a critical role in ensuring the sustained expansion of drop-in biofuels and in accelerating their eventual commercial success.

References

Adegboye MF, Ojuederie OB, Talia PM et al (2021) Bioprospecting of microbial strains for biofuel production: metabolic engineering, applications, and challenges. Biotechnol Biofuels 14:5

Agency IE (2023) Transport biofuels. https://www.iea.org/reports/renewables-2022/transport-biofuels. Accessed 15 June 2023

Ajjawi I, Verruto J, Aqui M et al (2017) Lipid production in Nannochloropsis gaditana is doubled by decreasing expression of a single transcriptional regulator. Nat Biotechnol 35:647–652

Akhtar MK, Danapani H, Thiel K, Jones PR (2015) Microbial production of 1-octanol: a naturally excreted biofuel with diesel-like properties. Metab Eng Commun 2:1–5

Andrianantoandro E, Basu S, Karig DK et al (2006) Synthetic biology: new engineering rules for an emerging discipline. Mol Syst Biol 2:0028

Atsumi S, Higashide W, Liao JC (2009) Direct photosynthetic recycling of carbon dioxide to isobutyraldehyde. Nat Biotechnol 27(12):1177–1180

Bai FW, Anderson WA, Moo-Young M (2008) Ethanol fermentation technologies from sugar and starch feedstocks. Biotechnol Adv 26:89–105

Belcher J, Mclean KJ, Matthews S et al (2014) Structure and biochemical properties of the alkene producing cytochrome P450 OleTJE (CYP152L1) from the Jeotgalicoccus sp. 8456 bacterium. J Biol Chem 289:6535–6550

Ben-Iwo J, Manovic V, Longhurst P (2016) Biomass resources and biofuels potential for the production of transportation fuels in Nigeria. Renew Sustain Energ Rev 63:172–192

Bentley GJ, Jiang W, Guamán LP et al (2016) Engineering Escherichia coli to produce branched-chain fatty acids in high percentages. Metab Eng 38:148–158

Bioenergy IEA (2019) Drop-in biofuels: the key role that co-processing will play in its production. IEA Bioenergy. www.ieabioenergy.com/wp-content/uploads/2019/09/Task-39-Drop-in-Biofuels-Full-Report-January-2019.pdf. Accessed 21 May 2023

Bohre A, Dutta S, Saha B et al (2015) Upgrading furfurals to drop-in biofuels: an overview. ACS Sustain Chem Eng 3:1263–1277

Bouman EA, Lindstad E, Rialland AI et al (2017) State-of-the-art technologies, measures, and potential for reducing GHG emissions from shipping – a review. Transp Res D Transp Environ 52:408–421

Bourdenx B, Bernard A, Domergue F et al (2011) Overexpression of Arabidopsis ECERIFERUM1 promotes wax very-long-chain alkane biosynthesis and influences plant response to biotic and abiotic stresses. Plant Physiol 156:29–45

Bruder S, Moldenhauer EJ, Lemke RD et al (2019) Drop-in biofuel production using fatty acid photodecarboxylase from Chlorella variabilis in the oleaginous yeast Yarrowia lipolytica. Biotechnol Biofuels 12:1–13

Cai W, Zhang W (2018) Engineering modular polyketide synthases for production of biofuels and industrial chemicals. Curr Opin Biotechnol 50:32–38

Ceroni F, Algar R, Stan G-B et al (2015) Quantifying cellular capacity identifies gene expression designs with reduced burden. Nat Methods 12:415–418

Chang W-C, Song H, Liu H-W (2013) Current development in isoprenoid precursor biosynthesis and regulation. Curr Opin Chem Biol 17:571–579

Cheong S, Clomburg JM, Gonzalez R (2016) Energy- and carbon-efficient synthesis of functionalized small molecules in bacteria using non-decarboxylative Claisen condensation reactions. Nat Biotechnol 34:556–561

Choi YJ, Lee SY (2013) Microbial production of short-chain alkanes. Nature 502:571–574

Chuck CJ, Donnelly J (2014) The compatibility of potential bioderived fuels with Jet A-1 aviation kerosene. Appl Energy 118:83–91

Ciliberti C, Biundo A, Albergo R et al (2020) Syngas derived from lignocellulosic biomass gasification as an alternative resource for innovative bioprocesses. Processes 8:1567

Daniell J, Köpke M, Simpson SD (2012) Commercial biomass syngas fermentation. Energies 5: 5372–5417

Das M, Patra P, Ghosh A (2020) Metabolic engineering for enhancing microbial biosynthesis of advanced biofuels. Renew Sustain Energ Rev 119:109562

Davison J (2002) Towards safer vectors for the field release of recombinant bacteria. Environ Biosaf Res 1(1):9–18

De Jong S, Van Stralen J, Londo M et al (2018) Renewable jet fuel supply scenarios in the European Union in 2021–2030 in the context of proposed biofuel policy and competing biomass demand. GCB Bioenergy 10:661–682

Dellomonaco C, Clomburg JM, Miller EN et al (2011) Engineered reversal of the β-oxidation cycle for the synthesis of fuels and chemicals. Nature 476:355–359

Dimitriadis A, Natsios I, Dimaratos A et al (2018) Evaluation of a hydrotreated vegetable oil (HVO) and effects on emissions of a passenger car diesel engine. Front Mech Eng 4
Dunn BJ, Khosla C (2013) Engineering the acyltransferase substrate specificity of assembly line polyketide synthases. J R Soc Interface 10:20130297
Dutta K, Daverey A, Lin J-G (2014) Evolution retrospective for alternative fuels: first to fourth generation. Renew Energy 69:114–122
Fan L, Liu J, Nie K et al (2013) Synthesis of medium chain length fatty acid ethyl esters in engineered Escherichia coli using endogenously produced medium chain fatty acids. Enzym Microb Technol 53:128–133
Fang L, Fan J, Luo S et al (2021) Genome-scale target identification in Escherichia coli for high-titer production of free fatty acids. Nat Commun 12:4976
Frederix M, Mingardon F, Hu M et al (2016) Development of an E. coli strain for one-pot biofuel production from ionic liquid pretreated cellulose and switchgrass. Green Chem 18:4189–4197
Fu J, Huo G, Feng L et al (2016) Metabolic engineering of Bacillus subtilis for chiral pure meso-2,3-butanediol production. Biotechnol Biofuels 9:90
Gaeta-Bernardi A, Parente V (2016) Organic municipal solid waste (MSW) as feedstock for biodiesel production: a financial feasibility analysis. Renew Energy 86:1422–1432
Garg A, Xie X, Keatinge-Clay A et al (2014) Elucidation of the cryptic epimerase activity of redox-inactive ketoreductase domains from modular polyketide synthases by tandem equilibrium isotope exchange. J Am Chem Soc 136:10190–10193
Gollakota A, Kishore N, Gu S (2018) A review on hydrothermal liquefaction of biomass. Renew Sustain Energ Rev 81:1378–1392
Gopalan K, Raikova S, Smith CR et al (2020) The impact of biodiesel and alternative diesel fuel components on filter blocking through accelerated testing on a novel high pressure common rail non-firing rig. Fuel 282:118850
Guo W, Sheng J, Zhao H et al (2016) Metabolic engineering of Saccharomyces cerevisiae to produce 1-hexadecanol from xylose. Microb Cell Fact 15:24
Gupta A, Reizman IMB, Reisch CR et al (2017) Dynamic regulation of metabolic flux in engineered bacteria using a pathway-independent quorum-sensing circuit. Nat Biotechnol 35: 273–279
Hagen A, Poust S, Rond TD et al (2016) Engineering a polyketide synthase for in vitro production of adipic acid. ACS Synth Biol 5:21–27
Halfmann C, Gu L, Gibbons W et al (2014) Genetically engineering cyanobacteria to convert CO2, water, and light into the long-chain hydrocarbon farnesene. Appl Microbiol Biotechnol 98: 9869–9877
Haushalter RW, Kim W, Chavkin TA et al (2014) Production of anteiso-branched fatty acids in Escherichia coli; next generation biofuels with improved cold-flow properties. Metab Eng 26: 111–118
Howard TP, Middelhaufe S, Moore K et al (2013) Synthesis of customized petroleum-replica fuel molecules by targeted modification of free fatty acid pools in Escherichia coli. Proc Natl Acad Sci 110:7636–7641
Initiative SS (2019) The role of sustainable biofuels in the decarbonisation of shipping. Technical report. www.ssi2040.org. Accessed 24 May 2023
Inui M, Kawaguchi H, Murakami S et al (2005) Metabolic engineering of Corynebacterium glutamicum for fuel ethanol production under oxygen-deprivation conditions. J Mol Microbiol Biotechnol 8:243–254
Jiang W, Jiang Y, Bentley GJ et al (2015) Enhanced production of branched-chain fatty acids by replacing β-ketoacyl-(acyl-carrier-protein) synthase III (FabH). Biotechnol Bioeng 112:1613–1622
Jiang GZ, Yao MD, Wang Y et al (2017) Manipulation of GES and ERG20 for geraniol overproduction in Saccharomyces cerevisiae. Metab Eng 41:57–66
Jiang W, Gu P, Zhang F (2018) Steps towards 'drop-in' biofuels: focusing on metabolic pathways. Curr Opin Biotechnol 53:26–32

Jin C, Yao M, Liu H et al (2011) Progress in the production and application of n-butanol as a biofuel. Renew Sustain Energ Rev 15:4080–4106

Jovanovic I, Jones SB, Santosa DM, et al (2010) A survey of opportunities for microbial conversion of biomass to hydrocarbon compatible fuels. https://www.osti.gov/servlets/purl/1013947. Accessed 25 May 2023

Jung J, Lim JH, Kim SY et al (2016) Precise precursor rebalancing for isoprenoids production by fine control of gapA expression in Escherichia coli. Metab Eng 38:401–408

Kang M-K, Zhou YJ, Buijs NA et al (2017) Functional screening of aldehyde decarbonylases for long-chain alkane production by Saccharomyces cerevisiae. Microb Cell Fact 16:74

Kararzos S, Van Dyk JS, McMillan JD et al (2017) Drop-in biofuel production via conventional (lipid/fatty acid) and advanced (biomass) routes. Part I. Biofuels Bioprod Biorefin 11:344–362

Karatzos S, McMillan JD, Saddler JN (2014) The potential and challenges of drop-in biofuels. Report for IEA Bioenergy Task 39. https://www.ieabioenergy.com/wp-content/uploads/2018/02/Task-39-Drop-in-Biofuels-Report-FINAL-2-Oct-2014-ecopy.pdf. Accessed 21 May 2023

Kargbo H, Harris JS, Phan AN (2021) "Drop-in" fuel production from biomass: critical review on techno-economic feasibility and sustainability. Renew Sustain Energ Rev 135:110168

Kusakabe T, Tatsuke T, Tsuruno K et al (2013) Engineering a synthetic pathway in cyanobacteria for isopropanol production directly from carbon dioxide and light. Metab Eng 20:101–108

Ladygina N, Dedyukhina EG, Vainshtein MB (2006) A review on microbial synthesis of hydrocarbons. Process Biochem 41:1001–1014

Lennen RM, Pfleger BF (2012) Engineering Escherichia coli to synthesize free fatty acids. Trends Biotechnol 30:659–667

Lim JH, Jung GY (2017) A simple method to control glycolytic flux for the design of an optimal cell factory. Biotechnol Biofuels 10:160

Liu R, Zhu F, Lu L et al (2014) Metabolic engineering of fatty acyl-ACP reductase-dependent pathway to improve fatty alcohol production in Escherichia coli. Metab Eng 22:10–21

Liu D, Xiao Y, Evans BS et al (2015a) Negative feedback regulation of fatty acid production based on a malonyl-CoA sensor–actuator. ACS Synth Biol 4:132–140

Liu Q, Wu K, Cheng Y et al (2015b) Engineering an iterative polyketide pathway in Escherichia coli results in single-form alkene and alkane overproduction. Metab Eng 28:82–90

Liu J, Zhou W, Fan S et al (2019) Study of the co-production of butanol and hydrogen by immobilizing Clostridium acetobutylicum CICC8012. Energy Procedia 158:1879–1884

Liu J, Fan S, Xiao Z (2021) Enhanced coproduction and trade-off of the hydrogen and butanol in the coupled system of pervaporation and repeated-cycle fixed-bed fermentation. Ind Crop Prod 161: 113172

Love J (2022) Microbial pathways for advanced biofuel production. Biochem Soc Trans 50:987–1001

Lü J, Sheahan C, Fu P (2011) Metabolic engineering of algae for fourth generation biofuels production. Energy Environ Sci 4:2451–2466

Madavi TB, Chauhan S, Keshri A et al (2022) Whole-cell biocatalysis: advancements toward the biosynthesis of fuels. Biofpr 16(9):859–876

Malik S, Shahid A, Liu C-G et al (2021) Developing fourth-generation biofuels secreting microbial cell factories for enhanced productivity and efficient product recovery; a review. Fuel 298: 120858

Mańkowska M, Pluciński M, Kotowska I (2021) Biomass sea-based supply chains and the secondary ports in the era of decarbonization. Energies 14:1796

Mendez-Perez D, Alonso-Gutierrez J, Hu Q et al (2017) Production of jet fuel precursor monoterpenoids from engineered Escherichia coli. Biotechnol Bioeng 114:1703–1712

Menendez-Bravo S, Comba S, Sabatini M et al (2014) Expanding the chemical diversity of natural esters by engineering a polyketide-derived pathway into Escherichia coli. Metab Eng 24:97–106

Mukhopadhyay A (2015) Tolerance engineering in bacteria for the production of advanced biofuels and chemicals. Trends Microbiol 23:498–508

Mund NK, Liu Y, Chen S (2022) Advances in metabolic engineering of cyanobacteria for production of biofuels. Fuel 322:124117

Niehus X, Crutz-Le Coq A-M, Sandoval G et al (2018) Engineering Yarrowia lipolytica to enhance lipid production from lignocellulosic materials. Biotechnol Biofuels 11:11

Niu F-X, He X, Wu Y-Q et al (2018) Enhancing production of pinene in Escherichia coli by using a combination of tolerance, evolution, and modular co-culture engineering. Front Microbiol 9: 1623

Ohtake T, Pontrelli S, Laviña WA et al (2017) Metabolomics-driven approach to solving a CoA imbalance for improved 1-butanol production in Escherichia coli. Metab Eng 41:135–143

Oliver NJ, Rabinovitch-Deere CA, Carroll AL et al (2016) Cyanobacterial metabolic engineering for biofuel and chemical production. Curr Opin Chem Biol 35:43–50

Osman AI, Hefny M, Abdel Maksoud MIA et al (2021a) Recent advances in carbon capture storage and utilisation technologies: a review. Environ Chem Lett 19:797–849

Osman AI, Mehta N, Elgarahy AM et al (2021b) Conversion of biomass to biofuels and life cycle assessment: a review. Environ Chem Lett 19:4075–4118

Patel A, Arora N, Mehtani J et al (2017) Assessment of fuel properties on the basis of fatty acid profiles of oleaginous yeast for potential biodiesel production. Renew Sustain Energ Rev 77: 604–616

Peralta-Yahya PP, Ouellet M, Chan R et al (2011) Identification and microbial production of a terpene-based advanced biofuel. Nat Commun 2:483

Peralta-Yahya PP, Zhang F, DelCardayre SB et al (2012) Microbial engineering for the production of advanced biofuels. Nature 488:320–328

Pfleger B, Pitera D, Smolke C et al (2006) Combinatorial engineering of intergenic regions in operons tunes expression of multiple genes. Nat Biotechnol 24:1027–1032

Plugge CM (2017) Biogas. Microb Biotechnol 10:1128–1130

Qiu Y, Tittiger C, Wicker-Thomas C et al (2012) An insect-specific P450 oxidative decarbonylase for cuticular hydrocarbon biosynthesis. Proc Natl Acad Sci 109:14858–14863

Rabe P, Schmitz T, Dickschat JS (2016) Mechanistic investigations on six bacterial terpene cyclases. Beilstein J Org Chem 12:1839–1850

Ramzi AB, Baharum SN, Bunawan H et al (2020) Streamlining natural products biomanufacturing with omics and machine learning driven microbial engineering. Front Bioeng Biotechnol 8

Reddy GK, Leferink NGH, Umemura M et al (2020) Exploring novel bacterial terpene synthases. PLoS One 15:e0232220

Rodriguez GM, Tashiro Y, Atsumi S (2014) Expanding ester biosynthesis in Escherichia coli. Nat Chem Biol 10:259–265

Rui Z, Li X, Zhu X et al (2014) Microbial biosynthesis of medium-chain 1-alkenes by a nonheme iron oxidase. Proc Natl Acad Sci 111:18237–18242

Sánchez OG, Birdja YY, Bulut M et al (2019) Recent advances in industrial CO2 electroreduction. Curr Opin Green Sustain Chem 16:47–56

Savage DF, Way J, Silver PA (2008) Defossiling fuel: how synthetic biology can transform biofuel production. ACS Chem Biol 3:13–16

Schirmer A, Rude MA, Li X et al (2010) Microbial biosynthesis of alkanes. Science 329:559–562

Schmitz AC, Hartline CJ, Zhang F (2017) Engineering microbial metabolite dynamics and heterogeneity. Biotechnol J 12:1700422

Sebesta J, Peebles CA (2020) Improving heterologous protein expression in Synechocystis sp. PCC 6803 for alpha-bisabolene production. Metab Eng Commun 10:e00117

Shears J (2019) Is there a role for synthetic biology in addressing the transition to a new low-carbon energy system? Microb Biotechnol 12:824–827

Shen CR, Liao JC (2008) Metabolic engineering of Escherichia coli for 1-butanol and 1-propanol production via the keto-acid pathways. Metab Eng 10:312–320

Sims R, Taylor M, Saddler J et al (2008) From 1st-to 2nd-generation biofuel technologies. International Energy Agency (IEA) and Organisation for Economic Co-Operation and

Development, Paris, pp 16–20. https://www.osti.gov/etdeweb/biblio/21589371. Accessed 20 May 2023
Smith S, Tsai S-C (2007) The type I fatty acid and polyketide synthases: a tale of two megasynthases. Nat Prod Rep 24:1041–1072
Solaymani S (2019) CO2 emissions patterns in 7 top carbon emitter economies: the case of transport sector. Energy 168:989–1001
Solem C, Dehli T, Jensen PR (2013) Rewiring Lactococcus lactis for ethanol production. Appl Environ Microbiol 79:2512–1518
Sorigué D, Légeret B, Cuiné S et al (2017) An algal photoenzyme converts fatty acids to hydrocarbons. Science 357:903–907
Speight JG (2014) The chemistry and technology of petroleum. CRC Press, New York
Stein UH, Abbasi-Hosseini M, Kain J et al (2022) Influence of gas-release strategies on the production of biohydrogen and biobutanol in ABE fermentation. Biofuels 13:9–15
Stevens DC, Wagner DT, Manion HR et al (2016) Methyltransferases excised from trans-AT polyketide synthases operate on N-acetylcysteamine-bound substrates. J Antibiot 69:567–570
Sukovich DJ, Seffernick JL, Richman JE et al (2010) Widespread head-to-head hydrocarbon biosynthesis in bacteria and role of OleA. Appl Environ Microbiol 76:3850–3862
Sun X, Atiyeh HK, Huhnke RL, Tanner RS (2019) Syngas fermentation process development for production of biofuels and chemicals: a review. Bioresour Technol Rep 7:100279
Tan Z, Yoon JM, Nielsen DR et al (2016) Membrane engineering via trans unsaturated fatty acids production improves Escherichia coli robustness and production of biorenewables. Metab Eng 35:105–113
Tan Z, Khakbaz P, Chen Y et al (2017) Engineering Escherichia coli membrane phospholipid head distribution improves tolerance and production of biorenewables. Metab Eng 44:1–12
Tarasava K, Lee SH, Chen J et al (2022) Reverse β-oxidation pathways for efficient chemical production. J Ind Microbiol Biotechnol 49
Tee TW, Chowdhury A, Maranas CD, Shanks JV (2014) Systems metabolic engineering design: fatty acid production as an emerging case study. Biotechnol Bioeng 111:849–857
Tong K, Gleeson MJ, Rong G et al (2014) Optimal design of advanced drop-in hydrocarbon biofuel supply chain integrating with existing petroleum refineries under uncertainty. Biomass Bioenergy 60:108–120
Torella JP, Ford TJ, Kim SN et al (2013) Tailored fatty acid synthesis via dynamic control of fatty acid elongation. Proc Natl Acad Sci 110:11290–11295
Tse TJ, Wiens DJ, Reaney MJT (2021) Production of bioethanol—a review of factors affecting ethanol yield. Fermentation 7:268
Van Der Kroft DFA (2020) The potential of drop-in biofuels for the maritime industry: a MILP optimization approach to explore future scenarios. Dissertation, Delft University of Technology
Van Der Kroft DFA, Pruyn JFJ (2021) A study into the availability, costs and GHG reduction in drop-in biofuels for shipping under different regimes between 2020 and 2050. Sustainability 13: 9900
Wagner DT, Stevens DC, Mehaffey MR et al (2016) α-Methylation follows condensation in the gephyronic acid modular polyketide synthase. Chem Commun 52:8822–8825
Walls LE, Rios-Solis L (2020) Sustainable production of microbial isoprenoid derived advanced biojet fuels using different generation feedstocks: a review. Front Bioeng Biotechnol 8
Weber C, Farwick A, Benisch F et al (2010) Trends and challenges in the microbial production of lignocellulosic bioalcohol fuels. Appl Microbiol Biotechnol 87:1303–1315
Whitaker WB, Jones JA, Bennett RK et al (2017) Engineering the biological conversion of methanol to specialty chemicals in Escherichia coli. Metab Eng 39:49–59
Winkler M (2018) Carboxylic acid reductase enzymes (CARs). Curr Opin Chem Biol 43:23–29
Wojcik EZ, Singleton C, Chapman LN et al (2017) Plant biomass as biofuels. eLS 1:1–11. https://doi.org/10.1002/9780470015902.a0023716
Wu J, Cheng S, Cao J et al (2019) Systematic optimization of limonene production in engineered Escherichia coli. J Agric Food Chem 67:7087–7097

Xiao Y, Bowen CH, Liu D et al (2016) Exploiting nongenetic cell-to-cell variation for enhanced biosynthesis. Nat Chem Biol 12:339–344

Yan S, Dong D (2018) Improvement of caproic acid production in a Clostridium kluyveri H068 and methanogen 166 co-culture fermentation system. AMB Express 8:175

Youngquist JT, Schumacher MH, Rose JP et al (2013) Production of medium chain length fatty alcohols from glucose in Escherichia coli. Metab Eng 20:177–186

Yuzawa S, Keasling JD, Katz L (2016) Insights into polyketide biosynthesis gained from repurposing antibiotic-producing polyketide synthases to produce fuels and chemicals. J Antibiot 69:494–499

Yuzawa S, Deng K, Wang G et al (2017) Comprehensive in vitro analysis of acyltransferase domain exchanges in modular polyketide synthases and its application for short-chain ketone production. ACS Synth Biol 6:139–147

Zargar A, Bailey CB, Haushalter RW et al (2017) Leveraging microbial biosynthetic pathways for the generation of 'drop-in' biofuels. Curr Opin Biotechnol 45:156–163

Zhang F, Ouellet M, Batth TS et al (2012) Enhancing fatty acid production by the expression of the regulatory transcription factor FadR. Metab Eng 14:653–660

Zhang L, Liang Y, Wu W et al (2016) Microbial synthesis of propane by engineering valine pathway and aldehyde-deformylating oxygenase. Biotechnol Biofuels 9:80

Chapter 7
Drop-In Biofuel Production Using Fungi

Suparna Sen, Chandana Malakar, Bapan Banik, Naorem Bela Devi, and Kannan Pakshirajan

Abstract Global energy demand is increasing due to global population growth and rapid industrialization. This has resulted in a heavy reliance on finite fossil fuels, endangering energy security and raising environmental concerns. At the current consumption rate, fossil fuel is expected to exhaust within the next five decades. Moreover, geopolitical instability in regions with significant crude oil reserves creates volatility in energy supply and pricing, increasing energy scarcity and greenhouse gas emissions. Derived from non-fossil sources such as plant oils, biomass, and algae, drop-in biofuels offer identical chemical properties to traditional fuels, making them an attractive and practical option for sustainable energy transitions. However, large-scale commercialization faces hurdles related to feedstock availability, production costs, and competition with food production. To overcome these challenges, microbial fermentation emerges as a promising approach for renewable biofuel production. Leveraging genetic and metabolic engineering techniques, novel microbes can be tailored to utilize non-agricultural feedstock while maximizing fuel yields. Fungi, offering a wide range of bioactive compounds, including enzymes and lipids, play a crucial role in lignocellulose-based biofuel production and serve as invaluable biocatalysts in biotechnology. This chapter explores critical aspects of fungal-assisted drop-in biofuel production, delving into underlying mechanisms and evaluating prospects amid the ongoing fuel crisis. Continued research and policy support are essential to realizing the full potential

Bapan Banik and Naorem Bela Devi contributed equally with all other contributors.

S. Sen (✉) · N. B. Devi · K. Pakshirajan (✉)
Department of Biosciences and Bioengineering, Indian Institute of Technology Guwahati, Guwahati, Assam, India
e-mail: ssen23@rnd.iitg.ac.in; pakshi@iitg.ac.in

C. Malakar
Department of Animal Biotechnology, College of Veterinary Sciences, Assam Agricultural University, Guwahati, Assam, India

B. Banik
Drug Delivery Research Laboratory, Department of Pharmaceutical Sciences, Dibrugarh University, Dibrugarh, Assam, India

S. K. Brar et al. (eds.), *The Microbiology of the Drop-in Biofuel Production*, Biofuel and Biorefinery Technologies 15, https://doi.org/10.1007/978-3-031-61637-2_7

of drop-in biofuels and enabling sustainable and scalable solutions for the transportation sector.

Keywords Drop-in biofuel · Fungi · Microbial fermentation · Sustainable energy source

7.1 Introduction

Throughout the last few decades, there has been a remarkable surge in energy demand due to population inflation and industrialization. According to the UN's projections, it is anticipated that the human population will be 10 billion by 2055. With this continuous population growth and expanding industrial activities, the demand for energy has substantially increased, especially from finite fossil fuel sources like coal, oil, and natural gas. Presently, approximately 80% of the world's energy requirements are provided by the use of fossil fuels, which, at current rates, are expected to run out within the next 50 years. This growing demand poses a serious risk of depleting fossil fuel reserves. Moreover, most crude oil reserves are situated in economically and politically unstable regions, causing frequent fluctuations in the global supply chain and pricing, leading to energy scarcity and soaring prices. Concerns about global warming and climate change have grown as a result of the significant increase in the production of harmful greenhouse gases (GHGs) such as carbon dioxide, carbon monoxide, methane, and chlorofluorocarbons (CFCs) as a result of the conversion of petroleum into fuels (Heleniak 2020).

Based on data released by the International Energy Agency (IEA), about 66% of global oil usage was for transportation (Nitte 2023). This trend is predicted to continue, with the present daily consumption of 92 million barrels of oil likely to rise to 150 million barrels per day by 2050 (European Commission, 2021). Policymakers have been encouraged to implement several programs and milestones to decrease the usage of fossil fuels for transportation at this point. Until 2030, oil demand for transportation is expected to climb by 1.3% per year (Bello 2012). This demand is now growing steadily. In the next 20 years, the fleet of aircraft for both freighters and civil aviation is predicted to quadruple, according to a new forecast from Airbus for 2038 (Peiris 2019).

It is critical to embrace a change toward cleaner and sustainable energy sources to manage the growing demand for transportation fuel, as well as the effects of population growth, mobility, and related environmental and economic issues. Renewable energy options, including solar, wind, hydro, and geothermal power, hold the key to this transition. Additionally, enhancing energy efficiency measures is crucial to curb wastage and optimize energy usage. With the global population, mobility, and environmental concerns on the rise, the quest for alternative and eco-friendly transportation fuels has gained significance, leading to the development of biofuels. These biofuels are derived from the conversion of diverse biomass sources like vegetable oils, agricultural residues, and domestic wastes into viable and environmentally friendly products (Janda et al. 2012). Biofuels can be categorized based on their physical states, which include solid, liquid, or gaseous forms. Among them, liquid biofuels like bio-gasoline, biodiesel, bioethanol, and bio-jet fuel, as well as gaseous biofuels like biohydrogen, can be employed as transportation fuels to reduce dependency on finite fossil fuels.

One notable subset of biofuels is "drop-in" biofuels, which represent advanced alternatives virtually identical in functionality to traditional fossil fuels like gasoline, diesel, and aviation fuel. The remarkable aspect of drop-in biofuels is their seamless integration into existing engines and infrastructure, as they possess identical chemical properties to conventional fuels. No blending or engine modifications are required, making them an attractive and practical solution for sustainable energy transitions (Velvizhi et al. 2022). Drop-in biofuels, in contrast to traditional bioethanol, are derived from non-fossil sources such as plant oils, biomass, or algae and undergo advanced refining processes to attain the desired chemical structure. Drop-in biofuels, as opposed to bioethanol, are made up of a combination of diverse hydrocarbons (e.g., butanol, bisabolene, farnesane, and alkanes) characterized by functional properties such as distillation profile, viscosity, and acidity. To achieve climate goals, the transportation sector is projected to require significant volumes of drop-in biofuels, possibly reaching up to 426 million tonnes (Karatzos et al. 2014). These biofuels can be produced using three main technology platforms (oleochemical, thermochemical, and biochemical), along with a hybrid category. The sole drop-in biofuel now produced commercially relies on lipid hydro-treatment and is mostly utilized for renewable diesel and bio-gasoline, with minor quantities going to bio-jet fuels (Van Dyk et al. 2019). In response to the urgent need for greenhouse gas emission reduction and sustainable transportation, countries are progressively embracing drop-in biofuels as a viable solution. Governments and industries worldwide are actively investing in research, development, and production of these biofuels as an integral part of their renewable energy and climate change mitigation strategies (Watanabe et al. 2022). To facilitate their integration, supportive policies and regulations have been implemented, offering incentives such as blending mandates, tax benefits, and emission reduction targets. These measures encourage the incorporation of drop-in biofuels into conventional transportation fuel blends. As a result, concerted efforts are being made to introduce drop-in biofuels to the market, particularly in sectors like aviation and marine transportation, as a viable alternative to traditional fossil fuels (Ebadian et al. 2020).

Despite significant progress, the widespread production of drop-in biofuels confronts significant problems in addressing the demands of the transportation sector. Key issues include the availability of suitable feedstock, high production costs, and the potential competition with food production for land and resources, all of which hinder large-scale commercialization. To address these challenges, microbial fermentation emerges as an appealing approach for renewable biofuel production. Leveraging screening methods for novel microbes and employing genetic and metabolic engineering techniques hold promise in enhancing the efficiency and yield of drop-in biofuels. Through genetic engineering, microbes can be modified to utilize non-agricultural feedstock like switchgrass, minimizing conflicts with food crops for land usage while producing a wide array of fuels and commodity chemicals. These advancements in microbial technologies offer an encouraging path forward for sustainable and scalable drop-in biofuel production (Zargar et al. 2017). Fungi, encompassing both yeasts and filamentous fungi, have emerged as invaluable biocatalysts in biotechnology, serving as versatile biorefineries for a diverse range of biochemical processes. With more than 9000 bioactive compounds derived from cultivable fungi, including enzymes, organic acids, and biofuels, these microorganisms hold immense potential. Particularly significant is their role in lignocellulosic-based biofuel production. Their role is particularly important in the expensive pretreatment and enzymatic hydrolysis stages. Various studies have confirmed the suitability of different fungi for biofuel production or integration into the production process. Fungi can be utilized as a source for lipid production, a crucial element in biodiesel manufacturing. Additionally, they produce lipase enzymes that facilitate the transesterification of lipids into biodiesel. Using cutting-edge omics technologies such as genomics, transcriptomics, metabolomics, and metagenomics, as well as genetic engineering, advanced biofuels can now be produced using model organisms such as *Pichia pastoris* and *Saccharomyces cerevisiae*, advancing the concept of drop-in biofuels (Chen et al. 2011; Brat et al. 2012; Peralta-Yahya et al. 2011; Lee et al. 2012).

In this chapter, we explore the critical aspects of fungal-assisted drop-in biofuel production, highlighting the underlying mechanisms and assessing its future potential amid the prevailing fuel crisis (Fig. 7.1).

7.2 Overview of Fungi and Their Characteristics

The kingdom Fungi, which is part of the Eukarya domain, is noted for its enormous diversity, with at least 11 different groups, including seven phyla and four subphyla of the polyphyletic Zygomycota. These groups showcase a wide range of genetic attributes and morphologies. Fungi are one of the largest branches of the Tree of Life and consist of approximately 1.5 million species. The kingdom Fungi has a lineage tracing back to the Precambrian and Devonian eras. The three primary forms of fungi are unicellular yeasts, filamentous molds, and mushrooms. Depending on growth conditions, certain fungi exist as a combination of multicellular filamentous structures, known as hyphae, which together form the mycelium, or unicellular yeast structures, known as pseudohyphae. While true fungi lack flagella, their chytrid ancestors are unicellular organisms that utilize flagella for locomotion (Singara Charya 2015). This phenomenon is known as dimorphism, and this characteristic is exhibited by some pathogenic fungi. A fundamental feature of fungi is their unique cell walls, which distinguish them from higher plants and bacteria because of the presence of chitin. These cell walls play an important role in protecting fungi from osmotic pressure and environmental stresses while also giving shape to their cells. Furthermore, the fungal wall acts as a protective barrier, safeguarding fungal

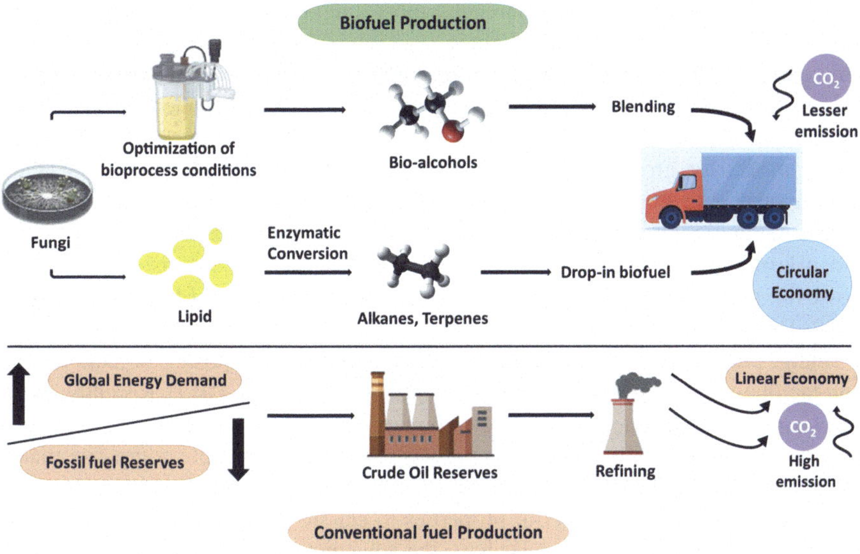

Fig. 7.1 Schematic illustrating conventional fuel production vs drop-in biofuel production using fungi

pathogens from specific fungicidal agents produced by their hosts (Bueno and Silva 2014). The composition of cell walls varies among fungal species, but in most fungi, the cell walls consist of five major components: (1/3)-β-glucan, (1/6)-β-glucan, (1/3)-α-glucan, chitin, and glycoproteins. The central core of the cell wall is a branched β 1,3, 1,6 glucan that connects to chitin through a β 1,4 linkage. Additionally, about 3–4% of the total glucan linkages are interchain β 1,6 glucosidic linkages (Gander 1974).

Fungi reproduce through both sexual and asexual means. Spores are frequently produced on specific structures known as ascus or basidium. The composition and quantity of these spores might vary throughout fungal groups and even during a single species' life cycle. The filamentous fungi's asexual life cycle begins with spore germination and the production of a vegetative hypha, which forms the mycelium and grows uniformly in all directions from a central point, resulting in a spherical colony. While true spheres are rare in nature, fungi can respond with rapid growth under favorable substrate conditions and external factors (Dongre et al. 2023). Unicellular yeasts, on the other hand, reproduce through fission or budding and might be spherical, oval, elliptic, elongated, or cylindrical. Yeasts can also develop filamentous structures on occasion. Their size ranges are normally between 3 and 10 mm in width and 5–30 mm in length. Various yeast species have different budding patterns, such as apical, lateral, bipolar, or multipolar, as well as morphological and physiological distinctions (Bueno and Silva 2014).

In terms of environmental preferences, most fungi thrive at temperatures ranging from 0 to 30 °C, with the ideal range typically falling between 20 and 30 °C. However, there are certain thermophilic species, such as those in the *Aspergillus* genus, that may thrive at temperatures as high as 50 °C, whereas others are psychrophilic, flourishing at temperatures as low as 10 °C. Furthermore, fungi prefer acidic media for growth, with a pH of 6 being best for the majority of species investigated. Metabolic stress might occur during growth in a slightly alkaline medium. Fungi release organic acids to acquire certain nutrients while also adjusting the pH of their surroundings. The pH range of growth media is rather broad; however, growth in a slightly alkaline media is generally slower (Almaguer Chávez 2022; Dix 2012).

Fungi play an important role in the ecosystem by decomposing organic matter and facilitating nutrient cycling and exchange (Spatafora et al. 2017). Recently, there has been a surge of interest in interdisciplinary research focused on using fungi to recycle agricultural and forestry by-products and trash to generate composite materials. These efforts have resulted in the development of novel fungal-based goods with comparable or even superior properties to classic petroleum-derived products. Additionally, these fungal products have the advantage of having a reduced or even negative carbon footprint and are fully biodegradable. Furthermore, fungal mycelia are anticipated to replace animal leather, surpassing both in terms of quality and price (Meyer et al. 2020). Adoption of these alternatives will have far-reaching consequences for the livestock, textile, and fashion industries. Furthermore, fungi stand out as critical components of the global carbon cycle among non-photosynthetic species (Barnett et al. 1990). Collectively, fungi possess the remarkable ability to break down nearly all naturally occurring biopolymers in addition to synthetic ones. This characteristic makes them highly promising in advancing alternative fuels, carbon sequestration, and the bioremediation of polluted ecosystems. However, to fully harness the benefits of fungi for humanity, a deeper understanding of their interactions within natural and synthetic communities is essential. Fungi play a vital role in synthesizing alcohols, enzymes, organic acids, and various pharmaceutical products, all of which are key to advancing modern technology (Meyer et al. 2016).

Mycotoxins, which include alkaloids and polyketides, are produced by a variety of fungi and can be toxic to both animals and humans. Certain fungal species pose significant threats as pathogens to humans and other animals, leading to substantial crop losses worldwide due to various fungal diseases and food spoilage. These impacts on agriculture can have far-reaching consequences for human food supplies and local economies. Fungi have a significant impact on human activities and ecological functioning, serving as decomposers, pathogens, and mutualistic symbionts. Furthermore, fungi play an important role as vital biological components of the global carbon cycle among non-photosynthetic species (Richard 2007; Murphy et al. 2006).

Based on variations in sexual reproduction, mycologists have traditionally classified fungi into four divisions (Ainsworth 2008): (1) Zygomycota (zygomycetes), (2) Ascomycota (sac fungi), (3) Basidiomycota (club fungi), and (4) Deuteromycota (fungi imperfecti). For years, yeasts have had a significant influence on society, playing a vital role in the production of bread, beer, and wine. They are extensively utilized in the brewing and baking industries, converting sugars into carbon dioxide and ethyl alcohol. One commonly used yeast, *S. cerevisiae*, possesses an enzyme complex that efficiently transforms sugars into alcohol and carbon dioxide. In the production of industrial alcohol, starch serves as the substrate instead of sugars (Walker and Stewart 2016).

Fungi display an extraordinary ability to produce unique enzymes that thrive under harsh conditions unsuitable for other higher organisms. These fungal enzymes can convert a variety of materials, including wood, plastic, paints, and jet fuel, into useful nutrients. Some of these enzymes have already found practical applications in the pulp and paper sector, as well as in biocatalysis for the synthesis of fine compounds. By incorporating these enzymes into industrial processes, it becomes feasible to carry out less energy-intensive procedures and generate fewer toxic by-products (Andlar et al. 2018). Furthermore, fungal enzymes are extremely versatile in their ability to degrade a wide spectrum of complex substances, making them particularly promising for bioremediation and the degradation of recalcitrant pollutants. Filamentous fungi excel as decomposers and have evolved to break down organic matter and polysaccharides present in plant biomass. They secrete an array of enzymes, including cellulases, amylases, pectinases, inulases, proteases, and lipases, which facilitate the hydrolysis of plant polysaccharides, proteins, and lipids. These exceptional capabilities render fungi indispensable in various industrial and environmental applications (Cherry and Fidantsef 2003).

Certain fungi, including *Aspergillus niger*, *A. oryzae*, *Trichoderma reesei*, and *Thermothelomyces thermophilus*, are renowned for their remarkable ability to enzymatically break down plant biomass into glucose and oligosaccharides. The by-products are subsequently absorbed by the fungal hyphae via specific sugar transporters. When coupled with their extraordinary capability for enzyme secretion, these fungi evolve into highly efficient factories for enzyme production. The resulting enzymes have widespread applications in various industries such as food, feed, detergent, pulp and paper, fuel, pharmaceuticals, and chemicals (Meyer et al. 2016).

There is ongoing research into the potential of *A. niger* to produce organic acids other than citric acids, such as itaconate and galactarate. Itaconate could be a sustainable substitute for petroleum-based polyacrylic acid, a critical precursor in the polymer industry used for absorbent polymers, polyester resins, and synthetic latex. Likewise, galactarate holds potential as a replacement for the currently used petroleum-based polyethylene terephthalate (PET), a common plastic. This line of research highlights the potential of fungi as flexible and environmentally friendly alternatives for enzyme production and the development of sustainable materials for a range of industrial applications (Karaffa and Kubicek 2019; Kuivanen et al. 2016). Several fungi are well known for producing natural compounds that function as antibiotics. Penicillin, an antibiotic produced by fungi, was first identified by Alexander Fleming in 1929. Another antibiotic found in *Penicillium griseofulvum*, griseofulvin, is used to treat skin conditions like athlete's foot and ringworm

Additionally, several fungal species have been employed as biocontrol agents, insecticides (*Ascherroni adeyroides*, *Isaria ferinosa*), or nematode control agents (*Heterodera* sp., *Globodera* sp.) (van den Berg 2011).

7.3 Biochemical Routes for the Production of Drop-In Biofuels

The majority of "drop-in" biofuels are produced through five prominent pathways: the α-keto acid pathway, non-decarboxylative Claisen condensation pathway, fatty acid pathway, isoprenoid pathway, and polyketides pathway (Fig. 7.2). These biofuels have similar properties to petroleum-based fuels and can be used in a variety of engines, including spark-ignition, compression-ignition, and gas-turbine engines. Among these processes, fatty acid biosynthesis has received a lot of attention because it produces short- and medium-chain alkanes, which are key components of gasoline and diesel fuel. Because of their branching and ring topologies, isoprenoid hydrocarbons have a high energy content, low water miscibility, and minimal danger of premature ignite, making them appealing alternatives to traditional diesel and jet fuel. Although the potential of polyketide synthases for biofuel applications has been explored to a lesser extent, they also hold promise as a biosynthetic approach for creating high-performance biofuels (Zargar et al. 2017).

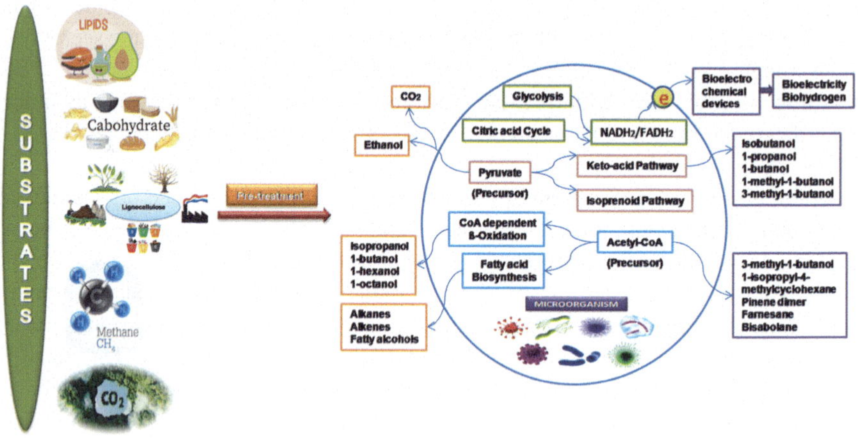

Fig. 7.2 An overview of metabolic pathways in microorganisms to produce drop-in biofuels

7.3.1 *Fatty Acid (FA) Pathway*

The FA pathway has been explored for producing various biofuel products such as alcohols, ketones, esters, and olefins. Medium- and short-chain alkanes synthesized through this pathway are viable alternatives to gasoline and diesel. Fatty aldehyde decarbonylase can be used to convert fatty aldehydes produced from free fatty acids, fatty-acyl-CoA, and a fatty-acyl carrier protein (ACP). Many enzymes have been utilized to directly convert fatty-acyl-ACPs or free fatty acids into fatty aldehydes, which can then be converted into alkanes. Examples of these enzymes are acyl carrier protein reductase (ACP-R) and carboxylic acid reductase (CAR). Fatty acid synthase complex FASII (found in bacteria) or multifunctional FASI synthase (found in eukaryotic cells) is responsible for the biosynthesis of fatty acids (Torella et al. 2013). By engineering the production of medium-chain fatty acids (C8-C12), the range of biofuels has been increased to include branched-chain fatty acids, which find application as biodiesel. This allows for the production of alkanes (Gajewski et al. 2017), alcohols (Youngquist et al. 2013), and esters (Fan et al. 2013) suitable for various biofuel applications. "Drop-in" biofuel production is made possible by the FA pathway's expanded chemical repertoire and its capability to modulate the α-functional group and FA chain length.

7.3.2 *Isoprenoid Pathway*

Terpenes and terpenoids, produced via the isoprenoid pathway, present a viable substitute for fuel. This process makes use of dimethylallyl pyrophosphate (DMAPP) and C5 isopentenyl pyrophosphate (IPP), which are products of the methylerythritol-4-phosphate (MEP) or mevalonate (MVA) pathway. Polyisoprenoid diphosphates are created by the repeated assembly of IPP or DMAPP. Terpene synthases or modifying enzymes can then convert these polyisoprenoid diphosphates into other terpenes or terpenoids.

Replacements for diesel and jet fuel include monoterpenes (C10), sesquiterpenes (C15), and their derivatives. Diverging the pathway at geranyl pyrophosphate (GPP) and farnesyl pyrophosphate (FPP) produced a range of terpenes and terpenoids with biofuel potential. Notably, pinene, limonene, hydrogenated products like farnesene and bisabolene, and novel candidates such as linalool, 1,8-cineole, and geraniol hold promising properties for fuel applications (Sarria et al. 2014; Peralta-Yahya et al. 2011). Amyris, a biotech company, successfully produced a substantial amount of farnesene, reaching 104.3 g/L, by introducing plant-derived farnesene synthases into the yeast *S. cerevisiae*. This farnesene product is marketed as Biofene (Paddon and Keasling 2014).

7.3.3 α-Keto Acid Pathway

Through the α-keto acid pathway, short-chain alcohols like 3-methyl-1-butanol, isobutanol, and 2-methyl-1-butanol are produced. In this process, cellular short-chain α-keto acids—intermediates in the production of branched-chain amino acids—are reduced and decarboxylated. Alcohol dehydrogenase and α-keto acid decarboxylase are two important enzymes in this process (Atsumi et al. 2008). LeuABCD enzymes' substrate specificity can be changed to allow for the recursive condensation of α-keto acids with acetyl-CoA and the release of CO_2, which results in longer acyl chains in the alcohols. Alcohols with medium-chain lengths (up to C8) or those with branched/aromatic terminals are produced using this technique (Marcheschi et al. 2012). Furthermore, the α-keto acids or the alcohols produced from them can be transformed into acetate-based esters or short-chain alkanes for use in biofuel applications (Rodriguez et al. 2014; Zhang et al. 2016a).

7.3.4 Polyketide Pathway

Polyketide synthases (PKSs) are molecular factories that produce various antibiotics and medicinal compounds. Type I PKSs, in particular, are attractive for biofuel production due to their ability to form saturated carbon skeletons. They offer an easily identifiable biosynthetic logic for tailored hydrocarbon production. Challenges arise, though, because PKSs have been observed to be slower than fatty acid synthases (FAS) (Shen et al. 2007). PKSs have been used to make biofuel, and iterative Type I PKSs have been expressed to generate esters and fatty alcohols. Researchers produced compounds like adipic acid by modifying chimeric Type I PKSs (Hagen et al. 2016). PKSs are being investigated by companies such as RHO Renewables to create cyclic aromatics for use as fuel additives (Junne and Kabisch 2017).

7.3.5 Claisen Condensation Pathway

The mechanism leading to non-decarboxylative Claisen condensation was first discovered in *Clostridium kluyveri* (Barker and Taha 1942). This process involves the condensation of acetyl-CoA molecules to form acetoacetyl-CoA, which is then reduced by β-reduction events to form butyrate. This system uses acetyl-CoA or other short-chain acyl-CoAs as primers to elongate the carbon chain on CoA, in contrast to the fatty acid biosynthesis pathway, which increases the carbon chain on acyl carrier protein (ACP) (Dellomonaco et al. 2011).

7.4 Fungi as Biofuel-Producing Cell Factories

Fungi have emerged as highly promising cell factories for biofuel production, thanks to their exceptional metabolic capabilities and efficient enzymatic machinery. As proficient decomposers, fungi demonstrate the ability to efficiently break down diverse organic materials, including lignocellulosic biomass and waste streams. Their capacity to accumulate substantial amounts of lipids and other valuable compounds positions them as attractive candidates for sustainable biofuel generation. With the advancements in genetic engineering and bioprocessing techniques, fungi can be optimized to further enhance their biofuel production potential, paving the way for renewable and environmentally friendly alternatives to conventional fossil fuels. This section highlights the role of fungi as biofuel-producing cell factories and underscores their significant potential in addressing the global energy challenges of our time.

7.4.1 Yeasts

Genetic engineering has played a pivotal role in advancing fungal biofuel production, particularly using yeasts. Yeasts are highly amenable to genetic manipulation, making them ideal candidates for engineering biofuel production pathways. Through genetic engineering, scientists can introduce or modify specific genes in yeasts to enhance their capacity to accumulate lipids, sugars, and other biofuel precursors. Several yeast species have been researched and designed to produce drop-in biofuels. Some of the commonly used yeasts for this purpose include (Table 7.1):

1. ***Saccharomyces cerevisiae* (Brat and Boles** 2013; **Guo et al.** 2016; **Morita et al.** 2017**):** *S. cerevisiae* is one of the most extensively researched and engineered yeasts for biofuel production. It is widely used to produce bioethanol via alcoholic fermentation of sugars produced from various feedstock such as sugarcane, corn, and lignocellulosic biomass. Furthermore, advancements in metabolic engineering have allowed scientists to modify *S. cerevisiae* to produce other drop-in biofuels beyond bioethanol.
2. ***Yarrowia lipolytica* (Darvishi et al.** 2017; **Blazeck et al.** 2013; **Xie** 2017**):** *Y. lipolytica* is a non-conventional yeast that can efficiently utilize fatty acids and lipids as substrates. *Y. lipolytica* has been engineered to produce drop-in biofuels like renewable diesel and aviation fuel from the lipid-rich feedstock.
3. ***Candida* species (Thangavelu et al.** 2021; **Somani et al.** 2018; **Lakshmi et al.** 2021**):** Various *Candida* species, such as *C. tropicalis* and *C. glabrata*, have been explored for the production of biodiesel (fatty acid methyl esters) from vegetable oils and waste oils.
4. ***Rhodotorula glutinis* (Dai et al.** 2007; **Saenge et al.** 2011; **Maza et al.** 2021**):** *R. glutinis* is a yeast species capable of producing lipids, and it has been studied for the production of biodiesel and other lipids-based biofuels.

Table 7.1 Drop-in biofuel production using various yeast species and their culture conditions followed

Yeast species	Culture conditions	Biofuel	Yield	References
Saccharomyces cerevisiae	Aerobic fermentation; D-xylose as substrate; synthetic medium	Isobutanol	1.36 ± 0.11 mg/L (0.16 ± 0.04 mg isobutanol per g D-xylose)	Brat and Boles (2013)
S. cerevisiae XF3	Batch and fed-batch fermentation; synthetic complete medium with 40 g/l xylose as substrate	Hexadecanol	1.2 g/L	Guo et al. (2016)
S. cerevisiae BY4742	Batch fermentation; synthetic dextrose medium	Isobutanol	93.2 ± 1.6 mg/L	Morita et al. (2017)
S. cerevisiae	Batch fermentation; synthetic medium with 10% glucose	Isobutanol	50 g/L	Ramli et al. (2017)
Yarrowia lipolytica	Batch fermentation; complete synthetic defined media (SD) contains 20 g/L glucose	Hexadecanol	636.89 mg/L intracellular	Wang et al. (2016a)
Y. lipolytica PO1f	Batch fermentation; YSC-LEU media 20 g/L glucose. YSC-URA media contained 0.77 g/L CSM-uracil	Pentane	4.98 mg/L	Blazeck et al. (2013)
Y. lipolytica Po1d	Minimal medium (YNB) with 2% glucose	TAG	5.6-fold respective increase in triacylglyceride (TAG) accumulation	Dulermo and Nicaud (2011)
Y. lipolytica	5 l stirred tank bioreactor; media: yeast nitrogen base (without amino acids and ammonium sulfate), ammonium sulfate, and 100 g/L glucose	FAEE, fatty alkanes, fatty alcohols, free fatty acids	142.5 mg/L FAEEs, 23.3 mg/L fatty alkanes, 2.15 g/L fatty alcohols, 9.67 g/L free fatty acids, and 66.4 g/ TAGs	Xu et al. (2016)
C. tropicalis ASY2	Batch fermentation; sago processing wastewater	FAME	86.56%	Thangavelu et al. (2021)

(continued)

Table 7.1 (continued)

Yeast species	Culture conditions	Biofuel	Yield	References
Saccharomyces cerevisiae	Batch fermentation; synthetic medium with 10% glucose	Isobutanol	50 g/L	Ramli et al. (2017)
R. glutinis 216	5-L automatic air-lift bioreactor; glucose as carbon source 100 g/L; yeast extract and peptone as nitrogen sources at, respectively, 8 and 3 g/L	Biodiesel	23.41 g/L	Dai et al. (2007)
R. glutinis TISTR 5159	Batch and semi-continuous fermentation; palm oil mill effluent (POME)	Biodiesel		Saenge et al. (2011)
Rhodotorula glutinis T13	Batch fermentation; GMY medium	Biodiesel	56.44%	Maza et al. (2021)
Pichia pastoris GS115	Minimal dextrose medium (MD), yeast extract peptone dextrose medium (YPD), buffered glycerol-complex medium (BMGY), and buffered methanol-complex medium (BMMY); 0.5–l stirred reactor	Methyl esters	84.4%	Jin et al. (2013)
P. pastoris	Buffered glycerol-complex medium (BMGY); batch fermentation	Methyl esters	83.14%	Huang et al. (2012)
Cutaneotrichosporon oleaginosus 20,509	YPD medium; shake flask fermentation	Biodiesel		Pham et al. (2021)

5. ***Pichia pastoris* (Jin et al.** 2013***;*** **Huang et al.** 2012***;*** **Dong et al.** 2020***):*** *P. pastoris* is widely employed in the synthesis of recombinant proteins, but it has also been investigated in the manufacture of biofuels, including biodiesel.
6. ***Cutaneotrichosporon oleaginosus* (Spagnuolo et al.** 2019***;*** **Pham et al.** 2021***):*** *C. oleaginosus* is another yeast species capable of accumulating lipids, making it a candidate for lipid-based biofuel production.

The choice of the yeast species used is determined by the type of biofuel and feedstock being used, as well as the metabolic engineering required to accomplish the appropriate biochemical pathways. Furthermore, breakthroughs in genetic engineering and synthetic biology can lead to the discovery and development of novel yeast strains with increased biofuel production capacities.

7.4.1.1 *Saccharomyces Cerevisiae* as a Workhorse for Drop-In Biofuel Production

The yeast *S. cerevisiae* emerges as the ideal candidate for biofuel production, given its extensive genetic information and versatile tools for manipulating metabolic pathways, and is generally regarded as safe (GRAS) for large-scale operation. In the current biofuel industry, *S. cerevisiae* serves as the workhorse, primarily used for ethanol production (Coyle 2010). Furthermore, through metabolic engineering, this yeast has been tailored to generate various drop-in biofuels, such as 1-butanol (Steen et al. 2008), isobutanol (Chen et al. 2011; Brat and Boles 2013), sesquiterpenoids like farnesene (Ubersax and Platt 2010), bisabolene (Peralta-Yahya et al. 2011), and amorphadiene (Paradise et al. 2008), as well as fatty acid ethyl esters (FAEEs) (biodiesel) (Buijs et al. 2013). These biofuels serve as outstanding alternatives: butanol and isobutanol are better gasoline substitutes than ethanol, farnesene, bisabolene, FAEEs are better diesel substitutes, and amorphadiene is a better jet fuel substitute.

Butanol

Butanol has a higher energy density, albeit lower than gasoline. It has better liquid-fuel qualities and is comparable to gasoline in certain aspects, with the potential to replace gasoline in current engines (Savage 2011). Though butanol has four isomers, 1-butanol and isobutanol stand out due to their benefits. These include a longer chain length, less hygroscopic nature, and the ability to blend with gasoline (85%), which makes them more desirable as biofuels. In addition to being a substitute for gasoline, isobutanol can be processed into isobutylene, which in turn can be converted into jet fuel (paraffinic kerosene) (Petrovič 2015).

Yeast species have shown promising potential for industrial applications due to their tolerance toward alcohol. However, cell toxicity and low tolerance toward isobutanol have been major obstacles in the production of this biofuel. Yeast is well documented as an ethanol producer and has found widespread use commercially owing to its high alcohol tolerance (greater than 2%) (Nor and Roshanida 2015).

Naturally occurring strains of *S. cerevisiae* yield low levels of isobutanol. To enhance the production of isobutanol, researchers have explored metabolic engineering techniques in host cells. One method includes restricting the entire pathway to the cytosol or the mitochondria, which results in higher isobutanol titers (Avalos et al. 2013; Matsuda et al. 2012). Additionally, experimental studies involving the overexpression of specific genes within the pathway have shown promising results in enhancing isobutanol production (Ida et al. 2015). The initial production of isobutanol in *S. cerevisiae* was achieved by enhancing the expression of specific genes involved in the mitochondrial valine biosynthesis pathway, leading to an increased yield of isobutanol from glucose (Weizmann 1919). Another study by Brat and Boles (2013) focused on overexpressing the valine biosynthetic pathway, ARO10-ADH2, in the cytosol and introducing a heterologous XI-XK pathway into a mitochondrial valine biosynthesis-deficient *S. cerevisiae* strain (Dilv2 Dilv3 Dilv5). This engineered strain produced isobutanol from xylose under aerobic culture conditions, demonstrating a yield of 0.16 mg/g xylose. This research marked the first successful combination of the xylose pathway and cytosolic re-localized valine pathway for isobutanol production. In one study, researchers enhanced isobutanol production by overexpressing four genes (ILV2, ILV5, ILV3, and BAT2), resulting in a significant increase in yield from 0.16 to 0.97 mg/g of glucose (Chen et al. 2011). Notably, commercial businesses like Douglas County, Colorado-based Gevo claim a theoretical yield of 97.5% when producing isobutanol using yeast.

Branduardi et al. (2013) described a novel method for using the glycine breakdown pathway in the yeast *S. cerevisiae* to produce butanol and isobutanol. They reported a titer of 92 mg/L of butanol and 58 mg/L of isobutanol with glycine as the substrate. They proposed that glycine is not just converted to CO_2 but is an efficient co-substrate for butanol formation, and butanol is produced from butyryl-CoA.

In addition to *S. cerevisiae*, various other yeast strains have been investigated for their potential to produce isobutanol. Among them are *P. pastoris*, *K. lactis*, and *Magnusiomyces magnusii*. When the endogenous amino acid biosynthetic pathway in *P. pastoris* was overexpressed, the isobutanol concentration increased significantly to 2.22 g/L (Siripong et al. 2018). Meanwhile, *K. lactis* was examined for the production of flavored volatile compounds, and it displayed an isobutanol titer of 8.6 mg/L during fermentation (Jiang 1993). It has been discovered in recent research that the yeast *M. magnusii* can also produce isobutanol; a wild-type strain of the yeast was able to produce 440 mg/L of the alcohol. Currently, three companies, namely Butalco, Butamax, and Gevo, are actively involved in the commercialization of isobutanol production (Lakshmi et al. 2021).

Fatty Acid (FA) Derivative

Energy-rich molecules, consisting of long hydrocarbon fatty-acyl chains, offer promising potential as precursors for a range of biofuels, including biodiesel. These hydrocarbons are seen as a sustainable substitute for conventional fossil diesel. The FAs are synthesized from acetyl-CoA and have an important effect as fundamental components in the production of triacylglycerols (TAGs) and other potential biofuels like fatty alcohols, FAMEs, FAEEs, and alkanes.

Certain microorganisms, such as yeasts and filamentous fungi (discussed in Sect. 9.4.2.1), have a remarkable ability to accumulate significant amounts of TAGs naturally. These accumulated TAGs can be utilized as valuable precursors for making biodiesel. Conventional *S. cerevisiae* strains do not, however, spontaneously assemble a lot of FAs or TAGs. To overcome this limitation, various metabolic engineering approaches have been employed. These approaches aim to enhance the carbon flux toward FA biosynthesis by increasing the availability of precursors and removing competing pathways while bypassing the host organism's regulatory network. Through these strategies, researchers aim to create more efficient and productive yeast strains for biodiesel production (Petrovič 2015).

Kalscheuer et al. (2004) conducted an early investigation for FAEE biosynthesis in *S. cerevisiae* using a wax ester synthase/acyl-coenzyme A: diacylglycerol acyltransferase (WS/DGAT) from *Acinetobacter calcoaceticus* ADP1, albeit with a poor yield. Shi et al. (2012) investigated the role of five distinct wax ester synthases derived from *Acinetobacter baylyi, Marinobacter hydrocarbonoclasticus, Rhodococcus opacus, Mus musculus,* and *Psychrobacter arcticus* in *S. cerevisiae* for the production of FAEE using palmitoyl-CoA as the fatty acid substrate and a variety of alcohols, from ethanol to octadecanol. Among the tested enzymes, the wax ester synthase derived from *M. hydrocarbonoclasticus* stood out as the most efficient while demonstrating the highest titer of FAEEs at 6.3 mg/L, with a further increase to 8.2 ± 1.1 mg/L with the overexpression of Acc1 (acetyl-CoA carboxylase), a key enzyme involved in the fatty acid synthesis. Valle-Rodriguez et al. (2014) inhibited the non-essential FA utilization pathways to boost FAEE. The mutant strain showed a significant threefold increase in FAEE production, reaching 17.2 mg/L as compared to the wild-type strain. The maximum known FAEE generation in *S. cerevisiae* is 520 mg/L when ethanol biosynthesis is increased, glycerol export is downregulated, and exogenous fatty acids are added (Yu et al. 2012).

For the first time, the yeast *S. cerevisiae's* heterologous production of long-chain alkanes was reported by Bujis et al. (2015). They reported that the combination of HFD1 deletion and the expression of an alkane biosynthetic pathway led to the enhanced production of the alkanes, i.e., tridecane, pentadecane, and heptadecane. With these results, evidence was provided for the production of long-chain alkanes in *S. cerevisiae*, a capability that was previously limited to bacteria, for the production of alkane-based drop-in biofuels.

Terpenoid Hydrocarbon

Pathway engineering has significantly advanced as a result of the potential isoprenoid (terpene/terpenoid) based drop-in components for a variety of fuels, from diesel to jet fuel. Terpenes and terpenoids encompass a vast array of over 40,000 structurally diverse molecules. Several terpenes have physical characteristics suggestive of desired properties, including low viscosities, flash and freezing temperatures, high energy densities, and high volumetric net heat of combustion. Sterols, monoterpenes, sesquiterpenes, and other non-steroidal terpenoid compounds are produced by different branches of this process. The carbon chain lengths of monoterpene (C10) and sesquiterpene (C15) molecules fall within the range of typical jet fuels. The isoprenoid pathway, which elongates carbon skeleton chains through the five-carbon units by condensing the building blocks isopentenyl pyrophosphate (IPP) and DMAPP, has attracted great interest as a biofuel production platform (Peralta-Yahya et al. 2011). Commercially, Amyris genetically modified *S. cerevisiae* to produce more amorphadiene and artemisinic acid (Westfall et al. 2012). In a study conducted by De Jong-Gubbels et al. (1995), all the MVA pathway genes were overexpressed, including the shortened HMG-CoA reductase gene, which was expressed in three copies. As a result, during fed-batch fermentation limited in ethanol, an amorphadiene titer of 37 g/L and yield of 18.67 Cmol% were obtained since this alcohol increases the amount of acetyl-CoA. Highly sought-after biodiesel precursors, bisabolene and farnesene, are sesquiterpenes that are produced from farnesyl pyrophosphate (FPP) by the enzymes bisabolene synthase (BS) and farnesene synthase (FS), respectively (Millo et al. 2014). Bisabolene can be chemically hydrogenated to produce fuel with characteristics akin to D2 diesel fuel. Significantly, in shake flasks, bisabolene has been successfully produced using *S. cerevisiae, achieving* titers greater than 900 mg/L (Özaydın et al. 2013).

7.4.1.2 Oleaginous Yeasts for Drop-in Biofuel Production

Microorganisms classified as "oleaginous" are those that can store more than 20% of their biomass as TAGs, typically in the form of intracellular vesicles (Ratledge 1993). Oleaginous yeasts have gained significant attention as promising candidates to produce drop-in biofuels as they possess a unique ability to accumulate large amounts of lipids, primarily in the form of TAGs. TAGs can serve as precursors for their conversion to biodiesel, green diesel, and jet fuel (Kruger et al. 2018; Nogué et al. 2018). In addition, the naturally high flux pathways for precursors used in fatty acid biosynthesis can drive oleaginous yeasts to undergo metabolic engineering to develop non-native compounds with improved fuel characteristics and potential utility as biofuels (Spagnuolo et al. 2019). Although significant advances have been made to engineer *S. cerevisiae* to produce a large quantity of fatty acids, oleaginous yeast continues to outperform the engineered non-oleaginous yeast (Ferreira et al. 2018) (Table 7.2). Among the divisions of oleaginous yeasts, two well-known groups that stand out for their exceptional ability to accumulate high amounts of lipids in the form of TAGs, classified based on their phylum, are Basidiomycota (*Rhodotorula toruloides* and *Cutaneotrichosporon curvatus*) and Ascomycota (*Yarrowia lipolytica* and *Lipomyces starkeyi*).

Eukaryotes store neutral lipids in specialized compartments called lipid bodies (LB) that form within a specific subdomain of the endoplasmic reticulum (ER). When the production of neutral lipids reaches a certain threshold, a bud forms and eventually develops into an LB, which has a neutral lipid core encased in a phospholipid monolayer embedded with different proteins involved in lipid metabolism, biosynthesis, and substrate trafficking. Triacylglycerol lipases and steryl ester hydrolases help the body release stored lipids from the LB when they are required. Free fatty acids (FFAs) are produced when these lipids are broken down, either as sources of energy or as building blocks for the production of membranes. The peroxisome for phospholipid oxidation or phospholipid biosynthesis can receive the hydrolyzed FAs.

The efficient production of microbial drop-in fuels from oleaginous sources relies on cost-effective fermentation, harvesting, and processing methods. This entails using inexpensive raw materials (Huang et al. 2013), as well as monitoring and adjustment of the culture media's C: N ratio. Additionally, the extraction of lipids

Table 7.2 Engineered strains for enhanced lipid and oleochemical (drop-in biofuel) production

Host organism	Titer (mg/L)	Biofuel produced	References
Rhodosporidium toruloides	261	Bisabolene (terpene)	Yaegashi et al. (2017)
	680	Bisabolene (terpene)	Yaegashi et al. (2017)
	14.94	1,8-cineole (monoterpene)	Zhuang et al. (2019)
	34.6	1,8-cineole (monoterpene)	Zhuang et al. (2019)
	8400	Fatty alcohols	Fillet and Adrio (2016)
	9500	Lipid (FAME)	Zhang et al. (2016c)
	16,400	Lipid (FAME)	Zhang et al. (2016b)
Yarrowia lipolytica	17.0	Alkanes	Xu et al. (2016)
	300–600	Free fatty acids	Rutter et al. (2015)
	5–90	α-Farnesene	Liu et al. (2019)
	212.7–955	β-Farnesene	Arnesen et al. (2020)
	539.38	β-Farnesene	Shi et al. (2021)
	170	Fatty alcohols	Wang et al. (2016b)
	5500	Lipid (FAME)	Qiao et al. (2017)
	8500	Lipid (FAME)	Friedlander et al. (2016)
Lipomyces starkeyi	800	Fatty alcohols	Wang et al. (2016b)

and their subsequent processing should be carried out cost-effectively (Carvalho et al. 2015). Significant advancements have been achieved in this field, as researchers have successfully created modified versions of *Trichosporon cutaneum*, *Candida lipolytica*, and acetyl-CoA synthase deletion mutants of *S. cerevisiae* capable of secreting lipids (Yagi et al. 1994; Scharnewski et al. 2008). Metabolic engineering approaches have been employed to broaden the spectrum of substrates utilized by oleaginous fungi, involving the functional expression of foreign genes. These metabolic engineering studies strive to unravel the molecular mechanisms responsible for lipid accumulation and enhance the yield of stored lipids.

Yaegashi et al. (2017) engineered a *Rhodosporidium toruloides* strain that could accumulate up to 261 mg/L of bisabolene, a terpene precursor to the biofuel bisabolane, by employing a novel biocompatible ionic liquid-treated corn stover hydrolysate on a bench scale. Furthermore, in a high-gravity fed-batch bioreactor using an alkaline-pretreated corn stover hydrolysate, the same strain achieved an even higher accumulation of 680 mg/L of bisabolene. Zhuang et al. (2019) investigated the potential of terpene synthases (TS) from diverse sources such as plants, bacteria, and fungi in *R. toruloides* for producing nine different monoterpenes. They discovered that eight of these TS were functional and capable of producing several monoterpenes, with 1,8-cineole being the most abundant at 14.94 ± 1.84 mg/L. The researchers further optimized the production of 1,8-cineole, reaching levels as high as 34.6 mg/L. Based on their findings, the authors suggested that 1,8-cineole could serve as a suitable petroleum blend stock or drop-in replacement fuel for application in spark-ignition engines.

Yarrowia lipolytica is a well-explored yeast model that demonstrates remarkable lipid production capabilities and has been genetically engineered to channel this flux into fatty acid-derived biofuels (Xu et al. 2016). *Y. lipolytica* has also been utilized for the production of various alkanes, offering a direct replacement for petroleum. This is achieved by expressing different enzymes in a heterologous manner that exploit the abundant FA species naturally present in the yeast (Blazeck et al. 2013). Bruder et al. (2019) reported the production of diesel-like, odd-chain 17- and 15-carbon-chain length alkanes and alkenes, respectively, by introducing a light-driven oxidase called CvFAP into the *Y. lipolytica*. By culturing the engineered yeast strains in glucose and the presence of light in a synthetic medium, the researchers achieved 10.9 mg/L of hydrocarbons, which was improved further to 58.7 mg/L under optimized conditions.

Recently, scientists have successfully engineered *Y. lipolytica* to produce significant amounts of farnesene by expressing three important genes related to the MVA pathway (acetyl-CoA acetyltransferase, HMG-CoA synthase, and HMG-CoA reductase) (Liu et al. 2019). The modifications led to a significant boost in farnesene production, reaching a maximum titer of 1.7 g/L. Further optimization of the fermentation conditions, including pH, dissolved oxygen, stirring speed, and high-density fed-batch fermentation led to an enhanced farnesene titer of 25.55 g/L, demonstrating the tremendous potential of *Y. lipolytica* as a robust platform for large-scale farnesene production.

In another investigation conducted by Shi et al. (2021), AaBFS was overexpressed in an engineered *Y. lipolytica* host. They successfully enhanced β-farnesene production by alleviating leucine auxotrophy, increasing the AaBFS copy number, disrupting the β-carotene pathway, and deleting DGA1 and DGA2, resulting in a production of 539.38 mg/L. By optimizing the fed-batch fermentation strategy, they further improved the β-farnesene titer up to 22.80 g/L.

7.4.2 *Filamentous Fungi*

Several filamentous fungi belonging to the genera *Aspergillus, Fusarium*, and *Penicillium* have been reported to produce volatile alcohols, terpenes, and alkanes, as drop-in fuel replacements (Spakowicz and Strobel 2015) (Fig. 7.3). The Ascocoryne genus of fungi has been found to produce various C8 compounds, C6 to C9 alkanes, and branched alcohols, which are potential replacements for gasoline (Griffin et al. 2010). Additionally, numerous fungi have been discovered to generate volatile terpene molecules, some of which are currently being explored as possible alternatives to jet or diesel fuels (Hassan et al. 2013).

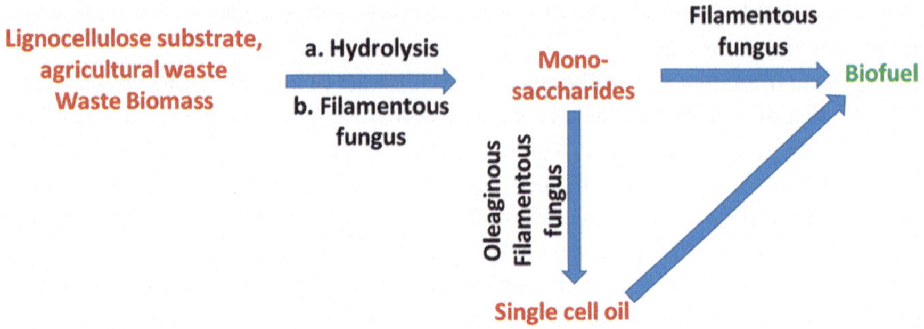

Fig. 7.3 Role of filamentous fungi in the generation of biofuel

7.4.2.1 Oleaginous Filamentous Fungi for Drop-In Biofuel Production

Filamentous fungi are believed to be efficient producers of lipids, thus indicating their efficacy in drop-in biofuel production technologies. Certain filamentous fungi belonging to the phyla Mucoromycota and Ascomycota can accumulate a significant amount of lipids in their cells and are considered to be favorable feedstock for biodiesel production (Mhlongo et al. 2021). Similar to oleaginous yeasts, key factors to take into account when choosing a filamentous fungi for biodiesel production encompass achieving a heightened overall lipid yield, particularly with elevated levels of TAGs, and ensuring an optimal composition of fatty acids (Bardhan et al. 2019). Various fungi such as *Aspergillus oryzae*, *Mortierella isabellina*, *Humicola lanuginosa*, and *Mortierella vinacea* are reported to be composed of 50–90% lipid per % dry weight (Subramaniam et al. 2010).

In oleaginous filamentous fungi, numerous enzymes are involved in the synthesis of lipids, such as acetyl-CoA carboxylase 1 (ACC1), desaturases, ATP-citrate lyase (ACL), adenosine monophosphate deaminase (AMPD), and malic enzymes (MEs) (Wang et al. 2019).

For the oleaginous microorganisms, deficiencies in critical nutrients such as nitrogen (N), phosphorus (P), sulfur (S), and vitamins and metals such as zinc (Zn), iron (Fe), and magnesium (Mg) cause a range of physiological and metabolic changes, including the diversion of carbon flow to the synthesis of lipids (Athenaki et al. 2018). Furthermore, the carbon-to-nitrogen (C/N) ratio within the culture plays a pivotal role in the lipid accumulation of oleaginous microorganisms. Research indicates that optimizing process parameters and culture conditions, including the C/N ratio, nitrogen source content, and oxygen concentration, can enhance microbial lipid accumulation. Innovative process configurations, such as solid-state or semi-solid-state fermentation methods, have been explored for this purpose (Kosa and Ragauskas 2011). Studies have demonstrated that increasing the C/N ratio in culture media significantly boosts lipid yields from oleaginous fungi (Fakas et al. 2007). At C/N ratios of 35, 44, and 57, *U. isabellina* ATHUM 2935 accumulated approximately 36%, 51.2%, and 64.3% of lipids, respectively, using rice hull hydrolysate (Economou et al. 2011). *Cunninghamella echinulata* ATHUM 4411 demonstrated lipid production exceeding 25% at a C/N ratio of 78 in an N-limited medium supplemented with xylose. The lipid content peaked at 57.7% with a higher C/N ratio of 285 (Fakas et al. 2009).

There are various characteristics of filamentous fungi that can be modified throughout the unit process consolidation. Rapid growth and the capacity to release a variety of hydrolytic enzymes are two of the most appealing characteristics of filamentous fungi (Zhao et al. 2017). Oils derived from single cells, particularly those produced by oleaginous microorganisms like filamentous fungi, are sought after as raw materials for biodiesel production owing to their high unsaturation levels. However, the high production costs connected with microbial single-cell oils (SCO) have prevented extensive industrialization. Over time, advancements such as the introduction of innovative fermentation technologies and the creation of versatile engineered oleaginous strains have enhanced the competitiveness of microbial lipid production (Abghari and Chen 2014). *Mucor circinelloides* was reported to be the first oleaginous fungus to produce polyunsaturated fatty acids (PUFAs) (Mhlongo et al. 2021). Improved lipids in oleaginous filamentous fungi can be achieved through metabolic engineering and optimization of culture conditions and media composition. Different sources, such as soil, herbaceous plants, and animal feed, have yielded the identification of numerous additional oleaginous filamentous fungi over time (Liu et al. 2010; Dey et al. 2011; Muniraj et al. 2017). *M. circinelloides* MU758 with a tendency to produce long-chain fatty acids (LCFAs) was genetically modified for enhanced production of medium-chain fatty acids (MCFAs). This was achieved by integrating a heterologous acyl-ACP thioesterase (TE) into the fatty acid synthase (FAS) complex and disrupting the acyl-CoA oxidase (ACOX) and/or acyl-CoA thioesterase (ACOT) genes in the β-oxidation pathway. Mutant strains (M-1, M-2, and M-3) exhibited significantly increased lipid production compared to the wild-type strain (WT) (Hussain et al. 2020). Because Mucormycetes (formerly Zygomycetes) and Ascomycota species are capable of high lipid accumulation, special attention has been paid to species like *Aspergillus niger*, *Cunninghamella echinulata*, *Thermomyces lanuginosus*, *Umbelopsis isabellina*, and *U. vinacea*.

To investigate the impact on lipid accumulation, the expression of genes encoding glucose 6-phosphate dehydrogenase (g6pd) and 6-phosphogluconate dehydrogenase (6pgd) from the PPP was amplified in the oleaginous fungus *M. circinelloides*. The findings revealed that overexpression of either g6pd or 6pgd led to a 20–30% increase in the lipid content of cell dry weight compared to the control strain. This effect was accompanied by elevated G6PD and 6PGD activities, as well as higher mRNA levels in the overexpressing strains. These results imply the significant role of G6PD and 6PGD as crucial NADPH providers, playing a key role in lipid accumulation in the oleaginous fungus *M. circinelloides* (Zhao et al. 2015).

In a study, *A. caespitosus* ASEF14 demonstrated a high lipid content (54% in synthetic medium (SM), 37.27% in sago processing wastewater (SWW) as its primary growth medium) with a fatty acid profile akin to commonly used vegetable oils in biodiesel production. The SWW-derived biodiesel properties align with established standards (CN: 82.32, IV: 30.53 g g^{-1}, KV: 2.62 mm^2 s^{-1}, CP: 4.66 ° C, PP: 11.90 °C), showcasing potential for *A. caespitosus* ASEF14 as a mycodiesel substitute for conventional diesel. Additionally, this approach efficiently decontaminates SWW, presenting a cost-effective and environmentally friendly alternative for diesel substitution (Srinivasan et al. 2021).

Cunninghamella echinulata biomass was reported to achieve 46.6% cellular lipids, consisting of 14% of γ-linolenic acid (Gema et al. 2002). The strain was found to exhibit efficient lipid accumulation in media supplemented with glucose as well as orange peel. A study on biodiesel production by *Fusarium oxysporum* NRC2017 revealed that the strain was capable of producing lipids while growing on a semi-solid fermentation medium consisting of agro-industrial by-products, namely rice, wheat bran, rice straw, corn stalk, sweet sorghum, and bagasse. The biodiesel produced by the strain NRC2017 consisted of 2.2% palmitoleic acid, 11.1% oleic acid, 5.6% linoleic acid, 3.0% α-linoleic acid, 2.5% eicosatrienoic acid, 6.8% eicosapentaenoic acid, 16.1% erucic acid, 4.2% margaric acid, and 3.9% behenic acid. Acylglycerides produced by various filamentous fungi can be used to produce biodiesel, jet fuel, etc. (Passoth 2017).

The production of the unsaturated fatty acid chain is also dependent on the age of the mycelia of the filamentous fungus. Filamentous fungi that produce monounsaturated fatty acids (MUFA) and saturated fatty acids (SFA) can be utilized for biodiesel production as the fatty acid profile of the biodiesel feedstock greatly influences melting point, cetane number, heating value/calorific value, kinematic viscosity, and oxidative stability of the generated biodiesel which in turn affect its performance as biodiesel (Sitepu et al. 2014). The tri-acyl glycerol produced by various oleaginous fungi can be transesterified into biodiesel. Oleaginous fungi such as *Aspergillus fumigatus, A. terreus, A. flavus, Trichoderma harzianum, Penicillium* sp., and *Fusarium graminearum* have been reported to grow on agro-wastes such as almonds husks, peanut husks, and sunflower husks to yield lipid-rich biomass that was used for the production of biodiesel (Shafiq and Chechan 2019). Sitepu et al. (2019) reported the direct production of biodiesel by using wet *Mucor plumbeus* biomass. The study revealed the direct transesterification of fatty acid to FAME, claiming 90% conversion efficiency to biodiesel.

Over the years, genetic manipulation of oleaginous filamentous fungi for increased lipid accumulation has involved recombinant DNA technology through RNA interference (RNAi), electroporation, and vector-mediated transformations (Gutiérrez et al. 2011; Takeno et al. 2005; Mhlongo et al. 2021). Unlike in oleaginous yeasts, where the recently developed clustered regularly interspaced short palindromic repeats (CRISPR)/associated protein 9 (CRISPR/Cas9) system for gene editing has been applied to improve fatty acid accumulation, there is a paucity of studies utilizing CRISPR/Cas9 to engineer oleaginous filamentous fungi for increased lipid accumulation (Yao et al. 2020). Utilizing current technologies, genetic alterations aimed at augmenting lipid accumulation in organisms that produce oil have been executed through diverse approaches. These include the overexpression of enzymes involved in fatty acid biosynthesis, upregulation of enzymes enhancing TAGs biosynthesis, modulation of related TAG biosynthesis pathways, partial inhibition of competing pathways for TAG, and the introduction of multiple genes to boost lipid accumulation.

In an attempt to augment FA and TAG production in *A. oryzae*, substituting the promoters for two fatty acid synthases and other relevant genes in the biosynthetic pathway with the highly expressed tef1 gene promoter increased the expression of fatty acid synthesis-related genes. This led to over a twofold increase in the production of FAs and TAGs. Although elevating the expression of ATP-citrate lyase and palmitoyl-ACP thioesterase genes had a positive impact, enhancing acetyl-CoA carboxylase expression showed no noticeable change in fatty acid levels. Quantitative real-time reverse transcription polymerase chain reaction confirmed increased expression of each gene, and the findings highlight that a simple upregulation of fatty acid synthase genes significantly boosts detectable fatty acid levels (Tamano et al. 2013).

Díaz-Fernández et al. (2019) explored waste industrial by-products as feedstock for microbial lipid production in engineered strains of the filamentous fungus *Ashbya gossypii*. A lipogenic xylose-utilizing strain was metabolically engineered to enhance lipid biosynthesis by alleviating regulatory mechanisms. Three genomic modifications were implemented: overexpressing a feedback-resistant acetyl-CoA carboxylase, expressing a truncated Mga2 regulator, and overexpressing an additional copy of DGA1 for diacylglycerol acyltransferase. The engineered strains performed effectively in culture media containing mixed formulations of corn-cob hydrolysates, sugarcane molasses, or crude glycerol, accumulating approximately 40% of cell dry weight (CDW) in lipid content using organic industrial wastes as feedstock.

7.4.3 Endophytic Fungi for Mycodiesel Production

Endophytic fungi are microorganisms that colonize the internal tissues of plant roots and shoots without causing any symptoms. They offer numerous benefits to host plants, including deterring pathogens and herbivores, enhancing tolerance to drought and stress, improving plant biomass, and increasing resilience in low soil fertility conditions (Ghimire et al. 2011). Over time, plants and endophytes have co-evolved mechanisms to control each other's growth. The host–endophyte relationship can lead to the development of genetic systems facilitating information transfer. Endophytes, in turn, have developed various tolerance mechanisms to the host's toxic metabolites, producing exoenzymes, mycotoxins, and extensive secondary metabolites, including terpenes, flavonoids, alkaloids, quinines, cyclohexanes, and hydrocarbons (Newman and Cragg 2015). Numerous secondary metabolites display valuable bioactivities, encompassing antimicrobial, anti-oxidant, anti-neoplastic, anti-leishmanial, anti-proliferative, and cytotoxic properties (Zhao et al. 2016; Wu et al. 2016; Kharwar et al. 2011).

Various endophytic fungi are reported to produce volatile organic compounds with industrial potential. Volatile organic compounds (VOCs) are carbon-based

Table 7.3 Endophytic fungi-derived volatile organic compounds with potential as drop-in biofuels

Endophytic fungi	Host plant	Volatile compound	References
Annulohypoxylon sp.	*Neolitsea pulchella*	1,8-cineole	Wang et al. (2017)
Daldinia sp.	*Myroxylon balsamum*	1,4 pentadiene, 1, 4-cyclohexadiene	Mends et al. (2012)
Nodulisporium sp.	*Thelypteris angustifolia*	2-pentanone, 1,8-cineole	Hassan et al. (2013)
Phoma sp.	*Larrea tridentata*	Sesquiterpenoids	Strobel et al. (2011)
Aspergillus nomius	*Eusideroxylon zwageri*	Hydrocarbons, alkyl halides, and alcohols	Azeez et al. (2016)
Fimetariella rabenhorstii	*Aquilaria sinensis*	Frabenol (sesquiter-pene alcohol)	Tao et al. (2011)
Lasiodiplodia sp. *Neofusicoccum* sp., *Botryosphaeria* sp., *Pseudofusicoccum* sp.	Plants in Caatinga Biome, Brazil	Sesquiterpenes	Oliveira et al. (2015)
Muscodor crispans	*Ananas ananassoides*	Propanoic acid	Mitchell (2008)
Ascocoryne sarcoides	*Eucryphia cordifolia*	Sesquiterpenes, alkanes	Griffin et al. (2010)

chemicals with low molecular weights and high vapor pressure, produced as part of the metabolic processes of living organisms (Table 7.3). Among these VOCs are terpenes, like pinene and bisabolene, which are currently under active research as potential drop-in biofuels. They are considered diesel and aviation fuel replacements due to their high energy densities (Wu et al. 2016). These high-energy-density volatile organic compounds are often referred to as mycodiesel (Strobel et al. 2008). The concept of mycodiesel was first reported by Gary Strobel and his research team, who observed the production of mycodiesel in an endophytic fungus called *Gliocladium roseum*. This fungus was isolated from the leaves of the ulmo tree (*Eucryphia cordifolia*) in the Andes mountain region of Patagonia (Strobel et al. 2008).

Endophytic fungi belonging to various Ascomycota lineages, particularly members of Xylariaceae, have shown the ability to produce hydrocarbons. Among them, the genus Muscodor, represented by *M. albus*, has captured the attention of mycologists due to its obligate endophytic nature, broad spectrum of antimicrobial activity, and potential for fuel production (Strobel et al. 2001).

Another example is *Myrothecium inundatum*, a fungal endophyte extracted from the herb *Acalypha indica* L., in Kalyani, India. This fungus demonstrated the production of several fuel-related hydrocarbons of interest under microaerophilic conditions. Notably, the fungus generated octane, 1,4-cyclohexadiene, 1-methyl, and cyclohexane (1-ethylpropyl). The peak production of VOCs was observed on day 15 during the experimental period (Paul et al. 2020). These findings highlight the potential of endophytic fungi in the production of fuel-related hydrocarbons and the importance of studying their diverse capabilities. In a study by Wu et al. (2016), terpene synthases from four endophytic fungi of the Xylariaceous family, renowned for their production of mycodiesel hydrocarbons, were characterized. These findings shed light on the fascinating potential of endophytic fungi in the realm of fuel-related hydrocarbon production.

Shaw et al. (2015) proposed an evolutionary connection between fungal terpene synthases and those present in plants and bacteria. Among the compounds identified was 1,8-cineole, produced by the endophytic *Hypoxylon* sp. In a separate investigation, *Hypoxylon* sp. (BS15) was isolated from the *Taxodium distichum* plant and cultivated on degraded cellulose and potato dextrose broth, demonstrating the potential to generate VOCs with significance as hydrocarbon fuels (Wang Kovaleff 2019). These findings underscore the intriguing capacity of fungal endophytes to produce valuable compounds with potential energy applications.

In the upper Napo region of the Ecuadorian Amazon, researchers have found an endophytic *Nodulisporium* sp. residing within *Myroxylon balsamum*. This remarkable fungus exhibits the ability to produce several compounds, including 1,4-cyclohexadiene, 1-methyl-, 1–4 pentadiene, and cyclohexene, 1-methyl-4-(1-methyl ethenyl), alongside specific alcohols and terpenoids of interest. These compounds show promising potential as fuels when cultivated under microaerophilic growth conditions (Mends et al. 2012).

Phomopsis sp. was discovered as an endophyte in *Odontoglossum* sp. (Orchidaceae) in the cloud forest of Northern Ecuador. *Phomopsis* sp. produced a distinctive blend of VOCs, including sabinene, a monoterpene only identified in higher plants. Major VOCs identified from the endophyte include 1-butanol, 3-methyl; benzeneethanol; 1-propanol, 2-methyl; and 2-propanone, as additive to gasoline (Singh et al. 2011). In the rainforests of Central America rainforest region, *Nodulisporium* sp. (*Hypoxylo*n sp.) was discovered as an endophyte residing within *Thelypteris angustifolia* (broadleaf leaf maiden fern). This exceptional fungus can produce a diverse array of ketones. Of particular interest, it also synthesizes cyclohexane propyl, a common ingredient found in diesel fuel when cultivated under suitable conditions (Hassan et al. 2013). In another study, *Nodulisporium* sp. (*Hypoxylon* sp.) was isolated from *Persea indica* (native to the Canary Islands), identified as producing a diverse range of VOCs, including 1,8-cineole, 1-methyl-1,4-cyclohexadiene, and tentatively identified (+)-.alpha.-methylene-.alpha.-fenchocamphorone, among others (Tomsheck et al. 2010). *G. roseum* emits antibiotic volatiles, primarily acetic acid esters of straight-chained compounds like hexyl, heptyl, octyl, sec-octyl, and decyl alcohols when grown on oatmeal-based agar in microaerophilic conditions, as determined by SPME-GC-MS analysis. This discovery, supported by other studies, indicated the fungus's metabolic capability to potentially produce Class 1 alkanes, common in diesel fuels (Strobel et al. 2008).

Fungi associated with biodiesel-producing plants (*Jatropha curcas*, *Pongamia pinnata*, *Sapindus mukorossi*, *Mesua ferrea*, *Terminalia bellerica*, *Cascabela thevetia*, and *Ricinus communis*) were screened for lipid production. The endophytic fungi isolated from the bio-diesel-producing plants, namely *Lasiodiplodia exigua* SPSRJ27, *Phomopsis* sp. SPSRJ28, *Pestalotiopsis microspora* SPSRJL35, and *Phomopsis* sp. SPSRJL36, exhibited lipid content exceeding 20% of their dry biomass, indicating oleaginous properties. Of these endophytic fungal isolates, the authors reported that *Phomopsis* sp. SPSRJ 28 was the most promising for biodiesel production. The study highlighted the potential of manipulating the growth of oleaginous fungal endophytes for biofuel production (Paul et al. 2020).

A screening process was conducted on 25 strains of filamentous fungi to assess their lipid production using the Nile red staining assay. Among these strains, *Penicillium commune* NRC 2016 stood out as a promising candidate for biodiesel production, as the lipids it produced were found to be suitable for this purpose. The fatty acid composition of the lipids was determined to be 20% palmitic acid, 4.65% stearic acid, 13.77% oleic acid, 32% linoleic acid, 1.75% α-linolenic acid, and 16.1% arachidonic acid (Hussein et al. 2017). In a study conducted in New Delhi, India, researchers investigated the endophytic fungus *Colletotrichum truncatum* of *Jatropha curcas*. Isolates EF9, EF10, and EF13 were found to produce lipids with significant potential as drop-in biofuels, and their hexane extracts were particularly promising in this regard (Kumar and Kaushik 2013).

7.5 Commercial Production of Drop-In Biofuels

Several companies are actively engaged in producing hydrocarbon-based biofuels through microbial synthesis, aiming to create "drop-in" replacements. These companies have also diversified their portfolios to include higher-value products such as platform molecules, pharmaceutical precursors, lubricants, and flavors, due to the volatile nature of oil prices.

Amyris: Utilizes sugar as a feedstock and *S. cere*visiae as the organism to produce diesel fuel, terpenes, and fine chemicals. Their product Biofene® is commercially available. Pathway: Terpenoid pathway. Partner: Total.

Genomatica: Uses sugar as a feedstock and *E. coli* as the organism to produce butadiene. Currently, at the pilot stage, they employ amino acid biosynthesis. Partners: Braskem, BASF.

Global Bioenergies: Sugar serves as the feedstock, and *E. coli* is the organism used to produce isobutene. The company has reached the commercial stage and employs amino acid biosynthesis. Partner: Audi.

Joule: Utilizes light and CO_2 with cyanobacteria as the organism to produce gasoline, diesel, and jet fuel. The company is at the commercial demonstration stage, utilizing fatty acid decarboxylation. Partner: Audi.

RHO Renewables: Uses sugar as a feedstock and *S. cere*visiae as the organism to produce monocyclic aromatic hydrocarbons as fuel additives. Currently in the pre-pilot stage, they employ polyketide synthases. Partner: Sasol.

REG Life Science (LS9): Sugar is used as the feedstock, and both *E. coli* and *S. cerevisiae* are employed to produce renewable chemicals and fuels. The company is currently in the research phase, using fatty acid decarboxylation. Partner: REG.

Sapphire Energy: Utilizes light and CO_2 with undisclosed algae and cyanobacteria as the organisms to produce gasoline, diesel, and jet fuel. The company is at the commercial demonstration stage, employing the terpenoid pathway. Partner: Phillips 66.

The success of these companies is often tied to the development of the bio-economy. During periods of high oil prices, there is increased support for research and start-ups in this field. However, when oil prices drop, the economic targets for these companies become more challenging to achieve within the expected time frames. Consequently, investments and research may shift toward alternative products during such periods. This cyclic pattern has been observed repeatedly over the years.

7.6 Future Perspectives

The future of drop-in biofuels looks promising, as advancements in microbial technologies and genetic engineering continue to enhance production efficiency and sustainability. As we move forward, several key areas deserve further attention and exploration:

Research and development: Continued investment in research and development is essential to discover novel microbes, optimize metabolic pathways, and improve the overall efficiency of drop-in biofuel production. Emphasizing the use of non-agricultural and low-cost feedstock (primary contributor to the production cost of drop-in fuel) can help mitigate competition with food crops and further reduce environmental impacts.

Scale-up and commercialization: The successful implementation of drop-in biofuels on a large scale will require overcoming technical and economic challenges. Collaborations between governments, industries, and research institutions are crucial to developing cost-effective production processes and establishing an extensive supply chain.

Policy and regulatory support: Governments worldwide should continue to offer supportive policies and regulations, such as blending mandates, tax incentives, and emission reduction targets, to encourage the adoption of drop-in biofuels and stimulate market growth.

Sustainability and environmental impact: As drop-in biofuels gain traction, it is imperative to assess their full life-cycle environmental impacts to ensure they deliver significant greenhouse gas reductions and contribute to a sustainable energy future.

Finally, drop-in biofuels hold great promise as a viable and sustainable solution to meet the world's growing transportation fuel demands while mitigating the environmental impacts of fossil fuel consumption. Continued efforts in research, development, and policy support are crucial to unlock the full potential of drop-in biofuels and paving the way for a cleaner and greener future in transportation.

7.7 Conclusion

The increasing energy demands of the rising population and industrialization, along with the depletion of finite fossil fuels and the negative consequences of greenhouse gas emissions, have prompted a change toward sustainable and environmentally acceptable transportation fuels. Drop-in biofuel development and adoption have emerged as potential fixes to these challenges. Drop-in biofuels, as opposed to traditional biofuels, have the advantage of seamless integration into existing engines and infrastructure, making them an appealing and feasible option for sustainable energy transitions.

Various drop-in biofuels, including renewable diesel, green diesel, and bio-gasoline, derived from non-fossil sources such as plant oils, biomass, and algae, have gained prominence in sectors like aviation, marine transportation, and rail. Governments and industries worldwide are actively investing in research, development, and production of these biofuels, supported by incentives and regulations to facilitate their integration into conventional transportation fuel blends. However, the widespread commercialization of drop-in biofuels faces several challenges, such as feedstock availability, high production costs, and competition with food production for land and resources. To overcome these obstacles, microbial fermentation emerges as a promising approach for renewable biofuel production. Microbe genetic and metabolic engineering has the potential to improve the efficiency and output of drop-in biofuels while minimizing conflicts with food crops for land usage. Fungi, including yeasts and filamentous fungi, have proven to be valuable biocatalysts in biotechnology, offering a diverse range of biochemical processes, enzymes, and bioactive compounds. Leveraging cutting-edge omics technologies and genetic engineering, model organisms like *Pichia pastoris* and *Saccharomyces cerevisiae* have paved the way for advanced biofuel production and the concept of drop-in biofuels.

Acknowledgments The authors would like to acknowledge the Department of Biosciences and Bioengineering, IIT Guwahati, Assam, India. SS would also like to acknowledge the financial support from SERB, DST, Govt. of India, in the form of a National post-doctoral fellowship (NPDF, PDF/2022/001568).

References

Abghari A, Chen S (2014) *Yarrowia lipolytica* as an oleaginous cell factory platform for production of fatty acid-based biofuel and bioproducts. Front Energy Res. https://doi.org/10.3389/fenrg.2014.000

Ainsworth GC (2008) Ainsworth & Bisby's dictionary of the fungi. CABI, London

Almaguer Chávez M (2022) Thermotolerance and adaptation to climate change. In: Frías-De-León (ed) The impact of climate change on fungal diseases, 1st edn. Springer, Cham, pp 37–71

Andlar M, Rezić T, Mardetko N, Kracher D, Ludwig R, Šantek B (2018) Lignocellulose degradation: an overview of fungi and fungal enzymes involved in lignocellulose degradation. Eng Life Sci 18(11):768–778

Arnesen JA, Kildegaard KR, Cernuda Pastor M, Jayachandran S, Kristensen M, Borodina I (2020) *Yarrowia lipolytica* strains engineered for the production of terpenoids. Front Bioeng Biotechnol. https://doi.org/10.3389/fbioe.2020.00945

Athenaki M, Gardeli C, Diamantopoulou P, Tchakouteu SS, Sarris D, Philippoussis A, Papanikolaou S (2018) Lipids from yeasts and fungi: physiology, production and analytical considerations. J Appl Microbiol 124(2):336–367

Atsumi S, Hanai T, Liao JC (2008) Non-fermentative pathways for synthesis of branched-chain higher alcohols as biofuels. Nature 451(7174):86–89

Avalos JL, Fink GR, Stephanopoulos G (2013) Compartmentalization of metabolic pathways in yeast mitochondria improves the production of branched-chain alcohols. Nat Biotechnol 31(4):335–341

Azeez LA, Muid S, Hasnul BM (2016) Identification of volatile secondary metabolites from an endophytic microfungus *Aspergillus nomius* KUB105. MJAS 20(4):751–759

Bardhan P, Gohain M, Daimary N, Kishor S, Chattopadhyay P, Gupta K, Chaliha C, Kalita E, Deka D, Mandal M (2019) Microbial lipids from cellulolytic oleaginous fungus *Penicillium citrinum* PKB20 as a potential feedstock for biodiesel production. Ann Microbiol 69(11):1135–1146

Barker H, Taha S (1942) *Clostridium kluyverii*, an organism concerned in the formation of caproic acid from ethyl alcohol. J Bacteriol 43(3):347–363

Barnett JA, Payne RW, Yarrow D (1990) Yeasts: characteristics and identification, 2nd edn. Cambridge University Press, Cambridge

Bello B (2012) Captured CO_2 to grow microalgae for bio-jet fuel production. UKCCSC Winter School, University of Cambridge, Cambridge, UK

Blazeck J, Liu L, Knight R, Alper HS (2013) Heterologous production of pentane in the oleaginous yeast *Yarrowia lipolytica*. J Biotechnol 165(3–4):184–194

Branduardi P, Longo V, Berterame NM, Rossi G, Porro D (2013) A novel pathway to produce butanol and isobutanol in *Saccharomyces cerevisiae*. Biotechnol Biofuels 6(1):1–12

Brat D, Boles E (2013) Isobutanol production from D-xylose by recombinant *Saccharomyces cerevisiae*. FEMS Yeast Res 13(2):241–244

Brat D, Weber C, Lorenzen W, Bode HB, Boles E (2012) Cytosolic re-localization and optimization of valine synthesis and catabolism enables increased isobutanol production with the yeast *Saccharomyces cerevisiae*. Biotechnol Biofuels 5(65):1–16

Bruder S, Moldenhauer EJ, Lemke RD, Ledesma-Amaro R, Kabisch J (2019) Drop-in biofuel production using fatty acid photodecarboxylase from *Chlorella variabilis* in the oleaginous yeast *Yarrowia lipolytica*. Biotechnol Biofuels 12(1):1–13

Bueno DJ, Silva JO (2014) Fungi. In: The fungal hypha. Elsevier, New York, pp 11–19

Buijs NA, Siewers V, Nielsen J (2013) Advanced biofuel production by the yeast *Saccharomyces cerevisiae*. Curr Opin Chem Biol 17(3):480–488

Buijs NA, Zhou YJ, Siewers V, Nielsen J (2015) Long-chain alkane production by the yeast *Saccharomyces cerevisiae*. Biotechnol Bioeng 112(6):1275–1279

Carvalho AK, Rivaldi JD, Barbosa JC, de Castro HF (2015) Biosynthesis, characterization and enzymatic transesterification of single cell oil of *Mucor circinelloides* – a sustainable pathway for biofuel production. Bioresour Technol 181:47–53

Chen X, Nielsen KF, Borodina I, Kielland-Brandt MC, Karhumaa K (2011) Increased isobutanol production in Saccharomyces cerevisiae by overexpression of genes in valine metabolism. Biotechnol Biofuels 49(1):1–12

Cherry JR, Fidantsef AL (2003) Directed evolution of industrial enzymes: an update. Curr Opin Biotechnol 14(4):438–443

Coyle WT (2010) Next-generation biofuels: near-term challenges and implications for agriculture. Diane Publishing, New York

Dai C-C, Tao J, Xie F, Dai Y-j, Zhao M (2007) Biodiesel generation from oleaginous yeast *Rhodotorula glutinis* with xylose assimilating capacity. AJB 6(18):2130–2134

Darvishi F, Fathi Z, Ariana M, Moradi H (2017) *Yarrowia lipolytica* as a workhorse for biofuel production. Biochem Eng J 127:87–96

De Jong-Gubbels P, Vanrolleghem P, Heijnen S, Van Dijken JP, Pronk JT (1995) Regulation of carbon metabolism in chemostat cultures of *Saccharomyces cerevisiae* grown on mixtures of glucose and ethanol. Yeast 11(5):407–418

Dellomonaco C, Clomburg JM, Miller EN, Gonzalez R (2011) Engineered reversal of the β-oxidation cycle for the synthesis of fuels and chemicals. Nature 476(7360):355–359

Dey P, Banerjee J, Maiti MK (2011) Comparative lipid profiling of two endophytic fungal isolates– *Colletotrichum* sp. and *Alternaria* sp. having potential utilities as biodiesel feedstock. Bioresour Technol 102(10):5815–5823

Díaz-Fernández D, Aguiar TQ, Martín VI, Romaní A, Silva R, Domingues L, Revuelta JL, Jiménez A (2019) Microbial lipids from industrial wastes using xylose-utilizing *Ashbya gossypii* strains. Bioresour Technol. https://doi.org/10.1016/j.biortech.2019.122054

Dix NJ (ed) (2012) Fungal ecology. Springer, Berlin, Germany

Dong C, Qiao J, Wang X, Sun W, Chen L, Li S, Wu K, Ma L, Liu Y (2020) Engineering *Pichia pastoris* with surface-display minicellulosomes for carboxymethyl cellulose hydrolysis and ethanol production. Biotechnol Biofuels 13(1):1–9

Dongre MA, Ashraf MS, Koshariya AK, Reddypriya P (2023) Plant pathology and plant diseases. AG Publishing House (AGPH Books), Bhopal, India

Dulermo T, Nicaud J-M (2011) Involvement of the G3P shuttle and β-oxidation pathway in the control of TAG synthesis and lipid accumulation in *Yarrowia lipolytica*. Metab Eng 13(5): 482–491

Ebadian M, van Dyk S, McMillan JD, Saddler J (2020) Biofuels policies that have encouraged their production and use: an international perspective. Energy Policy 147. https://doi.org/10.1016/j.enpol.2020.111906

Economou CN, Aggelis G, Pavlou S, Vayenas DV (2011) Single cell oil production from rice hulls hydrolysate. Bioresour Technol 102(20):9737–9742

Fakas S, Galiotou-Panayotou M, Papanikolaou S, Komaitis M, Aggelis G (2007) Compositional shifts in lipid fractions during lipid turnover in *Cunninghamella echinulata*. Enzyme Microb Technol 40(5):1321–1327

Fakas S, Makri A, Mavromati M, Tselepi M, Aggelis G (2009) Fatty acid composition in lipid fractions lengthwise the mycelium of *Mortierella isabellina* and lipid production by solid state fermentation. Bioresour Technol 100(23):6118–6120

Fan L, Liu J, Nie K, Liu L, Wang F, Tan T, Deng L (2013) Synthesis of medium chain length fatty acid ethyl esters in engineered *Escherichia coli* using endogenously produced medium chain fatty acids. Enzyme Microb Technol 53(2):128–133

Ferreira R, Teixeira PG, Gossing M, David F, Siewers V, Nielsen J (2018) Metabolic engineering of *Saccharomyces cerevisiae* for overproduction of triacylglycerols. Metab Eng Commun 6:22–27

Fillet S, Adrio JL (2016) Microbial production of fatty alcohols. World J Microbiol Biotechnol. https://doi.org/10.1007/s11274-016-2099-z

Friedlander J, Tsakraklides V, Kamineni A, Greenhagen EH, Consiglio AL, MacEwen K, Crabtree DV, Afshar J, Nugent RL, Hamilton MA (2016) Engineering of a high lipid producing *Yarrowia lipolytica* strain. Biotechnol Biofuels 9(77):1–12

Gajewski J, Pavlovic R, Fischer M, Boles E, Grininger M (2017) Engineering fungal de novo fatty acid synthesis for short chain fatty acid production. Nat Commun. https://doi.org/10.1038/ncomms14650

Gander J (1974) Fungal cell wall glycoproteins and peptido-polysaccharides. Annu Rev Microbiol 28(1):103–120

Gema H, Kavadia A, Dimou D, Tsagou V, Komaitis M, Aggelis G (2002) Production of γ-linolenic acid by *Cunninghamella echinulata* cultivated on glucose and orange peel. Appl Microbiol Biotechnol 58:303–307

Ghimire SR, Charlton ND, Bell JD, Krishnamurthy YL, Craven KD (2011) Biodiversity of fungal endophyte communities inhabiting switchgrass (*Panicum virgatum* L.) growing in the native tallgrass prairie of northern Oklahoma. Fungal Divers 47:19–27

Griffin MA, Spakowicz DJ, Gianoulis TA, Strobel SA (2010) Volatile organic compound production by organisms in the genus Ascocoryne and a re-evaluation of myco-diesel production by NRRL 50072. Microbiology. https://doi.org/10.1099/mic.0.041327-0

Guo W, Sheng J, Zhao H, Feng X (2016) Metabolic engineering of *Saccharomyces cerevisiae* to produce 1-hexadecanol from xylose. Microb Cell Fact 15:1–11

Gutiérrez A, López-García S, Garre V (2011) High reliability transformation of the basal fungus *Mucor circinelloides* by electroporation. J Microbiol Methods 84(3):442–446

Hagen A, Poust S, Rond TD, Fortman JL, Katz L, Petzold CJ, Keasling JD (2016) Engineering a polyketide synthase for *in vitro* production of adipic acid. ACS Synth Biol 5(1):21–27

Hassan SR-U, Strobel GA, Geary B, Sears J (2013) An endophytic Nodulisporium sp. from Central America producing volatile organic compounds with both biological and fuel potential. J Microbiol Biotechnol 23(1):29–35

Heleniak T (2020) Polar peoples in the future: projections of the Arctic populations. Nordregio, Stockholm

Huang D, Han S, Han Z, Lin Y (2012) Biodiesel production catalyzed by *Rhizomucor miehei* lipase-displaying *Pichia pastoris* whole cells in an isooctane system. Biochem Eng J 63:10–14

Huang C, Chen X-f, Xiong L, Ma L-l, Chen Y (2013) Single cell oil production from low-cost substrates: the possibility and potential of its industrialization. Biotechnol Adv 31(2):129–139

Hussain SA, Garcia A, Khan MAK, Nosheen S, Zhang Y, Koffas MA, Garre V, Lee SC, Song Y (2020) Increased accumulation of medium-chain fatty acids by dynamic degradation of long-chain fatty acids in *Mucor circinelloides*. Genes. https://doi.org/10.3390/genes11080890

Hussein AA, El Sayed O, Asker MS, Mohamed SS, Abdelhamid S (2017) Biodiesel production from local isolate *Penicillium commune* NRC 2016. JSRS 34:179–193

Ida K, Ishii J, Matsuda F, Kondo T, Kondo A (2015) Eliminating the isoleucine biosynthetic pathway to reduce competitive carbon outflow during isobutanol production by *Saccharomyces cerevisiae*. Microb Cell Fact 14(62):1–9

Janda K, Kristoufek L, Zilberman D (2012) Biofuels: policies and impacts. Agric Econ 58(8):372–386

Jiang J (1993) Identification of flavour volatile compounds produced by *Kluyveromyces lactis*. Biotechnol Tech 7(12):863–866

Jin Z, Han S-Y, Zhang L, Zheng S-P, Wang Y, Lin Y (2013) Combined utilization of lipase-displaying *Pichia pastoris* whole-cell biocatalysts to improve biodiesel production in co-solvent media. Bioresour Technol 130:102–109

Junne S, Kabisch J (2017) Fueling the future with biomass: processes and pathways for a sustainable supply of hydrocarbon fuels and biogas. Eng Life Sci 17(1):14–26

Kalscheuer R, Luftmann H, Steinbüchel A (2004) Synthesis of novel lipids in *Saccharomyces cerevisiae* by heterologous expression of an unspecific bacterial acyltransferase. Appl Environ Microbiol 70(12):7119–7125

Karaffa L, Kubicek CP (2019) Citric acid and itaconic acid accumulation: variations of the same story? Appl Microbiol Biotechnol 103:2889–2902

Karatzos S, McMillan JD, Saddler JN (2014) The potential and challenges of drop-in biofuels. Report for IEA Bioenergy Task 39

Kharwar RN, Mishra A, Gond SK, Stierle A, Stierle D (2011) Anticancer compounds derived from fungal endophytes: their importance and future challenges. Nat Prod Rep 28(7):1208–1228

Kosa M, Ragauskas AJ (2011) Lipids from heterotrophic microbes: advances in metabolism research. Trends Biotechnol 29(2):53–61

Kruger JS, Cleveland NS, Yeap RY, Dong T, Ramirez KJ, Nagle NJ, Lowell AC, Beckham GT, McMillan JD, Biddy MJ (2018) Recovery of fuel-precursor lipids from oleaginous yeast. ACS Sustain Chem Eng 6(3):2921–2931

Kuivanen J, Wang Y-MJ, Richard P (2016) Engineering *Aspergillus niger* for galactaric acid production: elimination of galactaric acid catabolism by using RNA sequencing and CRISPR/Cas9. Microb Cell Fact 15(210):1–9

Kumar S, Kaushik N (2013) Endophytic fungi isolated from oil-seed crop Jatropha curcas produces oil and exhibit antifungal activity. PLoS One. https://doi.org/10.1371/journal.pone.0056202

Lakshmi NM, Binod P, Sindhu R, Awasthi MK, Pandey A (2021) Microbial engineering for the production of isobutanol: current status and future directions. Bioengineered 12(2): 12308–12321

Lee W-H, Seo S-O, Bae Y-H, Nan H, Jin Y-S, Seo J-H (2012) Isobutanol production in engineered *Saccharomyces cerevisiae* by overexpression of 2-ketoisovalerate decarboxylase and valine biosynthetic enzymes. Bioprocess Biosyst Eng 35:1467–1475

Liu G-Q, Lin Q-L, Jin X-C, Wang X-L, Zhao Y (2010) Screening and fermentation optimization of microbial lipid-producing molds from forest soils. Afr J Microbiol Res 4(14):1462–1468

Liu Y, Jiang X, Cui Z, Wang Z, Qi Q, Hou J (2019) Engineering the oleaginous yeast *Yarrowia lipolytica* for production of α-farnesene. Biotechnol Biofuels 12(296):1–11

Marcheschi RJ, Li H, Zhang K, Noey EL, Kim S, Chaubey A, Houk K, Liao JC (2012) A synthetic recursive "+1" pathway for carbon chain elongation. ACS Chem Biol 7(4):689–697

Matsuda F, Kondo T, Ida K, Tezuka H, Ishii J, Kondo A (2012) Construction of an artificial pathway for isobutanol biosynthesis in the cytosol of *Saccharomyces cerevisiae*. Biosci Biotechnol Biochem 76(11):2139–2141

Maza DD, Viñarta SC, García-Ríos E, Guillamón JM, Aybar MJ (2021) *Rhodotorula glutinis* T13 as a potential source of microbial lipids for biodiesel generation. J Biotechnol 331:14–18

Mends MT, Yu E, Strobel GA, Hassan S, Booth E, Geary B, Sears J, Taatjes C, Hadi M (2012) An endophytic Nodulisporium sp. producing volatile organic compounds having bioactivity and fuel potential. J Pet Environ Biotechnol 331:14–18

Meyer V, Andersen MR, Brakhage AA, Braus GH, Caddick MX, Cairns TC, de Vries RP, Haarmann T, Hansen K, Hertz-Fowler C (2016) Current challenges of research on filamentous fungi in relation to human welfare and a sustainable bio-economy: a white paper. Fungal Biol Biotechnol 3(1):1–17

Meyer V, Basenko EY, Benz JP, Braus GH, Caddick MX, Csukai M, De Vries RP, Endy D, Frisvad JC, Gunde-Cimerman N (2020) Growing a circular economy with fungal biotechnology: a white paper. Fungal Biol Biotechnol 7(1):1–23

Mhlongo SI, Ezeokoli OT, Roopnarain A, Ndaba B, Sekoai PT, Habimana O, Pohl CH (2021) The potential of single-cell oils derived from filamentous fungi as alternative feedstock sources for biodiesel production. Front Microbiol. https://doi.org/10.3389/fmicb.2021.637381

Millo F, Bensaid S, Fino D, Marcano SJC, Vlachos T, Debnath BK (2014) Influence on the performance and emissions of an automotive Euro 5 diesel engine fueled with F30 from Farnesane. Fuel 138:134–142

Mitchell A (2008) *Muscodor crispans*, a novel endophyte from *Ananas ananassoides* in the Bolivian Amazon. Fungal Divers 31:37–43

Morita K, Nomura Y, Ishii J, Matsuda F, Kondo A, Shimizu H (2017) Heterologous expression of bacterial phosphoenol pyruvate carboxylase and Entner–Doudoroff pathway in *Saccharomyces cerevisiae* for improvement of isobutanol production. J Biosci Bioeng 124(3):263–270

Muniraj IK, Xiao L, Hu Z, Zhan X (2017) Screening and characterization of oleaginous fungi from Irish soil for growth under low carbon substrates. Int J Curr Microbiol Appl Sci 6:772–781

Murphy PA, Hendrich S, Landgren C, Bryant CM (2006) Food mycotoxins: an update. J Food Sci 71:R51. https://doi.org/10.1111/j.1750-3841.2006.00052.x

Newman DJ, Cragg GM (2015) Endophytic and epiphytic microbes as "sources" of bioactive agents. Front Chem. https://doi.org/10.3389/fchem.2015.00034

Nitte IS (2023) Global energy transition and its implications on energy security in nigeria: a critical review. EPRA Int J Clim Resour Econ Rev (CRER) 11(1):1–9

Nogué VS, Black BA, Kruger JS, Singer CA, Ramirez KJ, Reed ML, Cleveland NS, Singer ER, Yi X, Yeap RY (2018) Integrated diesel production from lignocellulosic sugars via oleaginous yeast. Green Chem 20(18):4349–4365

Nor AR, Roshanida AR (2015) Isobutanol production and alcohol tolerance by yeast wild strain. Adv Mat Res 1113:334–339

Oliveira FC, Barbosa FG, Mafezoli J, Oliveira MC, Camelo AL, Longhinotti E, Lima AC, Câmara MP, Gonçalves FJ, Freire FC (2015) Volatile organic compounds from filamentous fungi: a chemotaxonomic tool of the Botryosphaeriaceae family. J Braz Chem Soc 26(11):2189–2194

Özaydın B, Burd H, Lee TS, Keasling JD (2013) Carotenoid-based phenotypic screen of the yeast deletion collection reveals new genes with roles in isoprenoid production. Metab Eng 15:174–183

Paddon CJ, Keasling JD (2014) Semi-synthetic artemisinin: a model for the use of synthetic biology in pharmaceutical development. Nat Rev Microbiol 12(5):355–367

Paradise EM, Kirby J, Chan R, Keasling JD (2008) Redirection of flux through the FPP branch-point in *Saccharomyces cerevisiae* by down-regulating squalene synthase. Biotechnol Bioeng 100(2):371–378

Passoth V (2017) Lipids of yeasts and filamentous fungi and their importance for biotechnology. In: Sibirny A (ed) Biotechnology of yeasts and filamentous fungi. Springer, Cham, pp 149–204

Paul S, Bhagobaty R, Nihalani M, Joshi S (2020) Characterization of oleaginous endophytic fungi of biodiesel plants as potential biofuel minifactories. Biomass Bioenergy. https://doi.org/10.1016/j.biombioe.2020.105750

Peiris TC (2019) Future advances of aviation 36. https://www.doi.org/VIDU_36_4_6_15.pdf (https://www.nsf.gov.lk)

Peralta-Yahya PP, Ouellet M, Chan R, Mukhopadhyay A, Keasling JD, Lee TS (2011) Identification and microbial production of a terpene-based advanced biofuel. Nat Commun. https://doi.org/10.1038/ncomms1494

Petrovič U (2015) Next-generation biofuels: a new challenge for yeast. Yeast 32:583–593

Pham N, Reijnders M, Suarez-Diez M, Nijsse B, Springer J, Eggink G, Schaap PJ (2021) Genome-scale metabolic modeling underscores the potential of *Cutaneotrichosporon oleaginosus* ATCC 20509 as a cell factory for biofuel production. Biotechnol Biofuels 14(1):1–17

Qiao K, Wasylenko TM, Zhou K, Xu P, Stephanopoulos G (2017) Lipid production in *Yarrowia lipolytica* is maximized by engineering cytosolic redox metabolism. Nat Biotechnol 35(2): 173–177

Ramli N, Rahman R, Ngadi N, Samah R (2017) Optimisation of fermentation conditions for isobutanol production by *Saccharomyces cerevisiae* using response surface methodology. Chem Eng Trans 56:301–306

Ratledge C (1993) Single cell oils—have they a biotechnological future? Trends Biotechnol 11(7): 278–284

Richard JL (2007) Some major mycotoxins and their mycotoxicoses—an overview. Int J Food Microbiol 119(1–2):3–10

Rodriguez GM, Tashiro Y, Atsumi S (2014) Expanding ester biosynthesis in *Escherichia coli*. Nat Chem Biol 10(4):259–265

Rutter CD, Zhang S, Rao CV (2015) Engineering *Yarrowia lipolytica* for production of medium-chain fatty acids. Appl Microbiol Biotechnol 99:7359–7368

Saenge C, Cheirsilp B, Suksaroge TT, Bourtoom T (2011) Efficient concomitant production of lipids and carotenoids by oleaginous red yeast *Rhodotorula glutinis* cultured in palm oil mill effluent and application of lipids for biodiesel production. Biotechnol Bioprocess Eng 16:23–33

Sarria S, Wong B, Martín HG, Keasling JD, Peralta-Yahya P (2014) Microbial synthesis of pinene. ACS Synth Biol 3(7):466–475

Savage N (2011) Fuel options: the ideal biofuel. Nature 474:S9–S11. https://doi.org/10.1038/474S09a

Scharnewski M, Pongdontri P, Mora G, Hoppert M, Fulda M (2008) Mutants of *Saccharomyces cerevisiae* deficient in acyl-CoA synthetases secrete fatty acids due to interrupted fatty acid recycling. FEBS J 275(11):2765–2778

Shafiq S, Chechan R (2019) Influence of different natural media on production of myco-diesel. In: IOP conference series: earth and environmental science, vol 1. IOP Publishing, p 012054. https://doi.org/10.1088/1755-1315/388/1/012054

Shaw JJ, Berbasova T, Sasaki T, Jefferson-George K, Spakowicz DJ, Dunican BF, Portero CE, Narvaez-Trujillo A, Strobel SA (2015) Identification of a fungal 1,8-cineole synthase from *Hypoxylon* sp. with specificity determinants in common with the plant synthases. J Biol Chem 290(13):8511–8526

Shen B, Cheng Y-Q, Christenson SD, Jiang H, Ju J, Kwon H-J, Lim S-K, Liu W, Nonaka K, Seo J-W (2007) Polyketide biosynthesis beyond the type I, II, and III polyketide synthase paradigms: a progress report. In: Rimando A (ed) Polyketides. ACS Publications, New York

Shi S, Valle-Rodríguez JO, Khoomrung S, Siewers V, Nielsen J (2012) Functional expression and characterization of five wax ester synthases in *Saccharomyces cerevisiae* and their utility for biodiesel production. Biotechnol Biofuels 5(1):1–10

Shi T, Li Y, Zhu L, Tong Y, Yang J, Fang Y, Wang M, Zhang J, Jiang Y, Yang S (2021) Engineering the oleaginous yeast *Yarrowia lipolytica* for β-farnesene overproduction. Biotechnol J. https://doi.org/10.1002/biot.202100097

Singara Charya M (2015) Fungi: an overview. In: Bahadur B (ed) Plant biology and biotechnology: volume I: plant diversity, organization, function and improvement, 1st edn. Springer, New Delhi, India, pp 197–215

Singh SK, Strobel GA, Knighton B, Geary B, Sears J, Ezra D (2011) An endophytic *Phomopsis* sp. possessing bioactivity and fuel potential with its volatile organic compounds. Microb Ecol 61:729–739

Siripong W, Wolf P, Kusumoputri TP, Downes JJ, Kocharin K, Tanapongpipat S, Runguphan W (2018) Metabolic engineering of *Pichia pastoris* for production of isobutanol and isobutyl acetate. Biotechnol Biofuels 119(1):1–16

Sitepu IR, Garay LA, Sestric R, Levin D, Block DE, German JB, Boundy-Mills KL (2014) Oleaginous yeasts for biodiesel: current and future trends in biology and production. Biotechnol Adv 32(7):1336–1360

Sitepu EK, Jones DB, Zhang Z, Tang Y, Leterme SC, Heimann K, Raston CL, Zhang W (2019) Turbo thin film continuous flow production of biodiesel from fungal biomass. Bioresour Technol 273:431–438

Somani A, Smith D, Hegarty M, Fernandez-Fuentes N, Ravella SR, Gallagher JA, Bryant DN (2018) Draft genome assemblies of xylose-utilizing Candida tropicalis and *Candida boidinii* with potential application in biochemical and biofuel production. Genome Announc 6(7): 01594–01517

Spagnuolo M, Yaguchi A, Blenner M (2019) Oleaginous yeast for biofuel and oleochemical production. Curr Opin Biotechnol 57:73–81

Spakowicz DJ, Strobel SA (2015) Biosynthesis of hydrocarbons and volatile organic compounds by fungi: bioengineering potential. Appl Microbiol Biotechnol 99:4943–4951

Spatafora JW, Aime MC, Grigoriev IV, Martin F, Stajich JE, Blackwell M (2017) The fungal tree of life: from molecular systematics to genome-scale phylogenies. In: Spatafora (ed) The fungal kingdom. Wiley, New York, pp 1–34

Srinivasan N, Thangavelu K, Sekar A, Sanjeev B, Uthandi S (2021) *Aspergillus caespitosus* ASEF14, an oleaginous fungus as a potential candidate for biodiesel production using sago processing wastewater (SWW). Microb Cell Fact 20(1):1–23

Steen EJ, Chan R, Prasad N, Myers S, Petzold CJ, Redding A, Ouellet M, Keasling JD (2008) Metabolic engineering of *Saccharomyces cerevisiae* for the production of n-butanol. Microb Cell Fact 7:1–8

Strobel GA, Dirkse E, Sears J, Markworth C (2001) Volatile antimicrobials from *Muscodor albus*, a novel endophytic fungus. Microbiology 147(11):2943–2950

Strobel GA, Knighton B, Kluck K, Ren Y, Livinghouse T, Griffin M, Spakowicz D, Sears J (2008) The production of myco-diesel hydrocarbons and their derivatives by the endophytic fungus *Gliocladium roseum* (NRRL 50072). Microbiology 154(11):3319–3328

Strobel G, Singh SK, Riyaz-Ul-Hassan S, Mitchell AM, Geary B, Sears J (2011) An endophytic/pathogenic *Phoma* sp. from creosote bush producing biologically active volatile compounds having fuel potential. FEMS Microbiol Lett 320(2):87–94

Subramaniam R, Dufreche S, Zappi M, Bajpai R (2010) Microbial lipids from renewable resources: production and characterization. J Ind Microbiol Biotechnol 37(12):1271–1287

Takeno S, Sakuradani E, Tomi A, Inohara-Ochiai M, Kawashima H, Ashikari T, Shimizu S (2005) Improvement of the fatty acid composition of an oil-producing filamentous fungus, *Mortierella alpina* 1S-4, through RNA interference with Δ12-desaturase gene expression. Appl Environ Microbiol 71(9):5124–5128

Tamano K, Bruno KS, Karagiosis SA, Culley DE, Deng S, Collett JR, Umemura M, Koike H, Baker SE, Machida M (2013) Increased production of fatty acids and triglycerides in *Aspergillus oryzae* by enhancing expressions of fatty acid synthesis-related genes. Appl Microbiol Biotechnol 97:269–281

Tao M-H, Yan J, Wei X-Y, Li D-L, Zhang W-M, Tan J-W (2011) A novel sesquiterpene alcohol from *Fimetariella rabenhorstii*, an endophytic fungus of *Aquilaria sinensis*. Nat Prod Commun. https://doi.org/10.1177/1934578X1100600605

Thangavelu K, Sundararaju P, Srinivasan N, Uthandi S (2021) Bioconversion of sago processing wastewater into biodiesel: optimization of lipid production by an oleaginous yeast, *Candida tropicalis* ASY2 and its transesterification process using response surface methodology. Microb Cell Fact 20(1):1–18

Tomsheck AR, Strobel GA, Booth E, Geary B, Spakowicz D, Knighton B, Floerchinger C, Sears J, Liarzi O, Ezra D (2010) *Hypoxylon* sp., an endophyte of *Persea indica*, producing 1, 8-cineole and other bioactive volatiles with fuel potential. Microb Ecol 60:903–914

Torella JP, Ford TJ, Kim SN, Chen AM, Way JC, Silver PA (2013) Tailored fatty acid synthesis via dynamic control of fatty acid elongation. PNAS 110(28):11290–11295

Ubersax JA, Platt DM (2010, Dec 9) Genetically modified microbes producing isoprenoids. US Patent, 12/791,596

Valle-Rodríguez JO, Shi S, Siewers V, Nielsen J (2014) Metabolic engineering of *Saccharomyces cerevisiae* for production of fatty acid ethyl esters, an advanced biofuel, by eliminating non-essential fatty acid utilization pathways. Appl Energy 115:226–232

van den Berg MA (2011) Impact of the *Penicillium chrysogenum* genome on industrial production of metabolites. Appl Microbiol Biotechnol 92(1):45–53

Van Dyk S, Su J, Mcmillan J, Saddler J (2019) Potential synergies of drop-in biofuel production with further co-processing at oil refineries. Biofuels Bioprod Biorefin 13(3):760–775

Velvizhi G, Nair R, Goswami C, Arumugam SK, Shetti NP, Aminabhavi TM (2022) Carbon credit reduction: a techno-economic analysis of "drop-in" fuel production. Environ Pollut. https://doi.org/10.1016/j.envpol.2022.120507

Walker GM, Stewart GG (2016) *Saccharomyces cerevisiae* in the production of fermented beverages. Beverages 2:30. https://doi.org/10.3390/beverages2040030

Wang Kovaleff Y (2019) Novel fuel-producing fungi and methodologies for increasing fuel production. Dissertation, University of Central Florida

Wang G, Xiong X, Ghogare R, Wang P, Meng Y, Chen S (2016a) Exploring fatty alcohol-producing capability of *Yarrowia lipolytica*. Biotechnol Biofuels 9(1):1–10

Wang W, Wei H, Knoshaug E, Van Wychen S, Xu Q, Himmel ME, Zhang M (2016b) Fatty alcohol production in *Lipomyces starkeyi* and *Yarrowia lipolytica*. Biotechnol Biofuels 9(1):1–12

Wang K, Strobel GA, Yan D-H (2017) The production of 1,8-cineole, a potential biofuel, from an endophytic strain of *Annulohypoxylon* sp. FPYF3050 when grown on agricultural residues. J Sustain Bioenergy Syst 2:65–84

Wang L, Zhang H, Zhang Y, Song Y (2019) 13C metabolic flux analysis on roles of malate transporter in lipid accumulation of *Mucor circinelloides*. Microb Cell Fact 18(1):1–14

Watanabe MD, Cherubini F, Tisserant A, Cavalett O (2022) Drop-in and hydrogen-based biofuels for maritime transport: country-based assessment of climate change impacts in Europe up to 2050. Energy Convers Manag. https://doi.org/10.1016/j.enconman.2022.116403

Weizmann C (1919, Sep 9) Production of acetone and alcohol by bacteriological processes. US patent 1(315):585

Westfall PJ, Pitera DJ, Lenihan JR, Eng D, Woolard FX, Regentin R, Horning T, Tsuruta H, Melis DJ, Owens A (2012) Production of amorphadiene in yeast, and its conversion to dihydroartemisinic acid, precursor to the antimalarial agent artemisinin. PNAS. https://doi.org/10.1073/pnas.1110740109

Wu W, Tran W, Taatjes CA, Alonso-Gutierrez J, Lee TS, Gladden JM (2016) Rapid discovery and functional characterization of terpene synthases from four endophytic Xylariaceae. PLoS One. https://doi.org/10.1371/journal.pone.0146983

Xie D (2017) Integrating cellular and bioprocess engineering in the non-conventional yeast Y*arrowia lipolytica* for biodiesel production: a review. Front Bioeng Biotechnol. https://doi.org/10.3389/fbioe.2017.00065

Xu P, Qiao K, Ahn WS, Stephanopoulos G (2016) Engineering *Yarrowia lipolytica* as a platform for synthesis of drop-in transportation fuels and oleochemicals. PNAS 113(39):10848–10853

Yaegashi J, Kirby J, Ito M, Sun J, Dutta T, Mirsiaghi M, Sundstrom ER, Rodriguez A, Baidoo E, Tanjore D (2017) *Rhodosporidium toruloides*: a new platform organism for conversion of lignocellulose into terpene biofuels and bioproducts. Biotechnol Biofuels 10(1):1–13

Yagi T, Hatano A, Nakanishi T, Hatano T, Fukui S (1994) Extracellular production of palmitoleic triglycerides by a yeast, Trichosporon. J Ferment Bioeng 77(2):164–168

Yao F, Liu S-C, Wang D-N, Liu Z-J, Hua Q, Wei L-J (2020) Engineering oleaginous yeast *Yarrowia lipolytica* for enhanced limonene production from xylose and lignocellulosic hydrolysate. FEMS Yeast Res. https://doi.org/10.1093/femsyr/foaa046

Youngquist JT, Schumacher MH, Rose JP, Raines TC, Politz MC, Copeland MF, Pfleger BF (2013) Production of medium chain length fatty alcohols from glucose in *Escherichia coli*. Metab Eng 20:177–186

Yu KO, Jung J, Kim SW, Park CH, Han SO (2012) Synthesis of FAEEs from glycerol in engineered *Saccharomyces cerevisiae* using endogenously produced ethanol by heterologous expression of an unspecific bacterial acyltransferase. Biotechnol Bioeng 109(1):110–115

Zargar A, Bailey CB, Haushalter RW, Eiben CB, Katz L, Keasling JD (2017) Leveraging microbial biosynthetic pathways for the generation of 'drop-in' biofuels. Curr Opin Biotechnol 45:156–163

Zhang L, Liang Y, Wu W, Tan X, Lu X (2016a) Microbial synthesis of propane by engineering valine pathway and aldehyde-deformylating oxygenase. Biotechnol Biofuels 9(1):1–9

Zhang S, Ito M, Skerker JM, Arkin AP, Rao CV (2016b) Metabolic engineering of the oleaginous yeast *Rhodosporidium toruloides* IFO0880 for lipid overproduction during high-density fermentation. Appl Microbiol Biotechnol 100:9393–9405

Zhang S, Skerker JM, Rutter CD, Maurer MJ, Arkin AP, Rao CV (2016c) Engineering *Rhodosporidium toruloides* for increased lipid production. Biotechnol Bioeng 113(5):1056–1066

Zhao L, Tang X, Luan X, Chen H, Chen YQ, Chen W, Song Y, Ratledge C (2015) Role of pentose phosphate pathway in lipid accumulation of oleaginous fungus *Mucor circinelloides*. RSC Adv 5(118):97658–97664

Zhao K, Yu L, Jin Y, Ma X, Liu D, Wang X (2016) Advances and prospects of taxol biosynthesis by endophytic fungi. Sheng Wu Gong Cheng Xue Bao Chin J Biotechnol 32(8):1038–1051

Zhao C, Fang H, Chen S (2017) Single cell oil production by *Trichosporon cutaneum* from steam-exploded corn stover and its upgradation for production of long-chain α,ω-dicarboxylic acids. Biotechnol Biofuels 10(1):1–16

Zhuang X, Kilian O, Monroe E, Ito M, Tran-Gymfi MB, Liu F, Davis RW, Mirsiaghi M, Sundstrom E, Pray T (2019) Monoterpene production by the carotenogenic yeast *Rhodosporidium toruloides*. Microb Cell Fact 18:1–15

Chapter 8
Sustainable Aviation Fuel Production

Priscilla Zwiercheczewski de Oliveira,
Luciana Porto de Souza Vandenberghe, Gustavo Amaro Bittencourt,
Walter José Martinez Burgos, and Carlos Ricardo Soccol

Abstract Sustainable aviation fuel (SAF) is an alternative to traditional fossil-based jet fuel produced from renewable feedstocks, such as forestry and agricultural residues, used oils, and algae, where they are essential to avoid competition with food production and reduce the carbon footprint of the aviation sector. The SAF market is currently limited due to challenges in feedstock availability, production costs, and regulatory frameworks. SAF is now more expensive to produce than traditional jet fuel due to the high cost of specialized refining processes, in which production techniques have been developed, including Fischer–Tropsch, hydroprocessed esters and fatty acids, and alcohol-to-jet. Each method has advantages and disadvantages, and research is ongoing to improve production efficiency and reduce costs. SAF widespread adoption faces several challenges, including feedstock availability, high production costs, few regulatory frameworks, and the need to scale up production to meet the increasing demand for air travel. Addressing these challenges requires an intensive effort from the aviation industry, policymakers such as subsidies, incentives, and mandates can help accelerate investment in SAF and encourage its adoption, research and development labors can help improve production efficiency and reduce costs, and researchers promote development, adoption, and efficient implementation. SAF has the potential to be a critical tool in reducing greenhouse gas emissions from aviation, in order to achieve carbon-neutral emissions; it is a viable alternative to traditional fossil-based jet fuels.

Keywords Sustainable fuel · Feedstocks · Fischer–Tropsch · HEFA · Alcohol-to-Jet

---

P. Zwiercheczewski de Oliveira · L. Porto de Souza Vandenberghe (✉) ·
G. Amaro Bittencourt · W. J. Martinez Burgos · C. R. Soccol
Department of Bioprocess Engineering and Biotechnology, Federal University of Paraná, Curitiba, Paraná, Brazil
e-mail: lvandenberghe@ufpr.br

S. K. Brar et al. (eds.), *The Microbiology of the Drop-in Biofuel Production*, Biofuel and Biorefinery Technologies 15, https://doi.org/10.1007/978-3-031-61637-2_8

8.1 Introduction

According to the International Air Transport Association (IATA), climate change is an existential threat and a great challenge. So, environmental action remains at the top of the industry's agenda. At the 77th International Air Transport Association (IATA) Annual General Meeting in the USA, its airlines' members have committed to achieving net-zero carbon emissions generated from their operations, by 2050, which coincides with the objective of the Paris Agreement to limit global warming to 1.5 °C (SAF Grand Challenge Roadmap 2022).

IATA also projects approximately 10 billion people are expected to fly in 2050 (SAF Grand Challenge Roadmap 2022), which means that the corresponding consumption of aviation fuel will rise accordingly. Another very important point is the air cargo that, in 2021, generated $155.0 billion. The total number of flights reached 25.7 million in 2021, which is less than the average of the period from 2017 to 2021 which was 36.6, before the COVID-19 pandemic. However, numbers are continuously increasing with the normalization of passengers' transportation and air cargo with also the rising numbers of e-commerce with the consequent and significant increase of jet fuel consumption and CO_2 emissions. In this way, sustainable aviation fuels (SAF) bring a promising solution to reduce aviation emissions (ECEEE 2022). There are many types of SAFs including mainly e-kerosene, a synthetic fuel made of CO_2 (e.g., captured from the atmosphere) and green hydrogen. The production of SAF reached around 300 million liters in 2022. In 2021, SAF accounted for 0.04% of total jet fuel produced in the world and government policies. In this context, substantial production will be key for larger implementation across airlines in the world (Sharma 2022).

8.2 Why Green Aviation Fuels Are Exploding

Aircraft depend on fossil jet fuel to operate, leading to irreversible damage to the planet (ECEEE 2022). Since the aviation sector discharges approximately 2% of the globally generated CO_2, due to human-related activities, the search for environmental-friendly biojet fuels produced from renewable resources is highly increasing (Lim et al. 2021). SAFs offer then a promising alternative that will help to reduce aviation emissions. While demand reduction and energy efficiency are important steps to reduce aviation's climate impact, the European Union has proposed to increase the amount of SAFs used in aircraft, departing from the European bloc, with blending mandates that are currently being negotiated, where it is compulsory for all jet fuel supplied to EU airports to have a certain proportion of SAF (ECEEE 2022).

Among SAFs, e-kerosene is the most sustainable and holds one of the keys to aviation's decarbonization. Also known as electro-fuel or power-to-liquid (PtL) fuel, e-kerosene is a significant technological solution that will help reduce emissions. It is synthesized from hydrogen and CO_2 molecules, using green electricity such as wind or solar power. That is why the life cycle of e-kerosene can be near carbon-neutral. One advantage is direct air capture (DAC), which is the preferred pathway to obtain the carbon fraction for this fuel. So, the production process does not rely on (fossil) combustion technologies. The electricity used for the production process needs to be added, and producers need to guarantee the certificate that they can obtain additional green electricity or use their own green electricity production installations (ICCT 2022). However, e-kerosene production will need to be significantly increased. A sub-target of 0.7% by 2030 was proposed by the European Commission. However, specialists say that is too low to promote an impact on the market.

Other alternative and efficient technologies exist for other transport modes, such as electrification for road transport. Therefore, renewable hydrogen and additional renewable electricity for the creation of synthetic fuels will be developed (ECEEE 2022).

8.3 Feedstock for Sustainable Aviation Fuel Production

Many feedstocks have been evaluated for sustainable aviation fuel (SAF) production. The common characteristics between them are being renewable, and, in general, that they are biological in origin. Regarding the technology applied for feedstock conversion, some options can produce drop-in fuels that are well suited for the widely used design of aircraft engines. On the other hand, other strategies need additional infrastructure development, just in the case of the use of electricity, which requires batteries to store energy, electric motors to power airplane and charging stations to support them, hydrogen fuels that need the development of hydrogen storage and different transportation systems, and fueling stations required to support the aircraft. While it remains unclear which perspective strategy will lead to the highest overall environmental performance in the future, in terms of long-term investment decisions, the exploitation of biological feedstocks through technologic pathways already certified by ASTM D7566 is a viable option that could help address the reduction of aviation's carbon emissions in a short- to mid-term solution.

While the global annual biomass production is estimated as 100 trillion kilograms (Kandaramath Hari et al. 2015), a considerable advantage of its use is the worldwide availability of different biomass feedstocks. Especially for emergent countries, where unusable arable lands are available for crop production, the production of drop-in biofuels for aviation could offer economic profits. A diverse array of organic materials can be utilized to produce sustainable aviation fuels. This encompasses crops that are cultivated for energy or food/feed purposes, along with the by-products generated from forestry and agriculture. Additionally, materials such as used cooking oil, straw, wood waste, and municipal solid waste can also serve as feedstock for biojet fuel production.

Recently, the aviation initiative of ReFuelEU and its regulation proposal has also expanded the concept of sustainable aviation fuels to encompass feedstocks used for electro-fuels production, which are liquid hydrocarbons produced through electricity and renewable carbon inputs (Drünert et al. 2020; Eurocontrol 2021). Concerning this, water electrolysis, thermochemistry, or bio-photolysis are examples of various technologies that can be used to produce hydrogen from water, while carbon can be acquired from methods that are biomass-derived or not, including direct capture of carbon dioxide from the atmosphere or different point sources, and the use of biogenic carbon by syngas conversion through biomass gasification (Su-ungkavatin et al. 2023). However, it is imperative to consider the sustainability of different sources of electricity used, and the emissions generated during the production of hydrogen, while if it arises, for instance, from conventional steam reforming of natural gas instead of carbon-neutral alternatives, the original main target to reach sustainability in aviation transport systems is not reached at all.

In addition, the maturity of biofuel technologies produced from biomass feedstocks and their high compatibility with the existing fueling infrastructure make biofuels the outstanding pathway for sustainable aviation fuels (Su-ungkavatin et al. 2023). From the technology's viewpoint, any feedstock that possesses properties and traits aligned with their characteristics can be utilized; however, some worldwide available feedstocks present characteristics that are in conformity with some specific certified pathways (Ng et al. 2021). Traditional crops with carbohydrate-rich content, such as sugarcane, maize, and fodder beet, can be used in the direct sugar to hydrocarbon pathway to farnesane production and then its hydrotreatment to synthetic iso-paraffins. Most of the lignocellulosic biomass feedstock, such as agricultural and forestry residues and woody crops, can be processed in gasification-Fischer–Tropsch and alcohol-to-jet synthetic paraffinic kerosene technologies. In the case of oily feedstocks, lipids can be extracted/produced from algae, used cooking oil, oil seeds, and energy crops such as Jatropha and Camelina, in the hydroprocessed esters and fatty acids pathway (Zhang et al. 2020).

8.3.1 Carbohydrate-Rich Biomass

Although there are many plant species that can be used as biomass for bioenergy generation, only those that are grown extensively and processed in large-scale industries have the potential to support jet fuel production and blending in the near future (Escalante et al. 2022). For instance, sugarcane has been exploited by sugar mills across the world, mainly in Brazil, to produce ethanol and sugar from sugarcane juice in a synergic system (Vandenberghe et al. 2022). Another widely produced crop is corn, with 1112 million metric tons produced only in 2019. As the leading country of corn ethanol production, the USA produced 31.3% of the global corn, while its starch-rich composition is exploited for ethanol production (Li et al. 2022). Around 39% of the US corn production was used for ethanol production in the same year. Also, around 80% of annual ethanol production in China and 79% of annual ethanol production in Canada were feed with corn starch, due to its abundance and competitive price (Flach et al. 2021). Other first-generation feedstocks are used in ethanol production worldwide, such as wheat, cassava, rice, and barley, mainly in China and Europe (Flach et al. 2021). Sugars from industrial wastes such as waste molasses, corn syrup, and other by-products from food processing can also be used as feedstocks for aviation biofuels.

In this context, produced alcohols through established technologies can be used as intermediates after changes in their molecular structure through chemical processes, to attain characteristics similar to commercialized jet kerosene. The conversion of alcohols, such as methanol, ethanol, butanol, and/or long-chain fatty alcohols, through the alcohol-to-jet-fuel (AtJ) pathway, is a potential solution for the partial substitution of conventional kerosene in the aeronautical industry market. In July 2012, the US Air Force conducted a successful commercial-scale test of a technology that converts alcohol into jet fuel and utilized this fuel during a flight (Wang et al. 2016). The use of first-generation biorefinery feedstocks in biojet fuel systems can also occur with their blending with conventional kerosene. Other pathways also can be applied for the exploitation of first-generation feedstocks, such as the conversion of edible biomass to simple sugars, and the sub-sequential conversion of glucose into n-hexane by a succession of catalytic reduction steps (Dutta et al. 2023). Also, in the scope of hydroprocessing technologies, hydrotreated esters and fatty acids (HEFA) and hydroprocessed renewable jet (HRJ) can be applied (Zhang et al. 2020), which will be discussed in the next topics of this chapter.

However, the use of food crops for the production of SAF can lead to indirect changes in land use and have adverse effects on local environments through the use of irrigation, pesticides, and fertilizers. Additionally, it may cause competition with food production, impacting food prices (Ng et al. 2021). The direct land used transformation refers to direct changes specifically to the conversion of natural vegetation or tropical rainforests into areas used for cultivating the desired biomass, while indirect land use change occurs when there is a shift in land use in regions outside the boundary of the assessed system, as a direct consequence of compensating any deficit in market supply (Escalante et al. 2022). Both of these scenarios can significantly raise the greenhouse gas emissions (GHG) in the life cycle of biojet fuels, which could deviate from the original main target to reach sustainability in aviation transport systems (Staples et al. 2014). On the other hand, the use of food crops as energy feedstocks leads to resource competition in the food industry (Prieler et al. 2013). In addition, the established road transport sector is also competing for feedstock supply chain and policy support, making it challenging for the production of SAF, as there are currently no similar policies for it like those that exist for first-generation biofuels in many countries, such as Brazil, EU, and the USA (Ng et al. 2021). Hence, it is preferable to use raw materials that result in lower environmental impact and GHG emissions.

8.3.2 Lignocellulosic Biomass

The high recalcitrance of lignocellulosic biomass feedstocks, due to the presence of lignin and hemicellulose which prevents microbial attacks, makes its exploration for biofuels a more challenging process in comparison with first-generation feedstocks (Kumari and Singh 2018). Therefore, it requires the application of properly designed pretreatment steps, before its conversion through thermochemical or biochemical processes. However, the worldwide abundance of these materials and the higher GHG savings than food-based SAF pathways make its use a suitable solution and also a target to reach for aviation systems and scientific research. Some methods can be used to convert feedstock into biojet fuels, such as gasification coupled with Fischer–Tropsch synthesis, upgrading ethanol or isobutanol derived from the feedstock, and transforming alcohol into jet fuel (Su-ungkavatin et al. 2023).

Agriculture Residues

Despite the significant potential for agricultural waste to serve as an alternative feedstock for the production of biofuels and value-added chemicals, this opportunity has remained largely unexploited. The global generation of agricultural biomass is substantial, with wheat and rice straw, corn stover, and sugarcane bagasse being the primary agricultural wastes produced (Emmanouilidou et al. 2023). Thermochemical methods have been evaluated as a potential process to enable the utilization of agricultural biomass for biofuel production. There are two pathways that are commonly evaluated for the conversion of lignocellulosic biomass into biojet fuel. The first option is to convert the biomass into a bio-oil, which can then be further processed into kerosene. The second option involves converting the residual biomasses into gas, which can then be transformed into liquid fuels via gas-to-liquid (GtL) processes. However, efficient and economical biofuel production faces challenges such as energy market issues, a lack of supportive policies and financial incentives, and limited knowledge regarding equipment requirements (Emmanouilidou et al. 2023). The adoption of integrated technologies for biorefinery applications is a promising strategy for improving overall processing costs and environmental impact, based on techno-economic analysis and life cycle analysis, from the perspective of a circular economy.

Forestry

Forestry residues, such as sawdust, wood chips, bark, forest thinnings, logging slash, and residues from forest fires, are potential feedstocks for biojet fuel production (Ng et al. 2021). The availability of forestry residues as a feedstock for biojet fuel production varies by region and depends on factors such as the type of forestry operations being carried out and the local market demand for traditional forest products. For example, in regions with high levels of forest thinning and logging operations, there may be an abundance of forestry residues available for use as a feedstock for biojet fuel production. Cavalett and Cherubini (2018) introduced an approach for the analysis of forestry residues in Norway used for conversion into biojet fuels, based on life-cycle impact assessment methods. They found varying impacts on Sustainable Development Goals (SDGs) and highlighted the need for early environmental sustainability analysis, technological improvements, and policy mechanisms to address economic barriers and create synergies. Future development of SDG-specific indicators is necessary to manage the potential conflicting implications of biofuels.

Despite the possible advantages of using forestry residues as a feedstock for biojet fuel production, there are also some challenges associated. Since forestry residues are often spread out over large areas and may require specialized equipment for collection and transportation, the cost of harvesting and transporting these materials can be significant (Martinez-Valencia et al. 2021). The distinct needs of diverse regions and stakeholders prompt the adoption of diverse pathways, resulting in tailored solutions.

8.3.3 Oily Feedstocks

Although current levels of SAF production represent 0.004% of global jet fuel demand, at present, this production is primarily based on oily feedstocks, such as vegetable oil, animal fats, and used cooking oil, due to the high conversion efficiencies of the established hydroprocessing method, which is widely used for producing road transport fuel at large scales (Andrea Fantuzzi et al. 2023). Non-edible vegetable oils derived from sources such as *Jatropha curcas*, *Camelina sativa*, and *Pongamia pinnata* offer a sustainable solution (Atabani et al. 2013). Also, non-edible herbaceous oilseeds from the Brassica family, such as rapeseed, pennycress, and camelina, can be grown on marginal lands with minimal inputs in terms of fertilizers and agrochemicals and are more resilient to extreme weather conditions, making them suitable alternatives for oil production (Martinez-Valencia et al. 2021).

Camelina

Camelina is a non-food energy crop with an average oil content of 30–40%. As previously discussed, Camelina plant can grow on low fertile soils and marginal lands, while it maintains a high resistance to diseases and pests (Kandaramath Hari et al. 2015). Due to its minimal requiring inputs, it presents a promising solution for farmers looking to maximize their profits from their land while reducing the problems related to mono-cropping, as it can be grown as a rotational crop for wheat and cereals. Also, after oil extraction, its economic viability is further enhanced by the utilization of the remaining meal as animal feed.

Camelina oil, which costs around $0.40–0.70 per gallon, has high potential as a profitable and sustainable crop. The current market for the oil is mainly concentrated in the Northwest USA and Southern Canada, where in 2012, the USA produced 200 million gallons of camelina oil. Camelina-based biojet fuel is both more fuel-efficient and less harmful to the environment, and it is possible to produce up to 800 million metric gallons of oil from camelina each year (Shonnard et al. 2010).

Jatropha

Jatropha curcas is a non-edible, drought and pest-resistant multipurpose plant, ideal for biodiesel production and oilseed extraction, due to its high seed oil content, which is around 30–35% (Morgan et al. 2019). The plant is adaptable to degraded soil and is not in competition with food crops. It has a permanent pattern of high-yield oil production for up to 40 years with minimal moisture (Morgan et al. 2019). The plant meal left after oil extraction is rich in nitrogen, potassium, and phosphorous, making it useful as organic manure. Jatropha has great energy levels of around 18–19 MJ/kg, and a high seed dry weight of up to 55% (Kasim and Harvey 2011). Jatropha cultivation occurs in South Africa, South and Central America, and Southeast Asia, requiring irrigation for optimal growth.

Waste Oils

Animal fats, including beef tallow, pork lard, and chicken fat, are a type of lipid feedstock that has various applications. These fats are obtained as by-products from the meat industry, either through the rendering of inedible fats or from edible fats. Inedible fats, such as yellow grease and poultry fats, are used in soap processing, animal feed, and pet food (Badgett and Milbrandt 2020). On the other hand, edible tallow from cattle and lard from hogs are used in the production of food products and shortenings. Apart from food and cosmetic applications, animal fats are also used in the production of heat and power. They are a lower-cost alternative to vegetable oils and can be applied in power generation through combustion, pyrolysis, and gasification. However, only a portion of animal fats are considered eligible for use due to their quality.

Used cooking oil is another source of lipid feedstock, which is created during food preparation. Used cooking oil is discarded vegetable oil after its use for frying. It can be collected from commercial sources such as restaurants and some households. In 2019, around 32% of the total renewable and biodiesel consumed in the EU came from waste fats, oils, and greases, including animal fats, used cooking oil, and tall oil (Phillips et al. 2019). In addition, this type of waste accounted for approximately 2.5% of the total EU road diesel consumption (Phillips et al. 2019).

Algae

Algae is a third-generation feedstock with great potential to address fuel scarcity and SAF production, due to their high lipid content, fast carbon dioxide absorption, low land use, and rapid growth (Brennan and Owende 2010). Moreover, algae do not compete with food crops for resources, as they do not require land or freshwater to thrive. This feature makes algae particularly suitable for wastewater treatment, where they can produce large amounts of lipids and carbohydrates using sunlight and carbon dioxide.

Algae can be collected using various techniques, including centrifugal sedimentation, flotation, filtration, and flocculation. Dehydration, typically through solar exposure or low-pressure shelf drying, is the most common post-harvest method, as freezing temperatures are expensive (Tiwari et al. 2023). Third-generation biofuels are superior to other fuels in terms of cost and performance, with algae proving to have a higher production rate than first- and second-generation feedstock.

Algal cultivation uses brackish, salty, and wastewater instead of freshwater, which is an advantage over other crops. By-products from oil extraction, such as proteins, can be used as fertilizers. The application of photobiology is a low-cost, low-energy method for generating hydrogen from algae with lower GHG emissions than other refineries, while closed photobioreactor systems are more expensive, and energy-intensive (around $22.4/gal), but could reach higher efficiencies (Davis et al. 2011; Tiwari et al. 2023).

8.4 Advanced Technology for Sustainable Aviation Fuel Production

The aviation sector has been one of the significant contributors to greenhouse gas emissions (Said et al. 2022). According to Ng et al. (2021), the annual consumption of aviation fuel is approximately 300 Mt., which implies that each year around 859 Mt. of CO_2 is released into the atmosphere; then, the aviation industry contributes to approximately 2% of anthropogenic emissions of CO_2 released into the atmosphere. Therefore, there is an urgent need to make the transition from traditional aviation fuel based mainly on fossils with a high impact on the environment to alternative fuel, renewable, conveniently available, ecologically accessible, and with a smaller environmental footprint (Nouni et al. 2021). The main routes approved for the production of biofuel for aviation are Fischer–Tropsch, hydroprocessed esters and fatty acids, and alcohol-to-jet (European-Union-Aviation-Safety-Agency-EASA 2023).

8.4.1 Fischer–Tropsch

The Fischer—Tropsch (FT) process can be coupled with various biomass conversion processes such as liquefaction, pyrolysis, and gasification in order to produce the synthetic fuel using renewable sources. However, the only commercial and certified route for the production of jet fuel is gasification (Ng et al. 2021).

The FT process coupled with gassing is shown in Fig. 8.1. Initially, the biomass is pretreated and later it is gasified. The syngas is washed and packaged to be later made in the FT process, consisting of a set of reactions where hydrogen and carbon monoxide are converted into liquid hydrocarbons. The reactions are usually catalyzed by ruthenium, nickel, cobalt, and iron (Meurer and Kern 2021). The FT process produces hydrocarbons with carbon chains of different lengths, for example, light hydrocarbons (C_1–C_4) which can be used directly by turbines in power generation, naphtha (C_5–C_{10}) and kerosene (C_{10}–C_{16}) which can be blended with jet fuel, distillates (C_{14}–C_{20}) which can be refined into diesel oil, and waxes (C_{20+}) which can also be transformed into biodiesel via hydrocracking (Ng and Sadhukhan

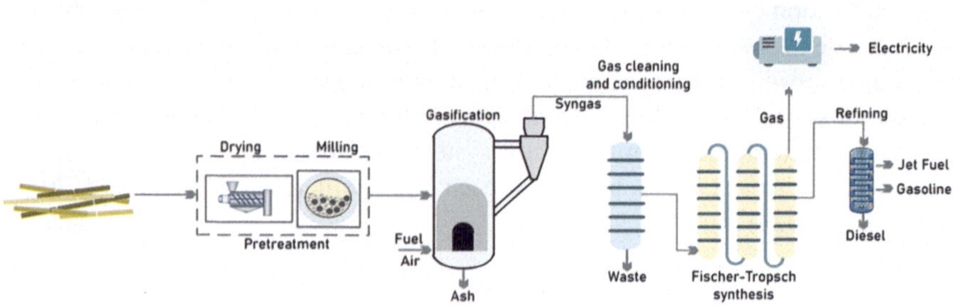

Fig. 8.1 Simplified Fischer–Tropsch process. Adapted from Geleynse et al. (2018), Ng et al. (2021)

2011). Hence, to produce a ton of FT liquid fuel, around 6 tons of biomass is needed (Ng et al. 2021).

Different technologies have been developed for the production of aviation biofuels using the FT process. For example, Guoru et al. (2022) in patent CN218609296U presented a utility model that provides a TF synthesis device. The system comprises two reduction reactors and each reactor contains a reduction cavity and a reduction feed inlet, whereas the FT synthesis reactor is provided with a synthesis cavity and a second air inlet. The reactors' reduction discharge ports are communicated with the inlet pipe and the outlet of the catalytic pipe. Kresnyak et al. (2022) presented a process for the production of synthetic hydrocarbons from biomass in patent WO2022236422A1. The process involves electrolysis of steam and CO_2 in a high-temperature co-electrolyzer to produce oxygen and hydrogen-rich synthesis gas. Next, the biomass is gasified under partial oxidation reaction conditions to generate hydrogen-poor synthesis gas. Subsequently, the synthesis gas is cooled to generate thermal energy; then the hydrogen-rich synthesis gas is mixed with the hydrogen-poor gas; this mixture reacts in an FT reactor where synthetic hydrocarbons and refinery gas are produced. Second, the authors of the patent said that the process produces renewable low-carbon hydrocarbons. Price et al. (2020) presented a process for the production of synthetic hydrocarbons from biomass in patent WO2021087618A1 similar to that exposed in patent WO2022236422A1. However, in patent WO2021087618A1, a portion of the water generated in the FT was recycled in the electrolysis process and the heat generated in the FT was also used in the drying of the biomass.

8.4.2 *Hydroprocessed Esters and Fatty Acids*

The hydroprocessed esters and fatty acids (HEFA) process is done in several steps shown in Fig. 8.2 and can use different types of vegetable oils and animal fats, including algae and microalgae oil.

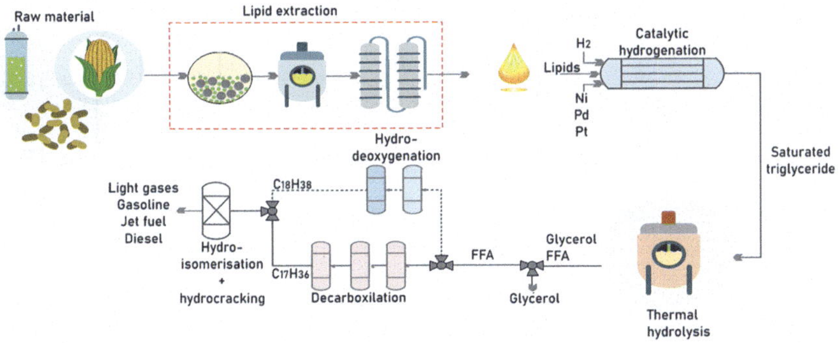

Fig. 8.2 Simplified process hydroprocessed esters and fatty acids. Adapted from Geleynse et al. (2018), Ng et al. (2021)

In the first step occurs the reactions to extract the fatty acids from the biomass, and the later reactions are isomerization and hydrocracking, that is, the rearrangement of molecules and reduction of the carbon chain length of molecules, respectively. Hydrocracking reactions are to obtain jet fuel that meets the jet fuel specification. The process begins with drying the biomass to subsequently extract the lipids. The oil is then catalytically hydrogenated to obtain saturated triglycerides. In hydrogenation, some catalysts such as nickel, palladium, and platinum are employed. The process is carried out at temperatures between 120 and 200 °C and pressures between 0.7 and 4.0 bar (Ng et al. 2021). Subsequently, the triglycerides are fractionated into a glycerol molecule and three free fatty acids (FFA). The latter is subjected to hydrodeoxygenation or decarboxylation processes to produce octadecane ($C_{18}H_{38}$) or heptadecane ($C_{17}H_{36}$), respectively. Finally, hydrocracking and hydroisomerization processes are performed to form compounds with 9–15 carbon atoms (Wang 2016; Monteiro et al. 2022).

8.4.3 Alcohol-to-Jet

The alcohol-to-jet (AtJ) process is specifically based on three catalytic reactions: (i) dehydration, (ii) oligomerization, and (iii) hydrogenation and a purification step (Geleynse et al. 2018). The AtJ involves the conversion of short-chain alcohols such as methanol, ethanol, and butanol into longer-chain hydrocarbons (C_8–C_{16}) (Ng et al. 2021). Dehydration of alcohols to alkenes takes place at high pressures (±14 bar) and temperatures around 350 °C. In this reaction, some catalysts such as ZSM-5 ($Na_nAl_nSi96–nO_{192}16·H_2O$) and some zeolites and resins can be used (Wang et al. 2016). The oligomerization process produces a length of carbon chains suitable for their subsequent fractionation in fuels after hydrogenation (Han et al. 2022). Methanol can be converted into jet fuel; the alcohol is placed in a fixed bed

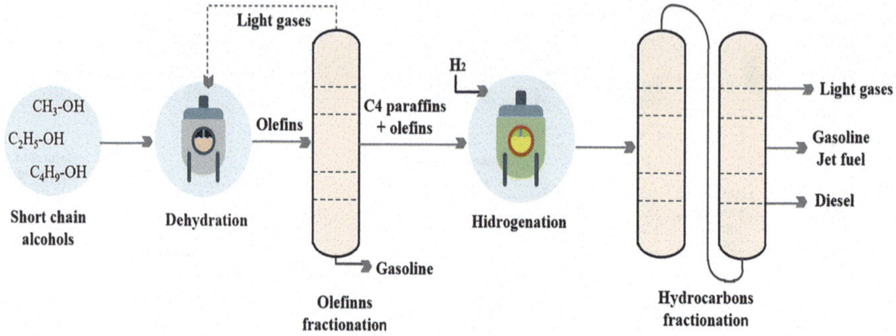

Fig. 8.3 Simplified alcohol-to-jet process. Adapted from Geleynse et al. (2018), Ng et al. (2021)

reactor at 482 °C and 14.5 Psi using the ZSM-5 catalyst. The main products generated in the reaction are olefins (56%, wt%) and gasoline (36% wt%). Subsequently, the olefins are fractionated into gasoline, olefins, paraffins, and light gases, the latter being recycled in the dehydration unit to improve yield. Subsequently, the olefins are processed in a second reactor at 400 °C, 14.5 Psi using the same ZSM-5 catalyst. The products of this unit are fractionated to finally produce kerosene, gasoline, and light gases as can be observed in Fig. 8.3 (Ng et al. 2021).

Different technological advances have been patented to produce aviation biofuels via AtJ. For example, patent WO2011140560A1 presents a process for the production of aviation fuel from alcohols obtained via fermentation, such as isobutanol. This alcohol is initially catalytically dehydrated and subsequently oligomerized to form higher molecular weight olefins such as C_8–C_{16}. Next, a part of the olefins is hydrogenated to obtain an aviation fuel mixture comprising C_{12}–C_{16} alkanes (Peters and Taylor 2011). In patent US20150247100A1, a process was presented to produce aviation fuel from synthesis gas, which was initially converted into alcohols. Subsequently, the alcohols were separated depending on the length of the carbon chain. Higher alcohols were transformed into paraffins through a dimethyl ether intermediate. This compound reacts with a catalyst to constitute aviation fuel grade aromatic hydrocarbons. Lower alcohols are dehydrated to form olefins which are then oligomerized and optionally hydrogenated and/or isomerized to form jet fuel line products (Bradin 2014). Cheiky et al. (2014) in patent WO2014031776A2 showed a system to produce renewable fuels, such as gasoline, diesel, and jet fuel from lignocellulosic biomass and even lipids. The system is made up of multiple processing stations and different catalysts coupled to a unit that produces volatile gases from biomass. The catalytic processing plants are programmed to maximize the conversion of biomass into jet fuel components. The system contains a gasifier to convert the carbonaceous material into a synthesis gas. The processing reactors can be a pyrolysis reactor, fixed bed, fluidized bed, or cyclonic reactor. The process temperature can range from 100 °C–1000 °C.

8.5 Market and Main Players of Sustainable Aviation Fuel

Certified in 2011, SAF still does not have a solid market due to fluctuations in the price of biomass, which are diverse (SABA 2023). However, as it is committed to sustainability, the SAF market is directly influenced by the carbon credits' policy. In summary, these credits exist as a mechanism to reduce greenhouse gas emissions, putting a price on carbon. The idea is that organizations or individuals can offset their emissions by buying credits that represent a reduction in carbon emissions elsewhere (Carbon Credits 2023). Such credits are often traded on carbon markets such as the European Union Emissions Trading System, the Chicago Climate Exchange, or the Clean Development Mechanism of the United Nations Framework Convention on Climate Change (United Nations 2009). The impact of the price of carbon credits can fluctuate based on supply and demand, as well as government policies and regulations. However, this mechanism has the potential to transform the aviation sector and create new opportunities for companies that can adapt to changing market conditions.

The cost of aviation fuel can vary widely depending on several factors such as the location, fuel type, and market conditions. Typically, the price of aviation fuel is quoted in cents per gallon or liters, and it is also influenced by some factors such as crude oil prices, refinery capacity, transportation costs, commercial taxes, and government regulations. Actually, the average global price for Jet-A aviation fuel is approximately $1.85–$2.00 USD per gallon. However, prices can vary widely across different regions. For example, in Europe, the price of Jet-A fuel can be as high as $2.50–$3.00 USD per gallon, while in the Middle East, it can be as low as $1.50 USD per gallon. The price of aviation fuel is typically higher at major airports and in remote areas where transportation costs are higher. It is important to note that the cost of aviation fuel is a significant expense for airlines, accounting for approximately 20–30% of their operating costs (Singh and Joshi 2022; PWC 2023). This is why airlines are constantly looking for ways to reduce their fuel consumption and explore alternative fuels, such as SAF.

In turn, the cost of SAF is currently higher than conventional fossil fuel, but the price also has fluctuations depending on the production process, feedstocks used, and market conditions. Generally, SAF production costs are higher due to the more complex production processes and the higher cost of feedstocks, such as waste oils, agricultural residues, and algae. Currently, the cost of SAF is estimated to be around $3.00–$6.00 USD per gallon, which is significantly higher than the market price for conventional aviation fossil fuel. However, the cost of SAF is expected to decrease as production scales up and technology advances, making it more cost-competitive with fossil fuels in the long term (PWC 2023; Singh and Joshi 2022).

The US Department of Energy's Office of Bioenergy Technologies (BETO) estimates that demand for jet fuel will increase over the next three decades, where 1 billion dry tons of biomass can be sustainably harvested each year in the USA, considered sufficient to produce 50–60 billion gallons of SAF. In terms of economic strategy, SAF-biomass crops can be profitable during the off-season, as SAF can be blended up to 50% with traditional jet fuel. Since 2011, more than 225,000 commercial flights have been powered by SAFs and 6 billion liters of SAFs are under current airline purchase agreements (BETO 2022).

The interest of airline corporations is linked to the amount of CO_2 absorbed by plants during biomass growth, considered equivalent to the amount of CO_2 returned to the atmosphere by a combustion engine. Currently, SAFs are more expensive than traditional jet fuel, about twice as much when using waste, and up to ten times as much when using synthetic fuels. Agreements signed by several airlines are considered beneficial to biomass suppliers, which allow financing the construction of new plants dedicated to the supply of SAF, which may help to reduce the cost of this fuel in the medium and long term (Aviation 2023).

Considering the foregoing, the SAF market is still relatively small due to slow growth at first; however, SAF production has increased significantly since 2019 when large companies started producing high amounts of SAF. With more organizations entering the market, the industry is expected to grow rapidly in the coming years due to increasing demand for reducing its carbon footprint (BCC Research 2022). According to the International Air Transport Association (IATA), approximately 300 million liters of SAF were produced in 2022, and around 17 billion US dollars were accounted for in SAF 2022 purchase transactions (IATA 2023). Still recovering due to the economic impacts caused by the COVID-19 pandemic, the growth of the SAF market is estimated from 219 million US dollars (in 2021) to 15,716 million US dollars by the year 2030, at a CAGR of 60.8% (Markets and Markets 2023). Currently, the main players in the SAF market include:

Neste: A Finnish company that is the largest producer of renewable diesel and the world's largest producer of SAF. They have been producing SAF for over ten years under the Neste MY Sustainable Aviation Fuel brand. Neste produces 100,000 tonnes of SAF annually and plans to increase its production capacity to 1.5 million tonnes per year (about 1.875 billion liters) by 2023. Neste's SAF is available at 12 airports in the USA and three in Europe. To define themselves as leaders in this space, Neste is focused on different partnerships to draw attention to their product (Neste 2023).

TotalEnergies: A French oil and gas company that is investing heavily in SAF production and aims to produce 10% of the world's SAF by 2030 (TotalEnergies 2023).

BP: A British multinational oil and gas company that is developing SAF production facilities in Europe and the USA (BP 2023).

Shell: A Dutch-British oil and gas company that is investing in SAF production and has committed to producing 2 billion liters of SAF per year by 2025 (Shell 2023).

Gevo: American renewable chemicals and advanced biofuel company founded in 2005 that is developing technology to produce SAF from renewable resources such as corn. The company plans to produce and sell 1 billion gallons of hydrocarbons by 2030, and they have focused its efforts on increasing its SAF production to reach a target of 55 million gallons per year (Gevo 2023).

World Energy: is one of North America's largest suppliers of biofuels, and their produced SAF is in active commercial use. The company was responsible for the world's first commercial SAF facility, in Paramount, California. Also, the next step is to convert its biodiesel facility in Texas to produce SAF, where additional 250 million gallons of SAF will count by 2025. The same production goal is intended in Paramount. Also, in California, the SAF facility plant in Los Angeles is expanding to produce 375 million gallons. These movements are part of the company's commitment to the US Grand Challenge program to produce, at least, 1 billion gallons of SAF by 2030 (World Energy 2023).

Fulcrum BioEnergy: is the first company to have a commercial plant producing low-carbon oil fuels from domestic waste as substrate. The plant located in Nevada is expected to produce 11 million gallons of renewable fuel from 175,000 tons of landfill waste. Three more plants are under development in Indiana, on US Gulf Coast, and in the Northwest of England (Fulcrum BioEnergy 2023).

LanzaJet: is a subsidiary of LanzaTech, a technology company involved in carbon recycling. The focus is on sustainable aviation fuels, diesel, and aviation decarbonization along with carbon recycling. Companies supporting LanzaJet are Suncor Energy (a Canadian energy company), Mitsui & Co. (Japanese trade and investment company), British Airways, Shell, and ANA (All Nippon Airways, a Japanese airline) (LanzaJet 2023).

SkyNRG: is an aviation fuel service that supplies, blends, and distributes SAF for airlines and its partner Shell Aviation. KLM Royal Dutch is one of the company's founding partners, also working with Boeing. The company aims to produce and supply 1 million metric tons of SAF by 2030 (SkyNRG 2023).

The aviation jet fuel market has space to expand and to receive new investments, especially in SAF. New advancements may come through the support for SAF suppliers and end users, coordination between commercial, civil, and military aviation purposes, and government support for the development of standards and benefit environmental programs. And the most important assortments for reducing SAF costs include improvement of the biomass technology conversion—mainly for lignocellulosic biomass—that can be profitable through existing production infrastructure and incentives for production across the supply chain; these adjustments can contribute to reducing the industrial risk (SAF Grand Challenge Roadmap 2022).

8.6 Global Standards, Policies, and Measures for Sustainable Aviation Fuel

Climate change is a fact and environmental actions remain at the top of discussions. During the 77th IATA Annual General Meeting in the USA, IATA member airlines defined the commitment to achieve net-zero carbon emissions from their operations by 2050 in order to limit global warming to 1.5 °C (SAF Grand Challenge Roadmap 2022). For the success of these actions, coordinated efforts of the entire industry (airlines, airports, air navigation services, manufacturers) and strong government support will be needed. In this way, the industry must progressively reduce its emissions and, concomitantly, absorb growing demand, which is very difficult in a developing world and growing population.

The resolution demands some important actions and definition of timelines involving all industry stakeholders such as (i) fuel-producing countries must develop and bring large-scale, cost-competitive, and sustainable SAF to the market; (ii) aircraft and engine manufacturers must produce more efficient airframe and propulsion technologies; and (iii) airport operators must provide adequate infrastructure to supply SAF in a cost-effective manner (SAF Grand Challenge Roadmap 2022).

The International Civil Aviation Organization's (ICAO) and the Carbon Offsetting and Reduction Scheme for International Aviation (CORSIA) defined some criteria that will promote the stabilization of international emissions in the short-to-medium term. IATA also reaffirmed all points with the net-zero resolution (SAF Grand Challenge Roadmap 2022). In 2019, CORSIA defined to purchase eligible emissions units, which are above a certain level, to offset the increase of CO_2 emissions. In 2020, the ICAO Council decided that 2019 emissions from international aviation would follow the CORSIA's baseline during the pilot phase (2021–2023). Actually, new definitions are being discussed for 2024–2035. Concomitantly, different actions and initiatives are progressing to reach climate goals such as investments for SAF production, but also new prototype electric and hydrogen aircraft. In this way, bringing sufficient SAF to market and transforming new technology into a reality will require enormous efforts from all stakeholders in the sector. Some concrete facts were reported. As an example, in 2021, 100 million liters of SAF were produced and used by airlines. Some 38 countries have accomplished SAF-specific policies so far. In 2022, it was estimated that more than 450,000 flights would operate using SAF with investments of $17 billion from airlines. New SAF production capacity is being discussed worldwide, and airlines and manufacturers are joining forces to develop hydrogen-powered aircraft or technical solutions to make alternative means of propulsion commercially viable.

In fact, to achieve net-zero by 2050 will require the elimination of emissions at the source with the implementation of carbon capture technologies. The strategy is to reduce as much CO_2 as possible using not only in-sector solutions, such as sustainable aviation fuels, new aircraft technology, more efficient operations, and infrastructure, but also developing zero-emissions energy sources, including electric and hydrogen power. Any emissions that cannot be eliminated at the source will be eliminated through out-of-sector options, such as carbon capture and storage and credible offsetting schemes (Fig. 8.4).

In July 2021, the European Commission's (EC) ReFuelEU Aviation proposal introduced blending mandates for sustainable aviation fuels (SAF) and sub-quotas of e-kerosene (Table 8.1). The EC has proposed an e-kerosene blending mandate of 0.7% by 2030, increasing to 28% in 2050, which is a good first step. However, other

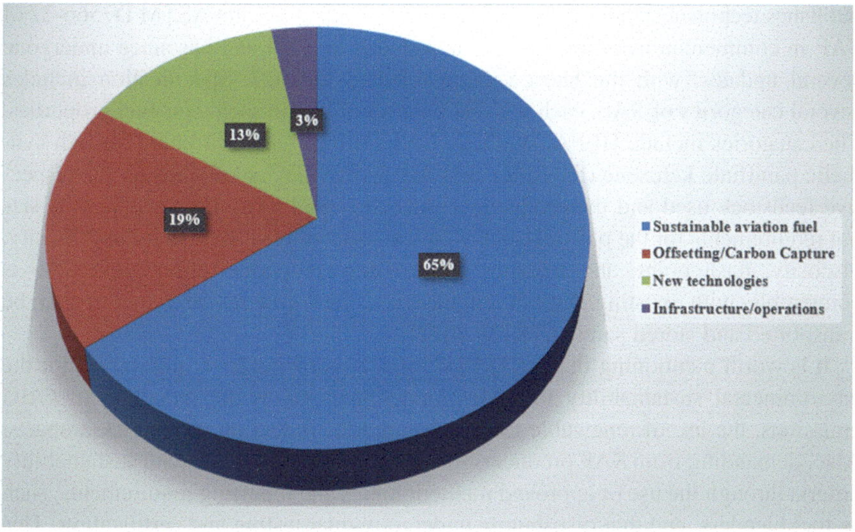

Fig. 8.4 Contribution for 2050 net-zero carbon. Source: IATA (2022)

Table 8.1 SAF's mandated actions

Year	2025	2030	2035	2040	2045	2050
Total mandated SAF actions (%)	2	5	20	32	38	63
Mandated actions for e-kerosene (%)	–	0.7	5	8	11	28

Source: ECEEE (2022)

policy incentives are needed to ensure the production of e-kerosene in Europe and set aviation on a pathway to zero emissions by 2050. A study showed that these sub-targets for e-kerosene can be revised, based on the current e-kerosene production plans in Europe. It was reported that there will be 0.16 Mt. of e-kerosene available in Europe by 2025, produced by 10 companies, and 1.83 Mt. in 2030, produced by at least 18 manufacturers. However, these predictions were based on favorable political conditions. So, blending mandates must be defined at high levels to support the market (ECEEE 2022). Some recommendations were made such as (i) blend mandates must be of at least 0.1% (0.05 Mt) in 2025 and at least 2% (1.00 Mt) in 2030; (ii) e-kerosene needs to be produced with renewable electricity and preferably with carbon sourced from direct air capture; and (iii) the synthetic fuel sub-target should also include renewable electricity and green hydrogen for direct use in aircraft, which will be used by novel propulsion technologies that will soon enter the market, such as electric, fuel cell, or hydrogen-powered aircraft. The market development of e-kerosene has great industrial potential in Europe's green market and know-how and must not be neglected.

The main standard to guide SAF is the ASTM D7566-22, where a set of specifications outline the requirements for the production, quality, and uses. Systematized by the ASTM International, a global organization that develops and publishes technical standards for a wide range of industries, the ASTM D7566-22 of SAF in commercial aviation was first introduced in 2009 and has since undergone several updates, with the latest version released in 2022. Specification includes several categories of SAF, each with its own set of requirements for fuel properties. The categories include HEFA, synthetic iso-paraffinic (SIP), Fischer–Tropsch synthetic paraffinic kerosene (FT-SPK), and AtJ, where they are defined by the respective feedstock used and the production processes employed. The specification sets out requirements for the physical and chemical properties of SAF, including density, viscosity, flash point, and freezing point. These properties ensure that SAF is compatible with existing aircraft engines and fuel systems and that it can be transported and stored safely (ASTM 2022).

It is worth mentioning that ASTM D7566-22 also includes requirements for the environmental sustainability of SAF production, such as the reduction of GHG emissions, the use of renewable feedstocks, and the reduction of land use impacts. Also, demanding from SAF producers to demonstrate compliance with sustainability criteria through the use of approved methodologies and reporting requirements, such as fuel blenders, and that distributors undergo regular testing and certification. This specification plays an important role in the adoption and use of SAF in commercial aviation, aiding to ensure its safety, quality, and performance, providing a level of certainty and transparency for airlines, fuel suppliers, and regulators, which can help to increase confidence in SAF and encourage its adoption. The most challenging barrier to its full implementation is the limited availability of SAF production facilities and distribution infrastructure. To overcome this challenge, governments and industry stakeholders are investing in the development of SAF logistics and distribution networks, as well as incentives to encourage SAF production and use (ASTM 2022).

8.7 Future Perspectives

Overall, SAF fuel offers a promising alternative to traditional fossil-based aviation fuels to support reducing aviation emissions in several ways that were discussed in this chapter. Concerning the lower carbon emissions, as SAF is made from renewable sources such as plant oils, agricultural waste, and algae biomass waste, which can reduce carbon emissions by up to 80% compared to traditional jet fuel. This is the advantage of using biomasses that have absorbed CO_2 during their growth phase, which is later released when the fuel is burned. SAF can, therefore, provide a significant reduction in aviation's carbon footprint. SAF can also reduce local air pollution, which is a significant concern in densely populated urban areas near airports. SAF can reduce the emission of particulate matter, sulfur oxides, and nitrogen oxides, which can have harmful effects on human health and the environment. The production of SAF is typically more sustainable than traditional fossil fuel production. SAF can be produced from waste products, reducing the need for land use, and other inputs associated with the production of traditional fuels. Moreover,

SAF is produced using a range of feedstocks, which can support local agriculture and provide new opportunities for rural communities. SAF has the potential to increase the efficiency of aircraft engines due to its higher energy density than traditional jet fuel, which means that aircraft can fly further on less fuel. This could aid in reduced global fuel consumption and emissions, making air travel more sustainable in the long term.

8.8 Conclusions

Although SAF has potential benefits to the environment, its implementation faces several challenges. Some aspects can be improved by bringing perspectives to SAF advancements, as the availability of sustainable feedstocks is one of the biggest challenges for the production of SAF since it may be limited, and their production may face technical and economic constraints. The production of SAF is currently more expensive than traditional fossil-based jet fuels due to the high cost of sustainable feedstocks and the need for specialized refining processes. SAF production costs need to be reduced significantly to compete with traditional fossil-based jet fuels. Further investment in research and development is required to improve production efficiency and bring down the cost. Scaling up SAF production to meet the increasing demand for air travel requires significant investment in infrastructure and technology. The current production capacity of SAF is limited, and the supply chain is not well established. Significant investment is required to scale up production, including developing the necessary infrastructure for feedstock collection, transport, and storage, as well as refining and distribution facilities. Ensuring that SAF meets environmental and safety standards is crucial to its adoption. The aviation industry has established standards for SAF, but further work is needed to establish a global certification system and to harmonize standards to ensure consistency and transparency. Creating a supportive policy and regulatory framework is necessary to promote the use of SAF. Policies such as subsidies, incentives, and mandates can help encourage investment in SAF and accelerate its adoption.

References

ASTM – American Society for Testing and Materials International Standards Organization (2022) ASTM D7566-22 standard specification for aviation turbine fuel containing synthesized hydrocarbons. https://www.astm.org/d7566-22.html. Accessed 30 Apr 2023

Atabani AE, Silitonga AS, Ong HC, Mahlia TMI, Masjuki HH, Badruddin IA, Fayaz H (2013) Non-edible vegetable oils: a critical evaluation of oil extraction, fatty acid compositions, biodiesel production, characteristics, engine performance and emissions production. Renew Sustain Energy Rev 18:211. https://doi.org/10.1016/j.rser.2012.10.013

Aviation Benefits Beyond Borders (2023). https://aviationbenefits.org/. Accessed 30 Apr 2023

Badgett A, Milbrandt A (2020) A summary of standards and practices for wet waste streams used in waste-to-energy technologies in the United States. Renew Sustain Energy Rev 117:109425. https://doi.org/10.1016/j.rser.2019.109425

BCC Research (2022) The leading companies in the sustainable aviation fuel industry. https://blog.bccresearch.com/10-leading-companies-in-the-sustainable-aviation-fuel-industry. Accessed 30 Apr 2023

BETO – Bioenergy Technologies Office (2022) Sustainable aviation fuel grand challenge roadmap: flight plan for sustainable aviation fuel report. https://www.energy.gov/eere/bioenergy/articles/sustainable-aviation-fuel-grand-challenge-roadmap-flight-plan-sustainable. Accessed 30 Apr 2023

BP (2023). https://www.bp.com/en/global/air-bp/aviation-fuel/sustainable-aviation-fuel.html. Accessed 30 Apr 2023

Bradin D (2014) Process for producing renewable jet fuel compositions. US patent 20150247100A1. 2022–2023

Brennan L, Owende P (2010) Biofuels from microalgae - a review of technologies for production, processing, and extractions of biofuels and co-products. Renew Sustain Energy Rev 14:557. https://doi.org/10.1016/j.rser.2009.10.009

Carbon Credits.com (2023). https://carboncredits.com/. Accessed 30 Apr 2023

Cavalett O, Cherubini F (2018) Contribution of jet fuel from forest residues to multiple sustainable development goals. Nat Sustain 1:799–807. https://doi.org/10.1038/s41893-018-0181-2

Cheiky M, Malyala R, Sills R (2014) System for making renewable fuels including gasoline, diesel, and jet fuel. International Application Published under the Patent Cooperation Treaty (PCT)-WO2014031776A2. 2011–2012

Davis R, Aden A, Pienkos PT (2011) Techno-economic analysis of autotrophic microalgae for fuel production. Appl Energy 88:3524–3531. https://doi.org/10.1016/j.apenergy.2011.04.018

Drünert S, Neuling U, Zitscher T, Kaltschmitt M (2020) Power-to-liquid fuels for aviation—processes, resources and supply potential under German conditions. Appl Energy 277:115578. https://doi.org/10.1016/j.apenergy.2020.115578

Dutta S, Madav V, Joshi G, Naik N, Kumar S (2023) Directional synthesis of aviation-, diesel-, and gasoline range hydrocarbon fuels by catalytic transformations of biomass components: an overview. Fuel 347:128437. https://doi.org/10.1016/j.fuel.2023.128437

ECEEE – European Council for an Energy Efficient Economy (2022) Analysis of green jet fuel production in Europe. https://www.eceee.org/all-news/news/analysis-of-green-jet-fuel-production-in-europe/. Accessed 30 Apr 2023

Emmanouilidou E, Mitkidou S, Agapiou A, Kokkinos NC (2023) Solid waste biomass as a potential feedstock for producing sustainable aviation fuel: a systematic review. Renew Energy 206:897–907. https://doi.org/10.1016/j.renene.2023.02.113

Escalante ESR, Ramos LS, Rodriguez Coronado CJ, de Carvalho Júnior JA (2022) Evaluation of the potential feedstock for biojet fuel production: focus in the Brazilian context. Renew Sustain Energy Rev 153:111716. https://doi.org/10.1016/j.rser.2021.111716

Eurocontrol (2021) Supporting European aviation. The EU's "Fit for 55" Package: what does it mean for aviation? https://www.eurocontrol.int/article/eus-fit-55-package-what-does-it-mean-aviation. Accessed 24 Apr 2023

European-Union-Aviation-Safety-Agency-EASA (2023) European aviation environmental report 2019. https://ec.europa.eu/transport/sites/transport/files2019-aviation-environmental-report.pdf. Accessed 25 Apr 2023

Fantuzzi A, Saenz PA, Cavazos NM, High M, Mai Bui A, Rutherford W, von Holstein I (2023) Low-carbon fuels for aviation. Imperial College of London Briefing Paper, pp 1–32. https://doi.org/10.25561/101834

Flach B, Lieberz S, Bolla S (2021) Biofuels annual

Fulcrum BioEnergy (2023). https://www.fulcrum-bioenergy.com/plants-overview. Accessed 30 Apr 2023

Geleynse S, Brandt K, Garcia-perez M, Wolcott M, Zhang X (2018) The alcohol-to-jet conversion pathway for drop-in biofuels: techno-economic evaluation. ChemSusChem 11:3728–3741. https://doi.org/10.1002/cssc.201801690

Gevo (2023). https://gevo.com/category/products/sustainable-aviation-fuel/. Accessed 30 Apr 2023

Guoru D, Nianyou D, Yongjie S, Feiyue Z, Angui Z, Jiang Q, Jiuzhou C, Hao H, Shun Y, Hua Y (2022) Fischer-Tropsch synthesis device. CN patent 218609296U. (April):156577

Han B, Goh H, Tung C, Chyuan H, Seljak T, Ng J, Tian B, Karmarkar S, Viktor J (2022) Recent advancements in catalytic conversion pathways for synthetic jet fuel produced from bioresources. Energy Convers Manag 251(8):114974. https://doi.org/10.1016/j.enconman.2021.114974

IATA – International Air Transport Association (2022) Annual review 2022 78th annual general meeting and world air transport summit, Doha, Qatar. https://www.iata.org/contentassets/c81222d96c9a4e0bb4ff6ced0126f0bb/annual-review-2022.pdf. Accessed 30 Apr 2023

IATA – International Air Transport Association (2023) Developing Sustainable Aviation Fuel (SAF). https://www.iata.org/en/programs/environment/sustainable-aviation-fuels/#tab-2. Accessed 30 Apr 2023

ICCT – The International Council on Clean Transportation (2022) Current and future cost of e-kerosene in the United States and Europe. https://theicct.org/wp-content/uploads/2022/02/fuels-us-europe-current-future-cost-ekerosene-us-europe-mar22.pdf. Accessed 30 Apr 2023

Kandaramath Hari T, Yaakob Z, Binitha NN (2015) Aviation biofuel from renewable resources: routes, opportunities and challenges. Renew Sustain Energy Rev 42:1234. https://doi.org/10.1016/j.rser.2014.10.095

Kasim FH, Harvey AP (2011) Influence of various parameters on reactive extraction of Jatropha curcas L. for biodiesel production. Chem Eng J 171:1373–1378. https://doi.org/10.1016/j.cej.2011.05.050

Kresnyak S, Crawford G, Suppiah S (2022) Process for producing synthetic hydrocarbons from biomass. International application published under the Patent Cooperation Treaty (PCT) WO2022236422A1

Kumari D, Singh R (2018) Pretreatment of lignocellulosic wastes for biofuel production: a critical review. Renew Sustain Energy Rev 90:877–891. https://doi.org/10.1016/j.rser.2018.03.111

LanzaJet (2023). https://www.lanzajet.com/. Accessed 30 Apr 2023

Li J, Zhao R, Xu Y, Wu X, Bean SR, Wang D (2022) Fuel ethanol production from starchy grain and other crops: an overview on feedstocks, affecting factors, and technical advances. Renew Energy 188:223. https://doi.org/10.1016/j.renene.2022.02.038

Lim JHK, Gan YY, Ong HC, Lau BF, Chen W-H, Chong CT, Ling TC, Klemeš JJ (2021) Utilization of microalgae for bio-jet fuel production in the aviation sector: challenges and perspective. Renew Sustain Energy Rev 149:111396. https://doi.org/10.1016/j.rser.2021.111396

Markets and Markets (2023) Sustainable Aviation Fuel (SAF) Market. https://www.marketsandmarkets.com/Market-Reports/sustainable-aviation-fuel-market-70301163.html#:~:text=The%20startup%20companies%20in%20the,Northwest%20Advanced%20Biofuels%20(Austria). Accessed 30 Apr 2023

Martinez-Valencia L, Camenzind D, Wigmosta M, Garcia-Perez M, Wolcott M (2021) Biomass supply chain equipment for renewable fuels production: a review. Biomass Bioenergy 148:106054. https://doi.org/10.1016/j.biombioe.2021.106054

Meurer A, Kern J (2021) Fischer–Tropsch synthesis as the key for decentralized sustainable kerosene production. Energies 14:1836

Monteiro RRC, Dos Santos IA, Arcanjo MRA, Cavalcante CL, de Luna FMT, Fernandez-Lafuente-R, Vieira RS (2022) Production of jet biofuels by catalytic hydroprocessing of esters and fatty acids: a review. Catalysts 12(2):1–24. https://doi.org/10.3390/catal12020237

Morgan TJ, Youkhana A, Turn SQ, Ogoshi R, Garcia-Pérez M (2019) Review of biomass resources and conversion technologies for alternative jet fuel production in Hawai'i and tropical regions. Energy Fuels 33:2699. https://doi.org/10.1021/acs.energyfuels.8b03001

Neste (2023). https://www.neste.com/products/all-products/saf. Accessed 30 Apr 2023
Ng KS, Sadhukhan J (2011) Techno-economic performance analysis of bio-oil based Fischer-Tropsch and CHP synthesis platform. Biomass Bioenergy 35(7):3218–3234. https://doi.org/10.1016/j.biombioe.2011.04.037
Ng KS, Farooq D, Yang A (2021) Global biorenewable development strategies for sustainable aviation fuel production. Renew Sustain Energy Rev 150:111502. https://doi.org/10.1016/j.rser.2021.111502
Nouni MR, Jha P, Sarkhel R, Banerjee C, Tripathi AK, Manna J (2021) Alternative fuels for decarbonisation of road transport sector in India: options, present status, opportunities, and challenges. Fuel 305:121583. https://doi.org/10.1016/j.fuel.2021.121583
Peters M, Taylor J (2011) Renewable jet fuel blendstock from isobutanol. International Application Published under the Patent Cooperation Treaty (PCT) WO2011140560A1. 2011–2012
Phillips S, Flach B, Lieberz S, Bolla S (2019) EU Biofuels Annual 2019
Price S, Kresnyak S, Crawford G (2020) Process for producing synthetic hydrocarbons from biomass. International application published under the Patent Cooperation Treaty (PCT) WO2021087618A1
Prieler S, Fischer G, van Velthuizen H (2013) Land and the food-fuel competition: insights from modeling. Wiley Interdiscip Rev Energy Environ 2:199–217. https://doi.org/10.1002/wene.55
PWC (2023) Sustainable aviation fuels cost less than you think. https://www.pwc.com/gx/en/issues/esg/the-energy-transition/sustainable-aviation-fuels-cost-less-than-you-think.html. Accessed 30 Apr 2023
SABA – Sustainable Aviation Buyers Alliance (2023) The state of the SAF market in 2022 and beyond. https://www.flysaba.org/2022/12/15/state-of-saf-market-2022-and-beyond/#:~:text=The%20Sustainable%20Aviation%20Fuel%20(SAF,benefit%20of%20communities%20near%20airports. Accessed 30 Apr 2023
SAF Grand Challenge Roadmap – Flight Plan for Sustainable Aviation Fuel (2022) U.-S. Department of Energy. https://www.energy.gov/sites/default/files/2022-09/beto-saf-gc-roadmap-report-sept-2022.pdf. Accessed 30 Apr 2023
Said Z, Nguyen TH, Sharma P, Li C, Ali HM, Nguyen VN, Pham VV, Ahmed SF, Van DN, Truong TH (2022) Multi-attribute optimization of sustainable aviation fuel production-process from microalgae source. Fuel 324:124759. https://doi.org/10.1016/j.fuel.2022.124759
Sharma A (2022) Sustainable aviation fuel production to hit 300 million litres in 2022: IATA. https://www.livemint.com/news/world/sustainable-aviation-fuel-production-to-hit-300-million-litres-in-2022-iata-11670413121177.html. Accessed 30 Apr 2023
Shell (2023). https://www.shell.com/business-customers/aviation/100years/flying-together/scale-saf.html. Accessed 30 Apr 2023
Shonnard DR, Williams L, Kalnes TN (2010) Camelina-derived jet fuel and diesel: sustainable advanced biofuels. Environ Prog Sustain Energy 29:382–392. https://doi.org/10.1002/ep.10461
Singh S, Joshi G (2022) How SAF can become cost competitive against conventional fuel. https://simpleflying.com/saf-cost-competitive-jet-fuel/?newsletter_popup=1. Accessed 30 Apr 2023
SkyNRG (2023). https://skynrg.com/. Accessed 30 Apr 2023
Staples MD, Malina R, Olcay H, Pearlson MN, Hileman JI, Boies A, Barrett SRH (2014) Lifecycle greenhouse gas footprint and minimum selling price of renewable diesel and jet fuel from fermentation and advanced fermentation production technologies. Energy Environ Sci 7:1545–1554. https://doi.org/10.1039/c3ee43655a
Su-ungkavatin P, Tiruta-Barna L, Hamelin L (2023) Biofuels, electrofuels, electric or hydrogen?: A review of current and emerging sustainable aviation systems. Prog Energy Combust Sci 96:101073. https://doi.org/10.1016/j.pecs.2023.101073
Tiwari R, Mishra R, Choubey A, Kumar S, Atabani AE, Badruddin IA, Khan TMY (2023) Environmental and economic issues for renewable production of bio-jet fuel: a global prospective. Fuel 332:125978. https://doi.org/10.1016/j.fuel.2022.125978
TotalEnergies (2023). https://aviation.totalenergies.com/en/fuels-and-services-aviation/sustainable-aviation-fuel. Accessed 30 Apr 2023

United Nations (2009) The state of play of the clean development mechanism, review of barriers and potential ways forward. United Nations Conference on Trade and Development, New York

Vandenberghe LPS, Valladares-Diestra KK, Bittencourt GA, Zevallos Torres LA, Vieira S, Karp SG, Sydney EB, de Carvalho JC, Thomaz Soccol V, Soccol CR (2022) Beyond sugar and ethanol: the future of sugarcane biorefineries in Brazil. Renew Sustain Energy Rev 167:112721. https://doi.org/10.1016/j.rser.2022.112721

Wang WC (2016) Techno-economic analysis of a bio-refinery process for producing hydro-processed renewable jet fuel from Jatropha. Renew Energy 95:63–73. https://doi.org/10.1016/j.renene.2016.03.107

Wang W, Tao L, Markham J, Zhang Y, Tan E, Batan L, Biddy M, Wang W, Tao L, Zhang Y, Tan E, Warner E, Biddy M (2016) Review of biojet fuel conversion technologies review of biojet fuel conversion technologies. National Renewable Energy Laboratory (NREL). https://doi.org/10.2172/1278318

World Energy (2023). https://www.worldenergy.net/. Accessed 30 Apr 2023

Zhang H, Fang Y, Wang M, Appels L, Deng Y (2020) Prospects and perspectives foster enhanced research on bio-aviation fuels. J Environ Manag 274:111214. https://doi.org/10.1016/j.jenvman.2020.111214

Chapter 9
Microbial Isoprenoids as Drop-In Biofuel

Xiaobing Yang, Qidou Gao, and Xiaochun Zheng

Abstract Sustainable development is the need of the hour, and it requires renewable energy resources having better combustion performance and compatibility with existing fossil fuel infrastructure. Isoprenoids are most abundant chemicals in plants. The isoprenoid-based biofuels represent a sustainable source for renewable energy in the transportation sector, particularly jet fuels. Terpenes including hemiterpenes, monoterpenes, and sesquiterpenes can be explored as gasoline, jet fuel, and diesel replacement, respectively. However, the production of terpenoid biofuels via sourcing plants is economically unviable to meet the ever-growing market while chemical synthesis route would suffer from environmental concerns. Metabolic engineering and synthetic biology offer promising approach to produce desired terpene fuels in microbes. To date, massive progress has been achieved to promote the terpenes' bioproduction efficiency of microbial cell factories toward economic viability. This chapter provides a comprehensive landscape in bioproduction of terpene fuels with microbes.

Keywords Drop-in biofuels · Terpenes · Sustainability · Microbial cell factory · Synthetic biology

9.1 Introduction

Urgent needs in sustainable development have stimulated efforts to advance the microbial production of eco-friendly and low-cost biofuels from renewable feedstocks. Currently, bioethanol is used as a major sustainable biofuel, especially for household vehicles (Staples et al. 2019). Nevertheless, two critical defects render it from being perfect: low-energy density and high hygroscopicity. Bioethanol has only 70% of the energy content of gasoline, which will reduce the endurance of vehicles. In addition, ethanol can corrode the automotive and distributing pipelines.

X. Yang (✉) · Q. Gao · X. Zheng
College of Enology, Northwest A&F University, Xianyang, Shaanxi, People's Republic of China
e-mail: yangxb@nwafu.edu.cn

S. K. Brar et al. (eds.), *The Microbiology of the Drop-in Biofuel Production*, Biofuel and Biorefinery Technologies 15, https://doi.org/10.1007/978-3-031-61637-2_9

Therefore, the research focus is now shifting to the development of advanced biofuels such as higher alcohols, fatty acid alkyl ester (FAAE), alkane, and terpene molecules (Balat 2009). Advanced biofuels have more properties of existing gasoline and diesel, and more comparable energy density and physicochemical properties than petroleum-based fuels. More importantly, they allow the direct use of the existing engine designs, fuel distribution, and storage systems. However, some advanced biofuels like FAEE or higher alcohols still lack the necessary high-density requirements (both physical and energy density) for use in special area, such as military jet and rocket fuels. The compact structures of isoprenoids endow them with potential as high-density fuels, especially, when there is ring unit to contribute high combustion heat. Short- and branched-chain terpenes and their derivatives are favorable alternatives to gasoline, while longer branched-chain or cyclic ones may be used as diesel and aviation fuel substitutes.

Isoprenoids are components of primary metabolites prevalent in plants, fungi, bacteria, and marine invertebrates, which account for over 90,000 compounds in nature (Hou and Dickschat 2021). They have found wide uses as drugs, fragrances, flavors, and drop-in biofuels. Terpenes are compounds derived from isoprene units (C_5H_8), whereas terpenoids are terpenes with additional functional groups, such as hydroxyl, epoxy, and carbonyl. Plants are the major natural sources, where the terpenes are most abundant and diverse. Nevertheless, sourcing isoprenoids from plants to produce commodity biofuels on a large scale remains economically unfeasible due to their exceedingly low natural concentrations in their original hosts. Besides, plant growth requires not only long cycle, massive land usage, and water resources but is also susceptible to climate change, making plant-based terpene production even more costly and difficult. In addition, the de novo chemical synthesis of terpenes suffers from low yield, undesired impurity, and high cost on catalysts due to the high complexity of terpene structures (Lapuerta et al. 2023).

Compared with plants, microbes are featured with multiple advantages such as fast growth, high carbon metabolism efficiency, versatile genetic enabling toolboxes, independence from climate and environmental divergence, and ease in up-scaling. Motivated by the rapid progress in metabolic engineering, the development of a microbial fermentation platform is gaining attention as an alternative route for efficient, sustainable, and greener terpene production. To date, microbes including bacteria, yeast, fungi, and cyanobacteria have been exploited for the total isoprenoid biosynthesis of drop-in biofuel from simple carbon sources such as sugars and glycerol (Fig. 9.1).

9.2 Categories of Isoprenoid-Derived Drop-In Biofuels

Terpenes are organic compounds derived from isoprene units (C_5H_8), whereas terpenoids are terpenes with additional functional groups, such as hydroxyl, epoxy, and carbonyl. In the past decade, diverse terpenoids have been investigated

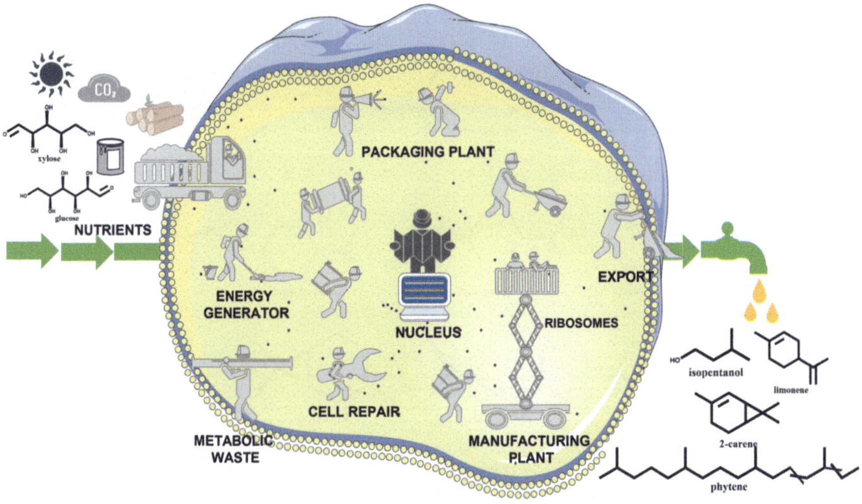

Fig. 9.1 Microbial cell factories produce biofuels

for their fuel-like properties. Their low hygroscopy, high energy density, and good fluidity at low temperatures place them as desirable transportation fuel alternatives (Hellier et al. 2013; Phulara et al. 2016). Terpenes including hemiterpenes, monoterpenes, sesquiterpenes, diterpenes, and triterpenes could be explored as renewable biofuels (Fig. 9.2). Generally, monoterpenes are viewed as biogasoline analogs and sesquiterpenes as replacement for conventional aviation fuels such as Jet A and Jet A-1, whereas sesquiterpenes and diterpenes could find application as biodiesels (Tippmann et al. 2013). The most promising examples include the monoterpenes limonene and β-phellandrene, the sesquiterpenes farnesene and bisabolene, and the diterpene phytoene, which can either be used as fuel additives or directly as replacements for gasoline, jet fuel, and diesel, respectively (Table 9.1) (Tippmann et al. 2013).

9.2.1 *Hemiterpene (C5) Biofuels*

Hemiterpenoids are derived from the C5-precursors IPP and DMAPP. Hemiterpenoids including prenol, isoprenol, isobutanol, and isopentanol are considered as gasoline substitutes (Niu et al. 2017).

Branched C5 alcohols exhibit superior energy content and octane numbers, typically ranging from 92 to 102, along with enhanced combustion characteristics and resistance to engine knock when compared to ethanol. These attributes make them promising candidates for serving as gasoline substitutes and anti-knock additives (Beller et al. 2015). All the hemiterpene biofuels can be biosynthesized from

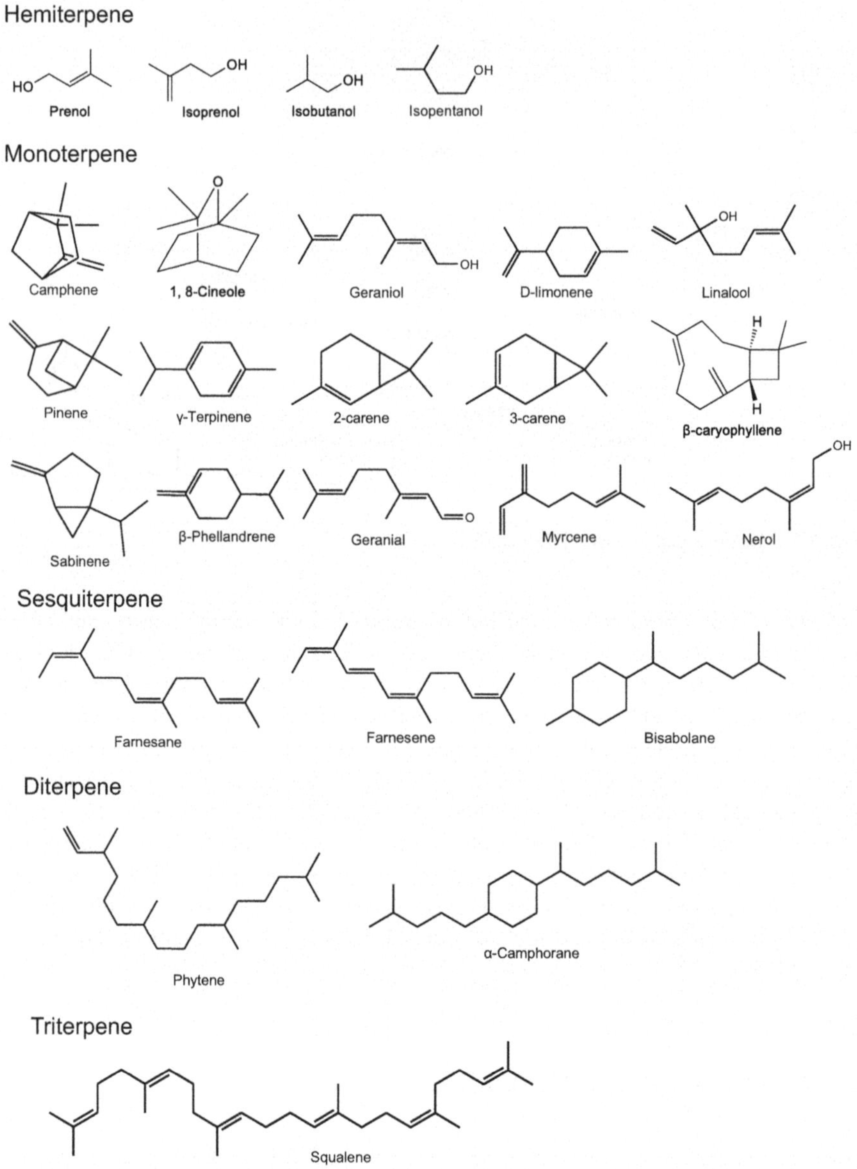

Fig. 9.2 Structure of partial terpenes suitable to be used as drop-in biofuels

the isoprenoid pathway. Alternatively, branched alcohols like isopentanol (isoamyl alcohol or 3-methyl-1-butanol) can be produced either by using a 2-keto acid decarboxylase module from a refactored leucine biosynthesis pathway or by hydrogenation of 3-methyl-3-butenol and 3-methyl-2-butenol (isopentenols) from

Table 9.1 Basic characteristics of terpenes suitable to be used as drop-in biofuels

	Compound	Formula	Density (g/mL)	Kinematic viscosity (mm^2/s, 40 °C)	Cetane number	References
	Non-oxygenated gasoline				85	Mack et al. (2014)
C5	Prenol	$C_5H_{10}O$	0.848 (25 °C)	/	96.5	Mack et al. (2014)
	Isoprenol	$C_5H_{10}O$	0.800 (25 °C)	/	98	Mack et al. (2014)
	Isobutanol	$C_4H_{10}O$	0.803 (25 °C)	/	106.5	Altinkurt et al. (2022)
	Isopentanol	$C_5H_{12}O$	0.809 (25 °C)	/	/	Runguphan et al. (2021)
C10	Camphene	$C_{10}H_{16}$	0.842 (25 °C)	34.96	29	Meylemans et al. (2012)
	1, 8-cineole	$C_{10}H_{16}$	0.992 (25 °C)	/	91	Zhuang et al. (2019)
	Geraniol	$C_{10}H_{18}O$	0.874 (25 °C)	7.652 (25 °C)	19.25	Gonçalves et al. (2018
	D-limonene	$C_{10}H_{16}$	0.841 (25 °C)	1.089 (25 °C)	18.9	Gonçalves et al. (2018)
	Linalool	$C_{10}H_{18}O$	0.857 (25 °C)	5.324 (25 °C)	16.45	Gonçalves et al. (2018)
	R-pinene	$C_{10}H_{16}$	0.855 (25 °C)	1.502 (25 °C)	23.13	Gonçalves et al. (2018)
	β-Pinene	$C_{10}H_{16}$	0.867 (25 °C)	1.766 (25 °C)	21.25	Gonçalves et al. (2018
	y-terpinene	$C_{10}H_{16}$	0.845 (25 °C)	1.008 (25 °C)	20.3	Gonçalves et al. (2018)
	3-carene	$C_{10}H_{16}$	0.857 (25 °C)	/	27	Yanowitz et al. (2017), Woodroffe et al. (2021)
	β-Caryophyllene	$C_{15}H_{24}$	0.900 (15 °C)	/	29	
C15	α/*β*-farnesene	$C_{15}H_{24}$	0.841 (20 °C)	/	32	Pahima et al. (2019)
	β-Bisabolane	$C_{15}H_{30}$	0.809 (25 °C)	/	32.2	Yanowitz et al. (2017) Gonçalves et al. (2018)

/, No overlay

isoprenoid pathway. Intriguingly, the cetane number of hemiterpene biofuels is much higher than that of others. In addition, esters of isoprenoid alcohols (C5, C10, and C15) have also the potential to be used as replacements for petroleum-based diesels.

9.2.2 *Monoterpene Biofuels*

Monoterpenes, a class of isoprenoids typically synthesized in the glandular structures of plants, are the main components of essential oils and are used in a wide variety of industrial applications including pharmaceuticals, insecticides, polymers, and fragrances (Jiang and Wang 2023). Interestingly, the low freezing point and high energy density (due to methyl branching and cyclic structures, respectively) of monoterpenes also make them good candidates for gasoline, diesel, and in particular, jet fuel substitutes. Monoterpene biofuels are produced from geranyl/Nerol diphosphate (G/NPP). Some monoterpenes are acyclic (such as myrcene and ocimene) or their derivatives (such as linalool and geraniol), but cyclic monoterpenes are more abundant in nature. Among these monoterpenes, the hydrogenated products of limonene, myrcene, and ocimene (2,6-dimethyloctane) display favorable fuel properties, and the hydroxylated compounds such as linalool and geraniol have been suggested as potential biofuel precursors. Besides, hydrogenated dimers of α- and β-pinene show similar high energy density to jet-fuel JP-10 (Lapuerta et al. 2023).

9.2.3 *Sesquiterpene*

In addition to their fragrance and medicinal values, sesquiterpenes are also candidates for advanced biofuels, among which farnesene, bisabolene, and their alkylated forms are most favored. Monocyclic sesquiterpenes, such as bisabolenes and their hydrogenation products (i.e., bisabolane), have been proposed as a diesel replacement. Probably due to the presence of the cyclic structure, bisabolane compares well to farnesane in that it has a lower cloud point (−78 °C), more favorable cold-temperature properties, and physical density (0.88 g/mL for bisabolane vs 0.77 g/mL for farnesane) (Peralta-Yahya et al. 2011). Acyclic sesquiterpenoids that explored for drop-in biofuels replacement include farnesene and isomeric alcohols such as nerolidol and farnesol. Farnesene is a collective name that refers to four stereoisomers of α-farnesene and two stereoisomers of β-farnesene. All these isomers have a branched hydrocarbon chain backbone and four double bonds in common, but they are differentiated by the location of the double bonds and the chirality of specific C–C bonds. Farnesane, the fully hydrogenated form of farnesene, is a diesel and jet-fuel replacement, which has an appropriate cetane number (CN) of ~58 (CN of D2 diesels ranges from 40 to 60) and better cold-temperature properties, with a lower cloud point (−78 °C) than that of D2 diesels (−3 °C) (Rude and Schirmer 2009). The most successful case was presented by Amyris, a US company, which has produced trans-β-farnesene under the name of "Biofene®" at 130 g/L with Brazilian sugarcane as the feedstock (Meadows et al. 2016).

9.3 Terpenes Modification Before Used as Biofuels

Terpenes and their derivatives, including blends, have emerged as the most promising alternatives to conventional aviation fuels. However, oxygen and double bonds in terpenes and terpenoids are not desired because they negatively impact energy density and storage stability. Thus, terpenes and terpenoids should be modified before fueling engines, and direct use must be avoided. In this respect, terpenes can be enhanced by chemical synthesis paths. Among various chemical processes, hydrogenation, isomerization, dimerization, alkylation, oxyfunctionalization, hydration, addition of alcohols, and acetoxylation are particularly attractive (Lapuerta et al. 2023). Typically, hydrogenation plays an important role in reducing or eliminating the unsaturation in terpenes and enhancing the stability. Oxyfunctionalization is a process in which an oxygen group is introduced, changing characteristics of the terpene for its use as fuel by transforming the terpene to ketones, aldehydes, esters, and ethers. For instance, oxyfunctionalized turpentine decreases particulate matter emissions, despite reducing its energy density, and low flash point can also be tackled using terpene oxyfunctionalization (Donoso et al. 2021; Garcia et al. 2020). Dimerization is effective in synthesizing high-energy density fuels. The dimers of pinene and camphene have shown comparable viscosities, densities, and net combustion heats to the missile fuel J-10 (Lapuerta et al. 2023).

9.4 Biosynthesis Pathway for Isoprenoids Biofuels

9.4.1 Canonical MEP and MVA Pathways

All isoprenoids are derived from the universal C5 precursors isopentenyl diphosphate (IPP) and dimethylallyl diphosphate (DMAPP). Typically, there are two routes to form IPP, the mevalonate (MVA) pathway and the methylerythritol phosphate (MEP) pathway (Fig. 9.3). Bacteria use the MEP pathway, except Actinobacteria, Bacteroidetes, Chloroflexi, Firmicutes, and Proteobacteria which have the classical MVA pathway or a variation thereof. Eukaryotes use the MVA pathway while plants also use the MEP pathway in plastids, which is also essential for survival. Most archaea use alternative MVA pathways, except *Sulfolobus, Metallosphaera*, and *Acidianus* which use the classical MVA pathway (Table 9.2). IPP is then converted to DMAPP and various isoprenoids.

In detail, the MEP pathway begins with the condensation of pyruvate and glyceraldehyde-3-phosphate (G3P) to form 1-deoxy-D-xylulose 5-phosphate (DXP). DXP is reduced to generate MEP by DXP reductoisomerase (DXR) using NADH, and MEP is activated with cytidine-triphosphate (CTP) to produce 4-diphosphocytidyl-2-C-methyl-D-erythritol (CDP-ME) by CPD-ME synthase (CMS). CDP-ME is phosphorylated by CDP-ME kinase (CMK) and cyclized to

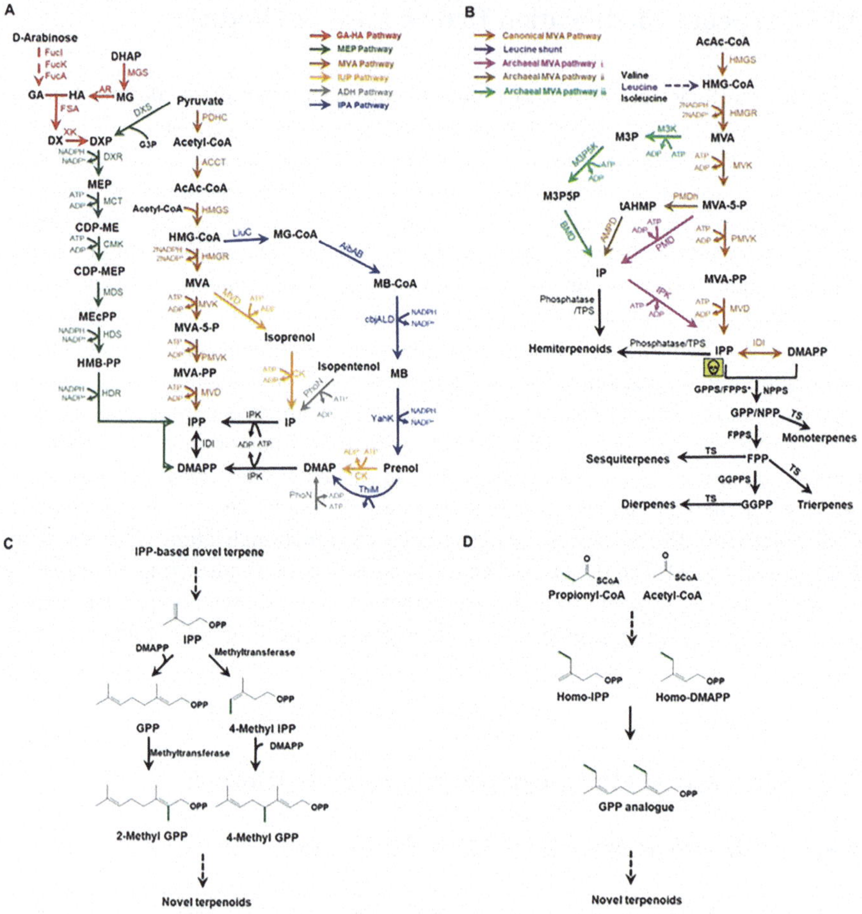

Fig. 9.3 Canonical and alternative pathways for terpenoid biosynthesis

2-C-methyl-D-erythritol 2,4-cyclodiphosphate (MEcPP) by MEcPP synthase (MCS). The cyclic diphosphate ring is opened and reductive dehydration is catalyzed by 4-hydroxy-3-methyl-butenyl diphosphate synthase (HDS) to produce 4-hydroxy-3-methyl-butenyl diphosphate (HMBPP). Finally, a reductive dehydration of HMBPP by HMBPP reductase (HDR) produces both IPP and DMAPP (Fig. 9.3). Since IPP and DMAPP are produced by the MEP pathway, the IPP delta-isomerase is not essential in organisms using the MEP pathway; many bacteria do not have a gene encoding this enzyme (Fig. 9.3). Notably, the MEP pathway has several regulation points. DXS catalyzes the most tightly regulated step. Isopentenyl pyrophosphate isomerase (IDI) is the second important regulation point and its overexpression doubled isoprene production (Table 9.3).

The MVA pathway originates from acetyl-CoA, which is formed through glycolysis and PDH or citrate pathway. Then, acetyl-CoA thiolase (ERG10) catalyzes

Table 9.2 Comparison of the biosynthesis of isoprenoid precursors and fatty acid

	MVA	MEP	IPA	FAS
Products	IPP	IPP and DMAPP	DMAPP	Palmitic acid
Starters	Acetyl-CoA	Pyruvate, glyceral-dehyde-3-phosphate	Acetyl-CoA	Acetyl-CoA
Cofactors	2 NADPH and 3 ATP	3 NADPH and 3 ATP	2 NADPH and 2 ATP	14 NADPH and 7 ATP
Mass yield on glucose	25.2%	30.2%	25.2%	35.6%
Steps	6	7	8	7
Key enzymes	HMGS, HMGR, MK, PMK, IDI	DXS, DXR, IspG, IspH, IDI	HMGS, LiuC, AjbAB, cbjALD, YahK, ThiM, IPK	ACC1, FAS
Natural distribution	Eukaryotes, fungi, algae, archaea, a few bacteria, and plant cytoplasm and mitochondria	Bacteria, algae, diatoms cyanobacteria, and plant chloroplasts	Artificially designed	All the organism

the condensation of two molecules acetyl-CoAs to acetoacetyl-CoA (AcAc-CoA), and AcAc-CoA is converted into 3-hydroxy-3-methylglutaryl-CoA (HMG-CoA) by the hydroxymethylglutaryl-CoA synthase (ERG13, HMGS), after which HMG-CoA is converted into mevalonate by HMG-CoA reductase (HMGR). IPP is generated by three additional reactions catalyzed by mevalonate kinase (MK, ERG12), phosphomevalonate kinase (PMK, ERG8), and mevalonate diphosphate decarboxylase (MVDD, ERG9), and is then isomerized into DMAPP by IDI. Finally, the IPP and DMAPP are condensed to form pyrophosphate precursors with different chain lengths, including GPP/NPP (C10), FPP(C15), and GGPP(C20). These precursors are further converted by terpene synthases (TSs) to yield diverse terpenes or backbone molecules.

9.4.2 *Alternatives to the Canonical Pathways*

9.4.2.1 Alternative Entry Points to the MEP and MVA Pathways

1. ***Alternative Pathways to DXP***
   DXS stands out as a highly regulated step in the MEP pathway. Exploring alternative routes to DXP might hold the potential to enhance carbon flux toward terpenoid biosynthesis (Ren et al. 2020). The structural similarity between DXP and other pentose phosphates has inspired the use of two *Escherichia coli* genes, *yajO* or *ribB* variant, to convert D-ribulose 5-phosphate from central metabolism

Table 9.3 Microbial production of terpenes that can be explored as drop-in biofuels

Biofuel		Host	Substrate and scale	Recovery	Titer	Strategy	References
C5	Isoprene	*S. cerevisiae*	Sucrose, Fed batch (5 L)	/	11.9 g/L, 52.2 mg/g	Strengthening of precursor supply; overexpression of MVD1 and IDI1; protein engineering	Yao et al. (2018)
		E. coli	Glucose, fed batch	/	60.0 g/L, 2 g/L/h, 0.8 g/g dcw	Overexpression of mvaE and mvaS, mvk, pmk, mvd, and idi, additional mvk, ispS, pgl genes, inducible Ptrc promoter	Isar et al. (2022)
		E. coli	Glucose, 300 L	/	25.2	Dual plasmid (pTS-sPt-MVA and pS-NA)	Lei et al. (2021)
		P. ananatis	Glucose, 1 L	Enclosed fermentation	25.0 g/L, 1.8% vol yield in 48 h	Metabolic switch (SWITCH-PphoC/pispSM), fed batch	Nitta et al. (2020)
		S. elongatus	CO_2, shake flask (25 mL)	/	1.26 g/L	Expression of Idi, ispS, dxs, ispG	Gao et al. (2016)
	Prenol	*E. coli*	Glycerol, shake flask (25 mL)	Enclosed fermentation	2.0 g/L	Designed upper IPA pathways, expression of *atoB*, *HMGS*, *aibA*, *aibB*, *cbjALD*, *liuC*, and *yahK*	Clomburg et al. (2019)
	Isoprenol	*E. coli*	Shake flask,	/	8.5 g/L, 85% of theoretical yield	MVA path, repression of yggV and accA, fed-batch fermentation	Wang et al. (2022a)
		S. cerevisiae	Shake flask	/	0.38 g/L	IPP bypath, deleting a promiscuous endogenous kinase, overexpressing promiscuous alkaline phosphatase	Kim et al. (2021)
	Isoprenyl acetate	*E. coli*	2-L bioreactor		28.0 g/L, 0.20 g/g	MVA pathway, IPP-bypass, thioester pathways, heterologous alcohol acyltransferases (ATF1, ATF2, and SAAT)	Carruthers et al. (2023)
	Isopentenol	*E. coli*	Glucose, shake flask	Ethyl acetate	1.1 g/L	Expression of MevTco, BBa1002, MKco, PMKco, PMDsc, NudB	Kang et al. (2016)

C10	Limonene	*E. coli*	Glycerol, stirred-tank reactor (3.1 L)	Diisononyl phthalate	3.6 g/L, 0.15 g/L/h	Expression of mevalonate pathway and the limonene synthase genes	Rolf et al. (2020)
		S. cerevisiae	Glucose, fed batch (3 L)	Dodecane	2.6 g/L	Combination of mitochondrial compartmentalization, push-pull-restrain strategy, precursor and cofactor supply, and optimization of batch feeding	Kong et al. (2023)
		S. cerevisiae			2.6 g/L	Dual regulation of peroxisomal compartmentalization and cytosol regulation of the biosynthesis path	Dusseaux et al. (2020)
		R. toruloides	Glucose, shake flask (small system)	Dodecane	0.39 g/L	Overexpression of LS, NPPS, HMGR; expression of EfMvaE and EfMvaS from *E. faecalis*, MmMK from *M. mazei*, and NPPS::LS	Liu et al. (2021)
		Y. lipolytica	Glycerol, fed batch (1.5 L)	Dodecane	0.165 g/L	Expression of XR, XDH, tNDPS1, tLS, and XK; overexpression of HMG1 and ERG12; utilization of xylose and lignocellulosic hydrolysate	Cheng et al. (2019)
	Linalool	*P. ananatis*	Glucose, fed batch (1 L)	IM	10.9 g/L, 0.051 g/g	Biphasic fermentation, dual-phase fed-batch fermentation	Nitta et al. (2021)
	1,8-cineole	*R. toruloides*	Hydrolysate of corn stover, 2 L	Dodecane	1.4 g/L	Expression of HMGR, MK, and PMK orthologs, fed-batch fermentation	Kirby et al. (2021)
	Citronellol	*S. cerevisiae*	Glucose, fed batch (5 L)	IM	8.3 g/L	Expression of CrIS, Erg20^{F96W}, protein scaffolds and protein fusion, degradation pathway gene ATF1	Jiang et al. (2021)
		C. glycerinogenes	Glucose, shake flask	Dodecane	12.8 mg/L	Enhancing the MVA pathway, overexpressed phosphatase and introduced NPPS	Ma et al. (2023)
	α-Pinene	*E. coli*	Glycerol, fed batch (1.3 L)	Dodecane	1.0 g/L	Expression of N-terminal truncated pinene synthase and AgGPPS, fine-tuning of IPTG concentration	Zhou et al. (2023a)

(continued)

Table 9.3 (continued)

Biofuel		Host	Substrate and scale	Recovery	Titer	Strategy	References
	Geraniol	*E. coli*	Glycerol, shake flasks	IM	13.2 g/L	MVA path, bioderivatization strategy	Wang et al. (2022b)
		S. cerevisiae	Glucose and ethanol, fed batch (1 L)	Dodecane	1.7 g/L	Expression of GES; deletion of OYE2 and ATF1; dynamic control of ERG20 expression; fed-batch cultivation	Zhao et al. (2017)
		C. glycerinogenes	Glucose (5 L)		1.09 g/L 47.5 mg/g DCW	Transcription factor-mediated ergosterol feedback system, modification of ergosterol-responsive promoters, optimization of transcription factor expression, xylose assimilation pathway, activating the pentose phosphate (PP) pathway	Zhao et al. (2023)
	Myrcene	*E. coli*	Glycerol, shake flask	IM	1.3 g/L	Rational enzyme modification biochemical process engineering	Wang et al. (2023)
	Nerol	*E. coli*	Glucose, shake flask	IM	0.97 g/L	Expression of NPPS, GmNES, NudJ	Lei et al. (2021)
	Sabinene	*E. coli*	Glucose and glycerol (2 L)	/	2.65 g/L	Biosynthetic pathway using both the MEP and MVA pathways combining the GPP synthase genes and sabinene synthase from *S. pomifera*	Zhang et al. (2014)
		S. cerevisiae	Glucose, shake flask	IM	0.2 g/L	Expression of t34SabS1, subcellular locations, AIM25	Jia et al. (2020)
C15	α-Farnesene	*Y. lipolytica*	Glucose, bioreactor (1 L)	Dodecane	25.6 g/L	Expression of α-farnesene synthase; overexpression of mevalonate pathway; promoter optimization of OptFSLERG20, Sc-tHMG1, and IDI; fed-batch cultivation; overexpression of Yl-HMGR and ERG19	Liu et al. (2019)
		P. pastoris	Glucose, 7 L	Dodecane	17.9 g/L, 0.107 g/L/h, 0.078 g/g	Promoter engineering, MVA biosynthesis module, and the cytosolic acetyl-CoA	Xu et al. (2023)

						supply module as well as to weaken FPP branch module, fed-batch fermentation	
		S. cerevisiae	Glucose, fed batch (5 L)	Dodecane	10.4 g/L	Overexpression of MVA pathway and Fsso, expression of HMGR and Fsso under the control of the GAL promoter, inactivation of lipid phosphate phosphatase	Wang et al. (2021b)
	β-Farnesene	*S. cerevisiae*	Cane syrup, 0.5 L	Poly-alpha-olefin	130.0 g/L, 0.173 g/g	Using non-native metabolic reactions to rewire central carbon metabolism, fed-batch fermentation	Meadows et al. (2016)
	β-Farnesene	*Y. lipolytica*	Oleic acid, 4 L	Dodecane	35.2 g/L, 0.17 g/g	Peroxisomal compartmentalization of MVA pathway, protein engineering of β-farnesene synthase and pathway engineering, fed-batch fermentation	Liu et al. (2022)
		E. coli	Glycerol, 5 L	Decane	10.3 g/L	Expression of atoB, HMGS, tHMGR, FPPS, E12, E8, MVD1, IDI, ispA, FG, fed-batch fermentation	Yao et al. (2020)
	α-Bisabolene	*R. toruloides*	Hydrolysate of corn Stover, bioreactor (2 L)	Dodecane	2.6 g/L	Expression of copies of BS; utilization of lignocellulosic hydrolysate	Kirby et al. (2021)
		Y. lipolytica	Waste cooking oils, shake flasks	Dodecane	1.06 g/L, 8.11 mg/L/h	Mitochondrial and cytosol dual regulation, introduction of MVA pathway and of multidrug efflux transporter, fermentation optimization	Zhu et al. (2022b)
		Rhodobacter capsulatus	Glucose, 5 L	Dodecane	9.8 g/L, 0.196 g/g	Promoter screening, improving the NADPH availability, strengthening the precursor supply, suppressing the side pathways, and introducing a heterologous mevalonate pathway	Zhang et al. (2021)
	β-Bisabolene	*Y. lipolytica*	Waste cooking oils, shake flasks	Dodecane	68.2 mg/L, 0.56 mg/L/h	Heterologous expression of β-bisabolene synthase, overexpression of endogenous	Zhao et al. (2021)

(continued)

Table 9.3 (continued)

Biofuel		Host	Substrate and scale	Recovery	Titer	Strategy	References
	γ-Bisabolene	*Y. lipolytica*	Waste cooking oils, shake flasks	Dodecane	20.2 mg/L, 0.17 mg/L/h	MVA pathway, and introduction of multidrug efflux transporters, fermentation optimization	
C30	Squalene	*S. cerevisiae*	Glucose, fed batch (5 L)	/	27.3 g/L, 650 mg/g	Expression of TPS1 and HSP 104, fermentation optimization	Zhang et al. (2023)
		A. mangrovei FB3	Glucose, shake flasks	/	21.2 g/L	Medium optimization and the treatment with terbinafine	Fan et al. (2010)
		C. glutamicum	Glucose, fed batch (5 L)	Dodecane	1.5 g/L	Expression of truncated ScSQS and HsSQS, knocking out pyc, idsA, crtE	Park et al. (2021).
		Pseudozyma sp. *SD301*	Glucose, fed batch (5 L)	/	2.4 g/L	C/N ratio, intermittent feeding, pH and temperature optimization	Song et al. (2015)

/, No overlay; *IM*, isopropyl myristate

to DXP. The second alternative pathway to DXP from pentose starts from D-arabinose. In this pathway, the exogenous D-arabinose is first degraded to glycolaldehyde and hydroxy-acetone, which are subsequently condensed by fructose-6-phosphate aldolase to produce 1-deoxyxylulose, which is then promiscuously phosphorylated by xylulose kinase to DXP. There are also other natural MEP pathway shunts, depending on S-adenosylmethionine (SAM)-based polyamine. In these routes, DXP is produced from 5′--methylthioadenosine, the dead-end product of universal polyamine metabolism, or from 5′-deoxyadenosine, a radical SAM byproduct (Rinaldi et al. 2022).

2. ***Leucine Shunt***

   To our knowledge, in most cases, leucine can be degraded to produce AcCoA and subsequently incorporated into the MVA pathway. However, an alternative shunt pathway is adopted in some organisms to convert leucine to terpenoids. As evidenced in the archaeon *Halobacterium salinarum* and the protozoan *Leishmania mexicana*, leucine is assimilated into sterols and terpenoid-based lipids, respectively (Rinaldi et al. 2022). These results inspired a novel route where leucine is converted to HMG-CoA, thus entering the MVA pathway (Fig. 9.3). Valine and isoleucine can also be incorporated into the MVA path via the leucine shunt, which stands as the novel pathway to synthesize isovaleryl coenzyme A (IV-CoA) for producing isobutene, a precursor of renewable fuels and chemicals, from glucose (Li et al. 2013). However, the generation and degradation of leucine should be finely tuned.

9.4.2.2 Archaeal MVA Pathways

Compared to the canonical MVA pathway, the archaeon originated pathways may offer important advantages in terpenoid production. By using the isopentenyl phosphate kinase (IPK) in the archaeal MVA pathways, the toxicity of IPP, in the MEP and MVA paths, can be reduced by balancing IP/IPP and DMAP/DMAPP ratios; the carbon flux between the MEP and MVA pathways can be regulated in a bacteria host to elevate terpenoid production. Other benefits of using archaeal variations are bypassing highly regulated enzymes (archaeal MVA pathway II) and reducing ATP consumption (archaeal MVA pathway III) (Rinaldi et al. 2022).

1. ***Archaeal MVA pathway I.*** The first archaeal alternative MVA pathway is featured with two enzymes: ATP-dependent M-5-P decarboxylase (PMD) and isopentenyl phosphate kinase (IPK) (Vannice et al. 2014). It was elucidated by analyzing the *Methanocaldococcus jannaschii* genome (Grochowski et al. 2006). In this pathway, isopentenyl phosphate (IP) is identified as the intermediate between M-5-P and IPP, and the IPK is recruited to convert IP to IPP (Fig. 9.3). This suggests that the order of decarboxylation and phosphorylation is interchanged in this archaeon. The second essential enzyme, ATP-dependent PMD, was identified in the Chloroflexi bacterium *Roseiflexus castenholzii* using in vitro assays (Dellas et al. 2013). For example, utilizing the upper MVA

pathway from *S. cerevisiae* in combination with *Herpetosiphon aurantiacus* PMD and IPK enzymes resulted in comparable isoprene production level to those achieved using the *S. cerevisiae* MVA pathway in *E. coli* (Rinaldi et al. 2022).

2. ***Archaeal MVA pathway II.*** The second Archaeal MVA pathway was identified in *Thermoplasma acidophilum* and closely related thermoplasmatales species (Vinokur et al. 2016). The pathway starts with the phosphorylation of MVA at 3-OH to produce mevalonate 3-phosphate (M-3-P) (Fig. 9.3). This step is performed by mevalonate 3-kinase (M3K), an enzyme homologous to decarboxylases of the MVA route. M3P is further phosphorylated to mevalonate 3,5-bisphosphate by mevalonate 3-phosphate 5-kinase. The phosphate group is removed during the decarboxylation performed by mevalonate biphosphate decarboxylase (BMD), producing IP. IP is phosphorylated by an IPK to produce IPP as in the archaeal MVA pathway I. It is worth mentioning that both decarboxylases share the same Asp/Lys/Arg catalytic triad in archaeal MVA pathways I and II; however, BMDs do not require ATP for activity since it lacks the ATP-binding pocket.
3. ***Archaeal MVA pathway III.*** The third alternative MVA pathway has been elucidated in the archaeon *Aeropyrum pernix*, where mevalonate 5-phosphate dehydratase and trans-anhydromevalonate 5-phosphate decarboxylase are responsible for converting M-5-P to IP through a new intermediate, trans-anhydromevalonate 5-phosphate (Fig. 9.4) (Hayakawa et al. 2018). In the last step, IP is converted to IPP by IPK as in the archaeal MVA pathways I and II. This pathway comprises enzymes that are not homologous to either mevalonate monophosphate or diphosphate decarboxylases and seems to be widely conserved among archaea. These enzymes do not require ATP. This might therefore create a more energy-efficient route to terpenoids. Considering the toxic and inhibitory nature of IPP, it is plausible that IPK could potentially mitigate the potentially adverse effects rose from its accumulation.

9.4.2.3 Artificially Designed Alternative Pathways

Notably, natural isoprenoid biosynthesis pathways are globally interacted with the cellular metabolism, which would hinder their rational engineering. To bypass the intricate regulation and achieve efficient accumulation of IPP and DAMPP, artificially designed alternative pathways including the isopentenol utilization pathway (IUP), the alcohol-dependent hemiterpene (ADH) pathway, and the isoprenoid alcohol (IPA) pathway could be recruited for terpenes overproduction (Ren et al. 2020; Rinaldi et al. 2022). The IUP and the ADH paths mentioned above require less energy and redox power input and give higher theoretical yield. However, they are not sustainable choices since the substrates, such as isopentenol, are petroleum derived. IPA route, on the other hand, can be transplanted into the platform microbes, as it has a direct linkage to the central metabolism, and operates with fewer ATP demands and comparable carbon conversion efficiency. In addition, like

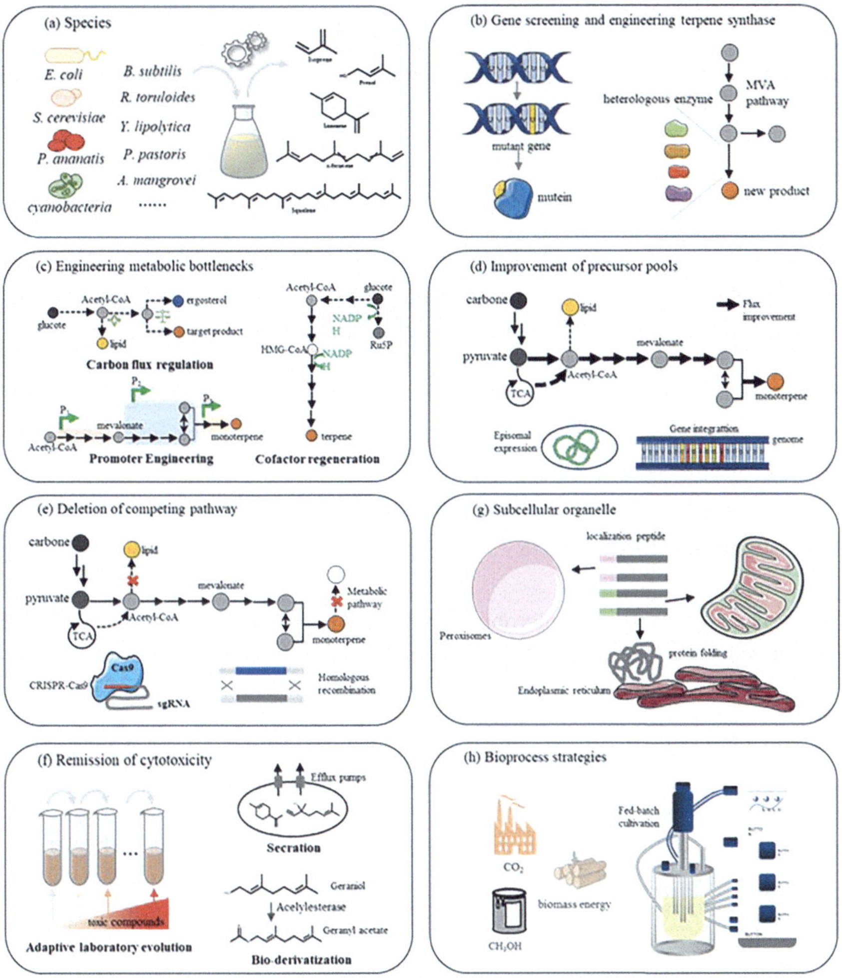

Fig. 9.4 Critical Issues for developing terpene fuel cell factory

other pathways, isoprenoids produced via the designed pathways are also highly dependent on the cofactors of NADPH and/or NADH (Rinaldi et al. 2022).

9.4.3 Pathways Outside the "Isoprene Rule"

The majority of terpenoids start from the C5 modules and follow the so-called isoprene rule, but recent discoveries allow the biosynthesis of terpenoids outside this rule (Rinaldi et al. 2022). These novel terpenoids might also find their proper place as biofuels in the future. By using a SAM-dependent methyl transferase, which catalyzes the electrophilic methylation of GPP, the C11 terpenoids like 2-methylisoborneol and 2-methylenebornane can be produced. The 2-methylisoborneol and 2-methylenebornane are responsible for the muddy smell in contaminated drinking water, and are derived from the 2-methylgeranyl diphosphate intermediate (Rinaldi et al. 2022). Other methyl transferases that are capable of methylating FPP have also been discovered and might play an important role in the synthesis of nonconventional terpenoids (Ozaki et al. 2018; Von Reuss et al. 2018). The second alternative shunt that out of the "isoprene rule" is the homomevalonate pathway, which is used by insects to produce C16, C17, or C18 terpenoid-derived hormones (Rinaldi et al. 2022). In this novel pathway, propionyl-CoA is condensed with Ac-CoA to produce an unusual ethyl branch in terpenoid skeletons (Fig. 9.3). This approach requires no additional ATP or the involvement of SAM.

9.5 Critical Issues for Efficient Bioproduction of Isoprenoids Fuels

The great achievements in microbial synthesis of terpenes biofuels share several common but effective strategies with other natural products. Here, the applied strategies for terpene fuel biosynthesis were briefly summarized: (i) choice of versatile chassis cell; (ii) key enzyme screening and engineering; (iii) modular pathway engineering; (iv) enzyme engineering, common enzyme fusion, direct evolution, structure-guided enzyme engineering, and N-terminal modification (truncation, fusion partners, membrane anchoring, etc.); (v) subcellular engineering strategies, compartmentalization of the biosynthesis routes in mitochondria, peroxisome, lipid droplet, and endoplasmic reticulum; (vi) blocking or downregulating competing pathways; (vii) efflux transporter engineering; and (viii) bioprocess optimization.

9.5.1 Choice of Versatile Microbial Chassis

The importance of a suitable platform cannot be overstated. With the rapid development in metabolic engineering and synthetic biology, microbial synthesis offers an alternative approach to sustainable production of high-value natural products. Microorganisms are potentially ideal platforms for industrial production of drop-in terpene biofuels (e.g., farnesene) from renewable resources due to their rapid growth and versatile metabolism (Meadows et al. 2016). So far, microbes including bacteria, yeast, and mold have been repurposed into cell factories for sustainable overproduction of terpenes (Fig. 9.4a).

The model bacterium *E. coli* and conventional yeast *Saccharomyces cerevisiae* which possess endogenous MEP and MVA pathways, respectively, are the most widely studied hosts for terpenes production, such as geraniol, limonene, and linalool (Jiang and Wang 2023). The two model hosts have much more clear genetic background, mature genetic engineering tools, and readily enabling genetic toolbox than other emerging nonconventional ones. However, they also have their own shortcomings as a workhorse. For example, *S. cerevisiae* suffers from insufficient supply of acetyl-CoA and limited capacity in utilizing pentoses from lignocellulosic biomass.

As a comparison, the unconventional yeasts, oleaginous yeast such as *Yarrowia lipolytica* and *Rhodotorula toruloides* and methylotrophic yeast *Pichia pastoris* and *Ogataea polymorpha*, are becoming potential candidates for the industrial production of terpenes due to their ever maturing enabling genetic toolbox (Ploessl et al. 2023). They usually possess outstanding metabolic characteristics, including the capacity to produce value-added compounds utilizing low-cost biomass feedstocks, such as corn stalk, crude glycerol and waste cooking oil, and wheat straw. Oleaginous yeasts naturally use various renewable carbon sources to accumulate triglycerides (TAG) from acetyl-CoA, ATP, and NADPH. As is shown in Table 9.2, acetyl-CoA, ATP, and NADPH are also precursor and cofactors used for terpenoids biosynthesis. Therefore, the carbon flux toward lipid production in oleaginous yeast can be diverted to support terpenoid production. The methylotrophic yeasts can use the one carbon substrate methanol to produce a portfolio of chemicals. Most recently, the nonconventional yeast *Ogataea polymorpha* was globally rewired for producing sesquiterpenoid β-elemene from glucose (4.7 g/L) (Ye et al. 2023). Along with the rapid expansion of biological elements and gene editing technologies, these yeasts are emerging as one of the main chassis for terpenes production.

9.5.2 Screening and Engineering Terpene Synthase

TSs are the key enzymes in synthesizing various terpene biofuels. Although platform microbes are equipped with the MEP or MVA pathway, few of them are endowed with desired terpenoid synthases. Therefore, once the host is determined, the next step is to screen specific TSs that convert pyrophosphate precursors into target products. Since most terpenoid biofuels are plant-sourced chemicals, TSs and modifying enzymes are mostly elucidated in plants and used in microbes (Fig. 9.4b).

Now, hundreds of TS (target sequences) have been reported and archived in the National Center for Biotechnology Information (NCBI) and other open-access resources. However, few of the plant-derived TSs have been clearly elucidated about their catalytic mechanism and most of them always show unexpected activity in microbes. Directed evolution is an alternative strategy to generate mutant TSs with desired properties when limited knowledge against structural and catalytic mechanism is available. With the aid of a high-throughput screening strategy, the selection process can be greatly accelerated. By coupling robotic liquid handling platforms to GC–MS and automated data extraction, an automatic screening pipeline for the identification of several monoterpene synthase mutants with altered products can be accomplished (Leferink et al. 2019).

In fact, few of the plant TPSs investigated are sufficiently active to support high-titer production of terpenes (Leferink et al. 2016). When eukaryotic proteins are expressed in *E. coli*, incorrect folding and inclusion bodies formation are always observed even after codon optimization and removal of eukaryotic elements such as N-terminal signal peptides (Rinaldi et al. 2022). A lack of eukaryote-specific post-translational modifications is responsible for less optimal catalytic activities or different product profiles of TSs, which might impact the final terpenoid titers. Thus, enhancing protein solubility in bacteria is particularly important for TPSs heterologous overexpression, especially after the precursor supply optimization. Otherwise, the strong precursor supply may result in the accumulation of toxic intermediates like IPP and FPP before efficiently uptaken by TPSs. In contrast, bacterial enzymes that are more similar to *E. coli* proteins would more likely be highly soluble, efficiently expressed, and catalytically active. For example, bacterial *Streptomyces clavuligerus* originated 1,8-cineole synthase (CinS) produces 23.4 mg/L 1,8-cineole, comparable to that of plant *Salvia fruticosa* CinS (23.6 mg/L) (Karuppiah et al. 2017).

While few TSs are clearly investigated, and TSs with desired properties are difficult to predict, they share structural similarities. First, many TSs possess the aspartic acid-rich structural motif DDXXD, locating at the entrance of the catalytic site, and use metal ions like Mg^{2+} as activators (Jiang and Wang 2023). Second, plant TSs usually use N-terminal signal peptides, rich in serine and threonine residues and less acidic amino acids, to direct the translocation of initial peptides (Gagat et al. 2013). Third, the N-terminal domain contains functional elements that usually act as a scaffold to facilitate the maturation of the catalytically active C-terminal domain (Ren et al. 2020). These common features could help us to rationally engineer the TS toward desired properties or to broaden the products spectrum. In addition, plant TSs show certain substrate promiscuity, which can be redressed for exploring new terpenoid products. As in the case of *Fusarium graminearum*, from which two clade III TSs FgJ07578 and FgJ07623 are identified. FgJ07578 is a chimeric protein consisting of the N-terminal terpene cyclase (TC) domain and a C-terminal prenyltransferase (PT) domain, while FgJ07623 contains only a TC domain. When they are combined with FPPS, GGPPS, and geranylfarnesyl pyrophosphate synthase (GFPPS) to form 6 different combinations, 50 terpenoids were produced, including 9 sesterterpenoids, 24 diterpenoids, and 17 sesquiterpenoids (Bian et al. 2017). The results above invite us to further explore the natural TS resource from the microbes.

9.5.3 Bottlenecks in the Terpenoid Biosynthesis Route

Microbial terpenes biosynthesis is highly regulated at genetic and metabolic levels. Of course, the MEP and MVA pathways are not always efficient in supporting the production of target chemicals. For the MEP pathway, key enzymes to be targeted include DXS, DXR, MCT, MDS, HDS, HDR, and IDI. In the MVA pathway, gate-keeping enzymes include acetyl-CoA C-acetyl transferase (ACCT), 3-hydroxy-3-methylglutaryl coenzyme A (HMG-CoA) synthase (HMGS), 3-hydroxy-3-methylglutaryl-CoA reductase (HMGR), mevalonate kinase (MK), mevalonate-5-phosphate kinase (MPK), and IDI (Ren et al. 2020). Nonetheless, the restrictive points of the metabolic flux might differ in a specific strain. Generally, in almost all the studies using the MVA pathway for isoprenoids production, the rate-limiting HMGR will be manipulated in three ways. The most common practice is removing the transmembrane domain to ameliorate its solubility. Second is to utilize strong constitutive promoters or high copy plasmids to improve the expression. Third is introducing other sourced HMGRs to strengthen the overall metabolic flux in the MVA path. When the MEP pathway is deployed, the overexpression of DXS, DXR, and HDR is a very common practice (Fig. 9.4c).

To be noted, the MEP pathway generally exhibits poor catalytic efficiency in the original host because the underlying enzyme functions and regulation mechanisms have not been thoroughly studied. As a result, even when coupled with reconstructed exogenous yeast MVA pathway, the MEP pathway can only provide limited amounts of IPP and DMAPP (Rinaldi et al. 2022). The highest titers reported in *E. coli* for most terpenoids production—especially the hemi-, mono-, and sesquiterpenoids—are obtained through heterologous rewiring of the MVA pathway (Table 9.3). GPPS/NPPS, FPPS, GGPPS, and TS are also regarded as knot-controlling toward efficient terpenoid biosynthesis. Specifically, GPPS/NPPS are critical for monoterpene, FPPS for sesquiterpene, and GGPPS for diterpene (Rinaldi et al. 2022).

In addition, regulating the allocation of metabolic flux is also critical for the biosynthesis of terpenes (Fig. 9.4c). For example, the conversion of acetyl-CoA to terpenes might interfere with internal metabolism like lipid biosynthesis and retard cell growth and terpene yield. The FPP is indispensible for biosynthesis of ergosterol and related chemicals, which are essential for maintaining cell health. In the meantime, FPP is also the precursor for the production of sesquiterpenes. Consequently, the distribution of acetyl-CoA and isopentenyl pyrophosphate is crucial for enabling efficient biosynthesis of terpenes and must be carefully fine-tuned.

9.5.4 The Supply of Precursors

The MVA pathway is more often employed than the MEP pathway for terpenoids production. To enhance the metabolic flux of the MVA pathway, it is necessary to strengthen the supply of cytosolic acetyl-CoA (Fig. 9.4d). There are at least 13 pathways reported for generating acetyl-CoA. There are 11 pathways using the Embden–Meyerhof–Parnas (EMP) pathway to produce redox cofactors and gain or lose ATP. The others function via the phosphoketolase (PK) will consume ATP; therefore, the PK-based ones must be combined with one of the 11 EMP-related pathways. Balance of these pathways can maximize the acetyl-CoA yield, thereby minimizing the overall cost of subsequent acetyl-CoA-derived molecules. For more details, please refer to the review by Zhu et al. (2022a).

When the MVA pathway is enhanced, the toxic IPP and its isomer DMAPP should be transformed into the final product. Hence, it is obligatory to redirect the carbon flux toward GPP or NPP for the high-level microbial production of C10, FPP toward C15, and GGPP to C20 (Fig. 9.4c). For C15 and C20 terpenes, the strategies are always screening the favorable FPPS and GGPPS. However, most microbial hosts do not have the specific GPP synthase (GPPS) or NPP synthase (NPPS) (Ren et al. 2020). Strategies for boosting GPP levels toward monoterpene biosynthesis are as follows: (1) engineering ERG20 (the FPPS in *S. cerevisiae*) to generate mutants like $ERG20^{K197E}$, $ERG20^{K197G}$, and $ERG20^{F96W\text{-}N127W}$, which have higher GPPS and minimal FPPS activity (Fig. 9.4b); (2) fusion of the ERG20 mutants with TS to reduce GPP/NPP flux leakage and facilitate mass transfer; (3) introduction of efficient FPPS mutant ($mFPS^{N144W}$) of heterologous sources; (4) dynamic regulation of the FPPS degradation; and (5) introducing a specific GPPS like the one from *Abies grandis* to avoid the generation of FPP (Ren et al. 2020).

When an NPP-based orthogonal biosynthetic pathway for terpene is established, and the feasible strategies for enhancing the target terpene biofuels are (1) screening enzymes with higher activity toward NPP, (2) subcellular compartmentalization, (3) protein fusion, and (4) protein engineering the catalytic efficiency and specificity of MTS toward NPP (Fig. 9.4).

9.5.5 Manipulation of the Cofactor NAD(P)H Balance

The introduction of terpene synthesis pathway usually leads to an imbalance in the NAD(P)H pool since it consumes ATP and NADPH to propel the biosynthesis process (Fig. 9.3). As ATP is efficiently produced in the central carbon metabolism, the generation of NAD(P)H should be finely manipulated to drive the metabolic flux toward the target biofuel product. There are three major sources that can be explored for NADPH generation: the PP pathway, the TCA cycle, and the NADH pool (Fig. 9.4c). Accordingly, the metabolic strategies for elevating NADPH supply are overexpression genes of the PP pathway such as Zwf1, tktA, and talB; genes of the TCA modules sucAB, gltA, and sdhABCD; and introduction and overexpression of efficient NADH kinase, which catalyzes the conversion of NADH to NADPH. In *E. coli,* the proton-translocating transhydrogenase PntAB exerts the same effect as NADH kinase (Rinaldi et al. 2022).

Importantly, the use of NADH-consuming enzymes should also be taken seriously to alleviate the high dependence upon NADPH and the resulting disturbance in cofactor pools. For example, the original HMGR in yeast always uses NADPH as the cofactor (Fig. 9.3), while the *Silicibacter pomeroyi* originated HMGR is NADH-dependent (Meadows et al. 2016). In addition, inactivating the rival pathway that consumes NADPH can also aid in the accumulation of NADPH. Take fatty acid biosynthesis as an example; it is a heavily ATP and NADPH demanding process. Tuning down the fatty acid biosynthesis while maintaining the basic cell growth should facilitate the redirection of cofactors toward terpene biofuel production.

9.5.6 Blocking the Competitive Pathway

In microorganisms, the isoprenoid synthetic pathway is related to many physiological pathways for cell growth, leading to many bypass and competitive pathways for target isoprenoid production. It is necessary to redirect the carbon flux from the bypass pathway and the competitive pathway toward improved isoprenoid production. The strategies often adopted for increasing target isoprenoid production via blocking the competitive pathway include deletion, downregulation, or dynamic control of the expression of flux-competing enzymes and utilization of enzyme inhibitors (Fig. 9.4e).

Deletion of the competing enzyme is a direct operation usually performed. In the case of geraniol production, more flux can be redirected to geraniol production by deleting yjgB in *E. coli*, and OYE2 and ATF1 in *S. cerevisiae*, which encode enzymes for geraniol degradation (Zhou et al. 2014). However, when the enzyme's deletion is lethal, downregulation strategies such as using weaker promoters, RNA interference, protein destabilization, and inhibitors on flux-competing enzymes are more feasible.

Take ERG9 (squalene synthase in *S. cerevisiae*) for example; it catalyzes the biosynthesis of squalene, which will be converted to an essential cell membrane component ergosterol and other molecules. Strategies applied should achieve a delicate balance between production of target terpenoids and squalene. When ERG9 involves, several approaches can be selected (Jiang and Wang 2023; Ren et al. 2020). First, replace the native promoter of ERG 9 with a weak promoter like pBTS1. Second, dynamic control of the bypass pathway with a glucose responding promoter like pHXT1. Third, control ERG9 degradation with the PEST peptide tag-activated endoplasmic reticulum-associated protein degrading machinery or N-degron-dependent protein degradation strategy. Notably, the protein degradation system requires no inducers, repressors, or specific conditions. Fourth but not last, utilize inhibitor against flux-competing enzymes. By using terbinafine, an ERG1 inhibitor, the content of ergosterol was kept at a low level and thereby reduced the loss of carbon from squalene.

9.5.7 Cytotoxicity Engineering

Despite all the efforts, the production of a certain terpenoid seems to reach a limit in final titer, whose further improvement might be hindered by the cytotoxicity. Particularly, the monoterpenes are highly toxic to microbial hosts. In addition, the intermediates IPP, FPP, MVA-5-P, and MVA-5-PP are also harmful to the microbial host. Thus, cytotoxicity engineering is expected to promote production in the future. In this part, we will talk about the toxicity mechanism and the relevant handling approaches (Fig. 9.4f).

9.5.7.1 Mechanisms of Toxicity

Currently, the toxicity of terpenes can be hard to predict and rather is assessed empirically. Though the cytotoxicity of terpene varies among the diverse research, several mechanisms have been proposed in recent years, including cell apoptosis triggering, DNA damage, protein prenylation inhibition, membrane disruption, cell wall integrity damage, and phase toxicity (Ren et al. 2020).

Terpenoids are hydrophobic chemicals that can accumulate in membranes system and increase membrane permeability, thus disrupting vital biological stability by allowing proton leakage and dissipating the proton motive force needed for energy production (Sikkema et al. 1994, 1995). A commonly used metric for hydrophobicity is the logarithm of the partition coefficient between octanol and water (log *P*). Hydrocarbons with log *P* between 1 and 4 partition easily into membranes and therefore can be toxic; above log *P* 4 water solubility also determines toxicity (Sikkema et al. 1995). The toxicity can also be affected by structural features. In general, hydroxyl groups and double bonds can increase toxicity. For example, *E. coli* grows as well with 25.23 g/L exogenously added limonene as a non-supplemented control, but oxidized forms of limonene such as terpineol and perillyl alcohol inhibit growth almost completely at concentrations of 4.67 and 4.79 g/L, respectively (Vermaas et al. 2018). Also, limonene is oxidized in vivo to a hydroperoxide that might cause oxidative damage, as evidenced by the finding that an endogenous alkyl hydroperoxidase AhpC (L177Q) variant alleviates toxicity (Chubukov et al. 2015). This additional toxicity might be due to DNA damage. RecA is a protein that activates SOS box genes for DNA repair and might confer resistance to geraniol as *Δ*recA strains are more geraniol sensitive than wild-type counterparts (Shah et al. 2013b).

9.5.7.2 Handling Approaches Against the Terpenoid Toxicity

Terpenoid production with microbes needs to address arguably the last step, i.e., removal of the end product from the cell, and even from the media, to handle end-product toxicity, to engineer subcellular organelle for synthesis products, and to achieve efficient product recovery and purification (Fig. 9.4f, g). Existing strategies to reduce the cytotoxicity of terpenoids include physical extraction, efflux engineering, compartment engineering, application of co-culture systems, in vivo derivation to lessen the toxicity, adaptive laboratory evolution, and cell-free systems. Of all the approaches, physical extraction and efflux engineering are always coupled to alleviate terpene toxicity, while adaptive laboratory evolution is widely used to strengthen the tolerance of host cells against the cytotoxicity of target product (Fig. 9.4f)

1. ***Physical Extraction***

   Terpenoid biofuels typically possess significant volatility and hydrophobic properties, upon which in situ product recovery (ISPR) depends, to mitigate cell exposure to the target product and its cytotoxic effect. Biocompatible organic solvents, condensers, and gas-stripping are widely used to recover volatile terpenoids from the medium and alleviate the evaporation loss during microbial production. In this regard, various solvents, including diisononyl phthalate, dibutyl phthalate, poly-α-olefin, and dodecane, have been used due to their biocompatibility, distribution coefficient, phase separation processes, and cost (Ren et al. 2020; Rinaldi et al. 2022). Organic overlays also help handle the toxicity of less volatile terpenoids, perhaps by displacing the equilibrium from the media and therefore reducing exposure of cells to the end products. Gas-stripping is more applicable to gaseous and volatile terpenoid products like isoprene. By using polyalpha-olefin as an overlay, 60 g/L isoprene was obtained with engineered *E. coli* (Whited et al. 2010).

2. ***Efflux Engineering***

   Efflux pumps have been applied to improve the host tolerance against several isoprenoid biofuels, such as limonene, geraniol, pinene, etc. (Fig. 9.4f) (Jiang and Wang 2023). The best studied efflux pump is multiprotein AcrAB-TolC. AcrAB-TolC is composed of AcrB, an ATP-binding cassette (ABC) transporter in the inner membrane, TolC, an outer membrane channel, and AcrA, an accessory protein in the periplasmic space that mediates interaction between AcrB and TolC to export the substrate through the two membranes. This is a native pump of *E. coli* that can be used to increase terpenoid tolerance and production level. The efflux efficiency toward monoterpenes can be enhanced through either directed evolution of AcrB (Rau et al. 2016) or fine-tuning its expression level in a direct or indirect way (Shah et al. 2013a, b). Indirectly, by overexpression of positive transcriptional factor MarA of AcrAB-TolC, the recombinant *E. coli* can grow on 8.9 g/L geraniol, which would otherwise inhibit growth completely (Jiang and Wang 2023). The increased resistance in *E. coli* might be explained by the reduction of the intracellular terpenoid concentrations from 12.9 to 6.3 μg/mg (Shah et al. 2013a).

   Functional expression of heterologous ABC transporters with high transport efficiency could also improve strains tolerance toward terpenes (Demissie et al. 2019). However, the benefits from the efflux engineering strategy can be intrinsically limited. On the one hand, the effects of pump engineering so far are limited, providing about fourfold titer increase at most. On the other hand, monoterpenes diffuse quickly across the bilayer cell membrane. If the terpene generation rate exceeds the pumps capacity, intracellular accumulation of terpenoid will also cause cytotoxicity (Hu et al. 2012). In addition, overexpression of efflux pumps could cause great changes in membrane composition and thus result in dysfunction of the cell membranes. Thus, the trade-off between biofuel toxicity and pump toxicity should be taken seriously (Turner and Dunlop 2015).

3. ***Compartment Engineering***

   With improvements in the isoprenoid production level, more research should be focused on intermediates cytotoxicity. The cellular accumulation of toxic intermediates is also observed when the heterogeneous pathway is intensified. MVA-5P, MVA-5PP, IPP, and FPP are proven to impair cell growth and influence product synthesis (Rinaldi et al. 2022). Unlike the efflux engineering strategy for the target isoprenoid, compartment engineering can be feasible in circumventing the toxicity of intermediates during terpene biofuels biosynthesis (Fig. 9.4g). The alternative IPP/DMAPP-bypass, which directly produces isopentenol from mevalonate or mevalonate monophosphate, can also be used to avoid the accumulation of IPP and DMAPP (Fig. 9.3). When the IPP/DMAPP-bypass is combined with subcellular compartmentalization strategy, the influence of the toxic intermediates on the cell metabolism should be further relieved. Importantly, metabolic pathways enclosed inside of a certain subcellular organelle would provide a subtle micro-environment, where enzymes, metabolites, and cofactors are enriched to enable higher reaction efficiency, and competing pathways are prevented (Fig. 9.4g).

4. ***Cell-Free System***

   The cell-free system that mixes the required enzymes in a reaction system can circumvent the toxicity of products and intermediates against chassis cells and avoid the comparative pathway in the host and facilitate the target product recovery (Korman et al. 2017). A cell-free system with 27 enzymes was designed for monoterpene (limonene, pinene, and sabinene) production. The rationally designed system can operate continuously for more than 5 days from a single addition of glucose. The final monoterpenes titers were 12.5, 14.9, and 15.9 g/L for limonene, pinene, and sabinene, respectively, and theoretical yields from glucose were 88.3%, 103.9%, and 94.5% , respectively. These results highlight the potential of cell-free approaches for producing terpenoid-derived drop-in biofuels.

9.5.8 Modulation of Culture Conditions

Although microbes can tolerate and survive a wide range of physicochemical parameters, such as osmotic pressure, pH, temperature, agitation, aeration, and so on, the culture conditions need to be optimized in order to achieve desirable titers of target terpenoids after genetic modification (Shukla and Phulara 2021). The physicochemical parameters, together with medium compositions, such as carbon and nitrogen source usage and concentrations, inducers, and cofactor supply, not only play a crucial role in the high-yield production of target terpenoids from an engineered host, but also reduce the generation of unwanted metabolites during fermentation and thus simplify product recovery. Furthermore, for the economical production of terpenoids, the low-cost lignocellulosic biomass can be explored together with the optimized culture conditions (Fig. 9.4h).

9.6 Microbial Production of Isoprenoids Fuels

9.6.1 Hemiterpene (C5) Biofuels

Hemiterpenoids include isoprene, isopentenols (prenol and isoprenol), and other chemicals, which can be used as a platform chemical for a variety of products, such as rubbers, adhesives, elastomers, pesticides, medicines, fragrances, and biofuels, among others.

9.6.1.1 Isoprene

Isoprene (C5H8) is the simplest isoprenoid, which is used to produce millions of tons of rubber annually and has been suggested as an alternative fuel (Rabinovitch-Deere et al. 2013). There is an increasing need for sustainable isoprene production at industrial scale. Metabolic engineering of microbial systems for isoprene production might be helpful to address this issue. The max titer (60 g/L) of isoprene was achieved with *E. coli* via the following manipulation: (1) plasmid-based introduction of the upper MVA pathway from *Enterococcus faecalis*, consisting of two genes (*mvaE* and *mvaS*), in concert with the *Populus alba* isoprene synthase, and (2) expression of the mevalonate kinase and phosphogluconolactonase gene and integration of the lower pathway of MVA pathway from *S. cerevisiae*. The final volumetric and cell productivity were 2 g/L/h and 0.85 g isoprene/g dry cell weight with a yield of 11% from glucose in 14-L fed-batch fermentation (Whited et al. 2010). By synergistic dual pathway of MVA and MEP, the isoprene yield was further enhanced to 0.267 g/g while the resulting production was 24.0 g/L with *E. coli* (Yang et al. 2016). Similar efforts have also been conducted with other microbial hosts like *P. ananatis*, *S. elongatus*, and *S. cerevisiae* (Table 9.3). The production of isoprene has been elevated to 11.9 g/L with a diploid *S. cerevisiae* strain via compartmentation of pathway and protein engineering. By using a mutant isoprene synthase ISPSLN, fortifying the supply of precursors in the cytoplasm and mitochondria, and redeploying the diphosphomevalonate decarboxylase (MVD1) and ID1, the results finally established *S. cerevisiae* as a potent isoprene producer (Yao et al. 2018).

9.6.1.2 Isoprenoid Alcohols

The production of isoprenoid alcohols has also been explored with microbes like *E. coli* and *S. cerevisiae* (Table 9.3). The precursor toward isoprenoid alcohols is generally insufficient in the target chassis, and these alcohols usually pose toxicity to the host. To improve the production of these alcohols, channeling the metabolic flux, engineering the cofactor supply, introducing the tolerance-enhancing genes, such as metR and mdlB, utilizing the organic overlay, and deriving the alcohols to less toxic esters proved effective. Usually, a cocktail of strategies should be utilized to achieve a remarkable progress. For example, by using the CRISPR/dCas9 interference (CRISPRi), genes involved in the phosphate intermediate hydrolysis (aphA, yieH, and yggV), carbon flux redirection (accA), and isoprenol tolerance (astE) pathways were identified. The combinatorial repression of yggV and accA permitted isoprenol titer of 3.63 g/L in a shake flask using batch cultivation, with a yield of 57% of the theoretical maximal pathway yield. Further, in the shake flask isoprenol titer was increased to 8.51 g/L, during which the isoprenol yield within the stationary growth phase could reach 85% of the theoretical maximal pathway yield (Wang et al. 2022a). Derivation of isoprenol to isoprenyl acetate further increased the product titer to 28.0 g/L (Carruthers et al. 2023).

9.6.2 *Monoterpene (C10) Biofuels*

Monoterpenoids are a plethora of C10 compounds derived from canonical precursor geranyl diphosphate (GPP) or neryl diphosphate (NPP), which are formed by condensing IPP and DMAPP under the catalyzation of GPP/NPP synthase (GPPS/NPPS). Monoterpenoids can be divided into three major subgroups based on their structural features: (1) acyclic monoterpenes, such as myrcene and ocimene; (2) monocyclic monoterpenes, such as limonene, menthol, and carvone; and (3) bicyclic monoterpenes, such as pinene, sabinene, and camphor. The application of monoterpenoids is expanded from traditional utilization as flavors, antibacterials, and insecticides, to high-density drop-in fuels. Particularly, monoterpenes like a-pinene, camphene, and limonene have been foreseen as raw material for the development of jet fuels (Lapuerta et al. 2023).

To date, microbial production of monoterpenes is less efficient. One of the factors contributing may be the catalysis sequence of the isoprenoid biosynthesis pathway in most organisms. A key feature is the sequential nature of FPPS, which is a bifunctional enzyme that catalyzes two successive steps of the pathway. Initially, it condenses DMAPP and IPP to form GPP, and subsequently it catalyzes the condensation of GPP with one more IPP molecule to generate FPP that is essential for FPP-derived molecules (Fig. 9.3). This facilitates the metabolic economy and cell viability, as microbes do not naturally produce other PP-derived compounds. However, the sequential reaction appears to limit the available GPP pool for monoterpenoid production. Overexpression of a GPP synthase is possible to increase GPP availability but usually shows only a moderate monoterpene increase as the increased GPP flux would be efficiently channeled by the endogenous FPPS to produce sterols and other essential isoprenoids. Further efforts concerning FPPS engineering are to downplay its FPP-synthesizing capacity and elevate the GPP preference without abolishing sterol synthesis.

9.6.2.1 Limonene

Limonene is derived from the direct precursor GPP/NPP. Microbes, including bacteria, yeasts, and molds, have been used for the biosynthesis of limonene. To improve limonene production, strategies include establishing orthogonal biosynthetic pathways, mitochondrial and peroxisomal compartmentalization, protein engineering like truncation, degradation, and fusion, site-directed mutation, and a combination have been adopted (Table 9.3).

In the context of limonene production using the bacterium *E. coli*, the most successful approaches often involve the optimization of both the heterologous MVA pathway and the orthogonal NPP pool. Site-directed mutation of key enzymes of the MVA pathway, for example the *Enterococcus faecalis* $HMGS^{A110G}$, and engineering the RBS of enzymes in the upstream MVA pathway led to successful limonene production with *E. coli*. The resultant limonene titer was 1.29 g/L by using fed-batch bioreactor (Wu et al. 2019).

Similarly, the NPP-based limonene synthesis has also been adopted in *S. cerevisiae* since the NPP would not be consumed by FPPS but efficiently converted by limonene synthases. Of course, to strengthen the limonene production, metabolic engineering strategies are always used in combination. Recently, there are two typical examples. One is the combinatorial manipulation of limonene synthesis in peroxisomes and cytosol, which obtained a titer of 2.58 g/L, the second highest with *S. cerevisiae* (Dusseaux et al. 2020). The other is the dual regulation of cytoplasmic and mitochondrial engineering for limonene production. First, the authors established an NPP-based limonene biosynthesis pathway by introducing an N-terminal truncated limonene synthesis from *Mentha spicata* and optimizing the expression of the rate-limiting tHMG1, $ERG20^{ww}$, and IDI1 in the MVA pathway. Then, by dynamic inhibition of ERG20-regulated competitive bypass and optimization of the copy number of tLimS, the titer of limonene was improved to 640.9 mg/L. Subsequently, they enhanced the limonene titer to 1097.4 mg/L by elevating acetyl-CoA and NADPH supply. Finally, dual regulation of the limonene synthesis in the mitochondria and cytosol resulted in 1586 mg/L in flask, and 2.63 g/L in a 3 L fermenter, the highest with *S. cerevisiae* (Kong et al. 2023). Two mechanisms may be responsible for the observed increase. One is the abundant availability of acetyl-CoA in mitochondria, which pushed the flux into the MVA pathway. The other is that metabolites derived from the mitochondrial MVA pathway may be more efficiently directed toward limonene synthesis. Especially, when ERG20 is relocated into the mitochondria, the FPP produced cannot be consumed to synthesize ergosterol, thus reducing the diversion of GPP.

Optimization of the fermentation conditions is vital for improving the limonene production titer. By optimizing the carbon source, inducer concentration, 368 mg/L limonene was produced by an engineered *E. coli* strain carrying an MVA pathway. The final titer was further improved to 3.6 g/L by employing the carbon-limited exponential feed strategy to prolong cells' exponential growth phase, supplying additional trace elements and using an inert organic phase (DINP) in a bioreactor (Rolf et al. 2020). This is the highest titer reported for microbial production of limonene.

The results above demonstrate that systematic rearrangement of the host metabolism and optimization of the fermentation system are essential and effective for promoting limonene biosynthesis. There are also interesting but not so successful examples for microbial limonene production with other hosts, such as *Yarrowia lipolytica, Rhodotorula toruloides*, and cyanobacteria (Ren et al. 2020; Rinaldi et al. 2022).

9.6.2.2 Linalool

Linalool is an acyclic monoterpenoid, existing as a pair of enantiomers, (S)-linalool and (R)-linalool, in many plant-derived essential oils (Rinaldi et al. 2022). Engineering microbes to produce linalool is an emerging alternative to plant extraction or chemical synthesis. Biosynthesis of linalool is specifically catalyzed by a linalool synthase (LIS) from GPP. The low catalytic activity of LIS and the shortage of precursor pools are considered the key factors for low titers in microbes. To improve the microbial production of linalool, the available options fall into the following: screening efficient LIS and GPPS, LIS truncation, HMGR truncation and overexpression, engineering the selectivity of FPPS, subcellular compartmentalization, and carbon flux redistribution. Currently, microbial production of linalool has been achieved with *E. coli*, *S. cerevisiae*, *P. ananatis*, etc.

In *E. coli*, strategies such as expressing GPPS from *Abies grandis* (*Ag*GPPS) and screening LIS from different resources, modifying ribosome-binding site (RBS) sequences, using fusion tags and peptidyl ligands GBD, SH3, and PDZ, and controlling carbon partition have been applied. Among which the use of peptidyl

ligands GBD, SH3, and PDZ and the manipulation of carbon flux redirection are demonstrated to be useful. LIS has been considered with lowest catalytic activity. By attaching GBD, SH3, and PDZ to the C-termini of EcIDI, AgGPPS, and bLIS, respectively, and manipulating the ratios of the scaffolds associated with each enzyme, linalool titer was improved to 187 mg/L, which was further pushed to 278 mg/L after optimizing the culturing conditions, such as adjusting induction temperature, inducer concentration, and carbon source concentration in the flasks. The final (R)-linalool titer was 1523 mg/L when the culture was scaled up to bioreactor level (Wu et al. 2021).

As in the case of dual manipulation of the linalool biosynthesis in the cytoplasm and mitochondria, though $ERG20^{F96W\text{-}N127W}$ was also introduced, the engineered *S. cerevisiae* strain gave merely 23 mg/L linalool under optimized carbon source (Zhang et al. 2020). Fortunately, however, the study presented by (Zhou et al. 2023b) shed light on the efficient bioproduction of linalool with *S. cerevisiae*. By rationally engineering the entrance of the substrate-binding pocket of linalool synthase ($t67OMcLIS_M$), the catalytic efficiency of this enzyme toward GPP was substantially improved. Specifically, the mutants $t67OMcLIS^{F447E}$ and $t67OMcLIS^{F447A}$, with decreased entrance hydrophobicity and steric hindrance, increased linalool production by 2.2 and 1.9 folds, respectively. Subsequently, cytoplasm and peroxisomes were harnessed to boost linalool production to a titer of 219.1 mg/L in shake-flask cultivation. Finally, the engineered diploid strain affected 2.6 g/L linalool in a 5 L bioreactor under a fed-batch fermentation.

A final optimization strategy is to control carbon partition between cell growth and linalool synthesis. This can be achieved by regulating flux distribution at the acetyl-CoA branch point, as this is a key intermediate participating in both the tricarboxylic acid (TCA) cycle (critical for cell growth) and the MVA pathway, which supplies the flux for isoprenoid synthesis. When the MVA pathway draws carbon flux for linalool production, it is important to prevent starvation of competing growth-critical pathways as well as optimally balance metabolite and energy demands for both functions.

By using a phosphate-sensing (PhoR/PhoB) system from *E. coli*, Hoshino et al. (2020) succeeded in separating the cell growth phase of *Pantoea ananatis* from the subsequent product-formation phase. With phosphate starvation-induced promoters, when phosphorylated isoprenoid intermediates were depleted, more carbon flux was directed at the acetyl-CoA node toward the MVA pathway in the engineered strain. This strategy was combined with screening for better LISs to construct superior linalool-producing strains. Synthases from *A. arguta* (AaLINS) and *Streptomyces clavuligerus* (bLIS) were selected for (S)-linalool and (R)-linalool biosynthesis, respectively. Co-expression of an ispA mutant from *E. coli* together with either of these two synthases resulted in 5.60 g/L (S)-linalool or 3.71 g/L (R)-linalool in test tube culture. According to the most recent report, the S-linalool titer was boosted to 10.9 g/L after a serial genetic operation with *P. ananatis*. First, the soluble expression level of AaLINS was increased by the N-terminal fusion of a halophilic β-lactamase from *Chromohalobacter* sp. Second, the precursor supply was elevated via recruiting the mevalonate pathway. Third, an aqueous-organic biphasic fermentation system was used with isopropyl myristate as the organic solvent for in situ extraction of cytotoxic and semi-volatile (S)-linalool. Fourth, "dual-phase" fed-batch fermentation was used (Nitta et al. 2021).

9.6.2.3 1,8-Cineole

The 1,8-cineole is another monoterpene that is a potential biofuel candidate; there are fewer reports on its microbial production when compared with limonene and linalool. Additionally, microbial production has been achieved in *E. coli*, *R. toruloides*, and cyanobacteria. By introducing an orthogonal MVA-based routes and engineering the FPPS (ispA) selectivity toward GPP, the obtained titer was only 653 mg/L with *E. coli* (Mendez-Perez et al. 2017). The biosynthesis routes were also established for 1,8-cineole in the oleaginous and carotenogenic *R. toruloides*. Recently, the titer has been improved to 1.4 g/L by overexpressing GPPS and MVA pathway genes (Kirby et al. 2021).

9.6.2.4 Pinene

Pinene is a bicyclic monoterpene with two structural isomers, α-pinene and β-pinene, which have attractive biological properties such as antiproliferative or anticarcinogenic potential and promising physicochemical properties for a candidate as jet fuel precursor (Raman et al. 2016).

Microbial production of this terpene has been tried in *E. coli, C. glycerinogenes*, and *S. cerevisiae*. However, the results are not unsatisfactory (Table 9.3), which might be owing to two factors. One is the high toxicity of the pinene to the host. The other is the poor catalytic activities of the pinene synthase. To improve the production of α-pinene, initial efforts focused on the tolerance enhancement of the producing organism. An evolved *E. coli* strain was obtained after adaptive laboratory evolution and the introduction of efflux pumps. Then, an *E. coli*–*E. coli* co-culture system was developed to balance the upstream MVA pathway and downstream pinene biosynthetic pathway. Metabolic burden caused by excessive protein expression could thus be alleviated in the co-culture system, and modular optimization could be achieved by fine-tuning the co-culture composition. A combination of these approaches led to the production of 167 mg/L α-pinene (Niu et al. 2018). Interestingly, a much higher pinene titer was achieved with a cell-free system by using crude lysates of *E. coli*. Optimization of the experimental parameters including medium composition, cofactor levels, and enzyme dosages resulted in the de novo production of 1.26 g/L pinene from glucose (Niu et al. 2020).

9.6.2.5 Geraniol

Geraniol is an acyclic monoterpenoid that is found in different plant essential oils. The rose-like scent made it a desirable ingredient in fragrances and cosmetics. Geraniol also has validated pharmacological properties, such as inhibiting tumor cell growth and possessing a cardio-protective effect (Jiang and Wang 2023; Rinaldi et al. 2022; Silva et al. 2022). For sustainable production of this alcohol, microbes including *S. cerevisiae*, *E. coli*, *Candida glycerinogenes*, *Methanococcus maripaludis*, and *Azospirillum brasilense* have been engineered. However, only a few of the reported production are at the g/L scale (Table 9.3).

As for limonene and linalool, *S. cerevisiae* and *E. coli* are the most investigated for geraniol biosynthesis. Strategies like selection of key enzymes (GPPS and GES), introduction of heterologous pathway and key enzymes, protein engineering, and bioderivation have been adopted to enhance geraniol production (Liu et al. 2021). Notably, the compartmentalized production of this alcohol in mitochondria and peroxisome has been proved effective, but limited in improving geraniol titer with *S. cerevisiae*. Though a sixfold improvement was observed in geraniol production when all enzymes involved in geraniol biosynthesis were targeted to mitochondria, the final titer was merely 43 mg/L (Yee et al. 2019). Likewise, peroxisomal relocation of GPPS and geraniol synthase (GES) could also be of limited benefits to geraniol production (Gerke et al. 2020).

Fusion of short leader peptide construct to the key enzymes like GES has proven to be preferable in enhancing the expression level. Additionally, the enhanced green fluorescence protein (eGFP)-based high-throughput screening is very helpful in identifying efficiently expressed GES with *E. coli*. After stepwise metabolic engineering, the best-performed strain, carrying GES with mutant leader peptide, produced 2.12 g/L geraniol in a shake-flask culture (Wang et al. 2021a). To avoid the internal interference, the prenol-based orthogonal route has been used to convert 3 g/L prenol into 2 g/L geranoids (geraniol and isomers/derivatives). Interestingly, the upper stream for prenol biosynthesis was linked to acetyl-CoA. The complete pathway produced 600 mg/L geranoids from 40 g/L glycerol in *E. coli* (Clomburg et al. 2019).

Despite all the efforts mentioned above toward sustainable microbial production of geraniol, the current yield and productivity of microbial systems are far below theoretical values. The toxicity of geraniol might account for part of the results. Bioderivatization provides a feasible strategy to alleviate the cytotoxicity of geraniol. By conversing geraniol to less toxic geranyl acetate, productivity can be significantly enhanced as in the following case. By combining systematic engineering, in vivo bioderivatization, geranyl acetate hydrolysis, and solid-phase adsorption, the titer of geraniol was finally elevated to 13.19 g/L, the highest ever reported in microbial production of monoterpenes, by fed-batch fermentation in a 10-L fermentor (Wang et al. 2022b).

9.6.2.6 Nerol

Nerol is an acyclic monoterpenoid alcohol that is widely used in cosmetics, agricultural, or pharmaceutical industries (Jiang and Wang 2023). The global demand for nerol is more than 5000 tons per year, but the annual production capacity worldwide is about 3000 tons, supplied mainly by the high-cost and eco-unfriendly chemical route. As an isomer of geraniol, it can be biosynthesized from GPP or NPP under the catalyzation of nerol synthase (NS), although the nonconventional *M. guilliermondii* was reported to produce nerol with a titer of 2.7 mg/L (Mo et al. 2021).

The first attempt to synthesize nerol from glucose in *E. coli* was implemented with NPP as the direct precursor. Co-expression of NS and truncated NPP synthase (tNPPS) led to 0.053 mg/L nerol. Overexpression of genes in the orthogonal MVA pathway and ERG10 enhanced carbon flux toward nerol by 30-fold in shake flask to

1.6 mg/L (Zong et al. 2019). To improve nerol production, metabolic and protein engineering strategies were coupled to enhance de novo nerol biosynthesis in *E. coli*. The strain harboring tNPPS and NS from *Glycine max* (GmNES) produced 55 mg/L nerol. When GmNES was replaced by an endogenous Nudix hydrolase (NudJ, a phosphatase for enhancing dephosphorylation of NPP), the nerol titer was improved to 262 mg/L. Optimization of induction conditions and initial glucose concentration further boosted the alcohol to 967 mg/L in shake flask, the highest nerol titer ever reported in microbes (Lei et al. 2021).

9.6.2.7 Sabinene

Sabinene is an important naturally occurring bicyclic monoterpene that can be used as flavorings, perfume additives, advanced biofuels, and so on. Up to now, this valuable terpene is commercially unavailable since there is no applicable manufacturing process (Cao et al. 2018). Microbial synthesis can be a promising route for its production. In fact, microorganisms including fungi, bacteria, and yeast have been investigated for sabinene production. The highest sabinene production titer via cell factory is 2.65 g/L (Zhang et al. 2014), a very remarkable result, considering the other efforts usually lead to titers below gram per liter. Additionally, cell-free production of sabinene was achieved with an encouraging titer of 15.9 g/L, and 94.5% of the theoretical yield (Korman et al. 2017).

Further improvement of the biosynthesis of sabinene still faces the following challenges. One is the lack of a specific and efficient sabinene synthase, the most important determinant for the purity of the final product and downstream processing input. The other is the cytotoxicity of sabinene against the microbial host. Efflux engineering may increase the product tolerance of the host strain, and biocompatible organic overlay may be applied to facilitate the alleviation of toxicity further by instant extraction of the resultant sabinene.

9.6.2.8 Myrcene

Myrcene is a promising block for producing the renewable diesel blendstock camphorane, hydrogenated camphorene (Staples et al. 2019). Microbial synthesis of myrcene is of remarkable interest due to its desirable physicochemical properties; however, achieving high biosynthetic titers remains a great challenge. The de novo biosynthesis of myrcene has also been achieved by metabolically engineered *E. coli*, where the MVA pathway was introduced and the GPPS and myrcene synthase (MS) were overexpressed. However, the resulted titer of myrcene was very low, which could be attributed to the inherently low enzymatic activity of MS or an imbalance in the metabolic network.

An innovative and efficient alternative is to convert monoterpenes that are acquired via microbial fermentation into myrcene. Recently, the myrcene titer was improved to 1.25 g/L from glycerol by introducing a truncated and mutant linalool dehydratase isomerase (tLDI*) to a geraniol producing *E. coli* strain (Wang et al. 2023). Initially, the LDI, a cofactor-independent linalool dehydratase isomerase (LDI) from the bacterium *Castellaniella defragrans*, possesses nominal activity for sequential catalysis of geraniol to linalool and the subsequent dehydration to myrcene in an anaerobic environment. To maintain and improve its anaerobic catalytic activity, rational enzyme modification and a series of biochemical process engineering were employed. By regulating the induction temperature and engineering the enzyme rationally, protein solubility and catalytic activity were greatly improved. The performance of the whole-cell biocatalyst was further improved via mild reducing agents and tuning the induction time. This anaerobic mineralization process is thermodynamically favorable to form myrcene from geraniol and reduce toxicity because monoterpene alcohols (geraniol/linalool) are more toxic to the host organism than monoterpenes (myrcene) (Engleder et al. 2019; Wojtunik-Kulesza 2022).

9.6.3 Sesquiterpene (C15) Biofuels

Sesquiterpenoids are one of the largest groups of isoprenoid natural products, which can be acyclic, monocyclic, bi-, or even tricyclic. They are 15 carbons in chain length close to the average of diesel (C16), but with a branched rather than a straight-chain structure. Among sesquiterpenoids, farnesol, farnesane, and bisabolane have been proposed as alternatives to conventional diesel fuels (Beller et al. 2015). To promote the sustainable production of these terpenes, metabolic engineering efforts have been focused on microbes including *E. coli*, *S. cerevisiae*, *Y. lipolytica*, *Pichia pastoris,* etc. The microbial production of farnesene is most successful among all the terpene fuels that both α-farnesene and β-farnesene are produced at 1010 g/L scale (Table 9.3). However, the microbial production of farnesol is below g/L that we will not go to details on this product.

9.6.3.1 Farnesene

Due to its intrinsic qualities of low hygroscopicity and high energy density, farnesene is considered an ideal biofuel precursor, particularly a potential to replace D2 diesel, which is a mixture of linear, branched, and cyclic alkanes with an average length of C16. Microbial production of farnesene has been achieved at 10 g/L scale with several microbial hosts, including *E. coli, Y. lipolytica, P. pastoris*, and *S. cerevisiae* (Table 9.3). The most successful case is β-farnesene production with an engineered *S. cerevisiae*. Meadows et al. constructed an artificial cytosolic acetyl-CoA biosynthetic pathway with a reduced ATP requirement, which contained *Dickeya zeae* aldehyde dehydrogenase (acylating) (ADA), *Leuconostoc mesenteroides* xylulose-5-phosphate-specific phosphoketolase (PK), and *Clostridium kluyveri* phosphor-transacetylase (PTA). Combining the artificial acetyl-CoA biosynthetic pathway with the NADH-dependent HMGR from *Silicibacter pomeroyi* in *S. cerevisiae* enhanced farnesene titer to 130 g/L in 200, 000-L bioreactor (Meadows et al. 2016).

The second worth mentioning is the production of a-farnesene with the methylotrophic yeast *P. pastoris* (Xu et al. 2023). An α-farnesene high-producing strain was obtained by systematically engineering *P. pastoris* XF22 by using appropriate promoters to control the expression of the key enzymes in MVA biosynthesis and the cytosolic acetyl-CoA supply as well as to weaken FPP branch. The final strain carrying an α-farnesene synthase (αFS) from *Malus domestica* was constructed to produce 17.92 g/L of α-farnesene with a productivity of 0.107 g/(L·h) and a yield of 0.078 g/g glucose after 168 h in fed-batch fermentation. Although the titer of α-farnesene is not the highest, the productivity and the yield are the highest in the literature (Xu et al. 2023).

9.6.3.2 Bisabolene

Bisabolane is a high-energy density monocyclic sesquiterpene with better cold properties and a potential to replace D2 diesel. In the past 5 years, microbes like *E. coli*, *Y. lipolytica*, *R. toruloides*, *Ruminococcus flavefaciens*, *Synechocystis* sp., *Chlamydomonas reinhardtii*, and *Rhodobacter capsulatus* have been refactored for its heterologous production with the aid of metabolic engineering. Fortunately, the feasibility of high-level production is demonstrated with the purple nonsulfur photosynthetic bacterium *R. capsulatus* (Zhang et al. 2021).

To obtain higher titers of bisabolene, engineering strategies such as promoter screening, improving the NADPH and precursor supply, down-tuning the side pathways, and introducing a heterologous MVA pathway have been used in *R. capsulatus*. The above strategies enabled a 35-fold increase in bisabolene titer, reaching 1089.7 mg/L from glucose in a shake flask. The engineered strain produced 9.8 g/L bisabolene with a yield of >0.196 g per gram glucose under the two-phase fed-batch fermentation, which corresponds to >78% of the theoretical yield (Zhang et al. 2021). The pioneering study demonstrates *R. capsulatus* as a promising platform for terpene-based advanced biofuel production.

9.6.4 Techno-Economic Assessment

Techno-economic assessment (TEA) is critical for evaluating the feasibility of microbial production of terpenoid biofuel at large scale. According to an estimated TEA for limonane, bisabolene, and epi-isozizaene, taking the estimated minimum selling price (MSP) and life-cycle greenhouse gas (GHG) emission into account, the production of target products from cellulose and hemicellulose is the main driver of cost mitigation (Baral et al. 2019). Through process optimization, the MSP of limonane can be reduced to $0.77 or $0.91/L Jet A. For bisabolane and epi-isozizaene, the MSPs were targeted to \$0.82 and \$0.73/L of Jet A, respectively, and GHG mitigation costs were within reasonable limits. It is estimated that additional efforts can further lower the target minimum price to \$0.66/L Jet A, such as metabolic and process optimization, lignin cost reduction, and using cheaper metal catalysts to improve jet-fuel yield and selectivity (Baral et al. 2019).

Take limonene as an example, the average market price is about \$7/kg and the global market size is about \$1.9 billion/year in the USA (Ren et al. 2020). The plant-based limonene production is at least an order of magnitude higher than the target \$2.50/gal biofuel sales price (Yang et al. 2020). Based on the reported limonene bioproduction data, the current microbial limonene would rocket to a forbidding \$465/kg. The main cost drivers of the microbial process include the product yield, feedstock input, capital investment, project length, aeration speed, and hurdle rates (Sun et al. 2020).

The additional extraction and recovery costs should be considered seriously. As for the example of β-farnesene biosynthesis with an engineered *S. cerevisiae* host in a 200,000-liter industrial bioreactor. For facilitating the downstream processing and minimizing the evaporation loss, the poly α-olefin was added as overlay during fermentation (Meadows et al. 2016). However, the process requires a significant amount of organic solvent, resulting in extra high cost. In addition, the large volume of solvent overlays in the fermentation broth also affects the dissolved oxygen level in the bioreactor, which could lower cell growth. A TEA of limonene production showed that the fermentation cost with solvent overlays is 4–6% higher than that without using an overlay (Wu and Maravelias 2018).

9.7 Conclusion

Rapid industrial advances have led to overuse of fossil fuel resources to meet the socioeconomic development and welfare of the growing global population. However, the fossil fuel manufacturing system has resulted in global climate and environmental divergences due to the imbalance in the carbon cycle. Additionally, worldwide concern on the availability of natural products and depleting oil reserves resulting in soaring oil prices has propelled widespread interest in unconventional renewable energy. Fortunately, microbial biosynthesis provides a promising alternative approach for the sustainable supply of terpenoids chemicals due to the rapid development in metabolic engineering and synthetic biology, which make it imperative to promote a bioeconomy, focused on efficient, green, sustainable, and low-carbon footprint energy technologies. In this regard, terpenes including hemiterpenes, monoterpenes, and sesquiterpenes can be explored as sustainable replacements to gasoline, jet fuel, and diesel, respectively.

9.8 Perspectives

Substantial product titer and economical productivity are the key factors for microbial isoprenoid-derived biofuels to achieve broad applications. Despite all the endeavors and progress, the microbial production of terpenoid biofuels is still challenging in most cases. Several points should be addressed to develop industrial-scale cell factories with high robustness, acceptable productivity, easy operation, and low cost. First, the discovery of isoprenoid synthases and rational design of pathways with impressive catalytic efficiency. Second, the exploration of novel hosts with strong tolerance against the isoprenoid biofuels. Third, the tactics against problems created by the volatility of isoprenoid biofuels on fermentation monitoring, product measurement, and downstream purification. Fourth, the balance between cell growth and product biosynthesis. Last but not the least, the scaling-up demonstration for assessing production process and cost. A well-established design-build-test-learn platform should significantly accelerate the research and development cycle. Combining the automated facilities and AI-driven data processing, it is possible to manufacture isoprenoid biofuels via synthetic biology in the future.

Acknowledgments This research was Supported by Chinese Universities Scientific Fund (2452018314) and Natural Science Foundation of Shaanxi Province (2020JM-177).

References

Altinkurt MD, Ozturk YE, Turkcan A (2022) Effects of isobutanol fraction in diesel-biodiesel blends on combustion, injection, performance and emission parameters. Fuel 330:125554

Balat M (2009) Bioethanol as a vehicular fuel: a critical review. Energy Sources A 31:1242–1255

Baral NR, Kavvada O, Mendez-Perez D et al (2019) Techno-economic analysis and life-cycle greenhouse gas mitigation cost of five routes to bio-jet fuel blendstocks. Energy Environ Sci 12: 807–824

Beller HR, Lee TS, Katz L (2015) Natural products as biofuels and bio-based chemicals: fatty acids and isoprenoids. Nat Prod Rep 32:1508–1526

Bian G, Han Y, Hou A et al (2017) Releasing the potential power of terpene synthases by a robust precursor supply platform. Metab Eng 42:1–8

Cao YJ, Zhang HB, Liu H et al (2018) Biosynthesis and production of sabinene: current state and perspectives. Appl Microbiol Biotechnol 102:1535–1544

Carruthers DN, Kim J, Mendez-Perez D et al (2023) Microbial production of high octane and high sensitivity olefinic ester biofuels. Biotechnol Biofuels Bioprod 16:1–17

Cheng BQ, Wei LJ, Lv YB et al (2019) Elevating limonene production in oleaginous yeast Yarrowia lipolytica via genetic engineering of limonene biosynthesis pathway and optimization of medium composition. Biotechnol Bioprocess Eng 24:500–506

Chubukov V, Mingardon F, Schackwitz W et al (2015) Acute limonene toxicity in *Escherichia coli* is caused by limonene hydroperoxide and alleviated by a point mutation in alkyl hydroperoxidase AhpC. Appl Environ Microbiol 81:4690–4696

Clomburg JM, Qian S, Tan ZG et al (2019) The isoprenoid alcohol pathway, a synthetic route for isoprenoid biosynthesis. Proc Natl Acad Sci USA 116:12810–12815

Dellas N, Thomas ST, Manning G et al (2013) Discovery of a metabolic alternative to the classical mevalonate pathway. Elife 2:e00672

Demissie ZA, Tarnowycz M, Adal AM et al (2019) A lavender ABC transporter confers resistance to monoterpene toxicity in yeast. Planta 249:139–144

Donoso D, Garcia D, Ballesteros R et al (2021) Hydrogenated or oxyfunctionalized turpentine: options for automotive fuel components. RSC Adv 11:18342–18350

Dusseaux S, Wajn WT, Liu YX et al (2020) Transforming yeast peroxisomes into microfactories for the efficient production of high-value isoprenoids. Proc Natl Acad Sci USA 117:31789–31799

Engleder M, Muller M, Kaluzna I et al (2019) Exploring *Castellaniella defragrans* linalool (De)hydratase-isomerase for enzymatic hydration of alkenes. Molecules 24:2092

Fan KW, Aki T, Chen F et al (2010) Enhanced production of squalene in the thraustochytrid *Aurantiochytrium mangrovei* by medium optimization and treatment with terbinafine. World J Microbiol Biotechnol 26:1303–1309

Gagat P, Bodył A, Mackiewicz P (2013) How protein targeting to primary plastids via the endomembrane system could have evolved? A new hypothesis based on phylogenetic studies. Biol Direct 8:18. https://doi.org/10.1186/1745-6150-8-18

Gao X, Gao F, Liu D et al (2016) Engineering the methylerythritol phosphate pathway in cyanobacteria for photosynthetic isoprene production from CO2. Energy Environ Sci 9:1400–1411

Garcia D, Bustamante F, Villa AL et al (2020) Oxyfunctionalization of turpentine for fuel applications. Energy Fuel 34:579–586

Gerke J, Frauendorf H, Schneider D et al (2020) Production of the fragrance geraniol in peroxisomes of a product-tolerant Baker's yeast. Front Bioeng Biotechnol 8:582052

Gonçalves D, Paludetti MF, Florido PM et al (2018) Physical behavior of the phases from the liquid-liquid equilibrium of citrus essential oils systems at 298.2 K. J Chem Eng Data 63:2718–2737

Grochowski LL, Xu HM, White RH (2006) Methanocaldococcus jannaschii uses a modified mevalonate pathway for biosynthesis of isopentenyl diphosphate. J Bacteriol 188:3192–3198

Hayakawa H, Motoyama K, Sobue F et al (2018) Modified mevalonate pathway of the archaeon *Aeropyrum pernix* proceeds via trans-anhydromevalonate 5-phosphate. Proc Natl Acad Sci USA 115:10034–10039

Hellier P, Al-Haj L, Talibi M et al (2013) Combustion and emissions characterization of terpenes with a view to their biological production in cyanobacteria. Fuel 111:670–688

Hoshino Y, Moriya M, Matsudaira A et al (2020) Stereospecific linalool production utilizing two-phase cultivation system in *Pantoea ananatis*. J Biotechnol 324:21–27

Hou AW, Dickschat JS (2021) Targeting active site residues and structural anchoring positions in terpene synthases. Beilstein J Org Chem 17:2441–2449

Hu FF, Liu JD, Du GC et al (2012) Key cytomembrane ABC transporters of *Saccharomyces cerevisiae* fail to improve the tolerance to D-limonene. Biotechnol Lett 34:1505–1509

Isar J, Jain D, Joshi H et al (2022) Microbial isoprene production: an overview. World J Microb Biot 38(7):122

Jia HJ, Chen TH, Qu JZ et al (2020) Collaborative subcellular compartmentalization to improve GPP utilization and boost sabinene accumulation in *Saccharomyces cerevisiae*. Biochem Eng J 164:107768

Jiang G, Yao M, Wang Y et al (2021) A 'push-pull-restrain' strategy to improve citronellol production in Saccharomyces cerevisiae. Meta Eng 66:51–59

Jiang H, Wang X (2023) Biosynthesis of monoterpenoid and sesquiterpenoid as natural flavors and fragrances. Biotechnol Adv 65:108151

Kang A, George KW, Wang G et al (2016) Isopentenyl diphosphate (IPP)-bypass mevalonate pathways for isopentenol production. Metab Eng 34:25–35

Karuppiah V, Ranaghan KE, Leferink NGH et al (2017) Structural basis of catalysis in the bacterial monoterpene synthases linalool synthase and 1,8-cineole synthase. ACS Catal 7:6268–6282

Kim J, Baidoo EEK, Amer B et al (2021) Engineering *Saccharomyces cerevisiae* for isoprenol production. Metab Eng 64:154–166

Kirby J, Geiselman GM, Yaegashi J et al (2021) Further engineering of *R. toruloides* for the production of terpenes from lignocellulosic biomass. Biotechnol Biofuels 14:1–16

Kong X, Wu YK, Yu WW et al (2023) Efficient synthesis of limonene in *Saccharomyces cerevisiae* using combinatorial metabolic engineering strategies. J Agric Food Chem 71:7752–7764

Korman TP, Opgenorth PH, Bowie JU (2017) A synthetic biochemistry platform for cell free production of monoterpenes from glucose. Nat Commun 8:15526

Lapuerta M, Tobio-Perez I, Ortiz-Alvarez M et al (2023) Heterogeneous catalytic conversion of terpenes into biofuels: an open pathway to sustainable fuels. Energies 16:2526

Leferink NGH, Jervis AJ, Zebec Z et al (2016) A 'plug and play' platform for the production of diverse monoterpene hydrocarbon scaffolds in *Escherichia coli*. ChemistrySelect 1:1893–1896

Leferink NGH, Dunstan MS, Hollywood KA et al (2019) An automated pipeline for the screening of diverse monoterpene synthase libraries. Sci Rep 9:11936

Lei DW, Qiu ZT, Wu JH et al (2021) Combining metabolic and monoterpene synthase engineering for de novo production of monoterpene alcohols in *Escherichia coli*. ACS Synth Biol 10:1531–1544

Li Y, Luxenburger E, Müller R (2013) An alternative isovaleryl CoA biosynthetic pathway involving a previously unknown 3-methylglutaconyl CoA decarboxylase. Angew Chem Int Ed 52(4):1304-1 308

Liu YH, Jiang X, Cui ZY et al (2019) Engineering the oleaginous yeast *Yarrowia lipolytica* for production of α-farnesene. Biotechnol Biofuels 12:1–11

Liu YR, Ma XQ, Liang H et al (2021) Monoterpenoid biosynthesis by engineered microbes. J Ind Microbiol Biotechnol 48:kuab 065

Liu YH, Zhang J, Li QB et al (2022) Engineering *Yarrowia lipolytica* for the sustainable production of β-farnesene from waste oil feedstock. Biotechnol Biofuels Bioprod 15:101

Ma TF, Zong H, Lu XY et al (2023) *Candida glycerinogenes*-promoted α-pinene and squalene co-production strategy based on α-pinene stress. J Agric Food Chem 71:5250–5260

Mack JH, Rapp VH, Broeckelmann M et al (2014) Investigation of biofuels from microorganism metabolism for use as anti-knock additives. Fuel 117:939–943

Meadows AL, Hawkins KM, Tsegaye Y et al (2016) Rewriting yeast central carbon metabolism for industrial isoprenoid production. Nature 537:694–697

Mendez-Perez D, Alonso-Gutierrez J, Hu QJ et al (2017) Production of jet fuel precursor monoterpenoids from engineered *Escherichia coli*. Biotechnol Bioeng 114:1703–1712

Meylemans HA, Quintana RL, Harvey BG (2012) Efficient conversion of pure and mixed terpene feedstocks to high density fuels. Fuel 97:560–568

Mo XY, Cai XH, Hui QY et al (2021) Whole genome sequencing and metabolomics analyses reveal the biosynthesis of nerol in a multi-stress-tolerant *Meyerozyma guilliermondii* GXDK6. Microb Cell Fact 20:1–11

Nitta N, Tajima Y, Katashkina JI et al (2020) Application of inorganic phosphate limitation to efficient isoprene production in *Pantoea ananatis*. J Appl Microbiol 128:763–774

Nitta N, Tajima Y, Yamamoto Y et al (2021) Fermentative production of enantiopure (S)-linalool using a metabolically engineered *Pantoea ananatis*. Microb Cell Fact 20:1–14

Niu FX, Lu Q, Bu YF et al (2017) Metabolic engineering for the microbial production of isoprenoids: carotenoids and isoprenoid-based biofuels. Synth Syst Biotechnol 2:167–175

Niu FX, He X, Wu YQ et al (2018) Enhancing production of pinene in *Escherichia coli* by using a combination of tolerance, evolution, and modular co-culture engineering. Front Microbiol 9:1623

Niu FX, Huang YB, Shen YP et al (2020) Enhanced production of pinene by using a cell-free system with modular cocatalysis. J Agric Food Chem 68:2139–2145

Ozaki T, Shinde SS, Gao L et al (2018) Enzymatic formation of a skipped methyl-substituted octaprenyl side chain of longestin (KS-505a): involvement of homo-IPP as a common extender unit. Angew Chem Int Ed 57:6629–6632

Pahima E, Hoz S, Ben-Tzion M et al (2019) Computational design of biofuels from terpenes and terpenoids. Sustain Energy Fuels 3:457–466
Park J, Kang DH, Woo HM (2021) Microbial bioprocess for extracellular squalene production and formulation of nanoemulsions. ACS Sustain Chem Eng 9:14263–14276
Peralta-Yahya PP, Ouellet M, Chan R et al (2011) Identification and microbial production of a terpene-based advanced biofuel. Nat Commun 2:483
Phulara SC, Chaturvedi P, Gupta P (2016) Isoprenoid-based biofuels: homologous expression and heterologous expression in prokaryotes. Appl Environ Microbiol 82:5730–5740
Ploessl D, Zhao YX, Shao ZY (2023) Engineering of non-model eukaryotes for bioenergy and biochemical production. Curr Opin Biotechnol 79:102869
Rabinovitch-Deere CA, Oliver JWK, Rodriguez GM et al (2013) Synthetic biology and metabolic engineering approaches to produce biofuels. Chem Rev 113:4611–4632
Raman V, Sivasankaralingam V, Dibble R et al (2016). α-Pinene-A high energy density biofuel for SI engine applications SAE Tech. 2001–2171.
Rau MH, Calero P, Lennen RM et al (2016) Genome-wide Escherichia coli stress response and improved tolerance towards industrially relevant chemicals. Microb Cell Fact 15:1–18
Ren YY, Liu SS, Jin GJ et al (2020) Microbial production of limonene and its derivatives: achievements and perspectives. Biotechnol Adv 44:107628
Rinaldi MA, Ferraz CA, Scrutton NS (2022) Alternative metabolic pathways and strategies to high-titre terpenoid production in *Escherichia coli*. Nat Prod Rep 39:90–118
Rolf J, Julsing MK, Rosenthal K et al (2020) A gram-scale limonene production process with engineered *Escherichia coli*. Molecules 25:1881
Rude MA, Schirmer A (2009) New microbial fuels: a biotech perspective. Curr Opin Microbiol 12: 274–281
Runguphan W, Sae-Tang K, Tanapongpipat S (2021) Recent advances in the microbial production of isopentanol (3-methyl-1-butanol). World J Microbiol Biotechnol 37(6):107
Shah AA, Wang C, Chung YR et al (2013a) Enhancement of geraniol resistance of *Escherichia coli* by MarA overexpression. J Biosci Bioeng 115:253–258
Shah AA, Wang C, Yoon SH et al (2013b) RecA-mediated SOS response provides a geraniol tolerance in *Escherichia coli*. J Biotechnol 167:357–364
Shukla V, Phulara SC (2021) Impact of culture condition modulation on the high-yield, high-specificity, and cost-effective production of terpenoids from microbial sources: a review. Appl Environ Microbiol 87(4):e02369–e02320
Sikkema J, Debont JAM, Poolman B (1994) Interactions of cyclic hydrocarbons with biological-membranes. J Biol Chem 269:8022–8028
Sikkema J, Debont JAM, Poolman B (1995) Mechanisms of membrane toxicity of hydrocarbons. Microbiol Rev 59:201–222
Silva GDE, Marques JND, Linhares EPM et al (2022) Review of anticancer activity of monoterpenoids: geraniol, nerol, geranial and neral. Chem Biol Interact 362:109994
Song XJ, Wang XL, Tan YZ et al (2015) High production of squalene using a newly isolated yeast-like strain *Pseudozyma sp* SD301. J Agric Food Chem 63:8445–8451
Staples O, Leal JH, Cherry PA et al (2019) Camphorane as a renewable diesel blendstock produced by cyclodimerization of myrcene. Energy Fuel 33:9949–9955
Sun CH, Theodoropoulos C, Scrutton NS (2020) Techno-economic assessment of microbial limonene production. Bioresour Technol 300:122666
Tippmann S, Chen Y, Siewers V et al (2013) From flavors and pharmaceuticals to advanced biofuels: production of isoprenoids in *Saccharomyces cerevisiae*. Biotechnol J 8:1435–1444
Turner WJ, Dunlop MJ (2015) Trade-offs in improving biofuel tolerance using combinations of efflux pumps. ACS Synth Biol 4:1056–1063
Vannice JC, Skaff DA, Keightley A et al (2014) Identification in haloferax volcanii of phosphomevalonate decarboxylase and isopentenyl phosphate kinase as catalysts of the terminal enzyme reactions in an archaeal alternate mevalonate pathway. J Bacteriol 196:1055–1063

Vermaas JV, Bentley GJ, Beckham GT et al (2018) Membrane permeability of terpenoids explored with molecular simulation. J Phys Chem B 122:10349–10361

Vinokur JM, Cummins MC, Korman TP et al (2016) An adaptation to life in acid through a novel mevalonate pathway. Sci Rep 6:39737

Von Reuss S, Domik D, Lemfack MC et al (2018) Sodorifen biosynthesis in the rhizobacterium serratia plymuthica involves methylation and cyclization of mep-derived farnesyl pyrophosphate by a sam-dependent c-methyltransferase. J Am Chem Soc 140:11855–11862

Wang X, Chen JM, Zhang J et al (2021a) Engineering *Escherichia coli* for production of geraniol by systematic synthetic biology approaches and laboratory-evolved fusion tags. Metab Eng 66: 60–67

Wang JH, Jiang W, Liang CJ et al (2021b) Overproduction of α-farnesene in *Saccharomyces cerevisiae* by farnesene synthase screening and metabolic engineering. J Agric Food Chem 69: 3103–3113

Wang J, Jiang T, Milligan S et al (2022a) Improving isoprenol production via systematic CRISPRi screening in engineered *Escherichia coli*. Green Chem 24:6955–6964

Wang X, Xiao LJ, Zhang XY et al (2022b) Combined bioderivatization and engineering approach to improve the efficiency of geraniol production. Green Chem 24:864–876

Wang X, Wang J, Zhang X et al (2023) Efficient myrcene production using linalool dehydratase isomerase and rational biochemical process in *Escherichia coli*. J Biotechnol 371-372:33–40

Whited GM, Feher FJ, Benko DA et al (2010) Technology update: development of a gas-phase bioprocess for isoprene-monomer production using metabolic pathway engineering. Ind Biotechnol 6:152–163

Wojtunik-Kulesza KA (2022) Toxicity of selected monoterpenes and essential oils rich in these compounds. Molecules 27:1716

Woodroffe JD, Lupton DV, Garrison MD et al (2021) Synthesis and fuel properties of high-energy density cyclopropanated monoterpenes. Fuel Process Technol 222:106952

Wu WZ, Maravelias CT (2018) Synthesis and techno-economic assessment of microbial-based processes for terpenes production. Biotechnol Biofuels 11:1–14

Wu J, Cheng S, Cao J et al (2019) Systematic optimization of limonene production in engineered Escherichia coli. J Agric Food Chem 67(25):7087–7097

Wu J, Wang X, Xiao LJ et al (2021) Synthetic protein scaffolds for improving R-(-)-linalool production in *Escherichia coli*. J Agric Food Chem 69:5663–5670

Xu JZ, Chen SL, Niu SQ (2023) Systematic modulating carbon metabolism to improve α-farnesene production in *Pichia pastoris*. Fuel Process Technol 247:107757

Yang C, Gao X, Jiang Y et al (2016) Synergy between methylerythritol phosphate pathway and mevalonate pathway for isoprene production in *Escherichia coli*. Metab Eng 37:79–91

Yang ML, Baral NR, Simmons BA et al (2020) Accumulation of high-value bioproducts in planta can improve the economics of advanced biofuels. Proc Natl Acad Sci USA 117:8639–8648

Yanowitz J, Ratcliff MA, McCormick RL et al (2017) Compendium of experimental cetane numbers NREL/TP-5400-67585. National Renewable Energy Lab (NREL), Golden, CO

Yao Z, Zhou PP, Su BM et al (2018) Enhanced isoprene production by reconstruction of metabolic balance between strengthened precursor supply and improved isoprene synthase in *Saccharomyces cerevisiae*. ACS Synth Biol 7:2308–2316

Yao P, You SP, Qi W et al (2020) Investigation of fermentation conditions of biodiesel by-products for high production of β-farnesene by an engineered *Escherichia coli*. Environ Sci Pollut Res 27:22758–22769

Ye M, Gao JQ, Zhou YJ (2023) Global metabolic rewiring of the nonconventional yeast *Ogataea polymorpha* for biosynthesis of the sesquiterpenoid β-elemene. Metab Eng 76:225–231

Yee DA, Denicola AB, Billingsley JM et al (2019) Engineered mitochondrial production of monoterpenes in *Saccharomyces cerevisiae*. Metab Eng 55:76–84

Zhang HB, Liu Q, Cao YJ et al (2014) Microbial production of sabinene—a new terpene-based precursor of advanced biofuel. Microb Cell Fact 13:1–10

Zhang Y, Wang J, Cao X et al (2020) High-level production of linalool by engineered Saccharomyces cerevisiae harboring dual mevalonate pathways in mitochondria and cytoplasm. Enzyme Microb Tech 134:109462

Zhang Y, Song X, Lai Y et al (2021) High-yielding terpene-based biofuel production in *Rhodobacter capsulatus*. ACS Synth Biol 10:1545–1552

Zhang YL, Wang WG, Wei WQ et al (2023) Regulation of ethanol assimilation for efficient accumulation of squalene in *Saccharomyces cerevisiae*. J Agric Food Chem 71:6389–6397

Zhao J, Li C, Zhang Y et al (2017) Dynamic control of ERG20 expression combined with minimized endogenous downstream metabolism contributes to the improvement of geraniol production in Saccharomyces cerevisiae. Microb Cell Fact 16:1–11

Zhao YK, Zhu K, Li J et al (2021) High-efficiency production of bisabolene from waste cooking oil by metabolically engineered *Yarrowia lipolytica*. Microb Biotechnol 14:2497–2513

Zhao C, Wang XH, Lu XY et al (2023) Metabolic engineering of candida glycerinogenes for sustainable production of geraniol. ACS Synth Biol 12:1836–1844

Zhou J, Wang CL, Yoon SH et al (2014) Engineering *Escherichia coli* for selective geraniol production with minimized endogenous dehydrogenation. J Biotechnol 169:42–50

Zhou YJJ, Wang JJ, Wang X et al (2023a) Efficient production of alpha-pinene through identifying the rate-limiting enzymes and tailoring inactive terminal of pinene synthase in *Escherichia coli*. Fuel 343:127872

Zhou PP, Zhou XQ, Yuan DD et al (2023b) Combining protein and organelle engineering for linalool overproduction in *Saccharomyces cerevisiae*. J Agric Food Chem 71:10133–10143

Zhu L, Zhang JZ, Yang JW et al (2022a) Strategies for optimizing acetyl-CoA formation from glucose in bacteria. Trends Biotechnol 40:149–165

Zhu K, Zhao BX, Zhang YH et al (2022b) Mitochondrial engineering of *Yarrowia lipolytica* for sustainable production of α-bisabolene from waste cooking oil. ACS Sustain Chem Eng 10: 9644–9653

Zhuang X, Kilian O, Monroe E, Ito M, Tran-Gymfi MB, Liu F, Davis RW, Mirsiaghi M, Sundstrom E, Pray T, Skerker JM, George A, Gladden JM (2019) Monoterpene production by the carotenogenic yeast Rhodosporidium toruloides. Microb Cell Fact 18(1):54

Zong Z, Hua QS, Tong XY et al (2019) Biosynthesis of nerol from glucose in the metabolic engineered *Escherichia coli*. Bioresour Technol 287:121410

Chapter 10
Microbial Alkane/Alkene as Drop-in Biofuel

Sumit Sharma, Shikha Singh, Saurabh Jyoti Sarma, and Satinder Kaur Brar

Abstract Microbial alkanes/alkenes are a special class of bio-hydrocarbons that can act as drop-in biofuels. It provides a bypass route for making an intermediate precursor of a commercial product. There is no complex processing required which is used in conventional petroleum refineries. Microbial processing also reduces GHG emissions. There is a variety of microbial alkanes/alkenes that can be used as drop-in biofuel (mostly heptadecane and 1-undecene). This chapter summarizes the types of alkane/alkenes, source microorganisms, biosynthetic pathways, engineering approaches, production improvements, suitable substrates, and the use of alkane/alkene as drop-in biofuel. Here, the potential utilization and future perspective of using bio-based alkane/alkene are also discussed.

Keywords Alkane/Alkene · Microorganism · Renewable resources · Bioprocessing · Drop-in biofuel

10.1 Introduction

The current era is moving toward bio-based energy resources and applications. Microbial alkanes/alkenes are the most promising replacement for fossil fuels because these act as drop-in biofuels and can be directly used in engines or other current mobilities. Microbial alkanes or alkenes are a class of hydrocarbons synthesized biologically through various microbial hosts. The microbial biosynthesis of these products showed a sustainable approach to biofuel production. Microbial alkanes/alkenes are distinctly defined from conventional biofuels such as bioethanol and biodiesel—currently blended with petroleum products. Bioethanol and biodiesel

S. Sharma · S. Singh · S. J. Sarma (✉)
Department of Biotechnology, School of Engineering and Applied Sciences, Bennett University, Greater Noida, India
e-mail: saurabh.sarma@bennett.edu.in

S. K. Brar
Department of Civil Engineering, Lassonde School of Engineering, York University, Toronto, ON, Canada

S. K. Brar et al. (eds.), *The Microbiology of the Drop-in Biofuel Production*, Biofuel and Biorefinery Technologies 15, https://doi.org/10.1007/978-3-031-61637-2_10

Table 10.1 Physicochemical properties of conventional jet propellent 5 (JP5) fuel vs blended drop-in bio-jet fuel farnesane (Zheng et al. 2022)

Property	Conventional aviation fuel JP5	Drop-in bofuel 36% farnesane +64% JP5
Density @ 15 °C(kg/m3)	826.8	807.7
Heat of combustion (MJ/kg)	43	43.29
Cetane number derived	47.9	47
Surface tension (mN/m)	28.4	27.7
Viscosity @ 20 °C (cSt)	6.5	8
Hydrogen content (mass %)	13.7	14.1
Aromatics (wt%)	20.59	13.61
Smoke point (mm)	20	25.2
Flash point (°C)	60	66

are still facing challenges with their implementation in currently running engines due to their hygroscopic property (Khuong et al. 2017; Bušić et al. 2018). These are prone to absorb water from moisture and lead to corrosion (Lou and Singh 2010; Khuong et al. 2017; Baena et al. 2021). These also need a proper blending ratio to match the similar ignition nature as petroleum hydrocarbons. Due to these facts, the microbial alkane/alkene gained the greatest interest and has similar properties to the hydrocarbon chain that matches the core chemical structure of petroleum hydrocarbons. These also do not need significant improvement while designing the engine and have no limitation of use in the current transport systems. There are several microbial hosts such as bacteria and cyanobacteria studied for microbial alkane biosynthesis (Brown et al. 2018).

Microbial alkane and alkene are categorized as drop-in biofuel for different types such as bio-gasoline, jet biofuel, and green fuels (Monteiro et al. 2022). Drop-in biofuels are produced by a bypass strategy in the form of liquid hydrocarbons that should have chemical as well as functional properties (Karatzos et al. 2017) such as flash point, boiling point, freezing point, carbon number, calorific value, heating value, energy density, H/C ratio, etc. and directly meet the final product demand. Recently, Zheng et al. (2022) studied the alternative jet fuel properties to identify lean blow out limits and represented the physicochemical properties of conventional fuel jet propellent 5 (JP5) and a 36% blend of drop-in bio-jet fuel farnesane with 64% JP5 that showed many similar characteristics represented in Table 10.1 (Zheng et al. 2022). The petroleum refinery process is the conventional way to produce alkane/alkene and is used as fuel in various forms, but it also imparts a higher GHG emission (Ma et al. 2023) and other toxic chemicals in the less sustainable process. Comparatively, microbial alkane/alkene biosynthesis is a sustainable and environmentally friendly process when renewable resources such as lignocellulosic biomass, algal biomass, or other sugar and fatty acid sources can be used (Azimov et al. 2021). These can be produced from cellulosic sugars using an oleaginous yeast where lipids are the major product (Caporusso et al. 2021). The lipid can be used as a

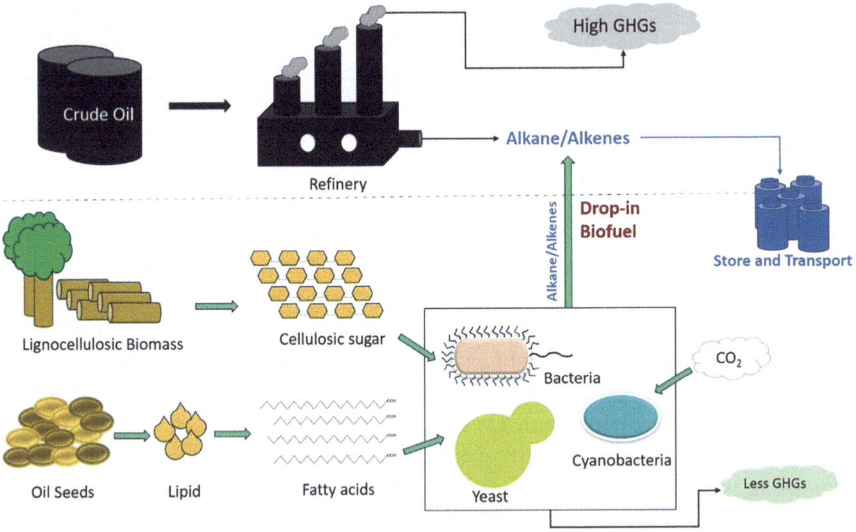

Fig. 10.1 Schematic representation of sustainable production of microbial alkane/alkene as drop-in biofuels

fatty acid source and then converted into microbial alkane/alkene by specialized microorganisms. An overview of microbial alkane/alkene production and its utilization as drop-in fuel against conventional processes is illustrated in Fig. 10.1. This chapter gives insights into the types and microbial biosynthesis of alkanes/alkenes, their production technologies, metabolic routes, and their application as drop-in biofuels.

Microbial alkanes/alkenes have similar functions as those derived from petroleum fuels because of their compatibility with the existing transport sector, especially jet fuels. These are specifically used in aviation where bioethanol or biodiesel may not be used. These also have applications in fragrances and other chemical reagents and largely find use as jet fuel. There is a range of equivalent properties required to meet as jet fuel, for example, very low freezing point (−40 °C) and low viscosity at low temperature. Unlike microbial alkane/alkenes, biodiesel freezes at 0 °C and is incompatible for aviation fuel. Aviation fuel is made up of C8–C16 hydrocarbon mixture and their boiling point ranges from 130 to 270 °C and is highly similar to microbial alkanes such as tridecane and tetradecane boiling points of 233.85 and 249.85 °C, respectively. The biochemical conversion using the fermentation process using cellulosic sugar is an advanced technology. However, the presence of a high content of oxygen (~40%) causes problems during combustion as oxygen consumes hydrogen to form water and lower energy density. The effective parameter to determine these values is the H/C ratio. Cellulosic sugars have very low H/C even close to 0 and are called poor feedstock for hydrocarbons while standard finished fuel should have H/C close to 2. Lipids are a simplified source and have higher H/C

1.6 and lower oxygen 11% (Karatzos et al. 2017). The external hydrogen supply is the best solution to overcome the oxygen content problem in renewable feedstocks.

10.2 Microbial Alkanes

Alkane biosynthesis has been studied specifically for the long-chain hydrocarbons. Most of the alkanes observed are decanes (tetra, penta, hexa, hepta, octa, and nonadecane) produced by photoautotrophic bacteria. Other than bacteria, several fungi and microalgae were also reported for alkane biosynthesis. Cyanobacteria are the major class of study for microbial alkane due to faster growth and life cycle and also contribute to direct CO_2 conversion into biofuels (Parveen and Yazdani 2022). Alkane production by recycling CO_2 is also a promising approach to reducing the environmental GHG (greenhouse gas) burden. Due to their halophilic nature, cyanobacteria can survive on seawater, brine water, high saline water, or wastewater coming from municipal areas. Therefore, its versatile nature is preferable for biofuel production. Cyanobacteria possess the ability to produce alkane/alkene via fatty acid biosynthetic pathway (Sun et al. 2015; Wang et al. 2020). The acetyl-CoA coming from glycolysis shifted to malonyl-ACP (acyl carrier protein) to forward the fatty acid cycle and produce fatty acid-acyl ACP. The fatty acyl ACP is converted into fatty aldehydes by acyl ACP reductase (AAR) which is a direct source of enzymatic conversion into alkane and alkene by aldehyde-deformylating oxygenase (ADO) enzyme. These are the two specific enzymes studied for electrostatic interaction where AAR attaches to the opening of ADO to facilitate the aldehyde into the ADO channel and then leads to the formation of alkanes (Chang et al. 2020). A schematic representation of the biosynthetic pathway for microbial alkane/alkene formation can be seen in Fig. 10.2, and the chemical structure of various alkanes and alkenes is illustrated in Fig. 10.3.

There are many cyanobacterial species identified for alkane formation. Few categories of the decane microbiologically synthesized by cyanobacteria are tetradecane (*Spirulina platensis)*, pentadecane (*Microcystis aeruginosa, Spirulina platensis, Synechococcus elongatus, Oscillatoria woronichinii*), hexadecane (*Nostoc commune, Synechococcus bacillaris, Trichodesmium erythraeum, Plectonema terebrans, Anacystis* sp.), heptadecane (*Synechococcus elongatus, Anabaena variabilis, Nostoc punctiforme, Gloeobacter violaceus*), octadecane (*Anacystis cyanea, Calothrix, Anacystis nidulans, Synechocystis UTEX 2470*), and nonadecane (*Calothrix* sp., *Microcystis aeruginosa, Spirulina platensis*) (Schirmer et al. 2010; Parveen and Yazdani 2022).

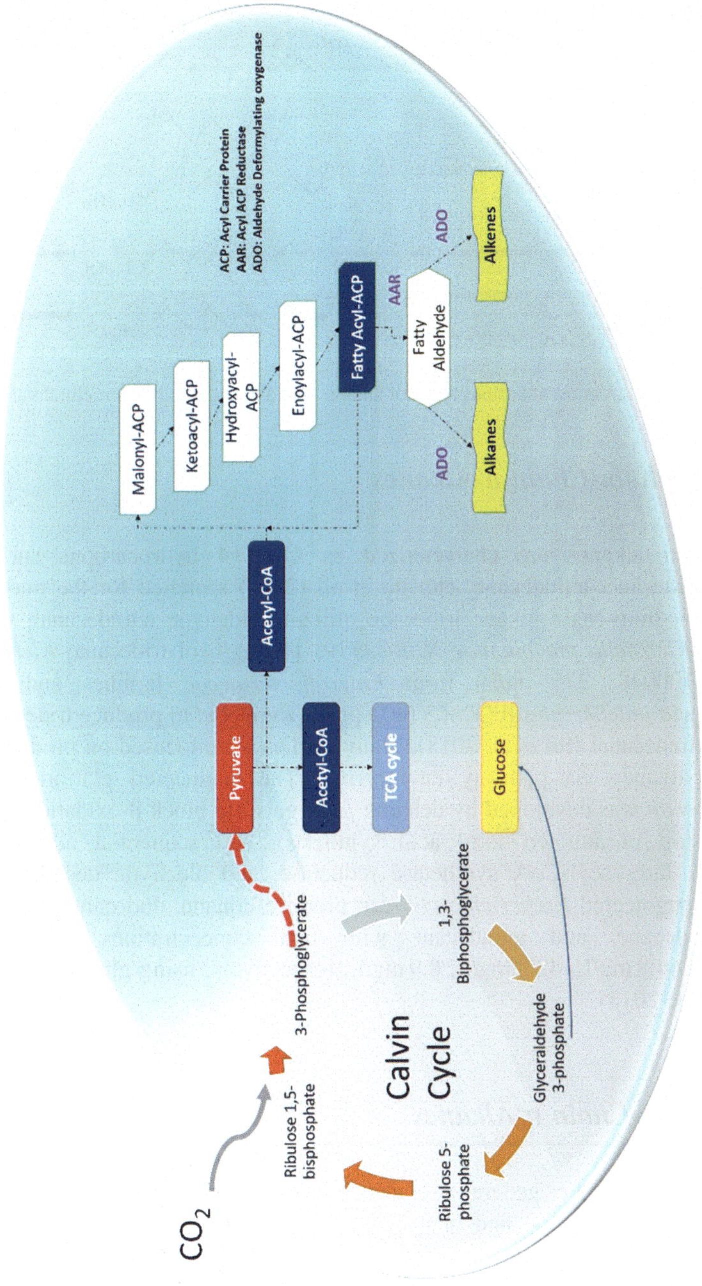

Fig. 10.2 Alkane/alkene biosynthetic pathway in a microbial host

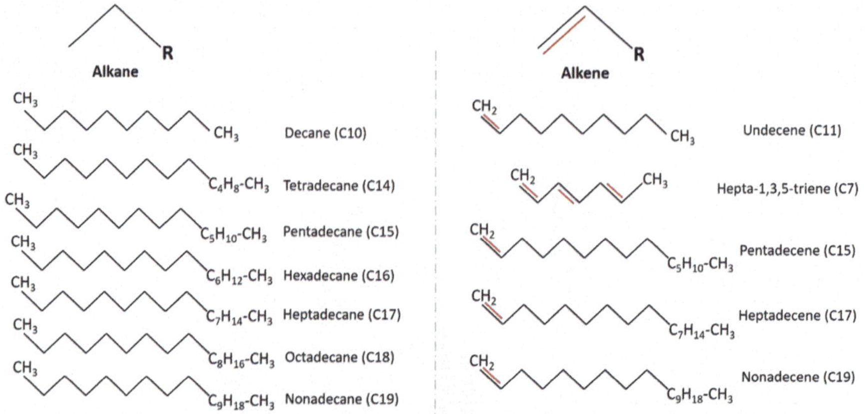

Fig. 10.3 Chemical structure of various types of alkanes and alkenes and their structural variations

10.2.1 Medium-Chain n-Alkanes

Medium-chain alkanes are characterized as C9–C14 hydrocarbons such as undecane, tridecane, tetradecane, etc. Ito et al. (2018) searched for the microbes producing medium-chain alkane tridecane utilizing aldehyde tetradecanal, where *Klebsiella (Klebsiella pneumoniae NBRC3319:* 116 μg/L of tridecane; *Klebsiella* sp. NBRC100048: 275 μg/L) from *Enterobacteriaceae* families and yeast *Metschnikowia pulcherrima* IFO0863 (402 μg/L) were able to produce tridecane at 1 mM of tetradecanal (Ito et al. 2018). Many studies were focused on producing a variety of alkanes via pathway engineering. The engineered platform using *Escherichia coli* was developed by deleting *fadE* gene (to block β-oxidation), *fadR* gene (to stop unsaturated fatty acid synthesis), and sequential addition of thioesterase, fatty acyl-CoA synthetase reductase, and aldehyde decarbonylase genes. The engineered *Escherichia coli* can produce nonane, dodecane, tridecane, 2-methyl-dodecane, and tetradecane with final concentrations 327.8 mg/L, 136.5 mg/L, 64.8 mg/L, 42.8 mg/L, 8.9 mg/L, respectively, using glucose substrate (Choi and Lee 2013).

10.2.2 Long-Chain n-Alkanes

Long-chain alkanes are generally considered for hydrocarbons with a chain length > C_{14}. Pentadecane and heptadecane are the most studied alkanes. The high efficiency of these alkanes was found in an engineered *Escherichia coli* strain BL21 (DE3) formed by introducing type-I fatty acid synthase for enhancing biosynthesis and catalase coexpression for removing by-products. The engineered *E. coli* BL21 (DE3) was able to produce 57 mg/L of pentadecane from 3% glucose

(Coursolle et al. 2015). The alkane biosynthesis and fatty acid metabolism were studied in *Saccharomyces cerevisiae* also. The engineered *Saccharomyces cerevisiae KB19* formed by *Hfd1* deletion and redox expression produce long-chain alkanes (pentadecane and heptadecane) with a cumulative titer of 22.0 1.4 μg/gDW (Buijs et al. 2015).

10.3 Microbial Alkenes

There are various long- and short-chain microbial alkenes synthesized from natural and engineered microbial hosts. *Pseudomonas* sp. can produce medium-chain alkene (1-undecene) by a specially conserved gene *UndA* found in different species of *Pseudomonas* such as *P. aeruginosa, P. fluorescens, P. putida, and P. syringae* (Rui et al. 2014). Some varieties of alkenes have been discovered by synthetic metabolic engineering. The engineered microbes *Pichia pastoris* and *Acinetobacter baylyi* have been identified for undecene (Luo et al. 2019; Cai et al. 2022), and *E. coli* has been constructed for hepta-1,3,5-triene (Messiha et al. 2021). The cyanobacteria *Synechococcus* sp. PCC7002 have been engineered for Ols (olefins) enzyme for the production of nonadecene (Mendez-Perez et al. 2011).

10.3.1 Medium-Chain n-Alkenes

Medium-chain hydrocarbons are prepared by fatty acid decarboxylation. These are the next-generation drop-in fuels produced by converting medium-chain fatty acids (C10–C14) into alkenes by fatty acid decarboxylase ($OleT_{JE}$), oxidative decarboxylase (*UndA* and *UndB*), and di-iron oxidases (*PfUndA*). 1-undecene is the most promising medium-chain alkene that can be prepared from lauric acid by the action of *UndA* (Wang and Chen 2023). It is a semi-volatile metabolite produced naturally by *Pseudomonas* via non-heme iron oxidase catalysis (Rui et al. 2014). In the production of 1-undecene, not only glucose but also ferulate (lignin-derived molecule) was utilized by the adapted strain of *A. baylyi* ADP1 (72 ± 7.5 μg/L of 1-undecene) (Luo et al. 2019).

10.3.2 Long-Chain n-Alkenes

Long-chain alkenes are generally categorized for hydrocarbons >C14 carbon. There are various applications of alkenes such as for epoxides (1,2-epoxytetradecane), alkanols (tetradecen-3-ol), and hydroxylated derivatives prepared from 1-tetradecene by unspecific peroxygenases (UPOs) from *Marasmius rotula* (Babot et al. 2022). The direct biosynthesis of long-chain alkenes was studied in

Macrococcus caseolyticus by identifying a new H_2O_2-independent P450 fatty acid (FA) decarboxylase ($OleT_{MC}$) able to produce C16 and C18 fatty acids (Lee et al. 2018). An engineered *E. coli* EcNN101 was also introduced by ($OleT_{MC}$) utilized glucose to produce 17.78 ± 0.63 mg/L of terminal alkenes (yield: 0.87 ± 0.03 mg/g) consisting of 82.7% ± 1.7% C15 alkene. Long-chain α-alkenes are produced by an engineered yeast *Pichia pastoris* using methanol as the sole carbon source and able to produce C15:1, C17:1, and C17:2 alkenes with a total titer of 1.6 mg/L (Cai et al. 2022). Not only yeast, eukaryotes like marine alga also took part in alkene biosynthesis. A ubiquitous marine alga *Emiliania huxleyi* was identified for long-chain alkenes (C31 to C38), especially dienes, trienes, and tetraenes (Volkman et al. 1980).

10.4 Microorganism and Substrate

The chemical synthesis of microbial alkane via the Fischer–Tropsch process is not suitable for the environment and is also less economical. Fatty acids are the major source of their biosynthetic pathway (Rahman et al. 2018) which can be derived from different types of substrates such as glucose, glycerol, lauric acid, acetate, methanol, and CO_2 using different types of microorganisms such as *E. coli, Acinetobacter baylyi, Pseudomonas* sp., *Saccharomyces cerevisiae, Pichia pastoris, Yarrowia lipolytica, Synechocystis* sp., and *Nostoc punctiforme* (Table 10.2).

Not only the natural microorganisms but biosynthesis was also majorly reported from engineered microbes, especially *Escherichia coli.* An engineered strain of *Acinetobacter baylyi* was used to produce bio-hydrocarbons from acetate. The approach was processed by continuing the acetate production from CO_2 by an acetogen *Acetobacterium woodii* that converts CO_2 and H_2 into acetate which is used as a substrate for alkane biosynthesis (Lehtinen et al. 2018). To expand the product profile, the *FabH2* enzyme from *Bacillus subtilis* incorporated into *Escherichia coli* to enhance the fatty acid initiation leads to fatty acid biosynthesis but leads to the altered alkanes C14 and C16 than C13, C15, and C17 (Harger et al. 2013).

Pathway engineering is the most recommended tool for the versatility of alkanes. During fatty acid metabolism, aldehyde reductase also produces by-products as a fatty alcohol. To assimilate the by-product, the synthetic alcohol to alkane route was engineered by alcohol dehydrogenase gene *PsADH* (able to oxidize short-chain alcohols (tetradecanol to tetradecenal) from *Pantoea* sp. strain 7-4. The gene was co-expressed with cyanobacterial aldehyde-deformylating oxygenase from *Nostoc punctiforme* PCC73102 in *Escherichia coli* (Sui et al. 2022). The coexpression helps in reducing the formation of by-products and converting them into the main product to enhance the total yield. So, there are several substrates considered for microbial alkane production.

Not only the bacteria but also yeast was also engineered to produce bio-hydrocarbons. Cai et al. (2022) worked on alpha-alkene, especially long-chain alkenes, and were able to develop a yeast *Pichia pastoris* using a combination of

Table 10.2 Classification of microorganisms able to produce microbial alkane/alkene from different substrates

Class	Organism name	Organism type	Substrate	Modification process	Hydrocarbon name	Hydrocarbon type	Titer	Extraction method	References
Bacteria	*Acinetobacter baylyi* ADP1	Engineered	Acetate	Transformation of an expression cassette cloned with *ADO* and *AAR*	Heptadecane	Alkane (C17)	540 μg/L	Chloroform–methanol extraction	(Lehtinen et al. 2018)
	Acinetobacter baylyi ADP1	Engineered	Glucose	Expression of *UndA* (*Pseudomonas putida*), *TesA* (*E. Coli*)	1-undecene	Alkene (C11)	694 ± 76 μg/L	SPME fiber-based adsorption/desorption	(Luo et al. 2019)
	Acinetobacter baylyi ADP1	Engineered	Ferulate (lignin-derived molecule	Expression of *UndA* (*Pseudomonas putida*), *TesA* (*E. coli*)	1-undecene	Alkene (C11)	72 ± 7.5 μg/L	SPME fiber-based adsorption/desorption	(Luo et al. 2019)
	E. coli XL1-Blue MR	Engineered	2,4,6-octatrienoic acid	Co-expressing the del_AdmH-AdmP construct With an *Fdc1*/*UbiX* expression construct	Hepta-1,3,5-triene	Alkene (C7)	78% (of 50 μM standard)	1 ml head-space volume was taken and injected in GC-MS	(Messiha et al. 2021)
	E. coli BL21 (DE3)	Engineered	Glycerol	Altering fattyacyl synthesis and elector transfer	Pentadecane, Heptadecene	Alkane (C15) Alkene (C17)	1.31 g/L	Ethyl acetate extraction	(Cao et al. 2016)
	E. coli BL21 Gold	Engineered	Lauric acid	Overexpression of *undA*	1-undecene	Alkene (C11)	0.06 Mol/UndA-Fe (II) second	SPME fiber (adsorption/desorption)	(Rui et al. 2014)
		Natural	Lauric acid	–	1-undecene	Alkene (C11)	15 ng/mL		(Rui et al. 2014)

(continued)

Table 10.2 (continued)

Class	Organism name	Organism type	Substrate	Modification process	Hydrocarbon name	Hydrocarbon type	Titer	Extraction method	References
	Pseudomonas aeruginosa PAO1							SPME fiber (adsorption/ desorption)	
	Pseudomonas syringae TLP2	Natural	Lauric acid	–	1-undecene	Alkene (C11)	70 ng/mL	SPME fiber (adsorption/ desorption)	(Rui et al. 2014)
	Pseudomonas putida F1	Natural	Lauric acid	–	1-undecene	Alkene (C11)	2 ng/mL	SPME fiber (adsorption/ desorption)	(Rui et al. 2014)
Yeast	*Saccharomyces cerevisiae* KB19	Engineered	Glucose	Elimination of *Hfd1* (hexadecenal dehydrogenase) and expression of a redox system	Tridecane, pentadecane, heptadecane	Alkanes (C13, C15, C17)	22.0 μg/ gDW	Freeze-dried cell Pellets in hexane	(Buijs et al. 2015)
	Pichia pastoris PC156	Engineered	Methanol	Combinatorial engineering (expression of *PfUndB* from *Pseudomonas fluorescens* Pf-5 and methanol-induced promoters (*PAOX1*, *PDAS2*, and *PCAT1*) by CRISPR–Cas9	1-pentadecene, 1-heptadecene, 1,8-heptadecadiene	Alkenes (C15:1, C17: 1 and C17:2)	1.6 mg/L	Chloroform–methanol (2: 1, v/v)	(Cai et al. 2022)

	Yarrowia lipolytica YLjbl-2	Engineered	Glucose	Expression of *CvFAP* (fatty acid photo-decarboxylase)	Heptadecane, Pentadecane, 8-heptadecene, 6,9-heptadecadiene	Alkanes C17: 0), (C15:0), and alkenes (C17:1), (C17:2)	1.47 g/L	Cell lysis (sodium hydroxide-methanol), acidified by H_2SO_4 and hexane extraction	(Li et al. 2020)
	Yarrowia lipolytica YLjbl-2	Engineered	Wheat straw hydrolysate	Expression of *CvFAP* (fatty acid photo-decarboxylase)	Heptadecane, Pentadecane, 8-heptadecene, 6,9-heptadecadiene	Alkanes C17: 0), (C15:0), and alkenes (C17:1), (C17:2)	242.04 mg/L	Cell lysis (sodium hydroxide-methanol), acidified by H_2SO_4 and hexane extraction	(Li et al. 2020)
	Yarrowia lipolytica YLjbl-2	Engineered	Acetic acid	Expression of *CvFAP* (fatty acid photo-decarboxylase)	Heptadecane, Pentadecane, 8-heptadecene, 6,9-heptadecadiene	Alkanes C17: 0), (C15:0), and alkenes (C17:1), (C17:2)	318.70 mg/L	Cell lysis (sodium hydroxide-methanol), acidified by H_2SO_4 and hexane extraction	(Li et al. 2020)
Cyanobacteria	*Synechocystis* 6803 F30	Engineered	CO_2	*FAR* and *FAD* genes overexpression	Heptadecane	Alkanes	~0.2% of dry cell weight	Chloroform–methanol extraction	(Hu et al. 2013)
		Engineered	CO_2		Heptadecane	Alkanes	26 mg/L (1.1% of		(Wang et al. 2013)

(continued)

Table 10.2 (continued)

Class	Organism name	Organism type	Substrate	Modification process	Hydrocarbon name	Hydrocarbon type	Titer	Extraction method	References
	Synechocystis sp. PCC6803 LX56			Overexpression of alkane bio-synthetic genes			dry cell weight)	Chloroform–methanol extraction	
	Nostoc punctiforme Np1710/11/5141	Engineered	CO_2	Expression of *aar/adc* genes and novel lipase gene encoded by *Npun_F5141*	Heptadecane	Alkanes	12.9% ± 0.30% of dry cell weight	Cell lysis and hexane-based extraction	(Peramuna et al. 2015)

engineering approaches such as expression optimization and co-factor engineering using *UndB*, *CamA*, and *CamB* genes. The engineered yeast was found to be able to produce 1.6 mg/L of (C15: 1-pentadecene, C17:1-heptadecene) alkenes from methanol as a non-food feedstock (Cai et al. 2022). It shows that there are lots of advancements going on for sustainable production of alkane/alkenes. Alkane/alkenes are chemically toxic to microorganisms and cause membrane integrity and growth inhibition. To reduce toxicity, various methods have been applied such as a two-phase culture system, engineering gene regulation, overexpressing heat show proteins or srRNA, and solvent-tolerant strain (Kang and Nielsen 2017). However, further research is still required to overcome this challenge.

10.5 Production and Improvements

Microbial alkanes/alkenes are mostly produced by the fed-batch fermentation methods (Cao et al. 2016) because they give a subsequent supply of nutrients to reduce carbon limitations and also help in improving microbial biomass (Lehtinen et al. 2018). During production, the microbial host is inoculated into a carbon-rich medium inside a bioreactor to develop microbial biomass to accumulate the produce inter- or extracellularly. In most cases, the product is intracellular, and cell lysis is required after fermentation that follows the solvent-based extraction for the purification and analysis of the product. There are several challenges arising during the large-scale production of alkane/alkenes, especially the titer and yield. Despite sufficient information available for alkane/alkene production from *E. coli* and *S. cerevisiae*, the engineered host produces lower than standard production at the industrial level. It often causes the scale-up issues of a biosynthetic product. While comparing the yield of natural ethanol in yeast to alkane/alkenes, there must be a 90% yield on the theoretical basis to meet the cost efficiency over ethanol. However, the downstream purification is less costly for alkane/alkenes than ethanol distillation. Overall, it sums up to improve the productivity at the upstream stage. There were several improvements established such as improving biosynthetic enzyme activity, precursor supply, adding co-factors, reducing toxicity tolerance, and decreasing feedback inhibition (Kang and Nielsen 2017).

10.5.1 Precursor Supply

The abundant supply of precursors such as acetyl-CoA and fatty acyl-CoA is very important. The supply of precursor depends on the host used; for example, yeast metabolizes acetyl-CoA in various cellular regions mitochondria, cytosol, peroxisomes, and nucleus. The yeast cytosol supply was improved by the co-expression of ATP-independent pyruvate dehydrogenase (*PDH*) gene, and lipoylation genes (*lplA* and *lplA2*) from *Enterococcus faecalis* fulfill acetyl-CoA supply in the cytosol

(Kozak et al. 2014). The strategy helps in minimizing the cost of ATP energy required for ATP-dependent pathways. The re-channeling of ethanol into acetyl-CoA was also helping the higher production of alkanes/alkenes. The overexpression of ethanol degradation genes (*ADH2, ALD6*) and acetyl-CoA synthetase acs_{SE}^{L641P} with wax ester synthase (*ws2*) genes (from *Marinobacter hydrocarbonoclasticus* DSM 8798) showed a three-fold increase (408 ± 270 μg FAEE $gCDW^{-1}$) in fatty acid ethyl ester production which indirectly relates to the alkane/alkene production yield in engineered *S. cerevisiae* (de Jong et al. 2014).

10.5.2 Co-Factor Supply

Recently, co-factor supply and its modification or replacement showed great improvement in alkane/alkene production. A newly identified cytochrome P450 member OleT is able to facilitate decarboxylation of free fatty acids into olefins using a co-factor H_2O_2 (Wise et al. 2018). H_2O_2 acts as an oxidant and electron donor. Other co-factors (NADPH and O_2) showed the H_2O_2-independent $OleT_{JE}$ activity. The enzyme showed improved decarboxylation of fatty acid (C_{12}–C_{20}) when NADPH and O_2 were used as co-factors and produced multiple alkanes. The stearic acid and palmitic acid converted into heptadecene and pentadecene (Liu et al. 2014). The $OleT_{JE}$ when coupled with metalloenzymes performs olefins hydrogenation to form paraffins (Monteiro et al. 2022). Therefore, many additional supplements improve the production of alkene from free fatty acids.

10.6 Effective Microbial Alkane/Alkene as Drop-in Biofuels

10.6.1 Farnesene

Farnesene is a successful example of the bio-based production of hydrocarbons used as precursors for jet fuel. Many advancements are going on for advanced biofuel and bioproducts (Pfleger and Takors 2023), where recombinant DNA technology plays an important role in metabolic engineering of a bacterium (*E. coli*) or yeast (*S. cerevisiae*) for the production of aviation fuels. Trans-β-farnesene is a renewable hydrocarbon and building block. It is a C-15 long-chain hydrocarbon produced by Amyris company via a sugar-based fermentation process using an engineered yeast (https://farnesene.net/). The yeast metabolic route was shifted from ethanol to hydrocarbon by rewiring the central metabolism (Meadows et al. 2016; Yu et al. 2016), which showed higher acetyl-CoA production at low ATP requirement. The process also lowers CO_2 emissions. The product efficiency was increased by 25%. There are many other advancements to improve farnesene from different substrates. The research progress showed that bio-jet fuel precursors can be produced from glucose, glycerol, waste cooking oil, and lignocellulosic biomass (Walls and Rios-

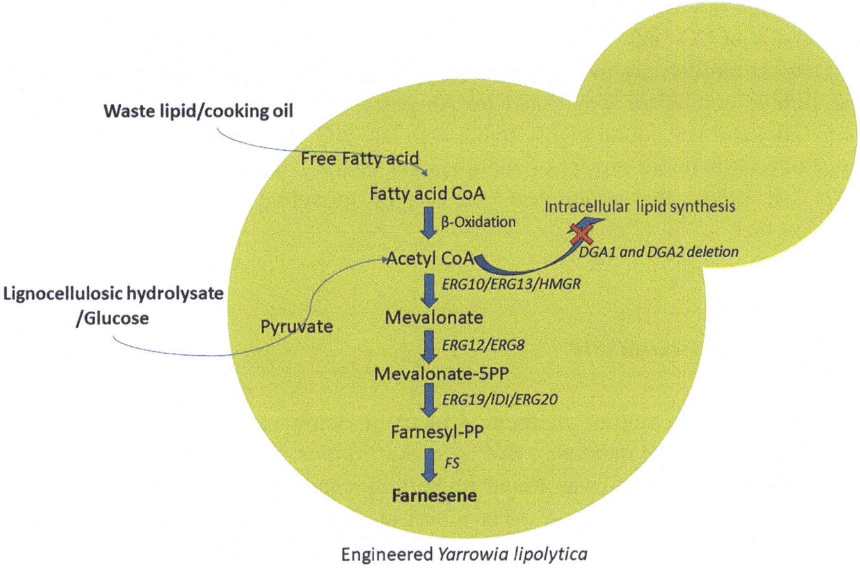

Fig. 10.4 Metabolic route in engineered yeast *Yarrowia lipolytica* for sustainable production of farnesene from waste lipid and glucose source (Adapted from Shaikh and Odaneth 2021; Liu et al. 2021, 2022; Bi et al. 2022)

Solis 2020) as well as α and β type of farnesene from CO_2 (Tang et al. 2021). *Yarrowia lipolytica* is the most prominent oleaginous yeast engineered to produce farnesene from glucose as well as lipids (Fig. 10.4). The production was enhanced by overexpression of pathway genes such as *ERG12 and ERG20* in lipid metabolism (Liu et al. 2021), as well as *ERG13, IDI, and HMGR* using glucose substrate (Shaikh and Odaneth 2021; Bi et al. 2022). Along with that, the deletion of lipid accumulating genes DGA1 and DGA2 also helped in decreasing lipid accumulation from 96.3 mg/g DCW to 24.1 mg/g DCW, and farnesene production was increased by 33.1.% from control (Liu et al. 2022). The biofuel synthesis from CO_2 was performed via an electrochemical system integration (Bi et al. 2023), where CO_2 was first converted into acetic and formic acid. Then, these organic acids were used by the engineered microbes to convert them into farnesene. A two-stage process enables the production of β-farnesene from CO_2. The other improvement in an engineered *E. coli* was found via a solar-driven photosynthetic system (Wang et al. 2022). The jet fuel farnesene was produced by implementing a mesoporous aluminum oxide that gives a long-lived electron in the presence of solar light. The electron enters into microbial cells and enhances the NAPDH level directly enhancing the rate of the MVA (mevalonate) pathway and leading to overproduction of farnesene by a factor of 100%. In another study, α-farnesene was produced from CO_2 using an engineered cyanobacterial species *Synechococcus elongatus* PCC 7942 (Lee et al. 2017). A heterologous farnesene synthase was expressed and the simultaneous introduction of the MEP (methylerythritol phosphate) pathway resulted in 4.6 ± 0.4 mg/L of α-farnesene from CO_2. Many advances in the direct conversion of CO_2 into biofuel contribute to reductions in GHG emissions.

Drop-in biofuels are mostly tested for aviation jet fuel replacement purposes. A gas turbine combustor was tested by Amyris's farnesane (hydrogenated form of fernesene) and was found highly stable with less NO and CO emissions and can be used directly without any prior modification in the combustor (Feser and Gupta 2021). It shows the importance and use of microbial alkenes as sustainable aviation fuel.

10.6.2 Botryococcene

Botryococcene is a kind of triterpene, structurally known as isoprenoid alkene. It is produced naturally in microalga *Botryococcus braunii* (Weiss et al. 2010). Around 76% of dry cell weight was found to be long-chain botryococcene ($C_{34}H_{58}$) in *Botryococcus braunii* Kützing (Maxwell et al. 1968). Botryococcene is the most promising hydrocarbon because it has similar physical properties to standard diesel fuel. The density (kg/cm3) and surface tension ($\times 10^{-3}$ N/m) were 0.835 and 29.7 of triterpenic hydrocarbon of *Botryococcus braunii* BOT-22 found similar to 0.826 and 28.2 of standard diesel fuel (Nagano et al. 2012). Recently, a new race B (Showa) strain of *Botryococcus braunii* holdspromise for producing a large number of alkenes in the range of C30 to C34, especially C33-1 botryococcene (Tatli et al. 2017). *Botryococcus braunii* is a special green alga with a thick cell wall and most alkenes form biofilm outside the cell and form aggregated colonies. To extract hydrocarbons, the in situ solvent-based extraction in a high shearing is preferred which resulted in productivity around 29.9 ± 8.5 mg L^{-1} day^{-1} using heptane at 415 s^{-1} shear rate (Jackson et al. 2019). Other than heptane, hexane (Furuhashi et al. 2013) and dimethyl ether (Kanda et al. 2013) also perform well. Other than the solvent extraction method, an electrical field is applied to weaken the cell walls. An electro-extraction process was tested for 80 nanoseconds using a pulsed electric field of 20–65 kV/cm which depicts an effective low-cost approach and causes less damage to cells (Guionet et al. 2017). It concludes that *Botryococcus braunii* is a rich source of long-chain alkene Botryococcene produced at a very high rate per dry cell weight.

10.7 Challenges and Bottlenecks During Microbial-Based Alkane/Alkenes Production

Alkane/alkene production via microbial route is a sustainable and green approach to contribute to fueling the transport system. It is still facing challenges at the commercial level such as low yield like 1–2 g/L, using metabolic engineering microbes, limited substrates, and post-modifications or refinements for matching actual standards. The technology needs a larger number of fermenters, and cost-consuming scale-up, real-time monitoring, and working within a CO_2 environment are the major industrial-level challenges. The strength of the substrate also affects the microbial stability as well as production rate. It showed that there is still more improvement and advancement needed to compete the conventional technologies. The major bottleneck for the whole process is the microorganism of interest, i.e., a more robust microorganism that is high-yielding and versatile for using a variety of substrates. Pathway engineering is the more promising approach for making these kinds of robust microorganisms that can withstand different environments. Toward the long-term sustainable goal, commercial growth needs to be addressed by using renewable feedstocks such as lignocellulosic biomass-derived sugars or non-edible or waste oil-derived lipids. However, these technologies are still at the lab or pilot scale and facing several scale-up challenges. Conventional chemical catalysis is cost-consuming, is economically weak, and has high catalyst cost, short lifespan, and less recyclability. Whether the microbial platform is more advantageous, but it is immature enough to commercialize for a very good amount of production.

10.8 Perspectives

The effectiveness of the biological production of alkanes/alkenes gives an alternative method for developing an environmentally friendly approach that also reduces GHG emissions. It concludes that microbial alkanes/alkenes are sustainable biofuels for commercial transportation systems and can be used as bio-jet fuel, bio-gasoline, and drop-in forms. There are more advancements needed to increase its yield so that sustainable commercial production can be achieved.

10.9 Conclusions

Microbial alkanes/alkenes are a category of bio-hydrocarbon produced microbiologically through a sustainable route. These act as drop-in biofuels because of their structural similarity and use for commercial vehicles. These are produced by different microorganisms such as bacteria, cyanobacteria, and yeast mostly engineered for higher product formation via engineering or altering the metabolic pathway. The alkane as heptadecane and alkene as 1-undecene are majorly found as bioproduct metabolized by using various substrates. The variety of substrates studied for microbial alkane/alkene production are CO_2, glucose, fatty acid, acetate, lignocellulosic hydrolysate, lauric acid, and methanol.

Acknowledgments The authors are thankful to Bennett University for providing the necessary infrastructure to complete this study.

References

Azimov U, Okoro V, Hernandez HH (2021) Recent progress and trends in the development of microbial biofuels from solid waste—A review. Energies 14:6011. https://doi.org/10.3390/EN14196011

Babot ED, Aranda C, Kiebist J et al (2022) Enzymatic epoxidation of long-chain terminal alkenes by fungal peroxygenases. Antioxidants 11:522. https://doi.org/10.3390/ANTIOX11030522/S1

Baena LM, Vásquez FA, Calderón JA (2021) Corrosion assessment of metals in bioethanol-gasoline blends using electrochemical impedance spectroscopy. Heliyon 7:e07585. https://doi.org/10.1016/J.HELIYON.2021.E07585

Bi H, Xv C, Su C et al (2022) β-Farnesene production from low-cost glucose in lignocellulosic hydrolysate by engineered Yarrowia lipolytica. Fermentation 8:532. https://doi.org/10.3390/FERMENTATION8100532/S1

Bi H, Wang K, Xu C et al (2023) Biofuel synthesis from carbon dioxide via a bio-electrocatalysis system. Chem Catalysis 3:100557. https://doi.org/10.1016/J.CHECAT.2023.100557

Brown S, Loh J, Aves SJ, Howard TP (2018) Alkane biosynthesis in bacteria. Biogenesis of Hydrocarbons:1–20. https://doi.org/10.1007/978-3-319-53114-4_7-1

Buijs NA, Zhou YJ, Siewers V, Nielsen J (2015) Long-chain alkane production by the yeast Saccharomyces cerevisiae. Biotechnol Bioeng 112:1275–1279. https://doi.org/10.1002/BIT.25522

Bušić A, Mardetko N, Kundas S et al (2018) Bioethanol production from renewable raw materials and its separation and purification: a review. Food Technol Biotechnol 56:289. https://doi.org/10.17113/FTB.56.03.18.5546

Cai P, Li Y, Zhai X et al (2022) Microbial synthesis of long-chain α-alkenes from methanol by engineering Pichia pastoris. Bioresour Bioprocess 9:1–8. https://doi.org/10.1186/S40643-022-00551-1/FIGURES/4

Cao YX, Xiao WH, Zhang JL et al (2016) Heterologous biosynthesis and manipulation of alkanes in Escherichia coli. Metab Eng 38:19–28. https://doi.org/10.1016/J.YMBEN.2016.06.002

Caporusso A, Capece A, De Bari I (2021) Oleaginous Yeasts as Cell Factories for the Sustainable Production of Microbial Lipids by the Valorization of Agri-Food Wastes. Fermentation 7:50. https://doi.org/10.3390/FERMENTATION7020050

Chang M, Shimba K, Hayashi Y, Arai M (2020) Electrostatic interactions at the interface of two enzymes are essential for two-step alkane biosynthesis in cyanobacteria. Biosci Biotechnol Biochem 84:228–237. https://doi.org/10.1080/09168451.2019.1677142

Choi YJ, Lee SY (2013) Microbial production of short-chain alkanes. Nature 502(7472):571–574. https://doi.org/10.1038/nature12536

Coursolle D, Lian J, Shanklin J, Zhao H (2015) Production of long chain alcohols and alkanes upon coexpression of an acyl-ACP reductase and aldehyde-deformylating oxygenase with a bacterial type-I fatty acid synthase in E. Coli. Mol BioSyst 11:2464–2472. https://doi.org/10.1039/C5MB00268K

de Jong BW, Shi S, Siewers V, Nielsen J (2014) Improved production of fatty acid ethyl esters in Saccharomyces cerevisiae through up-regulation of the ethanol degradation pathway and expression of the heterologous phosphoketolase pathway. Microb Cell Factories 13:1–10. https://doi.org/10.1186/1475-2859-13-39/TABLES/3

Feser J, Gupta A (2021) Performance and emissions of drop-in aviation biofuels in a lab-scale gas turbine combustor. J Energy Resour Technol Trans ASME 143(4):042103. https://doi.org/10.1115/1.4048243

Furuhashi K, Saga K, Okada S, Imou K (2013) Seawater-cultured Botryococcus braunii for efficient hydrocarbon extraction. PLoS One 8:e66483. https://doi.org/10.1371/JOURNAL.PONE.0066483

Guionet A, Hosseini B, Teissié J et al (2017) A new mechanism for efficient hydrocarbon electro-extraction from Botryococcus braunii. Biotechnol Biofuels 10:1–9. https://doi.org/10.1186/S13068-017-0724-1/FIGURES/9

Harger M, Zheng L, Moon A et al (2013) Expanding the product profile of a microbial alkane biosynthetic pathway. ACS Synth Biol 2:59–62. https://doi.org/10.1021/SB300061X/SUPPL_FILE/SB300061X_SI_001.PDF

Hu P, Borglin S, Kamennaya NA et al (2013) Metabolic phenotyping of the cyanobacterium Synechocystis 6803 engineered for production of alkanes and free fatty acids. Appl Energy 102:850–859. https://doi.org/10.1016/J.APENERGY.2012.08.047

Ito M, Kambe H, Kishino S et al (2018) A search for microorganisms producing medium-chain alkanes from aldehydes. J Biosci Bioeng 125:87–91. https://doi.org/10.1016/J.JBIOSC.2017.08.002

Jackson BA, Bahri PA, Moheimani NR (2019) Repetitive extraction of botryococcene from Botryococcus braunii: a study of the effects of different solvents and operating conditions. J Appl Phycol 31:3491–3501. https://doi.org/10.1007/S10811-019-01883-W/METRICS

Kanda H, Li P, Yoshimura T, Okada S (2013) Wet extraction of hydrocarbons from Botryococcus braunii by dimethyl ether as compared with dry extraction by hexane. Fuel 105:535–539. https://doi.org/10.1016/J.FUEL.2012.08.032

Kang MK, Nielsen J (2017) Biobased production of alkanes and alkenes through metabolic engineering of microorganisms. J Ind Microbiol Biotechnol 44:613–622. https://doi.org/10.1007/S10295-016-1814-Y

Karatzos S, van Dyk JS, McMillan JD, Saddler J (2017) Drop-in biofuel production via conventional (lipid/fatty acid) and advanced (biomass) routes. Part I. Biofuels Bioprod Biorefin 11:344–362. https://doi.org/10.1002/BBB.1746

Khuong LS, Masjuki HH, Zulkifli NWM et al (2017) Effect of gasoline–bioethanol blends on the properties and lubrication characteristics of commercial engine oil. RSC Adv 7:15005–15019. https://doi.org/10.1039/C7RA00357A

Kozak BU, van Rossum HM, Luttik MAH et al (2014) Engineering acetyl coenzyme a supply: functional expression of a bacterial pyruvate dehydrogenase complex in the cytosol of Saccharomyces cerevisiae. MBio 5:10–1128. https://doi.org/10.1128/MBIO.01696-14

Lee HJ, Lee J, Lee SM et al (2017) Direct conversion of CO2 to α-Farnesene using metabolically engineered Synechococcus elongatus PCC 7942. J Agric Food Chem 65:10424–10428. https://doi.org/10.1021/ACS.JAFC.7B03625

Lee JW, Niraula NP, Trinh CT (2018) Harnessing a P450 fatty acid decarboxylase from macrococcus caseolyticus for microbial biosynthesis of odd chain terminal alkenes. Metab Eng Commun 7:e00076. https://doi.org/10.1016/J.MEC.2018.E00076

Lehtinen T, Virtanen H, Santala S, Santala V (2018) Production of alkanes from CO2 by engineered bacteria. Biotechnol Biofuels 11:1–11. https://doi.org/10.1186/S13068-018-1229-2/TABLES/1

Li J, Ma Y, Liu N et al (2020) Synthesis of high-titer alka(e)nes in Yarrowia lipolytica is enabled by a discovered mechanism. Nature Communications 11(1):1–13. https://doi.org/10.1038/s41467-020-19995-0

Liu Y, Wang C, Yan J et al (2014) Hydrogen peroxide-independent production of α-alkenes by OleT JE P450 fatty acid decarboxylase. Biotechnol Biofuels 7:1–12. https://doi.org/10.1186/1754-6834-7-28/FIGURES/6

Liu Y, Wang Z, Cui Z et al (2021) α-Farnesene production from lipid by engineered Yarrowia lipolytica. Bioresour Bioprocess 8:1–12. https://doi.org/10.1186/S40643-021-00431-0/TABLES/2

Liu Y, Zhang J, Li Q et al (2022) Engineering Yarrowia lipolytica for the sustainable production of β-farnesene from waste oil feedstock. Biotechnol Biofuels Bioprod 15:1–15. https://doi.org/10.1186/S13068-022-02201-2/TABLES/2

Lou X, Singh PM (2010) Role of water, acetic acid and chloride on corrosion and pitting behaviour of carbon steel in fuel-grade ethanol. Corros Sci 52:2303–2315. https://doi.org/10.1016/J.CORSCI.2010.03.034

Luo J, Lehtinen T, Efimova E et al (2019) Synthetic metabolic pathway for the production of 1-alkenes from lignin-derived molecules. Microb Cell Factories 18:1–13. https://doi.org/10.1186/S12934-019-1097-X/TABLES/1

Ma S, Lei T, Meng J et al (2023) Global oil refining's contribution to greenhouse gas emissions from 2000 to 2021. The Innovation 4:100361. https://doi.org/10.1016/J.XINN.2022.100361

Maxwell JR, Douglas AG, Eglinton G, McCormick A (1968) The Botryococcenes—hydrocarbons of novel structure from the alga Botryococcus braunii, Kützing. Phytochemistry 7:2157–2171. https://doi.org/10.1016/S0031-9422(00)85672-1

Meadows AL, Hawkins KM, Tsegaye Y et al (2016) Rewriting yeast central carbon metabolism for industrial isoprenoid production. Nature 537(7622):694–697. https://doi.org/10.1038/nature19769

Mendez-Perez D, Begemann MB, Pfleger BF (2011) Modular synthase-encoding gene involved in α-olefin biosynthesis in Synechococcus sp. strain PCC 7002. Appl Environ Microbiol 77:4264–4267. https://doi.org/10.1128/AEM.00467-11

Messiha HL, Payne KAP, Scrutton NS, Leys D (2021) A biological route to conjugated alkenes: microbial production of Hepta-1,3,5-triene. ACS Synth Biol 10:228–235. https://doi.org/10.1021/ACSSYNBIO.0C00464/ASSET/IMAGES/LARGE/SB0C00464_0003.JPEG

Monteiro RRC, da Silva SSO, Cavalcante CL et al (2022) Biosynthesis of alkanes/alkenes from fatty acids or derivatives (triacylglycerols or fatty aldehydes). Biotechnol Adv 61:108045. https://doi.org/10.1016/J.BIOTECHADV.2022.108045

Nagano S, Yamamoto S, Nagakubo M et al (2012) Physical properties of hydrocarbon oils produced by Botryococcus Braunii: density, kinematic viscosity, surface tension, and distillation properties. Procedia Environ Sci 15:73–79. https://doi.org/10.1016/J.PROENV.2012.05.012

Parveen H, Yazdani SS (2022) Insights into cyanobacterial alkane biosynthesis. J Ind Microbiol Biotechnol 49:kuab075. https://doi.org/10.1093/JIMB/KUAB075

Peramuna A, Morton R, Summers ML (2015) Enhancing alkane production in cyanobacterial lipid droplets: a model platform for industrially relevant compound production. Life 5:1111–1126. https://doi.org/10.3390/LIFE5021111

Pfleger BF, Takors R (2023) Recent progress in the synthesis of advanced biofuel and bioproducts. Curr Opin Biotechnol 80:102913. https://doi.org/10.1016/J.COPBIO.2023.102913

Rahman Z, Nawab J, Sung BH, Kim SC (2018) A critical analysis of bio-hydrocarbon production in bacteria: current challenges and future directions. Energies 11:2663. https://doi.org/10.3390/EN11102663

Rui Z, Li X, Zhu X et al (2014) Microbial biosynthesis of medium-chain 1-alkenes by a nonheme iron oxidase. Proc Natl Acad Sci USA 111:18237–18242. https://doi.org/10.1073/PNAS.1419701112/SUPPL_FILE/PNAS.1419701112.SAPP.PDF

Schirmer A, Rude MA, Li X et al (2010) Microbial biosynthesis of alkanes. Science 329:559–562. https://doi.org/10.1126/SCIENCE.1187936

Shaikh KM, Odaneth AA (2021) Metabolic engineering of Yarrowia lipolytica for the production of isoprene. Biotechnol Prog 37:e3201. https://doi.org/10.1002/BTPR.3201

Sui YA, Kishino S, Maruyama S et al (2022) Utilizing alcohol for alkane biosynthesis by introducing a fatty alcohol dehydrogenase. Appl Environ Microbiol 88:e01264-22. https://doi.org/10.1128/AEM.01264-22/SUPPL_FILE/AEM.01264-22-S0001.PDF

Sun X, Shen X, Jain R et al (2015) Synthesis of chemicals by metabolic engineering of microbes. Chem Soc Rev 44:3760–3785. https://doi.org/10.1039/C5CS00159E

Tang R, Wen Q, Li M et al (2021) Recent advances in the biosynthesis of Farnesene using metabolic engineering. J Agric Food Chem 69:15468–15483. https://doi.org/10.1021/ACS.JAFC.1C06022/ASSET/IMAGES/MEDIUM/JF1C06022_0005.GIF

Tatli M, Naik MT, Okada S et al (2017) Isolation and characterization of cyclic C33 Botryococcenes and a Trimethylsqualene isomer from Botryococcus braunii race B. J Nat Prod 80:953–958. https://doi.org/10.1021/ACS.JNATPROD.6B00934/SUPPL_FILE/NP6B00934_SI_002.PDF

Volkman JK, Eglinton G, Corner EDS, Forsberg TEV (1980) Long-chain alkenes and alkenones in the marine coccolithophorid Emiliania huxleyi. Phytochemistry 19:2619–2622. https://doi.org/10.1016/S0031-9422(00)83930-8

Walls LE, Rios-Solis L (2020) Sustainable production of microbial isoprenoid derived advanced biojet fuels using different generation feedstocks: a review. Front Bioeng Biotechnol 8:599560. https://doi.org/10.3389/FBIOE.2020.599560/BIBTEX

Wang Y, Chen SL (2023) Mechanism for the synthesis of medium-chain 1-alkenes from fatty acids catalyzed by binuclear iron UndA decarboxylase. J Catal 420:123–133. https://doi.org/10.1016/J.JCAT.2023.02.021

Wang W, Liu X, Lu X (2013) Engineering cyanobacteria to improve photosynthetic production of alka(e)nes. Biotechnol Biofuels 6:1–9. https://doi.org/10.1186/1754-6834-6-69/TABLES/2

Wang L, Chen L, Yang S, Tan X (2020) Photosynthetic conversion of carbon dioxide to oleochemicals by cyanobacteria: recent advances and future perspectives. Front Microbiol 11: 526198. https://doi.org/10.3389/FMICB.2020.00634/BIBTEX

Wang J, Chen N, Bian G et al (2022) Solar-driven overproduction of biofuels in microorganisms. Angew Chem Int Ed 61:e202207132. https://doi.org/10.1002/ANIE.202207132

Weiss TL, Chun HJ, Okada S et al (2010) Raman spectroscopy analysis of botryococcene hydrocarbons from the green microalga Botryococcus braunii. J Biol Chem 285:32458–32466. https://doi.org/10.1074/jbc.M110.157230

Wise CE, Hsieh CH, Poplin NL, Makris TM (2018) Dioxygen activation by the biofuel-generating cytochrome P450 OleT. ACS Catal 8:9342–9352. https://doi.org/10.1021/ACSCATAL.8B02631/SUPPL_FILE/CS8B02631_SI_001.PDF

Yu Vivian Y, Chang MC (2016) High-yield chemical synthesis by reprogramming central metabolism. Nat Biotechnol 34(11):1129–1129. https://doi.org/10.1038/nbt.3723

Zheng L, Boylu R, Cronly J et al (2022) Experimental study on the impact of alternative jet fuel properties and derived cetane number on lean blowout limit. Aeronaut J 126:1997–2016. https://doi.org/10.1017/AER.2022.33

Chapter 11
Downstream Processing of Drop-In Biofuels

Júlio C. de Carvalho, Susan G. Karp, Walter J. Martínez-Burgos, Thamarys Scapini, Rafael N. Barros, Adriane B. P. Medeiros, Marcela M. Albuquerque, Adenise L. Woiciechowski, and Carlos R. Soccol

Abstract The world urgently requires decarbonizing its energy matrix. However, areas such as industry and transportation will likely rely on high-energy-density liquid fuels in the near future. Drop-in biofuels offer a sustainable solution that seamlessly integrates with existing pipelines, supply chains, and current applications. Examples include bioethanol from various sources (such as sucrose, starch, or lignocellulosic materials), biomethane, biohydrogen, and biohythane. Biodiesel (from plant or microbial sources), lipid-derived hydrocarbons, and pyrolysis gas or gasohol also qualify as drop-in biofuels to some extent. Some highly energetic molecules like higher alcohols, ketones, or ammonia do not fit this category, as they necessitate extensive engine modifications. However, high-energy molecules cannot be considered fuels until they undergo concentration and purification, achieved through a series of processing steps known as downstream processes.

This chapter explores the fundamentals of downstream processing for biofuels after their biosynthesis. It includes a material balance for the more technologically mature biofuels, allowing readers to estimate the potential fuel yield from a given amount of biomass.

Keywords Purification · Energy density · Bioethanol · Biodiesel · Biomethane · Biohydrogen

11.1 Introduction

Despite the accelerated electrification of vehicles, most human transportation still requires high-density, fluid fuels. Biofuels are a clean alternative to the fossil fuels used today. However, its recovery from plant biomass and microorganism cultures

J. C. de Carvalho (✉) · S. G. Karp · W. J. Martínez-Burgos · T. Scapini · R. N. Barros · A. B. P. Medeiros · M. M. Albuquerque · A. L. Woiciechowski · C. R. Soccol
Department of Bioprocess Engineering and Biotechnology, Federal University of Paraná, Curitiba, PR, Brazil
e-mail: jccarvalho@ufpr.br; susan.karp@ufpr.br; walter.burgos@ufpr.br; scapini.thamarys@ufpr.br; adrianebpm@ufpr.br; adenise@ufpr.br; soccol@ufpr.br

S. K. Brar et al. (eds.), *The Microbiology of the Drop-in Biofuel Production*, Biofuel and Biorefinery Technologies 15, https://doi.org/10.1007/978-3-031-61637-2_11

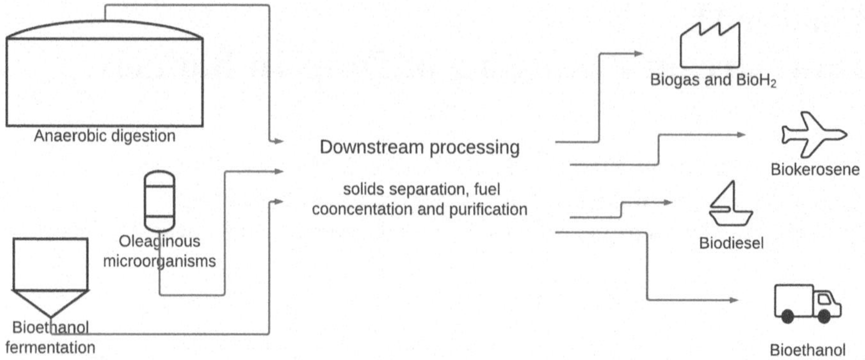

Fig. 11.1 Overall processing of the three main downstream processing used on drop-in biofuel production

may require separation and purification steps—the downstream processes. Sustainable bioprocesses use residual biomasses, high-productivity microorganisms in fermentation or lipid biosynthesis, or high-yielding methane or biohydrogen producers (Fig. 11.1). However, it is the downstream processes that guarantee the ultimate quality and applicability of the fuel.

11.1.1 Pollution and the Need for Biofuels

Today, the atmospheric carbon dioxide concentration is 50% higher than pre-industrial levels (Stein 2022) and averaged 413 ppm of CO_2 over continents in 2021 (Zhang et al. 2022). About 65% of the greenhouse gases emissions consist of carbon dioxide derived from the combustion of fossil fuels (Gür 2022). Carbon emissions continue rising, and never before was fossil fuel production so high—at about 130TWh of coal, oil, and gas (IEA 2022). Although this value could be reaching a peak, as optimistically observed by some, that could be an artifact of the recent COVID pandemic and its impact on the global economy. The global population is still rising, which may pressure fuel demand. At the same time, it is becoming clear that global warming is a reality and that there is a dire need for decarbonization of the world energy matrix.

Enter the, inherently renewable, biofuels. In a sense, the energy in biofuels is *solar energy*. This energy is stored as chemical energy in the biomass of photosynthetic organisms, and these compounds are then processed into fuels for use in electricity, heat generation, and transportation. Today, bioenergy makes up 10.5% of the total energy consumed, and biofuels comprise 3.7% of the energy used in the transportation sector—while 95% of the fuels are still fossil. The IEA (the International Energy Agency) projects biofuel production to grow by 80–200% by 2030 and its share in transportation to grow to between 5.1 and 8.5%. These scenarios are based on governmental policies worldwide (IEA 2022). Biofuels are a real solution to reduce our GHG emissions.

11.1.2 Biomolecules that Can Be Considered Drop-In Biofuels

Most of the existing fleet of vehicles already uses fossil fuels. Therefore, practical biofuels are those that can enter the existing distribution pipelines without adaptations—these are drop-in biofuels. In principle, any biomolecule can be a fuel; however, its energy content and physical properties are crucial considerations. The energy content is higher if carbon is reduced, as shown in Table 11.1 for one-carbon molecules. Actually, the enthalpy of combustion of a fuel correlates very well with its consumption of oxygen, yielding about 418 kJ/mol of O_2 (Schmidt-Rohr 2015, 2020).

Therefore, reduced-carbon molecules such as lipids and alcohols have the highest energy content and could be used as biofuels. However, many biomolecules easily produced in bioprocesses, such as biobutanol, butanediol, acetone, or residual glycerin, have properties incompatible with existing engines, with viscosities, volatilities, and flash points that require engine adaptation (Cortez et al. 2021; Poulikidou et al. 2018). These are not considered drop-in biofuels in this text.

Biohydrogen, although relatively new when viewed as an urban transportation fuel, *can* be considered a drop-in biofuel because it can be injected into some existing natural gas lines, increasing the combustion enthalpy of the mixture and "greening" the fossil fuel to an extent. Biohydrogen can already be considered a drop-in biofuel because it has an infrastructure being built (Apostolou and Xydis 2019; Moreno-Benito et al. 2017); the fuel, once purified, is identical to the gas produced by other methods.

11.1.3 Purity Is Key

A critical aspect of drop-in biofuels is their compliance with existing regulations. Besides having high energy content, they must have similar viscosity, density, flash point, or sulfur content characteristics as their synthetic counterparts. In general, the purer the fuel, the better—however, purifying fuels is also costly. Standardization details for each of the fuels can be found in technical standards by ISO (International Standards Organization), ANSI (American National Standards Institute), or ABNT (Brazilian National Standards Organization), among others (Gonçalves et al. 2022; Rehnlund 2008; van den Oever et al. 2022).

Most biofuels are reasonably pure compounds or mixtures of similar compounds. After biosynthesis, these compounds or their precursors must be concentrated and

Table 11.1 Energy of combustion for one-carbon molecules (compiled from NIST 2010; O'Neil 2013)

Compound	Methane	Methanol	Formaldehyde	Carbon	Formic acid	Carbon monoxide	Carbon dioxide
Minimal formula	CH_4	H_3COH	H_2CO	C	HCOOH	CO	CO_2
Carbon oxidation state	−4	−2	0	0	+2	+2	+4
Oxygen (O_2) required, mols/mol fuel	2	1.5	1	1	0.5	0.5	0
Combustion enthalpy, kJ/mol	830	727	526	393	254	283	0

separated from water to reach an adequate heating value and purity. Purification may require the removal of macromolecules to reduce viscosity and removing organic acids and sulfur to ensure that the fuel burns cleanly and to avoid corrosion. Removing free suspended solids or crystallizable material is necessary to avoid injectors clogging (Atelge et al. 2021; Karimi et al. 2021; Vedachalam et al. 2022). This must be guaranteed at a low cost by the downstream processing. The following sections present an overview of biofuel production's most common separation processes, followed by a closer look at the specificities of downstream processing of liquid and gaseous drop-in biofuels.

11.2 Biofuel Downstream Processing Basics

Each biofuel has its specificities and, thus, requires specific downstream processing steps. However, it can be argued that there are three main kinds of drop-in biofuels: solid, liquid, and gaseous, and this is reflected in their downstream processing.

Solid biofuels, such as raw microbial biomass or dry sludges, require a low water content and maybe a pelletizing process, but no further processing—they are inherently drop-in. Biomass can be fed to burners that use other solid fuels with little adaptation to accommodate the air-to-fuel ratio. However, the exhaust gas may have a higher content of sulfur oxides resulting from burning the protein in the biomass. Solid biofuels produced via bioprocesses are uncommon because they usually have lower heating value than simple and readily available lignocellulosic biomass (Manandhar et al. 2022; Obernberger et al. 2006).

Liquid biofuels produced by microorganisms can be further divided into hydrosoluble and hydrophobic. The hydrosoluble fuels are present in fermentation broths and can be separated via distillation, as in bioethanol production. At the same time, hydrophobic components such as lipids remain in the biomass and must be extracted using a suitable solvent. Since microbial fats and oils have a high viscosity, they must be chemically converted into lower-viscosity fuels through reactions such as transesterification or decarboxylation. These steps also segregate undesired fractions, such as inorganic compounds and proteins. Drying of the liquid fuels may be required (Segovia-Hernández and Sánchez-Ramírez 2022; Zhou et al. 2022).

Gaseous biofuels produced in anaerobic reactors naturally separate from culture media and can be used as is in some instances. However, removing contaminants and upgrading, i.e., removing inert fractions, can be beneficial—usually through absorption or adsorption. Pyrolysis gas has a more complex composition and can be trickier to purify, but involves essentially the same separation processes (Mallouppas et al. 2023; Zhang et al. 2023).

The following section describes the role of the essential operations necessary to allow access to biofuel precursors (biomass separation and drying), to convert lipids into biofuels (extraction and chemical reaction), to concentrate diluted biofuels and to recover solvents (distillation), and to purify and dehydrate fuels.

11.2.1 Microbial Biomass Separation

The whole fermentation broth *can* be fed into distillation columns for ethanol production because ethanol is in the liquid and not in the cells and separates with or without cells in the mixture. However, there may be advantages in recovering the biomass in multiproduct biorefineries. Cell separation is mandatory to process microbial biomass into some biofuels, such as lipid-based biodiesel or biokerosene, or for anaerobic digestion of biomass.

Removing cells from a fermentation broth is essentially a solid–liquid separation and can be done by filtration, sedimentation, or centrifugation (de Carvalho et al. 2019). However, industrial biofuel production involves enormous volumes, so filtration can be impractical because of the resistance of the biomass cake that would be formed. Sedimentation is usually impractical because it is slow and does not concentrate the biomass, except for microalgae production—where cultivation times are long, initial biomass concentrations are low, and, thus, a preliminary flocculation and sedimentation step does not interfere with throughput and is economical because of the low energy input (Sahoo et al. 2017); the thickened cell suspension can then be processed similarly to typical fermentation broths.

The most common separation process for broth clarification and biomass recovery is centrifugation; for example, sugarcane bioethanol industries use disk-stack centrifuges that give concentrated cell slurries of yeast, a "yeast cream." A particular aspect of this process in the bioethanol industry is that yeast cream can be acid-treated (Sekoai et al. 2019) to reduce bacterial contamination and re-inoculated in bioreactors to increase the ethanol yield.

11.2.2 Biomass Drying

Biomass can be converted into bioenergy without drying in processes such as anaerobic digestion of algal biomass, where energy is recovered as biomethane, or in the production of third-generation ethanol, where algal polysaccharides can be hydrolyzed and converted into bioethanol. However, biomass drying is a requisite for recovering lipid-based biofuels because lipid extraction or conversion is more efficient with low water contents in the biomass (Show et al. 2015).

Many methods can be used to dry microbial biomass, from batch tray drying to continuous spouted bed or spray drying. However, biofuels have a low added value, so drying in biofuel production must be economical. This translates to equipment capable of processing huge volumes because of the economy of scale or equipment that uses leftover or low-cost heat, such as solar drying. Considering the cost of drying and the fact that microbial lipids can be tailored to have a high content of nutritional, polyunsaturated fatty acids, many of the algal biodiesel endeavors of the last decades shifted to nutritional oils (de Carvalho et al. 2022).

11.2.3 Distillation

Distillation is the chief operation for recovery of bioethanol: because the alcohol is much more volatile than water, boiling the fermented broth containing ethanol gives vapors enriched in the fuel. If this is done sequentially, the end product is 95% ethanol. In practice, industrial distillation is done in large, efficient towers where many internal plates form sequential stages. Concentrated ethanol is the top product and an aqueous residue (vinasse) is the bottom product (Gavahian et al. 2019). Distillation is a physicochemically complex but mature technology that can also be used in the production of higher alcohols such as butanediol (although in this case, the alcohol would be part of the bottom of the distillation column), and to recover methanol, refine biodiesel, or even purify glycerin after its recovery in biodiesel production.

11.2.4 Lipid Extraction

Lipids in oleaginous microorganisms are extracted using solvents such as hexane (for triacylglycerols, TAGS) or more polar solvents such as chloroform for free fatty acids (FA). Analytical methods such as Bligh-Dyer or Folch, where extraction must be complete, allow for solvent loss—usually methanol. In industry, it may be better to have a lower recovery but less solvent use in solid–liquid extraction, and good recovery of the solvent used. The process involves contacting the dry biomass with the solvent of choice in mixing tanks, followed by centrifugation or filtration, or contacting the solvent with a packed bed of biomass solids in an extraction column (de Carvalho et al. 2022; Yellapu et al. 2018). The lipid solution, called *miscella* in industry, proceeds to solvent evaporation and oil refining.

11.2.5 Lipid Conversion

Microbial lipids come in many sizes, with the most prevalent FAs ranging from 14 to 18 carbons. These lipids can be converted into fatty acid methyl or ethyl esters (biodiesel) by esterification or transesterification. In these reactions, the glycerol in TAGs, or the OH group that is part of the carboxyl in FA, is exchanged by a short-chain alcohol, giving an overall smaller molecule compared to TAGs, reducing viscosity, and increasing volatility (Yellapu et al. 2018). However, other well-known reactions, such as decarboxylation or hydrodeoxygenation, are gaining importance. They may lead to hydrocarbon mixtures that can be processed using fractional distillation, similar to oil today (Loe et al. 2019).

11.2.6 Gaseous Fuels Purification

Gaseous fuels such as biogas and biohydrogen have a relatively low solubility in water and separate naturally from the anaerobic digestion mixture. However, these gases usually carry a significant fraction of CO_2 and smaller fractions of water, nitrogen, H_2S, and volatile fatty acids (VFAs). Carbon dioxide, water, and nitrogen are incombustible and, therefore, inert as fuel, reducing the energy content of the gas, while H_2S and VFAs are acidic and can cause corrosion problems. The case of H_2S is worse because it smells terrible and is toxic, and on burning, it generates sulfur oxides (SOx), which combine with water to form mineral acids corrosive to smoke-stacks (Franco-Morgado et al. 2018; Rocher-Rivas et al. 2022; Yang et al. 2012). These compounds cannot be freely liberated to the atmosphere for obvious reasons. It is straightforward to absorb these acid oxides by passing the fuel through an alkaline solution in a scrubber or an absorption column. The process requires an alkali such as sodium hydroxide, and similarly to drying, the process benefits from the economy of scale. More recent technology for separating gases, especially CO_2 from methane or hydrogen, uses molecular sieves, in which the molecules permeate or adsorb differentially into a nanoporous adsorbent based on their size. This technology is also gaining space in the separation of water.

11.2.7 Dehydration (for Novel Ethanol Plants, Biodiesel, and for Gases)

Removing the extra 5% of water from the azeotrope that forms on distillation is necessary to produce anhydrous ethanol. This was usually done using a third component, such as toluene, which would form another azeotrope with ethanol, free of water, in a second distillation. After condensation, the ethanol and toluene separate naturally. However, it is increasingly more common today to use water adsorbents such as silica gel or molecular sieves, where the hydrated ethanol loses its water to the adsorbent, and the saturated adsorbent can be recovered by heating (Plesu Popescu et al. 2021; Taghizadeh-Alisaraei et al. 2019). Gaseous fuels can be similarly dehydrated, although much of the water can be separated by simple compression and condensation of the excess water—compression may be necessary anyway for gas storage. Biodiesel may also require removing the low amount of water that can be present to improve stability. In this case, the fuel has a boiling point higher than that of water—therefore, it is possible to separate it by simply heating and applying a vacuum.

The following section delves more specifically into the downstream processes for the most important fluid biofuels.

11.3 Drop-in Biofuels

11.3.1 First-Generation Bioethanol, Biodiesel, and Biogas

First-generation biofuels are those derived from food crops rich in starch, sugars, or lipids (Bonatto et al. 2020; Saladini et al. 2016; Sengupta et al. 2020). Biogas, on the contrary, does not compete with food crops—it is a biofuel produced through the anaerobic digestion of different biomass sources, ranging from agricultural to urban wastes. However, it is a long-known fuel (Bond and Templeton 2011), older than biodiesel and bioethanol. With modern practices but the same principle, biogas production is increasingly popular (Farhang et al. 2023; Jain et al. 2019; Primmer 2021).

Over the years, incentives for the large-scale implementation and technological advancement of first-generation biofuels have given them an important role in the production of energy from renewable sources. They are reinforced by green and bio-based conversion technologies. Ethanol and biodiesel are often called "drop-in" due to their ability to reduce the use of petroleum-based fuels. These molecules can replace other petrochemical-based fuels without causing technical or environmental damage (IEA Bioenergy 2014, 2019; Saladini et al. 2016; Sengupta et al. 2020).

The advancement and integration into the market are already well-established, with significant progress in promoting drop-in biofuels. For instance, in Brazil, it is compulsory to blend 25–27% anhydrous ethanol into gasoline throughout the country (Portaria MAPA No 75 2015). Moreover, India has achieved a 10% ethanol blend by 2022, expected to reach 20% by 2025 (India 2022). Canada is implementing Clean Fuel Regulations to reduce GHG emissions by 13% by 2030 (Government of Canada 2023). The USA created the Inflation Reduction Act, which supports biofuel production and investments estimated at 9.4 billion dollars by 2031 (Government of Canada 2023).

This strategy extends to biodiesel, which is blended with diesel at a rate of 12% in Brazil, with the goal of increasing it to 15% by 2026 in accordance with the country's energy transition objectives (Resolução CNPE No 16, de 29 de Outubro de 2018 2023). In the USA, this blend goes as high as 20%, while in other regions like Canada, China, the European Union, and Chile, the proportion ranges from 2 to 10% (Lane 2016).

Given that first-generation biofuels are already well established on an industrial scale, they are poised to predominantly meet the rising market in this sector, with ethanol and biodiesel experiencing the most rapid growth (IEA 2023). Currently, Brazil and the USA are the world's largest producers of biodiesel and bioethanol, together accounting for 17% and 84% of global production in 2017, respectively (IEA 2019). Considering the downstream processes, significant advancements have been made in overcoming technical barriers and enhancing the separation efficiencies of these drop-in biofuels, facilitating large-scale production of these bio-based products. In the case of ethanol production, which is sourced globally from sugar cane, corn, and sugar beet, these processes have been widely scaled and are well established. As an example, the residual juice generated after ethanol distillation, known as vinasse, exhibits a high chemical oxygen demand, ranging from 27.5 to 300 kg/m^3, and a biochemical oxygen demand of up to 47,000 kg/m^3. Moreover, this downstream waste stream from ethanol production contains substantial concentrations of potassium, nitrogen, calcium, magnesium, and phosphorus, along with an acidic pH ranging from 4 to 5 (Deh Kiani et al. 2022).

The adaptation of the yeasts that produce ethanol allows fermentation efficiencies of over 95%. This means that the fermented broth collected in the downstream processes of the reactors is characterized by a high ethanol content, residual raw material in concentrations of less than 5% of the initial (~20–30% w w^{-1}), yeast cells (which are generally recycled to the fermentation vats, but not in their entirety, and a small concentration may be dragged into the vinasse), and other by-products (Basso et al. 2011; Sengupta et al. 2020).

In Brazilian production processes utilizing sugar cane as an input to produce 1G ethanol, it is estimated that after the distillation stage for ethanol recovery, 10–15 l of vinasse are produced per liter of ethanol produced (Basso et al. 2011).

The characteristics of vinasse generated in ethanol production make it an attractive substrate for biogas production (Deh Kiani et al. 2022; Fuess et al. 2017). This approach is already being implemented on a large scale, such as the food and energy company Raízen, which has one of the largest biogas plants in the world, converting vinasse and residual cake from the centrifugation of sugarcane ethanol production into electricity and fertilizer (Raizen 2023). Using industrial and agricultural waste substrates for biogas production is considered a suitable technological approach to waste management, aligning environmental stewardship with energy recovery through the controlled conversion of organic matter into biogas (Fuess et al. 2017).

In the context of downstream processes after anaerobic digestion, additional steps may be required to remove phosphorus and nitrogen from the residual digestate when the main objective is waste treatment. This can be achieved through processes such as nitrification, denitrification, and chemical precipitation (Lorick et al. 2020; Macura et al. 2019). Regarding biogas utilization, energy generation through burning is the most applied method. This involves passing the gas through a filter to remove moisture, followed by burning.

During the marketing stage, downstream processes are expected to yield products with a high degree of purity. For anhydrous fuel ethanol produced in Brazil for blending with gasoline, the product should have specific characteristics, including a minimum alcohol content of 99.3%, a maximum residual water content of 0.7%, maximum total acidity of 30 mg/L, maximum electrical conductivity of 300 μS/m, maximum hydrocarbon content of 3% v/v, maximum methanol content of 0.5% v/v, and maximum chloride, sulfate, iron, sodium, and copper content of 1, 4, 5, 2, and 0.07 mg/kg, respectively (Dispõe Sobre as Especificações Do Etanol Combustível e Suas Regras de Comercialização Em Todo o Território Nacional 2022). Ethanol blending with gasoline helps achieve the country's energy transition goals without any technical or environmental damage. About biodiesel, after transesterification and purification, the product for sale must have a water and sediment content of less than 0.05% and a viscosity of between 1.9 and 6.0. Biodiesel is more viscous than petroleum diesel, but it does not affect engine performance. It even enhances lubrication and prolongs engine life (Lotero et al. 2005; NREL 2023).

As the biodiesel production process is well established, most biodiesel companies have integration stages for marketing other products, such as glycerol, which is produced during the transesterification of biodiesel (approximately 40–70%). Glycerol can be purified and marketed for other purposes. This residue may also contain ethanol, methanol, water, and oils, with the content of these compounds varying based on the raw material used and the production and integration strategies (Almutairi et al. 2023; Barreiros et al. 2020; Lin et al. 2011; Lotero et al. 2005; Sun et al. 2023).

It is important to emphasize that first-generation biofuels play a vital role in the global energy transition. The partial replacement of petrochemical-based molecules with bioethanol and biodiesel has yielded promising environmental and economic outcomes. Although these are well-established processes, considerations must be made, particularly regarding the cost of raw materials and the use of agricultural food crops for energy production. In this context, significant efforts have been made to incorporate second-generation fuels (as discussed in the following topic). However, this doesn't diminish the importance of first-generation drop-in biofuels, which have demonstrated the feasibility of large-scale production with stability to meet market demand. New strategies will continue to address gaps and optimize production, expanding into new markets and facilitating the energy transition.

11.3.2 Advanced Bioethanol: Second and Third Generations

11.3.2.1 Second-Generation Bioethanol

Second-generation ethanol is produced from non-food raw materials, such as agricultural residues (straw, sugar cane bagasse, rice husks), forest residues, wood residues, and other lignocellulosic biomasses (Patel and Shah 2021). The second-generation ethanol production process is more complex than the first generation. It involves the breakdown of cellulose and hemicellulose, components of lignocellulosic biomass, into fermentable sugars. This often requires chemical and enzymatic treatments to release the sugars (Gandam et al. 2022).

The composition of broths fermented from lignocellulosic materials, such as agricultural residues, wood, and other biomasses rich in cellulose and hemicellulose, is complex and varies depending on the source of raw material, the pretreatment process, and the microorganism used in fermentation (Kucharska et al. 2018). Residual sugars such as glucose, xylose, arabinose, and other less common sugars can be found in small quantities or even in considerable quantities, depending on the conversion efficiency of the microorganism used (Periyasamy et al. 2022). Lignocellulosic compounds such as lignin can inhibit the action of microorganisms in fermentation. For this reason, it is removed or modified during the pretreatment process to increase fermentation efficiency. However, some pretreatment methods

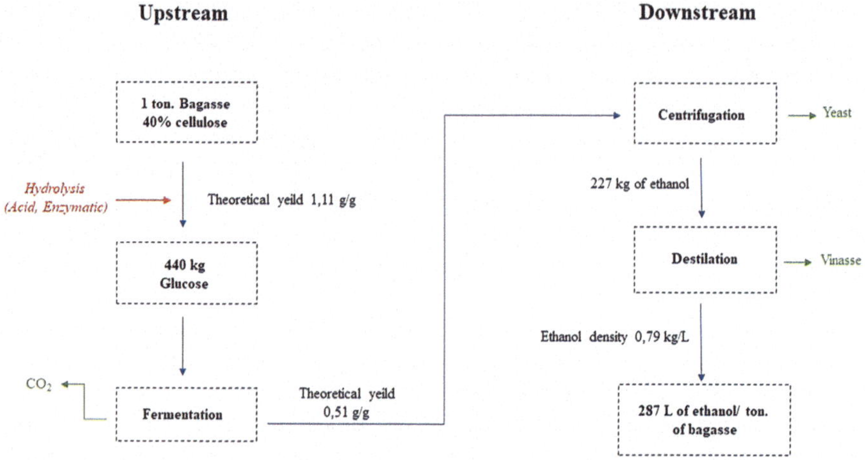

Fig. 11.2 Upstream and downstream in optimal conditions for second-generation ethanol production from bagasse

have proven less efficient, leading to the presence of lignin in the culture medium (Das et al. 2023). Water is the main component of fermented wort and is used to dissolve sugars and other soluble components of lignocellulosic materials. Small amounts of organic acids, such as acetic acid and formic acid, may be present in lignocellulosic materials and may be released during the pretreatment process (Liu et al. 2022).

Wort fermented from lignocellulosic materials may also contain impurities, such as residues from products used in pretreatment. These residues may include acids, bases, enzymes or other chemicals, toxic compounds, or unwanted by-products such as higher alcohols (Mussatto and Roberto 2004). Additional nutrients, such as nitrogen, vitamins, and minerals, can be added to the culture medium to provide the necessary nutrients for the yeast or fermentative bacteria. Whatever is not used by the microorganism remains in the medium at the end of fermentation. Unwanted microbiological contaminations, such as bacteria or wild yeast, may be present in the wort at the end of fermentation, competing with the desired strains of microorganisms and negatively affecting the fermentation process (Sekoai et al. 2019). The main product of must fermentation from lignocellulosic materials is ethanol, which, under perfect conditions, should reach a yield of 0.51 g/g of substrate (Fig. 11.2) (Paulova et al. 2015). Ethanol has intrinsic characteristics, such as boiling point, solidification, flash point, auto-ignition, and calorific value or energy potential. Its boiling point (transformation into a gas) is 78.4 °C, while its solidification point (transformation into a solid) is −114.3 °C, average values valid for a pressure of 1 atm. Its flash point starts at a temperature of 13 °C, the point at which it emits vapors that ignite on contact with a heat source. Its auto-ignition point is 363 °C; at this temperature, combustion occurs even without direct contact with a heat source. The calorific value of hydrated ethanol fuel is 5380 (kcal/l). This value corresponds to approximately 70% of the calorific value of gasoline, 8325 (kcal/l) (Whim and Johnson 1996).

Ethanol is sold for various purposes, mainly as fuel (anhydrous ethanol fuel and hydrated ethanol fuel). Ethanol quality standards may vary depending on local regulations and the specific application. It is essential to consult the rules and regulations in force in each region or country. Second-generation ethanol does not differ from first-generation ethanol when it comes to physical-chemical aspects and must present the same specifications to comply with the regulations of each region or country, such as ANP (Agência Nacional do Petróleo, Gás Natural e Biocombustíveis) RESOLUTION N° 907 (ANP 2022), which regulates the characteristics of ethanol in Brazil. Second-generation ethanol production involves more complex processes compared to first-generation ethanol, which is obtained from food crops (Ayodele et al. 2020). Raw materials for second-generation ethanol have low cost and good availability in agricultural regions. In contrast, ethanol produced from lignocellulosic biomass generally requires more rigorous purification processes to remove impurities such as organic acids, aldehydes, and other unwanted compounds (Nanda et al. 2014). Pretreatment is a crucial and often expensive step in the process. The development of more efficient and economical pretreatment technologies is essential. The economic viability of cellulosic ethanol is a challenge, but as technological advances and process improvements continue, second-generation ethanol production can become increasingly competitive and sustainable.

11.3.2.2 Third-Generation Bioethanol

The third generation of biofuels, or third-generation ethanol, seeks to overcome the limitations of previous generations, focusing on the production of ethanol from micro- and macroalgae (Tan et al. 2020). There are several reasons to consider using third-generation ethanol. Third-generation ethanol is produced from non-food biomass, which can be grown in non-agricultural areas, minimizing competition with food production (Müller et al. 2023). This makes third-generation ethanol production more sustainable from the point of view of land and natural resource use (Aron et al. 2020). Some varieties of algae are highly efficient at capturing atmospheric carbon and can contribute to reducing the concentration of carbon dioxide in the atmosphere (Onyeaka et al. 2021).

The downstream process for third-generation ethanol is similar to that for second-generation ethanol due to the nature of ethanol production. Significant differences occur in the initial phases, such as cultivation, harvesting, and raw material hydrolysis technology; third-generation ethanol involves the cultivation and harvesting of algae, while second-generation ethanol generally begins with harvesting from lignocellulosic waste (Devi et al. 2023). Although both generations may involve hydrolysis, the specific technology and enzymes used may differ depending on the biomass composition and process conditions (Vasić et al. 2021). Both processes use microorganisms to carry out the fermentation of sugars. This may involve the use of genetically modified yeast, bacteria, or microorganisms that convert sugars into ethanol.

The composition of the fermentation broth fermented from algae in the production of third-generation ethanol may vary depending on the species of algae used, the cultivation conditions, and the conversion process (Behera et al. 2015). Algae generally contain carbohydrates, such as sugars (such as glucose, sucrose, and galactose) and starches (Behera et al. 2015). When not used in their entirety for the production of ethanol, the presence of these carbohydrates in the residual form in the fermented must occur. Essential nutrients such as nitrogen, phosphorus, and potassium are necessary for the growth of algae and the functioning of fermenting microorganisms and are also found in fermented wort. In addition to the nutrients mentioned above, most may contain trace elements and vitamins that are essential for the growth and metabolism of algae and microorganisms involved in fermentation (Das et al. 2023). Water is an essential component of wort, as it serves as a reaction medium and helps maintain the viability of microorganisms. In some cases, buffering agents or pH regulators, such as acids and salts, may be added to maintain the proper pH. Depending on the algae cultivation process and the biomass pretreatment steps, broth may also contain impurities, such as lipids, proteins, and other unwanted substances (Pôjo et al. 2021). It is important to highlight that the exact composition of broth can vary according to the algae species, the algae growth stage, the cultivation technology, the pretreatment technology, and the fermentation conditions. Furthermore, the selection of fermenting microorganisms and the specific enzymes used can also influence the composition of the wort and the efficiency of ethanol production.

The characteristics and quality parameters pre-established for third-generation ethanol are the same as those established for first- and second-generation ethanol and must meet the standards and specifications of each region or country. The production technology is in the research, testing, and development phase and has not yet reached large-scale commercial production, being limited to bench- and pilot-scale tests. There are significant challenges to be overcome, such as developing efficient algae strains, optimizing cultivation processes, and making large-scale production economically viable (Musa et al. 2019). One of the most critical aspects of techno-economics is the assessment of the production cost of third-generation ethanol. This includes costs associated with the cultivation of algae, pretreatment of biomass, the fermentation process, and separation and purification of ethanol, among others (Wiatrowski et al. 2022). The scale of production is an important factor to consider. It must be assessed whether large-scale production is viable, taking into account the ability to meet market demand and the costs associated with increasing the scale of production. The environmental impact of the production process, such as the reduction of greenhouse gas emissions and the efficiency in the use of natural resources, is a great advantage compared to previous generations (Müller et al. 2023). Research into third-generation ethanol continues to hold promise due to the potential benefits this approach can offer in terms of sustainability, efficiency, and reduced carbon emissions. However, there are significant challenges to overcome.

11.3.3 Second-Generation Biodiesel (Microbial Biodiesel)

The downstream process to produce biodiesel of microbial origin begins with separating the biomass (microalgae or yeast) from the fermentation broth. Generally, the operations used to separate the biomass are centrifugation, filtration, or flocculation (Fig. 11.2). The latter is the most recommended due to its low costs and large volumes that can be treated despite being slower than centrifugation filtration (Goh et al. 2019). When centrifuges are used to separate biomass, they use revolutions above 3500 rpm, which implies high energy demand (Javed et al. 2022). The characteristics of the supernatant fermentation broth vary depending on the type of microorganism. In processes with microalgae and yeast, the nitrogen content in the residual supernatant is zero (a condition that forces them to accumulate carbon in the form of fatty acids) (Goh et al. 2019).

The residual medium in microalgae contains some residual salts and carbon sources, while the pH variation is minimal. In the case of yeast fermentation broth, some traces of the carbon source remain, around 5 g/L (Martinez-Burgos et al. 2023; Soccol et al. 2017), and the pH undergoes small changes, generally decreasing around one unit between 4 and 5.

After harvesting, the microbial biomass can be washed to remove some impurities from the fermentation broth or dried directly (the moisture content in the centrifuged biomass is around 90%). The drying process is also a critical variable as it can mainly affect polyunsaturated fatty acids (the most important in biodiesel production). At this stage, large amounts of input energy are also required. According to Javed et al. (2022), drying temperatures should not exceed 70 °C. The oil from the microbial cells is then extracted. Different techniques can be employed. However, there is no standard technique or process. Currently, processes are focused on reducing solvent consumption, increasing extraction yield, reducing operation time, and not altering the properties of the lipid fraction (Goh et al. 2019). According to Sierra et al. (2017), the primary strategy to improve the extraction of lipids on a large scale is the alteration of the cell wall since the process of passive diffusion through the cell wall is slow. Therefore, changes in the cell wall and membrane facilitated the access of solvents to intracellular lipid fractions (Goh et al. 2019). These methods can be mechanical and non-mechanical, such as high-pressure homogenizers, microbead mills, and ultrasound-assisted extraction. These processes reduce the use of solvents and process time. However, mechanical methods can have high biomass losses and low selectivity toward lipid material (Goh et al. 2019; Peralta-Ruiz et al. 2013).

Non-mechanical methods include treatments with acidic, enzymatic chemical compounds and ionic liquids. Abbassi et al. (2014) reported a recovery of the lipidic fraction of 53% using hydraulic pressing for the fractionation of *Nannochloropsis oculata*. However, the recovery of the lipid fraction increased to 95% after freezing in liquid N_2 because the cell wall gets brittle and that improves the lipid separation. Other authors (Zhang et al. 2016) evaluated treatment with ultrasonic assistance in *Trichosporon oleaginosus*, which, in addition to reducing the processing time by

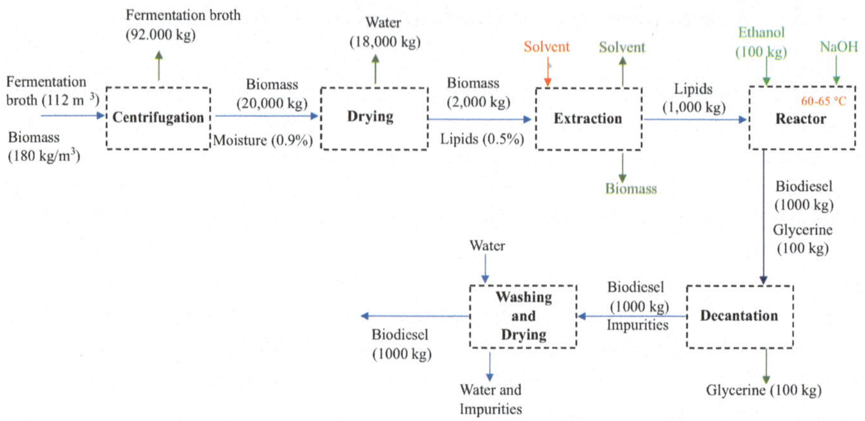

Fig. 11.3 Downstream process of microbial biodiesel production

23 h, obtained an extraction of 95% of the lipid fraction. Soxhlet extractions with conventional solvents are also highly efficient. For example, Escorsim et al. (2018) reported an extraction efficiency of the lipid fraction of 92% of *Acutodesmus obliquus* biomass in Soxhlet at 60 °C for 12 h and using ethanol-hexane (1:2) as solvent.

After extraction, the transesterification reaction takes place between triglycerides of microbial origin and short-chain alcohol (mainly methanol or ethanol) in the presence of a catalyst, and biodiesel is produced. Traditionally, sodium hydroxide and potassium hydroxide have been used as catalysts. However, some lipase enzymes can also be used. Strong bases are used in a maximum proportion of 1.0% with respect to the amount of oil (Khan et al. 2009). The process is developed at temperatures between 60 °C and 65 °C, and the time for the reaction to be completed takes around 90 min (Fukuda et al. 2001). Stoichiometrically, in transesterification, 3 moles of alcohol are needed for each mole of triglycerides, which produces 1 mole of glycerol and 3 moles of methyl esters. However, on an industrial scale, according to Khan et al. (2009), 6 moles of alcohol are used for each mole of triglycerides. This alcohol excess improves the reaction rate and shifts equilibrium. Under these conditions, the biodiesel yield is above 98% (Fukuda et al. 2001). Subsequently, biodiesel is separated from glycerin by decantation because glycerin is denser and immiscible with biodiesel (Thiru et al. 2011). In some cases, it is necessary to wash the biodiesel with water to remove traces of soap. Finally, the biodiesel is dried until the water is completely removed. The production process and mass balances for biodiesel production are shown in Fig. 11.3. The process was carried out on a calculation basis of 1 ton of microbial oil. Furthermore, a lipid content in the biomass of 50% and a process efficiency of 100% were considered.

The physicochemical parameters of microalgae and yeast biodiesels comply with the specifications established by ASTM, which indicates the sustainability of

Table 11.2 Physicochemical parameters of microbial biodiesels

Parameters	Microalgal biodiesel	Yeast biodiesel	ASTM limits
Acid number [mg KOH/g]	0.49	0.47	0.8
Biodiesel density (at 15 °C) [kg/m^3]	853	879	N/A
Viscosity (at 40 °C) [mm^2/s]	5.52	5.061	1.9–6.0
Flash point [°C]	140	>120	100–170
Glycerin [mass %]	155	–	240
Cu strip corrosion	1	1	3 max

Adapted from Faried et al. (2017) and Thiru et al. (2011)

microbial oils to produce biofuel. Table 11.2 presents the physicochemical characteristics of biodiesels obtained from microalgae and yeast. The production of biodiesel of microbial origin can be economically competitive compared to biodiesel produced from plant origin. According to Singh et al. (2022) the production cost of biodiesel from vegetable oil is approximately US$ 0.81/L, while the cost of biodiesel of microbial origin is US$ 0.76/L. Soccol et al. (2017) showed that the highest costs in the production of microbial biodiesel are due to energy expenditure, which is approximately US$0.504/L. In addition, they indicated that the operations that required the most energy were the fermentation itself, the centrifugation, and drying. Other technical and economic research has also shown that the production of microbial biodiesel is economically profitable. For example, Patil (2010) concluded that the cost of producing biodiesel from *Lipomyces starkeyi* is US$0.78/L. Huang et al. (2013) showed that the cost of producing biodiesel from microalgae is around US$1.11/L.

11.3.4 Biomethane

The downstream or purification processes in the production of biogases, unlike other fermentative processes in which the metabolite remains in the fermentation broth, can be started when the production of the metabolite begins, that is, almost in parallel with the production process. This is possible since the gas being produced can be stored in other containers without interrupting the process. This presents an advantage for the biogas production process as high pressures negatively affect biogas production (Ram et al. 2020).

In anaerobic digestion, in addition to biogas, a liquid effluent is also obtained, containing significant amounts of some organic acids, such as acetic acid, butyric acid, and formic acid, which are intermediate metabolites generated in the production of hydrogen (Martinez-Burgos et al. 2021a; do Nascimento Junior et al. 2021; Sydney et al. 2019).

The composition of biogas depends on several factors, including the type of inoculum, the type of effluent and residue, and the process conditions. According to De Carvalho et al. (2023) biogas is mainly composed of methane (maximum 70%)

and (30–40%) carbon dioxide. Other gases are in traces, such as hydrogen sulfide, hydrogen (Martinez-Burgos et al. 2020; Rosa et al. 2020), water vapor, nitrogen, oxygen, ammonia, and siloxanes (Awe et al. 2018; Das et al. 2022). Contaminants significantly affect equipment where biogas is used as fuel and affect the calorific value of biogas. For example, CO_2 is recalcitrant, decreasing the density of biogas as well as its calorific value (Awe et al. 2018). In the case of hydrogen sulfide, it is highly corrosive to metal parts of engines, power generators, turbines, pumps, etc., reducing the useful life of the equipment. Therefore, these contaminants must be removed from the biofuel before it is used. The concentration of CH_4 in biogas must be between 95 and 99% and CO_2 between 1.0% and 5%, and H_2S must be eliminated or be in undetectable concentrations (Schenk et al. 2008). The purity required for biomethane varies between each country. In Sweden, Switzerland, and Spain, the minimum required concentration of biomethane is 97%, 96%, and 95%, respectively. Regarding biomethane contaminants in these countries, in the case of H_2S, the permitted concentration must be lower than 5 ppm, 5 ppm, and 15 ppm, respectively.

CO_2 Removal The removal of this compound can be done through different techniques or processes, including physical absorption, chemical absorption, and pressure swing adsorption (PSA), among others. The first technique is based on the solubility of gaseous compounds in liquid washing solutions. According to Awe et al. (2018), water is the main selective absorbent on an industrial scale. According to Wang et al. (2020), CO_2 has low solubility in water; however, its solubility is 25 times larger than the solubility of CH_4, so the gas mixture has to be recirculated several times in the process (Wang et al. 2020). Furthermore, according to Awe et al. (2018) the binding forces of CO_2 and H_2S have a greater tendency to polarity, meaning they can be separated in the same process, while methane is more non-polar. In general, gases are washed in columns filled with Pall rings to increase mass transfer (gas-liquid) (Wang et al. 2020). The water used in the adsorption process is regenerated in a desorption column. Organic solvents, such as polyethylene glycol, can also be used in CO_2 removal processes. This solvent can simultaneously remove other gases, such as H_2S and water vapor.

In the second technique, "chemical absorption," the transfer principle is similar to the first technique. However, the reaction occurs between solvents such as alkanol amines and CO_2 (Maile et al. 2017). Amines are largely selective in removing CO_2. Therefore, up to 99% of methane can be recovered. However, before carrying out chemical treatments with amines, H_2S must be removed. Alkaline solutions such as NaOH, KOH, $Fe(OH)_3$, or ferric complex of EDTA can also be used (Awe et al. 2018; Maile et al. 2017) with which CO_2 reacts to form other compounds such as sodium carbonate.

The third technique that can be used for biomethane purification, PSA, begins with drying the desulfurized gas. Subsequently, the gas passes through a series of 4 vertical columns, which are filled with different materials. The process order is adsorption, depressurization, desorption, and pressurization. In the pressurization column, CO_2 and other contaminants such as H_2S, N_2, and O_2 are removed by selective absorption. Generally, the materials used in this process are activated

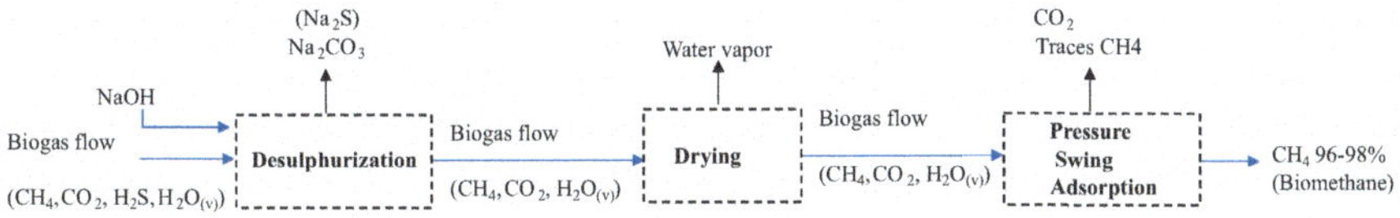

Fig. 11.4 Biogas purification process

carbon, silica gel, some zeolites, and synthetic resins. According to Bauer et al. (2013) and Ryckebosch et al. (2011) the recovery of biomethane with this technique is between 96 and 98% and a maximum of 4.0% losses.

H_2S Removal H_2S along with other compounds, such as ammonia, must be removed from the biogas at the beginning of the purification process as these can affect the CO_2 removal system. During combustion, these compounds can be transformed into NO_X or SOx (Abdeen et al. 2016). According to Awe et al. (2018), H_2S in biogas can reach concentrations of up to 10,000 ppm. To remove this compound, physical and chemical processes are used. Generally, the process is carried out in liquid contactors using aqueous solutions to transform H_2S into elemental sulfur or in the form of metallic sulfide (Maia et al. 2014). The process can be a single pass. However, in this condition, the water consumption is excessive. To improve the removal of H_2S, it is necessary to use some reagents such as NaOH, KOH, and $Fe(OH)_3$, among others. These blackings are used for PSA systems. H_2S can be eliminated instantly in the biodigester by adding iron ions (Fe^{2+} and Fe^{3+}), which react with S^{2-} to produce insoluble FeS (Ryckebosch et al. 2011). Semi-permeable membranes can also be used to remove H_2S and CO_2. They are permeable to these compounds while impermeable to biomethane. In Fig. 11.4, a general outline of a typical downstream process for the purification of biogas into biomethane is presented. In the first stage, desulfurization is carried out, and part of the CO_2 is also removed in the form of sodium carbonate. The biogas is then dried, and finally, the PSA is carried out where the biomethane reaches contractions between 96 and 98%.

In addition to the basic technologies discussed here, upgrading biogas to biomethane, membrane separation, cryogenic purification, and some biological methods can also be viable alternatives. In terms of cost, the choice of purification technology will depend more on the biogas production capacity than the type of technology itself.

11.3.5 Biohydrogen and Biohythane

Dark fermentation is the most studied biological method to obtain biohydrogen ($bioH_2$) due to a series of advantages presented by this process, such as its independence from light, the possibility of continuous production, use of waste as a substrate, lower energy costs, and greater ease scaling and control (Alavi-Borazjani et al. 2019). In this process, bioH_2 is produced by strict or facultative anaerobic bacteria such as of the genera *Clostridium, Enterobacter, Escherichia, Klebsiella, Citrobacter, and Bacillus* (Argun et al. 2008; Cheng et al. 2022) in their pure forms or consortium, which carry out the oxidation of organic matter present in waste from various sources, such as wastewater from industrial processing (do Nascimento Junior et al. 2021; Martinez-Burgos et al. 2021a, b), food waste (Ding et al. 2022), sewage sludge (Tena et al. 2020; Yang et al. 2019), animal husbandry waste (Sillero et al. 2022), and others.

The gas from dark fermentation is composed of around 30% H_2 and 60% carbon dioxide (CO_2), in addition to smaller impurities such as nitrogen (N_2), hydrogen sulfide (H_2S), and water vapor (Bakonyi et al. 2013; D'Silva et al. 2023; Deheri and Acharya 2022; Dunikov and Blinov 2020). Furthermore, bioH_2 yields obtained in this process are very variable due to the variability of process parameters, such as the wide range of substrates and microorganisms that can be used, temperature, pH, and other factors. Strategies such as pretreatment of raw materials, use of nanoparticles, and optimization of the process have been proposed to maximize bioH_2 production (do Nascimento Junior et al. 2021; Martinez-Burgos et al. 2021b). However, a theoretical maximum of 4 mol of hydrogen/mol of glucose is estimated (Sasidhar et al. 2022). Furthermore, during dark fermentation, a significant fraction of the substrate is converted into volatile fatty acids such as butyric, propionic, and acetic acids, which decrease the pH of the culture medium (Chandra and Venkata Mohan 2014; Davila-Vazquez et al. 2008) inhibiting bioH_2-producing microorganisms and, consequently, negatively impacting their yields, resulting in an effluent that is still quite rich in energy due to incomplete fermentation of the substrate (Mozhiarasi et al. 2023).

To overcome this problem, several studies have used effluent from dark fermentation rich in volatile fatty acids to produce biomethane (bioCH_4), which is also widely applied in the chemical and transport industries (Deheri and Acharya 2022; Hans and Kumar 2019; Kumari and Das 2016; O-Thong et al. 2016). In this second stage, microorganisms of genera such as *Methanoculleus*, *Methanosarcina*, *Methanomassiliicoccus*, and *Methanobrevibacter* (Meena et al. 2020) carry out the oxidation of these organic acids producing biogas composed of biomethane (50–75%), CO_2 (25–45%) (Deheri and Acharya 2022), and traces of H_2, N_2, water vapor, and H_2S (Chuichulcherm et al. 2017), presenting, as in the production of biohydrogen, quite variable yields, with a theoretical yield of 4 mol of bioCH_4 in the presence of 4 mol of H_2 and 1 mol of CO_2 (Khan et al. 2021).

After fermentation, the gases from the first and second reactors need to be purified to remove CO_2, as well as trace elements such as water vapor and H_2S (Fig. 11.5) (Bolzonella et al. 2018). Subsequently, the gases from the two stages, bioH_2, and bioCH_4, can be used separately or can be mixed, resulting in a gas known as biohythane (Ding et al. 2022; Meena et al. 2020; Mozhiarasi et al. 2023). Biohythane is made up of around 60% CH_4, followed by 30% CO_2 and 10% H_2 (Meena et al. 2020). In this way, biohythane can be produced as an alternative to overcome the energy limitations of bioH_2 production by dark fermentation, achieving a higher

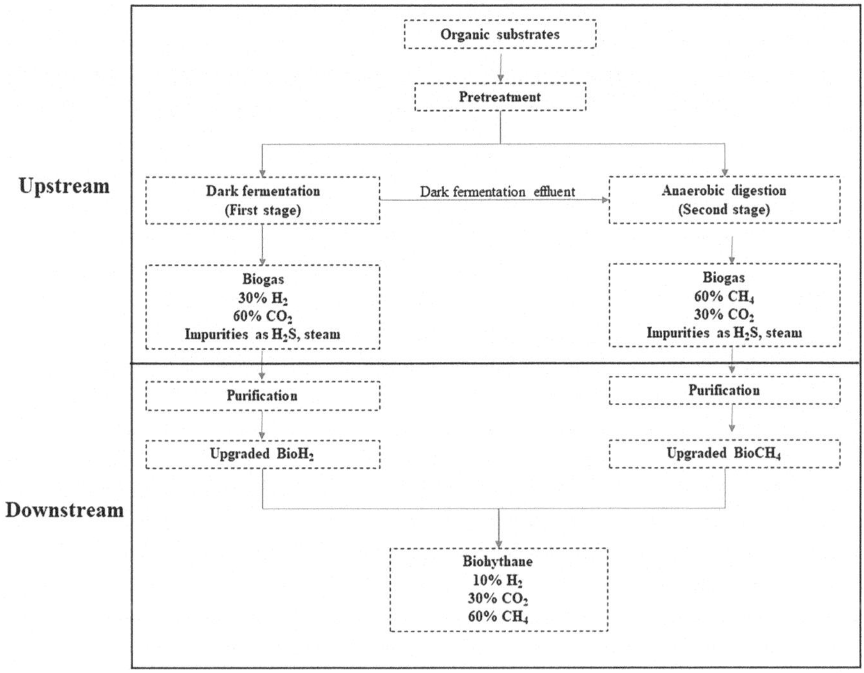

Fig. 11.5 Overall processes in the production and upgrading of biomethane into biohythane

energy conversion compared to single-stage anaerobic digestion (Mozhiarasi et al. 2023). This mixture improves the energy efficiency of the fuel and reduces the emission of pollutants such as CO_2 due to the presence of bioH_2, in addition to not requiring special storage conditions and modifications to the ignition systems used in the combustion of natural gas (Mozhiarasi et al. 2023; Shanmugam et al. 2021).

For biohydrogen to be converted into electrical energy in fuel cells, this gas must reach purity levels greater than 99%, according to the standard ISO 14687:2019, since the presence of other compounds damages the electrolysis cells (Cheng et al. 2022). Furthermore, the use of hydrogen in industrial applications such as hydrogenation also requires high levels of purity (Wang et al. 2006). Regarding the gas from the second reactor, it requires purification to remove CO_2 and impurities, improving the purity of the bioCH_4, so that it can later also be used in the production of the biohythane mixture (Deheri and Acharya 2022). Hydrogen and methane can be purified through a variety of chemical and physical methods, some of them being adsorption by pressure or temperature oscillation, cryogenic distillation, and membrane purification (Du et al. 2021). Despite the high efficiency of methods such as pressure swing adsorption and cryogenic distillation (Cheng et al. 2022), due to their high cost, they are most commonly used in the purification of hydrogen and methane from large-scale production processes of steam reforming and coal gasification, often becoming economically unviable depending on the amount of gas that needs to be purified. Therefore, they are not applicable on smaller scales of bioH_2 and bioCH_4 production (Dunikov et al. 2016).

Thus, membrane purification has proven to be a promising method for purifying bioH_2 and bioCH_4, due to its simpler operation, compactness and lightness, high energy efficiency, possibility of application in small-scale processes, lower production costs capital, operation and maintenance, lower environmental impact, and other benefits (Bernardo et al. 2020). Thus, much research has focused on the development of these membranes, which can be polymeric membranes, carbon, palladium, and membranes of metallic organic structures (MOFs) (Du et al. 2021). He et al. (2021) investigated the technical feasibility of using a two-stage carbon membrane system for bioH_2 purification, performing a process simulation along with cost estimation to evaluate operational expenses (OPEX) and capital expenses (CAPEX). According to the study, this purification system can achieve hydrogen purity levels of approximately >99.5% with a recovery of 90%, presenting a cost of US$ 0.06/m^3.

Technical-economic analysis studies are essential to reveal the commercial viability of a biorefinery. In the bioH_2 and bioCH_4 economy, in order to understand the global cost of their production, the main costs to be considered are the cost of the raw material, its pretreatment, bioreactors, gas purification, and operation/maintenance/labor costs. Mahmod et al. (2021) carried out a technical-economic analysis of the biohydrogen and methane production process in two stages using palm oil effluent as a substrate in a plant with a capacity of 700 m^3 to produce bioH_2 and 7.000 m^3 to produce bioCH_4. The temperature used to produce both phases was 50 °C and HRT of 1 and 10 days for bioH_2 and bioCH_4, respectively. The technical economic analysis demonstrated a total capital investment cost of 4.766.000 dollars, presenting an annual operating cost of 7.215.000 USD/year, annual revenue of 1.686.790 USD/kg of bioH_2 or bioCH_4/year, a unit cost of production of 4.28 USD/kg of bioH_2 or bioCH_4, and projected payback period of 8.01 years.

11.3.6 Pyrolysis Fuels

The thermochemical conversion of organic matter through pyrolysis generates pyrolysis oil (or bio-oil), biochar, and incondensable gases. Fast pyrolysis, usually operated at 425–600 °C for a few seconds, favors the formation of bio-oil (ca. 60%), which can be a drop-in biofuel in industrial boilers or engines. Fast pyrolysis crude oil presents a dark brown color and high viscosity, and its physicochemical characteristics demand downstream operations or upgrading when the intended application involves long-term storage or blending with conventional fuels (Schmitt et al. 2021). Crude fast pyrolysis oil is composed, in great part, of oxygenated compounds including carboxylic acids (ca. 10%), ketones, aldehydes, sugars (ca. 7% each), alcohols (ca. 1%), and water (15–30%), and of aromatic compounds including phenolics, furans, guaiacols, and partially depolymerized lignin (or pyrolytic lignin, ca. 30%). These lignin oligomers tend to repolymerize during storage, altering viscosity and burning properties. The presence of ashes and pyrolysis coke also influences the physicochemical characteristics of bio-oil (Lachos-Perez et al. 2023; Schmitt et al. 2021).

The oxygen content in crude bio-oil is typically between 35 and 50%, while that of heavy fossil oil is around 1%. The presence of oxygen in high proportion increases the polarity and reduces the miscibility with petroleum-based fuels, makes the oil more reactive, and contributes to its low energy density as compared to petroleum-derived oil. The high heating value (HHV) of bio-oil ranges between 16 and 20 MJ/kg, against 40 MJ/kg of fossil oil (Lachos-Perez et al. 2023; Schmitt et al. 2021). The downstream methods for bio-oil upgrading can be divided according to two main goals, namely, improving the physicochemical properties of bio-oil without significantly altering its primary chemical structures, which can be achieved by filtration, emulsification, and esterification (solvent addition), and promoting chemical reactions to reduce the oxygen content and increase the hydrocarbon content, through hydrotreatment, catalytic cracking, or electrochemical upgrading. Sometimes, the combination of two or more methods is required to achieve the desired properties (Kumar and Strezov 2021). Table 11.3 presents the main downstream methods applied to bio-oil upgrading for use as a drop-in biofuel.

The successful upgrading of pyrolysis bio-oil will result in enhanced stability during storage, reduced corrosiveness, increased HHV, and enhanced miscibility with conventional fuels. However, the upgradation strategy has an important impact on the technoeconomic analysis of bio-oil production for use as a drop-in biofuel. Each method raises technoeconomic challenges that may hinder the large-scale application of bio-oil. The bio-oil upgradation stage significantly contributes to the value of the minimum fuel selling price, which is within the range of 6–8 $/gasoline gallon equivalent (Lachos-Perez et al. 2023). The cost of bio-oil upgraded by hydrotreatment can be in the range of $ 0.74–1.80/L, higher than the expected cost for a commercially viable liquid fuel (Kumar and Strezov 2021).

11.4 Future Perspectives

Biofuels are critical to the decarbonization of the energy matrix. The developments in biomass production and processing technologies are essential to allow biofuels to expand without competing with food usage, and this requires the extensive use of lignocellulosic biomasses. However, the fermentation of lignocellulosic tends to produce more dilute fuels because of the presence of inhibiting compounds in the fermentation broths. This makes the economics of downstream processing even more important because the recovery must be of low energy input and, therefore, low-cost but still produce a high-quality fuel that complies with the existing legislation. Although disruptive technologies can always appear, the next decade shall see technology adoption that already exists in small-scale or other industries. Some of these trends are:

Table 11.3 Main downstream methods applied to bio-oil upgrading for use as a biofuel

Downstream method	Upgrading mechanism	Main objectives
Filtration	Membrane separation of solids (1–10 μm) from bio-oil in liquid phase or hot vapor phase	Removal of solid biochar particles and alkali metals, prevention of repolymerization
Emulsification	Emulsification with conventional fuels such as diesel and biodiesel with the aid of a surfactant	Increase of stability and improvement of physicochemical properties
Solvent addition/ esterification	Addition of polar solvents to bio-oil (ethanol, isopropanol, methanol, acetone, ethyl acetate, N, N-dimethylformamide) to promote esterification or acetalization	Increase of stability and decrease of viscosity
Hydrotreatment	Treatment with hydrogen in the presence of a metal catalyst (metal sulfides, noble metals, transition metals), at 200–500 °C and 3–30 MPa, for oxygen removal through hydrodeoxygenation reaction, producing hydrocarbons and water as by-product	Removal of oxygen from oxygenated compounds, increase of hydrocarbons content and HHV
Electrochemical upgrading	Electrocatalytic hydrogenation to convert carbonyl containing compounds to hydrocarbons, usually in an electrolytic membrane fuel cell that utilizes the protons that result from the oxidation of water	Removal of oxygen from oxygenated compounds, increase of hydrocarbons content and HHV, reduction of acidity
Catalytic cracking	Dehydration, cracking of hydrocarbons, and production of larger molecules via C–C bonding using a zeolite catalyst, at >350 °C and 0.7–14 MPa	Removal of oxygen in the form of water and CO_2, generation of light products
Hydrocracking	Dehydration, cracking of hydrocarbons, and production of larger molecules via C–C bonding using hydrogen at >400 °C	Removal of oxygen in the form of water and CO_2, generation of light products

Source: Adapted from Kumar and Strezov (2021), Lachos-Perez et al. (2023), Schmitt et al. (2021)

- In bioethanol production, membrane separations for ethanol and waste stream concentration and dehydration may increase the process competitiveness. Concentrated vinasse can be used for co-generation in boilers, compensating for downstream energy requirements.
- Microbial biodiesel may become popular, probably from yeasts grown on carbohydrates but possibly also from microalgae. In both cases, low-cost biomass drying is necessary, requiring process development. Another route that deserves investigation in large-scale processing of oleaginous biomass is extraction from wet biomass with novel solvent systems, where the energy requirement can be significantly lower than in drying-extracting processes.

- Gaseous fuels are rising, and economical purification and upgrading of biomethane is essential. While alkalis and iron ores can be used for biogas purification, it is possible that developments in compression and fractionation systems can lead to purification with minimal use of chemicals. The possible injection of synthetic natural gas (from biogas) in the existing pipelines will require purified methane.
- In biohydrogen production, coupling the VFA-rich streams to posterior biomethane production of VFA recovery steps is in its infancy, which can become a source of organic acids in large-scale processes. However, the economical separation of VFAs from the complex effluents from anaerobic digestion requires developing low-cost adsorption systems.
- Pyrolysis, gasification, and decarboxylation of raw biomasses and purified lipids are some of the most promising routes to hydrocarbon fuels.

11.5 Conclusion

The transition from fossil to biofuels is simultaneously essential and promising. However, widespread transition hinges on the efficacy of downstream processing, determining the quality and applicability of the biofuel. The potential of a biofuel to be a drop-in alternative depends on its energy content, physical properties, and compliance with existing fuel standards. Ensuring this compatibility is crucial, as it dictates the fuel's usability in existing processes and vehicles while ensuring a low environmental impact. This is especially challenging as the process must balance the need for high-quality fuel production with economic viability and integration into zero-waste biorefineries. Future trends in biofuel production indicate a shift toward innovative processing techniques that minimize competition with food resources and reduce the environmental impact. Adopting existing small-scale technologies in larger biofuel production systems is expected, emphasizing low-energy, low-cost recovery processes.

Acknowledgments The authors acknowledge CAPES for its support through the PROEX-PPGEBB program, and CNPq for the BIOGEN project, process 440138/2022-1.

References

Abbassi A, Ali M, Watson IA (2014) Temperature dependency of cell wall destruction of microalgae with liquid nitrogen pretreatment and hydraulic pressing. Algal Res 5:190–194. https://doi.org/10.1016/j.algal.2013.12.008

Abdeen FRH, Mel M, Jami MS, Ihsan SI, Ismail AF (2016) A review of chemical absorption of carbon dioxide for biogas upgrading. Chin J Chem Eng 24(6):693–702. https://doi.org/10.1016/j.cjche.2016.05.006

Alavi-Borazjani SA, Capela I, Tarelho LAC (2019) Dark fermentative hydrogen production from food waste: effect of biomass ash supplementation. Int J Hydrog Energy 44(48):26213–26225. https://doi.org/10.1016/j.ijhydene.2019.08.091

Almutairi AW, Abomohra A, Elsayed M (2023) A closed-loop approach for enhanced biodiesel recovery from rapeseed biodiesel-based byproducts through integrated glycerol recycling by black soldier fly larvae. J Clean Prod 409:137236. https://doi.org/10.1016/j.jclepro.2023.137236

ANP (2022) RESOLUÇÃO ANP N° 907, DE 18 DE NOVEMBRO DE 2022 - DOU DE 23-11-2022, pp 1–16

Apostolou D, Xydis G (2019) A literature review on hydrogen refuelling stations and infrastructure. Current status and future prospects. Renew Sust Energ Rev 113:109292

Argun H, Kargi F, Kapdan IK, Oztekin R (2008) Biohydrogen production by dark fermentation of wheat powder solution: effects of C/N and C/P ratio on hydrogen yield and formation rate. Int J Hydrog Energy 33(7):1813–1819. https://doi.org/10.1016/j.ijhydene.2008.01.038

Aron NSM, Khoo KS, Chew KW, Show PL, Chen W-H, Nguyen THP (2020) Sustainability of the four generations of biofuels - a review. Int J Energy Res 44:9266–9282. https://doi.org/10.1002/er.5557

Atelge MR, Senol H, Djaafri M, Hansu TA, Krisa D, Atabani A, Eskicioglu C, Muratçobanoğlu H, Unalan S, Kalloum S (2021) A critical overview of the state-of-the-art methods for biogas purification and utilization processes. Sustainability 13(20):11515

Awe OW, Zhao Y, Nzihou A, Minh DP (2018) A review of biogas utilisation, purification and upgrading technologies. HAL Open Sci 8(2):267–283

Ayodele BV, Alsaffar MA, Mustapa SI (2020) An overview of integration opportunities for sustainable bioethanol production from first- and second-generation sugar-based feedstocks. J Clean Prod 245:118857. https://doi.org/10.1016/J.JCLEPRO.2019.118857

Bakonyi P, Nemestóthy N, Bélafi-Bakó K (2013) Biohydrogen purification by membranes: an overview on the operational conditions affecting the performance of non-porous, polymeric and ionic liquid based gas separation membranes. Int J Hydrog Energy 38(23):9673–9687. https://doi.org/10.1016/j.ijhydene.2013.05.158

Barreiros T, Young A, Cavalcante R, Queiroz E (2020) Impact of biodiesel production on a soybean biorefinery. Renew Energy 159:1066–1083. https://doi.org/10.1016/j.renene.2020.06.064

Basso LC, Basso TO, Rocha SN (2011) Ethanol production in Brazil: the industrial process and its impact on yeast fermentation. In: Bernardes MADS (ed) Biofuel production-recent developments and prospects, 1st edn, InTech, pp 85–100. https://doi.org/10.5772/17047

Bauer F, Hulteberg C, Persson T, Tamm D (2013) Biogas upgrading—review of commercial technologies. http://www.sgc.se/ckfinder/userfiles/files/SGC270.pdf

Behera S, Singh R, Arora R, Sharma NK, Shukla M, Kumar S (2015) Scope of algae as third generation biofuels. Front Bioeng Biotechnol 2:1–13. https://doi.org/10.3389/fbioe.2014.00090

Bernardo G, Araújo T, da Silva Lopes T, Sousa J, Mendes A (2020) Recent advances in membrane technologies for hydrogen purification. Int J Hydrog Energy 45(12):7313–7338. https://doi.org/10.1016/J.IJHYDENE.2019.06.162

Bolzonella D, Battista F, Cavinato C, Gottardo M, Micolucci F, Lyberatos G, Pavan P (2018) Recent developments in biohythane production from household food wastes: a review. In: Bioresource technology, vol 257. Elsevier, pp 311–319. https://doi.org/10.1016/j.biortech.2018.02.092

Bonatto C, Camargo AF, Scapini T, Stefanski FS, Alves SL, Müller C, Fongaro G, Treichel H (2020) Biomass to bioenergy research: current and future trends for biofuels. In: Gupta V, Treichel H, Kuhad R, Rodriguez-Couto S (eds) Recent developments in bioenergy research, 1st edn. Elsevier, pp 1–17. https://doi.org/10.1016/B978-0-12-819597-0.00001-5

Bond T, Templeton MR (2011) History and future of domestic biogas plants in the developing world. Energy Sustain Dev 15(4):347–354. https://doi.org/10.1016/j.esd.2011.09.003

Chandra R, Venkata Mohan S (2014) Enhanced bio-hydrogenesis by co-culturing photosynthetic bacteria with acidogenic process: augmented dark-photo fermentative hybrid system to regulate volatile fatty acid inhibition. Int J Hydrog Energy 39(14):7604–7615. https://doi.org/10.1016/j.ijhydene.2014.01.196

Cheng D, Ngo HH, Guo W, Chang SW, Nguyen DD, Deng L, Chen Z, Ye Y, Bui XT, Hoang NB (2022) Advanced strategies for enhancing dark fermentative biohydrogen production from biowaste towards sustainable environment. In: Bioresource technology, vol 351, Elsevier. https://doi.org/10.1016/j.biortech.2022.127045

Chuichulcherm S, Kasichan N, Srinophakun P, Saisriyoot M, Thanapimmetha A (2017) The use of ozone in a continuous cyclical swing mode regeneration of Fe-EDTA for a clean biogas process from a swine farm waste. J Clean Prod 142:1267–1273. https://doi.org/10.1016/j.jclepro.2016.06.181

Cortez L, Franco TT, Valença G, Rosillo-Calle F (2021) Perspective use of fast pyrolysis bio-oil (FPBO) in maritime transport: the case of Brazil. Energies 14(16):4779

D'Silva TC, Khan SA, Kumar S, Kumar D, Isha A, Deb S, Yadav S, Illathukandy B, Chandra R, Vijay VK, Subbarao PMV, Bagi Z, Kovács KL, Yu L, Gandhi BP, Semple KT (2023) Biohydrogen production through dark fermentation from waste biomass: current status and future perspectives on biorefinery development. Fuel 350:128842. https://doi.org/10.1016/J.FUEL.2023.128842

Das J, Ravishankar H, Lens PNL (2022) Biological biogas purification: recent developments, challenges and future prospects. J Environ Manag 304:114198. https://doi.org/10.1016/j.jenvman.2021.114198

Das N, Jena PK, Padhi D, Mohanty MK, Sahoo G (2023) A comprehensive review of characterization, pretreatment and its applications on different lignocellulosic biomass for bioethanol production. Biomass Convers Biorefinery 13:1503–1527. https://doi.org/10.1007/s13399-021-01294-3

Davila-Vazquez G, Arriaga S, Alatriste-Mondragón F, De León-Rodríguez A, Rosales-Colunga LM, Razo-Flores E (2008) Fermentative biohydrogen production: trends and perspectives. Rev Environ Sci Biotechnol 7(1):27–45. https://doi.org/10.1007/s11157-007-9122-7

de Carvalho JC, Sydney EB, de Souza Kirnev PC, Medeiros ABP, Soccol CR (2019) Technologies for separation and drying of algal biomass for varied applications. In: Ravishankar GA, Ambati RR (eds) Handbook of algal technologies and phytochemicals: volume II phycoremediation, biofuels and global biomass production. CRC Press

de Carvalho JC, Goyzueta-Mamani LD, Karp SG, Aulestia DTM, Sydney EB, Fanka LS, Pandey A, Soccol CR (2022) Downstream processing and formulation of microbial lipids. Biomass Biofuels Biochem 2022:261–287. https://doi.org/10.1016/B978-0-323-90631-9.00007-7

De Carvalho JC, de Souza Vandenberghe LP, Sydney EB, Karp SG, Magalhães AI, Martinez-Burgos WJ, Medeiros ABP, Thomaz-Soccol V, Vieira S, Letti LAJ, Rodrigues C, Woiciechowski AL, Soccol CR (2023) Biomethane production from sugarcane vinasse in a circular economy: developments and innovations. Fermentation 9(4):1–27. https://doi.org/10.3390/fermentation9040349

Deh Kiani MK, Parsaee M, Ardebili SMS, Reyes IP, Fuess LT, Karimi K (2022) Different bioreactor configurations for biogas production from sugarcane vinasse: a comprehensive review. Biomass Bioenergy 161:106446. https://doi.org/10.1016/j.biombioe.2022.106446

Deheri C, Acharya SK (2022) Purified biohythane (biohydrogen+biomethane) production from food waste using CaO2+CaCO3 and NaOH as additives. Int J Hydrog Energy 47(5):2862–2873. https://doi.org/10.1016/j.ijhydene.2021.10.232

Devi A, Bajar S, Sihag P, Ud Z, Sheikh D, Singh A, Kaur J, Bishnoi NR, Pant D (2023) A panoramic view of technological landscape for bioethanol production from various generations of feedstocks. Bioengineered 14(1):81–112. https://doi.org/10.1080/21655979.2022.2095702

Ding L, Cheng J, Qiao D, Li H, Zhang Z (2022) Continuous co-generation of biohydrogen and biomethane through two-stage anaerobic digestion of hydrothermally pretreated food waste. Energy Convers Manag 268:116000. https://doi.org/10.1016/j.enconman.2022.116000

Dispõe sobre as especificações do etanol combustível e suas regras de comercialização em todo o território nacional, Pub. L. No. 907 (2022)
do Nascimento Junior JR, Zevallos Torres LA, Medeiros ABP, Woiciechowski AL, Martinez-Burgos WJ, Soccol CR (2021) Enhancement of biohydrogen production in industrial wastewaters with vinasse pond consortium using lignin-mediated iron nanoparticles. Int J Hydrog Energy 46(54):27431–27443. https://doi.org/10.1016/j.ijhydene.2021.06.009
Du Z, Liu C, Zhai J, Guo X, Xiong Y, Su W, He G (2021) A review of hydrogen purification technologies for fuel cell vehicles. Catalysts 11(3):1–19. https://doi.org/10.3390/catal11030393
Dunikov D, Borzenko V, Blinov D, Kazakov A, Lin CY, Wu SY, Chu CY (2016) Biohydrogen purification using metal hydride technologies. Int J Hydrog Energy 41(46):21787–21794
Dunikov D, Blinov D (2020) Extraction of hydrogen from a lean mixture with methane by metal hydride. Int J Hydrog Energy 45(16):9914–9926. https://doi.org/10.1016/j.ijhydene.2020.01.201
Escorsim AM, Vargas JVC, Mariano AB, Ramos LP, Corazza ML, Cordeiro CS (2018) Extraction of Acutodesmus obliquus lipids using a mixture of ethanol and hexane as solvent. Biomass Bioenergy 108:470–478. https://doi.org/10.1016/j.biombioe.2017.10.035
Farhang B, Ghaebi H, Javani N (2023) Techno-economic modeling of a novel poly-generation system based on biogas for power, hydrogen, freshwater, and ammonia production. J Clean Prod 417:137907. https://doi.org/10.1016/j.jclepro.2023.137907
Faried M, Samer M, Abdelsalam E, Yousef RS, Attia YA, Ali AS (2017) Biodiesel production from microalgae: processes, technologies and recent advancements. Renew Sustain Energy Rev 79 (May):893–913. https://doi.org/10.1016/j.rser.2017.05.199
Franco-Morgado M, Toledo-Cervantes A, González-Sánchez A, Lebrero R, Muñoz R (2018) Integral (VOCs, CO2, mercaptans and H2S) photosynthetic biogas upgrading using innovative biogas and digestate supply strategies. Chem Eng J 354:363–369. https://doi.org/10.1016/j.cej.2018.08.026
Fuess LT, de Araújo Júnior MM, Garcia ML, Zaiat M (2017) Designing full-scale biodigestion plants for the treatment of vinasse in sugarcane biorefineries: how phase separation and alkalinization impact biogas and electricity production costs? Chem Eng Res Des 119:209–220. https://doi.org/10.1016/j.cherd.2017.01.023
Fukuda H, Kondo A, Noda H (2001) Biodiesel fuel production by transesterification of oils. J Biosci Bioeng 92(5):405–416. https://doi.org/10.1016/S1389-1723(01)80288-7
Gandam PK, Chinta ML, Pabbathi NPP, Baadhe RR, Sharma M, Thakur VK, Sharma GD, Ranjitha J, Gupta VK (2022) Second-generation bioethanol production from corncob—a comprehensive review on pretreatment and bioconversion strategies, including techno-economic and lifecycle perspective. Ind Crops Prod 186:115245. https://doi.org/10.1016/J.INDCROP.2022.115245
Gavahian M, Munekata PES, Eş I, Lorenzo JM, Khaneghah AM, Barba FJ (2019) Emerging techniques in bioethanol production: from distillation to waste valorization. Green Chem 21(1):1171–1185. https://doi.org/10.1039/C8GC02698J
Goh BHH, Ong HC, Cheah MY, Chen WH, Yu KL, Mahlia TMI (2019) Sustainability of direct biodiesel synthesis from microalgae biomass: a critical review. Renew Sustain Energy Rev 107:59–74. https://doi.org/10.1016/j.rser.2019.02.012
Gonçalves DNS, Goes GV, de D'Agosto MA, La Rovere EL (2022) Development of policy-relevant dialogues on barriers and enablers for the transition to low-carbon mobility in Brazil. Sustainability 14(24):16405
Government of Canada (2023) Clean fuel regulations. https://www.canada.ca/en/environment-climate-change/services/managing-pollution/energy-production/fuel-regulations/clean-fuel-regulations.html
Gür TM (2022) Carbon dioxide emissions, capture, storage and utilization: review of materials, processes and technologies. Prog Energy Combust Sci 89:100965
Hans M, Kumar S (2019) Biohythane production in two-stage anaerobic digestion system. Int J Hydrog Energy 2019:17363–17380. https://doi.org/10.1016/j.ijhydene.2018.10.022

He X, Lei L, Dai Z (2021) Green hydrogen enrichment with carbon membrane processes: techno-economic feasibility and sensitivity analysis. Sep Purif Technol 276:119346. https://doi.org/10.1016/j.seppur.2021.119346
Huang C, Chen X, Xiong L, Chen X, Ma L, Chen Y (2013) Single cell oil production from low-cost substrates: the possibility and potential of its industrialization. Biotechnol Adv 31(2):129–139. https://doi.org/10.1016/j.biotechadv.2012.08.010
IEA (2019) How competitive is biofuel production in Brazil and the United States? https://www.iea.org/articles/how-competitive-is-biofuel-production-in-brazil-and-the-united-states
IEA (2022) World energy outlook 2022
IEA (2023) Biofuels. https://www.iea.org/energy-system/low-emission-fuels/biofuels
IEA Bioenergy (2014) The potential and challenges of drop-in biofuels. https://task39.sites.olt.ubc.ca/files/2014/01/Task-39-drop-in-biofuels-report-summary-FINAL-14-July-2014-ecopy.pdf
IEA Energy (2019) Drop-in biofuels: the key role that co-processing will play in its production. https://www.ieabioenergy.com/blog/publications/new-publication-drop-in-biofuels-the-key-role-that-co-processing-will-play-in-its-production/
India (2022) India has achieved the target of 10 percent ethanol blending, 5 months ahead of schedule. https://pib.gov.in/PressReleaseIframePage.aspx?PRID=1831289
Jain S, Newman D, Nzihou A, Dekker H, Le Feuvre P, Richter H, Gobe F, Morton C, Thompson R (2019) Global potential of biogas. World Biogas Association
Javed F, Rehman F, Khan AU, Fazal T, Hafeez A, Rashid N (2022) Real textile industrial wastewater treatment and biodiesel production using microalgae. Biomass Bioenergy 165 (August):106559. https://doi.org/10.1016/j.biombioe.2022.106559
Karimi S, Karri RR, Tavakkoli Yaraki M, Koduru JR (2021) Processes and separation technologies for the production of fuel-grade bioethanol: a review. Environ Chem Lett 19:2873–2890
Khan SA, Rashmi, Hussain MZ, Prasad S, Banerjee UC (2009) Prospects of biodiesel production from microalgae in India. Renew Sust Energy Rev 13:2361–2372. https://doi.org/10.1016/j.rser.2009.04.005
Khan MU, Lee JTE, Bashir MA, Dissanayake PD, Ok YS, Tong YW, Shariati MA, Wu S, Ahring BK (2021) Current status of biogas upgrading for direct biomethane use: a review. Renew Sustain Energy Rev 149. https://doi.org/10.1016/j.rser.2021.111343
Kucharska K, Rybarczyk P, Hołowacz I, Łukajtis R, Glinka M, Kamí Nski M (2018) Molecules pretreatment of lignocellulosic materials as substrates for fermentation processes. Molecules 23:2937. https://doi.org/10.3390/molecules23112937
Kumar R, Strezov V (2021) Thermochemical production of bio-oil: a review of downstream processing technologies for bio-oil upgrading, production of hydrogen and high value-added products. Renew Sustain Energy Rev 135. https://doi.org/10.1016/j.rser.2020.110152
Kumari S, Das D (2016) Biologically pretreated sugarcane top as a potential raw material for the enhancement of gaseous energy recovery by two stage biohythane process. Bioresour Technol 218:1090–1097. https://doi.org/10.1016/j.biortech.2016.07.070
Lachos-Perez D, Martins-Vieira JC, Missau J, Anshu K, Siakpebru OK, Thengane SK, Morais ARC, Tanabe EH, Bertuol DA (2023) Review on biomass pyrolysis with a focus on bio-oil upgrading techniques. Analytica 4(2):182–205. https://doi.org/10.3390/analytica4020015
Lane J (2016) Biofuels mandates around the world: 2016. https://www.biofuelsdigest.com/bdigest/2016/01/03/biofuels-mandates-around-the-world-2016/
Lin L, Cunshan Z, Vittayapadung S, Xiangqian S, Mingdong D (2011) Opportunities and challenges for biodiesel fuel. Appl Energy 88(4):1020–1031. https://doi.org/10.1016/j.apenergy.2010.09.029
Liu B, Liu L, Deng B, Huang C, Zhu J, Liang L, He X, Wei Y, Qin C, Liang C, Liu S, Yao S (2022) Application and prospect of organic acid pretreatment in lignocellulosic biomass separation: a review. Int J Biol Macromol 222:1400–1413. https://doi.org/10.1016/J.IJBIOMAC.2022.09.270

Loe R, Santillan-Jimenez E, Crocker M (2019) Upgrading of lipids to fuel-like hydrocarbons and terminal olefins via decarbonylation/decarboxylation. In: Chemical catalysts for biomass upgrading. Springer, pp 497–528. https://doi.org/10.1002/9783527814794.CH12

Lorick D, Macura B, Ahlström M, Grimvall A, Harder R (2020) Effectiveness of struvite precipitation and ammonia stripping for recovery of phosphorus and nitrogen from anaerobic digestate: a systematic review. Environ Evid 9(1):27. https://doi.org/10.1186/s13750-020-00211-x

Lotero E, Liu Y, Lopez DE, Suwannakarn K, Bruce DA, Goodwin JG (2005) Synthesis of biodiesel via acid catalysis. Ind Eng Chem Res 44(14):5353–5363. https://doi.org/10.1021/ie049157g

Macura B, Johannesdottir SL, Piniewski M, Haddaway NR, Kvarnström E (2019) Effectiveness of ecotechnologies for recovery of nitrogen and phosphorus from anaerobic digestate and effectiveness of the recovery products as fertilisers: a systematic review protocol. Environ Evid 8(1): 29. https://doi.org/10.1186/s13750-019-0173-3

Mahmod SS, Jahim JM, Abdul PM, Luthfi AAI, Takriff MS (2021) Techno-economic analysis of two-stage anaerobic system for biohydrogen and biomethane production from palm oil mill effluent. J Environ Chem Eng 9(4):105679. https://doi.org/10.1016/j.jece.2021.105679

Maia DCS, Cardosoa FH, Frare LM, Gimenes ML, Pereira NC (2014) Purification of biogas for energy use. Chem Eng Trans 37(August 2015):643–648. https://doi.org/10.3303/CET1437108

Maile OI, Muzenda E, Tesfagiorgis H (2017) Chemical absorption of carbon dioxide in biogas purification. Procedia Manuf 7:639–646. https://doi.org/10.1016/j.promfg.2016.12.095

Mallouppas G, Yfantis EA, Ioannou C, Paradeisiotis A, Ktoris A (2023) Application of biogas and biomethane as maritime fuels: a review of research, technology development, innovation proposals, and market potentials. Energies 16(4):2066

Manandhar A, Mousavi-Avval SH, Tatum J, Shrestha E, Nazemi P, Shah A (2022) Solid biofuels. In: Biomass, biofuels, biochemicals. Elsevier, pp 343–370

Martinez-Burgos WJ, Sydney EB, Paula DR, Medeiros ABP, Carvalho JC, Soccol VT, de Vandenberghe LPS, Woiciechowski AL, Soccol CR (2020) Biohydrogen production in cassava processing wastewater using microbial consortia: process optimization and kinetic analysis of the microbial community. Bioresour Technol 309:123331. https://doi.org/10.1016/j.biortech.2020.123331

Martinez-Burgos WJ, Junios JR, Medeiros ABP, Herrmann LW, Sydney EB, Soccol CR (2021a) Biohydrogen production from agro-industrial wastes using Clostridium beijerinckii and isolated bacteria as inoculum. Bioenergy Res 15:987–997. https://doi.org/10.1007/s12155-021-10358-1

Martinez-Burgos WJ, Sydney EB, de Paula DR, Medeiros ABP, de Carvalho JC, Molina D, Soccol CR (2021b) Hydrogen production by dark fermentation using a new low-cost culture medium composed of corn steep liquor and cassava processing water: process optimization and scale-up. Bioresour Technol 320:124370. https://doi.org/10.1016/j.biortech.2020.124370

Martinez-Burgos WJ, de Vandenberghe LPS, Karp SG, De Mello M, Soccol VT, Soccol CR (2023) Microbial lipid production from soybean hulls using Lipomyces starkeyi LPB53 in a circular economy. Bioresour Technol 372:128650. https://doi.org/10.1016/j.biortech.2023.128650

Meena RAA, Rajesh Banu J, Yukesh Kannah R, Yogalakshmi KN, Kumar G (2020) Biohythane production from food processing wastes—challenges and perspectives. In: Bioresource technology, vol 298, Elsevier. https://doi.org/10.1016/j.biortech.2019.122449

Moreno-Benito M, Agnolucci P, Papageorgiou LG (2017) Towards a sustainable hydrogen economy: optimisation-based framework for hydrogen infrastructure development. Comput Chem Eng 102:110–127

Mozhiarasi V, Natarajan TS, Dhamodharan K (2023) A high-value biohythane production: feedstocks, reactor configurations, pathways, challenges, technoeconomics and applications. In: Environmental research, vol 219. Academic Press. https://doi.org/10.1016/j.envres.2022.115094

Müller C, Scapini T, Rempel A, Abaide ER, Camargo AF, Nazari MT, Tadioto V, Bonatto C, Tres MV, Zabot GL, Colla LM, Treichel H, Alves SL (2023) Challenges and opportunities for third-

generation ethanol production: a critical review. Eng Microbiol 3(1):100056. https://doi.org/10.1016/J.ENGMIC.2022.100056
Musa M, Ayoko GA, Ward A, Rösch C, Brown RJ, Rainey TJ (2019) Factors affecting microalgae production for biofuels and the potentials of chemometric methods in assessing and optimizing productivity. Cells 8:851. https://doi.org/10.3390/cells8080851
Mussatto SI, Roberto IC (2004) Alternatives for detoxification of diluted-acid lignocellulosic hydrolyzates for use in fermentative processes: a review. Bioresour Technol 93(1):1–10. https://doi.org/10.1016/J.BIORTECH.2003.10.005
Nanda S, Mohammad J, Reddy SN, Kozinski JA, Dalai AK (2014) Pathways of lignocellulosic biomass conversion to renewable fuels. Biomass Convers Bioref 4:157–191. https://doi.org/10.1007/s13399-013-0097-z
NIST (2010) Standard reference database 69: NIST chemistry WebBook. National Institute of Standards and Technology
NREL (2023) Biodiesel handling and use guide. https://afdc.energy.gov/files/u/publication/biodiesel_handling_use_guide.pdf
O'Neil MJ (2013) The Merck index: an encyclopedia of chemicals, drugs, and biologicals. RSC Publishing
Obernberger I, Brunner T, Bärnthaler G (2006) Chemical properties of solid biofuels—significance and impact. Biomass Bioenergy 30(11):973–982
Onyeaka H, Miri T, Obileke KC, Hart A, Anumudu C, Al-Sharify ZT (2021) Minimizing carbon footprint via microalgae as a biological capture. Carbon Capture Sci Technol 1:100007. https://doi.org/10.1016/J.CCST.2021.100007
O-Thong S, Suksong W, Promnuan K, Thipmunee M, Mamimin C, Prasertsan P (2016) Two-stage thermophilic fermentation and mesophilic methanogenic process for biohythane production from palm oil mill effluent with methanogenic effluent recirculation for pH control. Int J Hydrog Energy 41(46):21702–21712. https://doi.org/10.1016/j.ijhydene.2016.07.095
Patel A, Shah AR (2021) Integrated lignocellulosic biorefinery: gateway for production of second generation ethanol and value added products. J Bioresour Bioprod 6(2):108–128. https://doi.org/10.1016/J.JOBAB.2021.02.001
Patil SY (2010) Lipid production from glucose and starch using Lipomyces starkeyi. Doctoral. Doctoral dissertation, University of Louisiana at Lafayette
Paulova L, Patakova P, Branska B, Rychtera M, Melzoch K (2015) Lignocellulosic ethanol: technology design and its impact on process efficiency. Biotechnol Adv 33(6):1091–1107. https://doi.org/10.1016/J.BIOTECHADV.2014.12.002
Peralta-Ruiz Y, González-Delgado A, Kafarov V (2013) Evaluation of alternatives for microalgae oil extraction based on exergy analysis. Appl Energy 101:226–236. https://doi.org/10.1016/j.apenergy.2012.06.065
Periyasamy S, Karthik V, Senthil Kumar P, Isabel JB, Temesgen T, Hunegnaw BM, Melese BB, Mohamed BA, Dai-Viet VN (2022) Chemical, physical and biological methods to convert lignocellulosic waste into value-added products. A review. Environ Chem Lett 20:1129–1152. https://doi.org/10.1007/s10311-021-01374-w
Plesu Popescu AE, Pellin JL, Bonet J, Llorens J (2021) Bioethanol dehydration and mixing by heterogeneous azeotropic distillation. J Clean Prod 320:128810. https://doi.org/10.1016/J.JCLEPRO.2021.128810
Pôjo V, Tavares T, Malcata FX (2021) Processing methodologies of wet microalga biomass toward oil separation: an overview. Molecules 26:641. https://doi.org/10.3390/molecules26030641
Portaria MAPA N° 75, Pub. L. No. 75 (2015)
Poulikidou S, Heyne S, Grahn M, Harvey S, Hansson J (2018) A comparative assessment of current and future fuels for the transport sector. In: Proceedings of the European Biomass Conference and Exhibition Proceedings, pp 14–17
Primmer N (2021) Biogas: pathways to 2030 – report. World Biogas Association
Raizen (2023) Raízen geo biogas. https://geobiogas.tech/projeto/raizen-geo-biogas/

Ram B, Ram P, Sharma K, Man I (2020) Effect of reduced gas pressure on yield of biogas and other physicochemical parameters. Int J Energy Environ Eng 0123456789:1–7. https://doi.org/10.1007/s40095-020-00351-3

Rehnlund B (2008) Outlook on standardization of alternative vehicle fuels

Resolução CNPE nº 16, de 29 de outubro de 2018, Pub. L. No. Resolução nº 3, de 20 de março de 2023, do Conselho Nacional de Política Energética-CNPE (2023)

Rocher-Rivas R, González-Sánchez A, Ulloa-Mercado G, Muñoz R, Quijano G (2022) Biogas desulfurization and calorific value enhancement in compact H2S/CO2 absorption units coupled to a photobioreactor. J Environ Chem Eng 10(5):108336. https://doi.org/10.1016/j.jece.2022.108336

Rosa D, Medeiros ABP, Martinez-Burgos WJ, Nascimento JR, Carvalho JC, Sydney EB, Soccol CR (2020) Biological hydrogen production from palm oil mill effluent (POME) by anaerobic consortia and Clostridium beijerinckii. J Biotechnol 323:17–23. https://doi.org/10.1016/j.jbiotec.2020.06.015

Ryckebosch E, Drouillon M, Vervaeren H (2011) Techniques for transformation of biogas to biomethane. Biomass Bioenergy 35(5):1633–1645. https://doi.org/10.1016/j.biombioe.2011.02.033

Sahoo NK, Gupta SK, Rawat I, Ansari FA, Singh P, Naik SN, Bux F (2017) Sustainable dewatering and drying of self-flocculating microalgae and study of cake properties. J Clean Prod 159:248–256

Saladini F, Patrizi N, Pulselli FM, Marchettini N, Bastianoni S (2016) Guidelines for energy evaluation of first, second and third generation biofuels. Renew Sustain Energ Rev 66:221–227. https://doi.org/10.1016/j.rser.2016.07.073

Sasidhar KB, Kumar PS, Xiao L (2022) A critical review on the two-stage biohythane production and its viability as a renewable fuel. Fuel 317:123449. https://doi.org/10.1016/j.fuel.2022.123449

Schenk PM, Thomas-Hall SR, Stephens E, Marx UC, Mussgnug JH, Posten C, Kruse O, Hankamer B (2008) Second generation biofuels: high-efficiency microalgae for biodiesel production. Bioenergy Res 1(1):20–43. https://doi.org/10.1007/s12155-008-9008-8

Schmidt-Rohr K (2015) Why combustions are always exothermic, yielding about 418 kJ per mole of O2. J Chem Educ 92(12):2094–2099

Schmidt-Rohr K (2020) Oxygen is the high-energy molecule powering complex multicellular life: fundamental corrections to traditional bioenergetics. ACS Omega 5(5):2221–2233

Schmitt CC, Fonseca FG, Fraga MC, Wisniewski A, Karp S, José ÁHM, Rodrigues RCLB, Moreira R, Hirayama DE, Raffelt K, Dahmen N (2021) Thermochemical and catalytic conversion technologies for the development of Brazilian biomass utilization. Catalysts 11(12). https://doi.org/10.3390/catal11121549

Segovia-Hernández JG, Sánchez-Ramírez E (2022) Current status and future trends of computer-aided process design, applied to purification of liquid biofuels, using process intensification: a short review. Chem Eng Process 172:108804

Sekoai PT, Mhlongo SI, Ezeokoli OT (2019) Progress in the development of methods used for the abatement of microbial contaminants in ethanol fermentations: a review. Rev Environ Sci Biotechnol 18:795–821. https://doi.org/10.1007/s11157-019-09511-2

Sengupta S, Bhattacharya D, Mukhopadhyay M (2020) Downstream processing of biofuel. In: Kuila A, Sharma V (eds) Genetic and metabolic engineering for improved biofuel production from lignocellulosic biomass. Elsevier, pp 47–62. https://doi.org/10.1016/B978-0-12-817953-6.00004-X

Shanmugam S, Mathimani T, Rene ER, Edwin Geo V, Arun A, Brindhadevi K, Pugazhendhi A (2021) Biohythane production from organic waste: recent advancements, technical bottlenecks and prospects. Int J Hydrog Energy 46(20):11201–11216. https://doi.org/10.1016/j.ijhydene.2020.10.132

Show K-Y, Lee D-J, Tay J-H, Lee T-M, Chang J-S (2015) Microalgal drying and cell disruption - recent advances. Bioresour Technol 184:258–266. https://doi.org/10.1016/j.biortech.2014.10.139

Sierra LS, Dixon CK, Wilken LR (2017) Enzymatic cell disruption of the microalgae Chlamydomonas reinhardtii for lipid and protein extraction. Algal Res 25:149–159. https://doi.org/10.1016/j.algal.2017.04.004

Sillero L, Solera R, Pérez M (2022) Effect of the hydraulic retention time on the acidogenic fermentation of sewage sludge, wine vinasse and poultry manure for biohydrogen production. Biomass Bioenergy 167:106643. https://doi.org/10.1016/J.BIOMBIOE.2022.106643

Singh S, Pandey D, Saravanabhupathy S, Daverey A, Dutta K, Arunachalam K (2022) Liquid wastes as a renewable feedstock for yeast biodiesel production: opportunities and challenges. Environ Res 207(October 2021):112100. https://doi.org/10.1016/j.envres.2021.112100

Soccol CR, Neto CJD, Soccol VT, Sydney EB, Costa ESF, Medeiros ABP, de Vandenberghe LPS (2017) Pilot scale biodiesel production from microbial oil of Rhodosporidium toruloides DEBB 5533 using sugarcane juice: performance in diesel engine and preliminary economic study. Bioresour Technol 223:259–268. https://doi.org/10.1016/j.biortech.2016.10.055

Stein T (2022) Carbon dioxide now more than 50 higher than pre-industrial levels. https://www.noaa.gov/news-release/carbon-dioxide-now-more-than-50-higher-than-pre-industrial-levels

Sun H, Yang M, Gao Z, Wang X, Wu C, Wang Q, Gao M (2023) Economic and environmental evaluation for a closed loop of crude glycerol bioconversion to biodiesel. J Biotechnol 366:65–71. https://doi.org/10.1016/j.jbiotec.2023.03.001

Sydney EB, Duarte ER, Martinez-Burgos WJ, Carvalho JC, Larroche C, Soccol CR (2019) Development of short chain fatty acid-based artificial neuron network tools applied to biohydrogen production. Int J Hydrog Energy 5:2175–5185. https://doi.org/10.1016/j.ijhydene.2019.10.128

Taghizadeh-Alisaraei A, Motevali A, Ghobadian B (2019) Ethanol production from date wastes: adapted technologies, challenges, and global potential. Renew Energy 143:1094–1110. https://doi.org/10.1016/j.renene.2019.05.048

Tan IS, Lam MK, Foo HCY, Lim S, Lee KT (2020) Advances of macroalgae biomass for the third generation of bioethanol production. Chin J Chem Eng 28(2):502–517. https://doi.org/10.1016/J.CJCHE.2019.05.012

Tena M, Luque B, Perez M, Solera R (2020) Enhanced hydrogen production from sewage sludge by cofermentation with wine vinasse. Int J Hydrog Energy 45(32):15977–15984. https://doi.org/10.1016/j.ijhydene.2020.04.075

Thiru M, Sankh S, Rangaswamy V (2011) Process for biodiesel production from Cryptococcus curvatus. Bioresour Technol 102(22):10436–10440. https://doi.org/10.1016/j.biortech.2011.08.102

van den Oever A, Costa D, Messagie M (2022) Comparison of BtL drop-in biofuels with transport fuels

Vasić K, Knez Ž, Leitgeb M (2021) Bioethanol production by enzymatic hydrolysis from different lignocellulosic sources. Molecules 26:753. https://doi.org/10.3390/molecules26030753

Vedachalam S, Baquerizo N, Dalai AK (2022) Review on impacts of low sulfur regulations on marine fuels and compliance options. Fuel 310:122243

Wang X, Chen R, Chen C, Wang Q (2006) Hydrogen storage properties of TixFe + y wt.% La and its use in metal hydride hydrogen compressor. J Alloys Compd 425(1–2):291–295. https://doi.org/10.1016/J.JALLCOM.2006.01.025

Wang H, Ma C, Yang Z, Lu X, Ji X (2020) Improving high-pressure water scrubbing through process integration and solvent selection for biogas upgrading. Appl Energy 276:115462. https://doi.org/10.1016/j.apenergy.2020.115462

Whim BP, Johnson PG (1996) Directory of solvents. Springer. https://doi.org/10.1007/978-94-009-1549-7

Wiatrowski M, Klein BC, Davis RW, Quiroz-Arita C, Tan ECD, Hunt RW, Davis RE (2022) Techno-economic assessment for the production of algal fuels and value-added products:

opportunities for high-protein microalgae conversion. Biotechnol Biofuels Bioprod 15(1):1–14. https://doi.org/10.1186/s13068-021-02098-3

Yang ZL, Zhang D, Caglayan O, Jenkinson ID, Bonsall S, Wang J, Huang M, Yan XP (2012) Selection of techniques for reducing shipping NOx and SOx emissions. Transp Res D Transp Environ 17(6):478–486. https://doi.org/10.1016/j.trd.2012.05.010

Yang G, Hu Y, Wang J (2019) Biohydrogen production from co-fermentation of fallen leaves and sewage sludge. Bioresour Technol 285:121342. https://doi.org/10.1016/j.biortech.2019.121342

Yellapu SK, Bharti, Kaur R, Kumar LR, Tiwari B, Zhang X, Tyagi RD (2018) Recent developments of downstream processing for microbial lipids and conversion to biodiesel. Bioresour Technol 256(November 2017):515–528. https://doi.org/10.1016/j.biortech.2018.01.129

Zhang X, Yan S, Dayal R, Drogui P, Surampalli RY (2016) Ultrasonication aided biodiesel production from one-step and two-step transesterification of sludge derived lipid. Energy 94: 401–408. https://doi.org/10.1016/j.energy.2015.11.016

Zhang L, Wang Z, Zhou W, Yang X, Zhao S, Li Q (2022) GOSAT mapping of global greenhouse gas in 2020 and 2021. Atmosphere 13(11):1814

Zhang Y, Liang Y, Li S, Yuan Y, Zhang D, Wu Y, Xie H, Brindhadevi K, Pugazhendhi A, Xia C (2023) A review of biomass pyrolysis gas: forming mechanisms, influencing parameters, and product application upgrades. Fuel 347:128461

Zhou Y, Remón J, Jiang Z, Matharu AS, Hu C (2022) Tuning the selectivity of natural oils and fatty acids/esters deoxygenation to biofuels and fatty alcohols: a review. Green Energy Environ 8(3): 722–743

Chapter 12
Recent Advances in Genetic and Metabolic Engineering for Drop-in Biofuel Production

Lachi Wankhede, Gaurav Bharadwaj, Rahul Saini, Carlos S. Osorio-Gonzalez, and Satinder Kaur Brar

Abstract This chapter explores recent genetic and metabolic engineering advances to produce drop-in biofuels. It provides a comprehensive overview of the techniques employed, including gene overexpression, gene cloning, gene therapy, and the knockout gene technique, which enhances microorganisms' ability to produce drop-in biofuels. Likewise, the metabolic engineering approach is also explored, focused mainly on cellular metabolic pathways manipulation to boost drop-in biofuel production, particularly isoprenoids, alkanes/alkenes, and jet fuels. Despite the challenges like complexity of cellular metabolic networks, significant progress has been made. The chapter highlights several investigations, demonstrating the potential of these technologies to produce sustainable and renewable drop-in biofuels. The future of genetic and metabolic engineering for drop-in biofuel production appears promising. It offers renewable alternatives to fossil fuels, integration into existing industrial processes, and the potential for economic growth and job creation in the biofuel industry. While challenges persist, the potential benefits make this field an active research and development area. Overall, this chapter serves as a comprehensive guide, providing insights into drop-in biofuel production's current state, challenges, and future directions through genetic and metabolic engineering.

Keywords Drop-in biofuels · Genetic engineering · Metabolic engineering · Isoprenoids · Alkanes/alkenes · Sustainable aviation fuel

12.1 Introduction

The global energy landscape is undergoing a significant transformation. Finding sustainable and renewable energy sources has become increasingly important as the world faces the dual challenges of rising energy needs and the need to reduce environmental impacts. As a result of increased demand for bio-based products,

L. Wankhede · G. Bharadwaj · R. Saini · C. S. Osorio-Gonzalez · S. K. Brar (✉)
Department of Civil Engineering, Lassonde School of Engineering, York University, Toronto, ON, Canada
e-mail: satinder.brar@lassonde.yorku.ca

S. K. Brar et al. (eds.), *The Microbiology of the Drop-in Biofuel Production*, Biofuel and Biorefinery Technologies 15, https://doi.org/10.1007/978-3-031-61637-2_12

the research and development efforts are now focused on commercially oriented research trends as the market will start to swell from 2–20% by 2032 (Cordova et al. 2020). Among the various alternatives, drop-in biofuels have emerged as a promising solution, offering a renewable and potentially carbon-neutral way to meet our energy needs. Drop-in biofuels, serving as a renewable substitute for fossil fuels, are derived from microorganisms or from the metabolic by-products of organic or food waste. In contrast to fossil fuels, which release carbon dioxide that has been stored for many years, the use of drop-in biofuels can lead to no overall increase in atmospheric carbon dioxide. This balance is maintained if the plants or biomass used for producing these biofuels is consistently regrown, as they absorb the same amount of carbon dioxide during their growth that is released when the biofuels are burned. During the last few decades, drop-in biofuels have gained significant attention. Drop-in biofuels, as direct substitutes for aviation fuel, diesel, and gasoline, possess the potential to exhibit similar or even higher energy densities, thereby offering a viable alternative without the need for significant modifications in existing fuel infrastructures. Some of the features that make drop-in biofuels better against non-renewable biofuels are the low or nil content of sulfur or polyaromatic hydrocarbons and no production of particulates or carbon monoxide after combustion, contributing to cleaner energy transition decreasing gas emissions (Saini et al. 2020; Xue et al. 2018). These fuels are fully compatible with existing infrastructure and engines, requiring no modification or adjustment. The ability to "drop" these biofuels into the existing infrastructure makes them an attractive and practical alternative to fossil fuels. Extensive efforts have been made to efficiently produce microbial strains with improved overall tolerances, broader substrate spectrums, and enhanced metabolic fluxes to convert biomass into drop-in biofuels (Ko et al. 2020; Ko and Lee 2018; Majidian et al. 2018). The production of drop-in biofuels, however, presents unique challenges, because they are typically derived from complex metabolic pathways in living organisms, requiring sophisticated techniques to enhance their production efficiency and yield. This is where genetic and metabolic engineering come into play.

In recent years, genetic and metabolic engineering have emerged as powerful tools for optimizing drop-in biofuel production and enhancing its viability as a sustainable energy solution. These fields encompass a range of techniques that enable the manipulation and fine-tuning of microbial strains and metabolic pathways to increase the yield of drop-in biofuels. Through genetic engineering, scientists can modify the genetic makeup of organisms to improve desired traits, such as yield increasing or tolerance to stress conditions. Metabolic engineering focuses on changing the metabolic pathways within organisms to redirect metabolic flux toward producing specific drop-in biofuel precursors (Nielsen and Keasling 2016; Wong et al. 2017). Genetic and metabolic engineering techniques offer several advantages in drop-in biofuel production. They enable the design and construction of microorganisms with enhanced capabilities to convert renewable resources, such as plant biomass and carbon dioxide, into valuable drop-in biofuels. These techniques also allow for optimizing metabolic pathways, enzyme activities, and regulatory

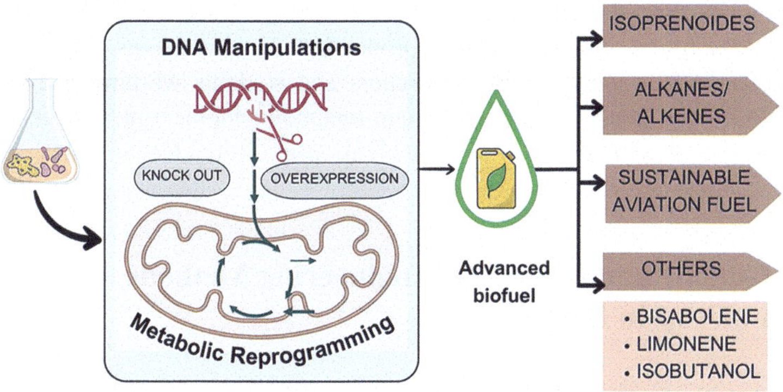

Fig. 12.1 Conceptual overview of the interplay between genetic and metabolic engineering in the production of advanced drop-in biofuels

mechanisms, leading to improved yield, productivity, and robustness of drop-in biofuel-producing microorganisms.

Genetic and metabolic engineering techniques present viable solutions to the limitations inherent in traditional drop-in biofuel production methods, such as suboptimal conversion efficiencies and competition with food supply chains. Pioneering work has focused on leveraging microbes like *Escherichia coli* and *Saccharomyces cerevisiae*, both of which naturally synthesize ethanol, a prominent but energy-deficient drop-in biofuel (Peralta-Yahya and Keasling 2010). However, ethanol's low energy content limits its future economic use. To counter this challenge, production of advanced drop-in biofuels, like long-chain alcohols and biofuels derived from isoprenoids and fatty acids, is being pursued. These drop-in biofuels exhibit physical characteristics that are more similar to those of fuels derived from petroleum (Peralta-Yahya and Keasling 2010). Metabolic engineering has further extended the applicability of drop-in biofuel production to alternative organisms like microalgae. Such platforms not only hold promise as a renewable energy source but also benefit from metabolic interventions to amplify biomass concentration, thereby mitigating the economic burden of downstream processes (Medipally et al. 2015). While the emergence of sophisticated genetic and metabolic engineering methodologies has significantly expanded the scope of drop-in biofuels, challenges remain. Key among them is the need for optimized production processes that include fine-tuned fermentation parameters, strategic nutrient allocation, and cost-effective downstream processing. Regulatory considerations and societal acceptance are also a factor in the large-scale deployment of genetically modified drop-in biofuel-producing organisms.

This chapter provides a comprehensive overview of recent genetic and metabolic engineering advances to produce drop-in biofuels specifically focusing on isoprenoids, alkanes/alkenes, sustainable aviation fuels, bisabolene, limonene, and isobutanol (Fig. 12.1). It explores the principles, techniques, and applications of genetic and metabolic engineering in drop-in biofuel production. Additionally, it discusses the challenges and future perspectives in the field, highlighting the potential for further advancements and the role of genetic and metabolic engineering in shaping the future of drop-in biofuel production. Finally, the insights provided in this chapter will be valuable for researchers and students working in bioenergy, renewable energy, genetic engineering, and metabolic engineering to accelerate the development and deployment of sustainable drop-in biofuel technologies.

12.2 Genetic and Metabolic Engineering Methods in Drop-in Biofuel

Genetic and metabolic engineering serve as pivotal techniques in the modern arena of drop-in biofuel generation. Genetic engineering plays a critical role by enabling targeted genomic manipulations by employing methods such as gene alteration, gene splicing, and gene therapy, where genetic engineering enhances the genomic landscape of microbes, thereby amplifying the drop-in biofuel output. Concurrently, metabolic engineering operates to systematically fine-tune existing enzymatic pathways. It deploys tactics such as enhancing carbon flux toward targeted end products, as evidenced by Liew et al. (2017), and to integrate non-native metabolic pathways to either create new drop-in biofuels or make the production process more efficient (Dusséaux et al. 2020). Further strategies, such as the elimination of metabolic routes that divert valuable precursors or produce unwanted by-products, have been pursued to optimize drop-in biofuel synthesis (Wess et al. 2019). Further dimensions of optimization have been researched through cofactor engineering to maintain cellular redox balance (Zhou et al. 2015), engineering of promoters for better gene expression (Madhavan et al. 2017), and the optimization of codons for effective protein translation (Kirby et al. 2021). Strategies that modulate cellular respiratory rates and improve toxic metabolite tolerance have also been adopted, thus enhancing drop-in biofuel production efficiencies (Kalnenieks et al. 2019). In summary, genetic and metabolic engineering operate synergistically yet distinctly, addressing the multifaceted complexities of microbial drop-in biofuel production. While the former genetic and metabolic engineering approaches modify genomes to introduce beneficial traits, recent efforts are increasingly centered on the rational manipulation of existing metabolic networks. Each methodology is indispensable and often synergistically deployed to navigate the intricacies of microbial drop-in biofuel production. In the subsequent sections, this chapter will explore the principles of genetic and metabolic engineering as they apply to a variety of advanced drop-in biofuels, examining the distinct challenges and opportunities each presents for optimization and scaling up production.

12.3 Isoprenoids

Isoprenoids, alternatively termed terpenoids, are a varied class of organic compounds that are sourced from isoprene. These compounds are present in plants, animals, and microorganisms, playing crucial roles in numerous biological processes (Birchfield and McIntosh 2020; Farrokh et al. 2019; Rahmat and Kang 2020). Isoprenoids are responsible for flavors, aromas, pigments, and medicinal properties (Heuston et al. 2012). As a result, they hold considerable commercial importance in industries such as pharmaceuticals, cosmetics, and food additives. The process of isoprenoid biosynthesis encompasses a sophisticated pathway known as either the mevalonate (MVA) pathway or the 2-C-methyl-D-erythritol 4-phosphate (MEP) pathway, which varies based on the organism (Goldstein and Brown 1990; Lichtenthaler 2000). Their biosynthesis involves complex pathways, and advancements in metabolic and genetic engineering offer opportunities for improved production. Technological advancements in both metabolic and genetic engineering offer interesting possibilities for elevating isoprenoid production. Genetic interventions, for example, provide targeted means to elevate the cellular yields of these compounds, often through gene overexpression strategies. Such strategies may encompass the insertion of additional gene copies or the alteration of gene-regulatory elements to maximize the enzymatic output pivotal for isoprenoid biosynthesis (Pulido et al. 2012; Spagnuolo et al. 2019).

In a study, Yang et al. developed a pathway in *Escherichia coli* for isoprene production by integrating an isoprene synthase (ispS) enzyme sourced from *Populus alba*. They enhanced isoprene production by increasing the expression of critical enzymes, namely, *dxs*, *dxr*, and *idi* as well as different gene orders within the pathway to optimize the process. The study revealed that *idi* played a vital role in controlling isoprene production. Likewise, they achieved the highest yield of isoprene at 2.727 mg g − 1 h − 1 (per dry weight) in *E. coli,* which was greater than the wild strain (Lv et al. 2013; Zhao et al. 2011). Another study by Yang et al. 2012 shows the efficiency of the MVA pathway for isoprene production. Two significant improvements were made: firstly, the "upper pathway" responsible for converting acetyl-CoA into MVA was switched from *Saccharomyces cerevisiae* to *Enterococcus faecalis* to increase MVA production. Secondly, a specific mutation was introduced in the *mvaS* gene to enhance MVA and further isoprene production. The final optimized strain YJM25 (*E. coli*), incorporating the modified MVA pathway and isoprene synthase gene derived *from Populus alba*, achieved an isoprene accumulation of 6.3 g/L after 40 h under fed-batch cultivation (Yang et al. 2012).

Likewise, Fiona et al. employed the *cyanobacterium Synechocystis* sp., along with a combination of isoprene synthase expression and mevalonic acid (MVA) pathway integration, leading to a notable 2.5-fold increase in isoprene yield. This achievement was made possible through the chromosomal integration of seven recombinant enzymes into the microorganism, a feat that marks a considerable advancement in the field of genetic engineering for microorganisms (Bentley et al. 2014). Using cyanobacteria, isoprene production was achieved by introducing the

IspS gene from *Pueraria montana* into *Synechocystis* sp. The co-expression of the mevalonate (MVA) pathway and the *IspS* gene in *Synechocystis* sp. 6803 led to an increase in isoprene production, as reported by Lindberg et al. (2010). In addition, Gao et al. conducted a study where they engineered the isoprene biosynthetic pathway in *Synechococcus elongatus*, enabling the production of isoprene directly from CO2. They opted for the methylerythritol phosphate (MEP) pathway due to its effectiveness in carbon utilization and precursor generation. The team introduced high-activity isoprene synthases from plants and amplified the dimethylallyl pyrophosphate to isopentenyl pyrophosphate (IPP) ratio by overexpressing IPP isomerase. This led to the engineered strain producing a notable amount of isoprene, reaching 1.26 g/L directly from CO_2 (Gao et al. 2016).CRISPR/Cas9 stands as a transformative gene editing technology, revolutionizing the realm of molecular biology. It is derived from a natural defense mechanism found in bacteria, where CRISPR sequences act as a genetic memory of past viral infections. CRISPR-associated protein 9 (Cas9) is an enzyme that, when guided by a small RNA molecule called a guide RNA (gRNA), can accurately target and modify specific genes within an organism's genome (Cong et al. 2013). In a recent study, researchers used CRISPR/Cas9 to develop a multiplex genome engineering approach in *Saccharomyces cerevisiae* for enhanced isoprenoid production. They simultaneously modified up to 5 genomic loci, resulting in a 41-fold increase in mevalonate titers, a key intermediate in isoprenoid biosynthesis. Genome resequencing confirmed the specificity of the CRISPR/Cas9 system, as no unintended genetic alterations were detected. This study showcases the power of CRISPR/Cas9 for targeted and efficient genome engineering, enabling the manipulation of multiple genes simultaneously. The findings highlight the potential of CRISPR/Cas9 in accelerating metabolic engineering efforts and optimizing the production of isoprenoids, offering opportunities for sustainable drop-in biofuel and biochemical production methods (Jakočiūnas et al. 2015).

Metabolic engineering aims to transform various organisms, from bacteria to plants, into efficient chemical factories for isoprenoid production. It has traditionally been applied in well-studied microorganisms like *E. coli* and yeast, but there is increasing interest in exploring novel hosts through genetic modification (Ward et al. 2018). Recently, a study by Zada et al. (2018) successfully leveraged the versatile activity of NudB, a native phosphatase in *Escherichia coli*, to generate a mixture of isoprenoid alcohols and their derivatives. By introducing NudB and the enzyme IspA, involved in farnesyl diphosphate (FPP) production, into a modified *E. coli* strain with a heterologous mevalonate (MVA) pathway, they achieved production levels of up to 1652 mg/L of a mixture of C5 to C15 alcohols. Furthermore, the addition of AphA, another endogenous phosphatase, in combination with NudB, led to an enhanced production yield of isoprenoid alcohol. In another approach, replacing IspA with a mutant variant, IspA*, which produced geranyl diphosphate (GPP), yielded 1418 mg/L of the isoprenoid mixture, with C10 alcohols as the primary component. This study demonstrates the successful application of metabolic engineering for producing isoprenoid alcohol mixtures, showcasing their potential as advanced drop-in biofuels and valuable chemicals. These engineering strategies

Table 12.1 Studies on genetic and metabolic engineering of isoprenoid drop-in biofuels

Microorganism	Genetic/Metabolic techniques employed	Isoprene yield	Reference
E. coli	Introduction of ispS from Populus alba, overexpression of dxs, dxr, and idi	2.727	(Lv et al. 2013)
E. coli	MVA pathway modification, mvaS gene mutation	6.3 g/L	(Yang et al. 2012)
Synechocystis sp.	Chromosomal integration of recombinant enzymes, isoprene synthase expression	–	(Bentley et al. 2014)
Synechocystis sp.	Expression of the MVA pathway and IspS from Pueraria montana	–	(Lindberg et al. 2010)
Synechococcus elongatus	MEP pathway optimization	1.26 g/L	(Gao et al. 2016)
Saccharomyces cerevisiae	Multiplex genome engineering, up to 5 loci modified		(Jakočiūnas et al. 2015)
E. coli	Introduction of NudB and IspA/IspA*, modified MVA pathway	1.652 g/L	(Zada et al. 2018)

offer a versatile platform for tailoring the synthesis of other isoprenoid mixtures by modulating the pathways of IPP, DMAPP, GPP, and FPP.

Likewise, metabolic engineering studies have focused on enhancing the production of isoprenoids within glandular trichomes (GTs) by manipulating the metabolic pathway involved. This pathway, known as the 1-deoxy-D-xylulose 5-phosphate (DXP) pathway, plays a crucial role in isoprenoid metabolism. By increasing the levels of precursor molecules, isopentenyl diphosphate (IPP) and dimethylallyl diphosphate (DMAPP), through improvements in the DXP pathway, substantial increases in isoprenoid production within G.T.s can be achieved. Crucial enzymes of the DXP pathway, including 1-deoxy-D-xylulose 5-phosphate synthase (DXS) and 1-deoxy-D-xylulose 5-phosphate reductoisomerase (DXR), have been pinpointed as targets for the product overexpression. Furthermore, studies have shown that individual overexpression of these enzymes can lead to a 350% boost in isoprenoid production (Mahmoud et al. 2021) (Table 12.1).

12.4 Alkanes/Alkenes

Alkanes and alkenes, desirable oleochemicals, also known as hydrocarbons, are essential components of drop-in biofuels. These organic compounds, made up of hydrogen and carbon atoms, are highly valued for their energy content and compatibility with current fuel infrastructure (Kallio et al. 2014). They can be obtained from various biological sources and processes, including the metabolic activities of specific microorganisms (Schirmer et al. 2010). Alkanes and alkenes hold particular significance in drop-in biofuels because they can be used as “drop-in” fuels, seamlessly integrated into existing engines and infrastructure without any modifications.

This quality makes them an attractive choice for the immediate implementation of more sustainable energy sources. However, producing these hydrocarbons through biological processes is a complex challenge. This is where genetic and metabolic engineering play a crucial role, allowing scientists to modify microorganisms like bacteria and yeast to produce it from renewable feedstocks like sugars or lignocellulosic biomass (Chuck and Donnelly 2014; Peralta-Yahya et al. 2012).

For example, the metabolic pathways of *E. coli* have been engineered to produce alkanes by introducing a gene from the cyanobacteria (Schirmer et al. 2010). The metabolic pathway for generating alkanes and alkenes involves transforming fatty acids into these hydrocarbons via a sequence of enzyme-mediated reactions. Fatty acyl-ACP or acyl-CoA, the end products of fatty acid synthesis, are converted into fatty aldehydes by an enzyme called fatty acyl-ACP/CoA reductase. Subsequently, these fatty aldehydes are transformed into alkanes or alkenes by another enzyme, known as aldehyde-deformylating oxygenase (ADO). The efficiency of this pathway and the production of alkanes and alkenes can be significantly improved in organisms like bacteria and yeast using genetic and metabolic engineering techniques. For instance, the oil-producing yeast *Lipomyces starkeyi* was genetically modified to generate long-chain fatty alcohols, which are the building blocks for alkanes/alkenes. This was achieved by incorporating a fatty acyl-CoA reductase gene (mFAR1) from *Mus musculus*. Two distinct versions of mFAR1 were optimized for *Y. lipolytica* or *S. cerevisiae* and inserted into *L. starkeyi*, along with specific promoters and transcriptional terminators from *L. starkeyi's* genomic DNA. The modified yeast effectively converted glucose, xylose, and glycerol into C16 and C18 fatty alcohols, especially when the carbon-to-nitrogen ratio was high (McNeil and Stuart 2018).

In another study, photosynthetic microorganisms like cyanobacteria and algae were genetically engineered to accumulate hydrocarbons, potentially substituting fossil fuel. The researchers introduced five synthetic metabolic pathways in these organisms. This led to efficient conversion of native acyl-ACP into free fatty acids (FFAs) and their subsequent transformation into corresponding alkenes and alkanes, yielding high results. (Yunus et al. 2018). In another study, a combined strategy of overexpression and deletion of genes has been carried out. This study demonstrates the successful metabolic engineering of *Escherichia coli* to produce n-alkanes. Researchers significantly increased alkane production by introducing two genes from *Synechococcus elongatus* and manipulating *E. coli* own genes to enhance fatty acid biosynthesis and reduce the substrate competition. The most effective strategy combined removal of the *yqhD* gene and overexpression of the *fadR* gene, resulting in an alkane production titer of 255.6 mg/L (Song et al. 2016). The genetic manipulation of the yeast species *Yarrowia lipolytica* for the production of fuel-like compounds and oleochemicals, including various types of alkanes, was carried out. The researchers directed acyl-CoA/acyl-ACP processing enzymes to different cellular compartments, enabling the production of specific fatty alkanes. Moreover, the researchers incorporated a lipoxygenase enzyme from soybean into the yeast. This enzyme is capable of converting naturally present linoleic acid into products such as13-oxo-cis-9, trans-11-tridecadienoic acid, and pentane, further diversifying the potential outputs of this bioengineered organism. Additionally, they adjusted alternative cytosolic acetyl-CoA pathways to enhance the yeast's lipid-producing capabilities. These strategies, which utilize the plentiful fatty acids present within the yeast, represent significant advancements toward creating a yeast-based biorefinery platform for sustainable alkene production (Blazeck et al. 2013; Spagnuolo et al. 2019; Xu et al. 2016). This study employed metabolic engineering techniques to produce alkanes and alkenes from lignin-derived molecules. Using *Acinetobacter baylyi* ADP1 as a chassis organism, the researchers established a cellular platform for synthesizing 1-undecene solely from ferulate. The engineered cells demonstrated efficient production and excretion of 1-undecene (Luo et al. 2019).

A combined computational modeling approach with pathway and process optimization was used to significantly boost the production of long-chain hydrocarbons, including alkenes, in *E. coli*. Where constraint-based metabolic modeling strategy was applied to pinpoint potential engineering targets in an *E. coli* strain that was already modified to express cyanobacterial enzymes for alkane production. The targets identified were primarily linked to the pentose phosphate pathway and the latter part of glycolysis; by applying these computational findings, which included optimizing the cyanobacterial enzymes' expression and modifying *E. coli* based on the identified targets, a remarkable 148.6-fold increase in alkene production was achieved, reaching a titer of 425 mg/L (Fatma et al. 2018).

The drop-in biofuel industry's use of alkanes/alkenes is crucial for combating climate change by lowering greenhouse gas emissions when compared to fossil fuels. According to a recent study, the production of alkanes/alkenes as part of drop-in biofuel production aligns with financial development and globalization, which are linked to lower carbon emissions (Ruza and Caro-Carretero 2022). This suggests that increasing alkane/alkene production for drop-in biofuels could contribute to further emission reductions. In addition, the production of alkanes/alkenes through genetic and metabolic engineering for drop-in biofuel applications is a rapidly evolving field with significant potential for future advancements (Spagnuolo et al. 2019) as ongoing research and development in this area are key to attain a sustainable and carbon-neutral future.

12.5 Sustainable Aviation Fuel

Sustainable aviation fuel (SAF) marks a notable progress in the aviation sector's efforts to diminish its carbon footprint and lessen the effects of climate change. SAFs are fuels derived from sustainable, non-fossil sources that, when used, can significantly reduce the overall carbon emissions of aviation compared to conventional jet fuel (Ashokkumar et al. 2022; Lee et al. 2021). Several commercially viable options exist for producing sustainable aviation fuels on a large scale. These include vegetable oils/animal fats (hydro processed), isoparaffinic fuel, biomass feed to fuel, jet fuel, and paraffinic kerosene aromatics. These methods offer sustainable alternatives that can be used to produce aviation fuels without relying on fossil fuels

(Farooqui et al. 2022). The aviation industry contributes approximately 2% of the global total carbon dioxide emissions. As global air travel continues to grow, so does the industry's potential impact on climate change. SAFs offer a promising solution to this challenge. They can potentially reduce aviation fuel's life cycle carbon emissions by up to 80% (Fan et al. 2020). The importance of SAFs extends beyond their environmental benefits. They also represent a significant opportunity for economic development and energy security. By diversifying the sources of aviation fuel it's possible to decrease dependence on fossil fuels and create new opportunities for sustainable economic growth. However, the widespread adoption of SAFs is not without its challenges. These include technical issues related to fuel compatibility and performance, economic factors such as cost and market demand and regulatory and policy considerations. Overcoming these challenges will require a concerted effort from all aviation industry stakeholders and continued research and development to enhance the efficiency and sustainability of SAF production.

One of the critical applications of genetic and metabolic engineering in SAF production is the development of strains capable of producing drop-in biofuels from non-traditional feedstocks. For example, researchers have engineered strains of *Escherichia coli* and *Saccharomyces cerevisiae* to produce drop-in biofuels from lignocellulosic biomass, a renewable resource that does not compete with food production (Lynd et al. 2008). The field of genetic and metabolic engineering for sustainable aviation fuel (SAF) production has seen significant advancements in recent years. Researchers are continually developing innovative strategies to improve the efficiency and sustainability of drop-in biofuel production. For example, the bacterium *Pseudomonas putida*, known for its metabolic versatility, as a host for producing isoprenol, a key precursor for the potential SAF compound which is 1,4-dimethylcyclooctane (Deepanwita et al. 2023). They employed computational modeling approaches, specifically bilevel optimization and constrained minimal cut sets, to identify gene knockout targets and refine the "IPP-bypass" pathway in *P. putida*. This approach yielded a significant increase in isoprenol production, reaching a titer of 3.5 gram per liter in *P. putida* under fed-batch cultivation conditions. In another study, Feng et al. (2022) employed a combination of engineering techniques to develop *Escherichia coli* cells proficient at efficiently synthesizing itaconic acid (Feng et al. 2022). They created 12 recombinant *E. coli* strains using various genetic constructs and combinations of plasmid. The strain *E. coli* BL-CAR470E-DS/A-CS engineered using enzyme evolution and protein scaffolds to bring together essential enzymes showed the highest titer. When subjected to optimized reaction conditions and a controlled feeding strategy, this strain achieved an itaconic acid titer of 51.79 g/L, with a notably high specific production rate of 9.42 g/g (DCW) at 48 h, surpassing other heterologous hosts. This study shows the power of combinatorial metabolic engineering in enhancing drop-in biofuel precursor production. Likewise, Wegat et al. (2022) conducted a comprehensive review of the metabolic processes and characteristics of methylotrophic yeasts and their application in producing a wide range of profitable products, from specialized chemicals to bulk chemicals (Wegat et al. 2022). They highlighted the potential of synthetic methylotrophy, where natural or artificial methanol assimilation pathways are integrated into biotechnologically relevant microorganisms like *Saccharomyces cerevisiae* and *Yarrowia lipolytica*. The authors emphasized that a thorough understanding of methylotrophic metabolism is the basis for microbial strain engineering, thus contributing to the development of a CO2-based circular bioeconomy.

Despite promising genetic and metabolic engineering advancements for sustainable aviation fuel (SAF) production, several challenges remain. One of these challenges is the need for high-cost inputs and complex processing steps, which can make drop-in biofuel production economically unviable compared to conventional jet fuel (Yang and Chen 2022). Additionally, drop-in biofuel production often requires food crops or arable land, raising concerns about food security and land use (Searchinger et al. 2008). Another challenge lies in the technical aspects of drop-in biofuel production. The engineered strains often have lower productivity and stability than their natural counterparts, and the drop-in biofuel products sometimes do not meet the strict quality standards for aviation fuels. Regulatory and policy challenges also exist. The lack of clear and consistent policies supporting the development and use of SAFs can hinder investment and slow down the pace of innovation (Wang et al. 2020).

Despite these challenges, the field also presents numerous opportunities. Advances in genetic and metabolic engineering techniques are continually improving the efficiency and sustainability of drop-in biofuel production. For example, the advancement of synthetic biology tools allows for the design of novel metabolic pathways and the enhancement of existing ones (Nielsen and Keasling, 2016). There is also growing interest in drop-in biofuel production using non-traditional feedstocks, such as lignocellulosic biomass, algae, and waste materials. These feedstocks do not compete with food production and can be more sustainable than traditional feedstocks (Chisti 2007). Furthermore, the increasing recognition of the need for decarbonization in the aviation sector is driving policy support for SAFs. This, in turn, can stimulate investment and innovation in the field.

The impact of these advancements on the aviation industry and climate change mitigation could be significant. The aviation industry is a major contributor to global carbon emissions, and developing SAFs is a crucial strategy for reducing these emissions. As SAF production becomes more efficient and sustainable, we expect to see greater adoption of SAFs in the aviation industry. Furthermore, the development of SAFs can contribute to the broader goal of a circular economy, where waste materials are turned into valuable resources. This reduces the environmental impact of waste and creates new opportunities for economic development (Yang and Chen 2022).

In conclusion, while challenges exist in genetic and metabolic engineering for SAF production, the opportunities for innovation and improvement are vast. Ongoing research and development in this area are essential to fully harness the potential of sustainable aviation fuels (SAFs) for a sustainable future in aviation. The intersection of genetic and metabolic engineering with the production of SAFs represents a dynamic and rapidly evolving field that holds great promise for the future of sustainable aviation and the broader goal of mitigating the impacts of climate change.

12.6 Others

12.6.1 Bisabolene

Bisabolene is a sesquiterpene, a class of hydrocarbons that have shown significant potential as advanced drop-in biofuels due to their high energy density and their ability to integrate seamlessly with current infrastructure. Bisabolene is considered a promising option because of its substantial energy content and the possibility of its production from renewable sources like biomass. However, the current methods of bisabolene production, which typically involve extraction from plant sources or chemical synthesis, are not sustainable or scalable. This has led to a growing interest in developing bio-based production methods for bisabolene.

By modification of microorganism's metabolic pathways to produce bisabolene from simple and complex sugars is a suitable solution. For example, a study by Junko Yaegashi et al. (2017) demonstrated bisabolene production in the engineered strains of *Rhodosporidium toruloides*. The researchers identified and optimized a biosynthetic pathway, leading to a significant increase in yield. Engineered *R. toruloides* (BIS3) has shown great promise using various carbon sources such as xylose, glucose, and p-coumaric acid. It was observed that p-coumaric acid utilization led to the highest levels of bisabolene production, probably due to its increased carbon content Surprisingly, when all three carbon sources were combined, complete utilization was achieved within 4 days, surpassing the capabilities of well-engineered strains like *E. coli* and *Saccharomyces cerevisiae* (Reider Apel et al. 2016). Moreover, genetic engineering techniques were employed to integrate codon-optimized expression modules for bisabolene synthase (BIS) into the genome of *R. toruloides*. The bisabolene synthase (BIS) gene was genetically engineered for optimal performance in *R. toruloides*. Codon optimization and gene synthesis were performed to adapt the genes specifically for the *R. toruloides* strain. The resulting constructs placed the genes under the control of the GAPDH promoter and NOS terminator. These engineered gene constructs were then integrated into the genome of *R. toruloides*. This genetic engineering approach allowed for improved expression and production of bisabolene in *R. toruloides*. This resulted in significantly higher bisabolene titers (294 mg/L) in contrast to *S. cerevisiae* that had been modified with high-copy number plasmids. Furthermore, the stability of bisabolene production *in R. toruloides* was demonstrated over an extended period of cultivation, with minimal variation (less than 16%) in titers over 24 days (Yaegashi et al. 2017). These findings underscore the promising potential of *R. toruloides* in efficiently producing bisabolene through genetic and metabolic engineering, paving the way for large-scale industrial production of terpene-based bioproducts.

Moreover, another study employed a metabolic engineering approach to improve bisabolene production in *Synechocystis* sp. PCC 6803. The researchers targeted genes related to biosynthesis, core carbon metabolism, amino acid biosynthesis, and ATP generation using the Hfq-MicC antisense RNA regulatory tool. Notably, the interference with chlorophyll biosynthesis and carotenogenesis pathways proved to be particularly effective in redirecting metabolic flux toward bisabolene production. These pathways are closely linked to the biosynthesis of terpenoids, and their manipulation resulted in improved carbon partitioning and increased availability of precursors for bisabolene synthesis. This study highlights the potential of metabolic engineering strategies, specifically gene knockdown techniques, to optimize and increase bisabolene production in *Synechocystis* (Rodrigues et al. 2023).

In a recent study, *Chlamydomonas reinhardtii,* a green microalga, was successfully engineered to produce (E)-α-bisabolene which is a sesquiterpene biodiesel precursor. They employed a combined approach of metabolic and genetic engineering to boost the synthesis of this sesquiterpenoid compound. Metabolic engineering techniques were utilized to increase the flow of the precursor FPP (farnesyl pyrophosphate) toward the desired product. Genetic engineering involved the fine-tuning of the isoprenoid metabolism by overexpressing terpene synthase and using amiRNA-based repression to suppress competing pathways. The integration of these approaches resulted in a remarkable 15-fold increase in bisabolene production. The engineered microalgae produced up to 10.3 mg of bisabolene per gram of dry cell weight within a span of 5 days. When cultivated in a mixotrophic environment, the yield reached up to 11.0 mg of bisabolene per liter over the course of 7 days. These findings signify a significant advancement in the development of sustainable and photosynthetic algal bio-factories for the production of valuable compounds like bisabolene (Wichmann et al. 2018).

12.6.2 Limonene

Limonene, a naturally occurring terpene, has drawn a lot of interest as a possible drop-in biofuel due to its high energy content and renewability. It is a versatile compound that can be derived from various renewable resources, including citrus fruits, pine trees, and engineered microorganisms (Liu et al. 2021; Pang et al. 2019; Rolf et al. 2020; Wu et al. 2019; Zhang et al. 2021). However, the commercial production of limonene as a drop-in biofuel is currently limited due to several challenges, including low natural abundance and the toxicity of limonene to microbial hosts. This has led to exploring genetic and metabolic engineering strategies to enhance limonene production.

Genetic engineering has become a highly effective approach for boosting limonene production. For instance, the mevalonate pathway in *Escherichia coli* was engineered by introducing a gene responsible for limonene synthase. The engineered *E. coli* strain produced 486 mg/L of limonene, a significant improvement (Rolf et al. 2020). In another study, the yeast *Saccharomyces cerevisiae* was engineered introducing a gene responsible for limonene synthase from *Mentha spicata* into *S. cerevisiae*. As a result, the modified yeast strain could produce limonene at a concentration of 8.2 mg/L. The method employed a comprehensive strategy, including the modification of the PDH bypass and the removal of specific genes not directly involved in the MVA pathway but influencing limonene synthesis.

Additionally, the research examined the synergistic impact of these genetic changes in conjunction with the expression of a particular gene, SeACSL641P. This approach exemplifies the effective use of genetic and metabolic engineering techniques to enhance the production capabilities of yeast for valuable substances such as limonene (Zhang et al. 2021).

Cyanobacteria have also been engineered for limonene production. A study performed by Lin et al. (2017), in which the researchers introduced a gene responsible for limonene synthase from *Citrus limon* into the cyanobacterium *Synechocystis* sp. PCC 6803. The engineered cyanobacterium was capable of producing limonene from CO_2, demonstrating the potential of photosynthetic microorganisms in sustainable limonene production. To further optimize limonene production, the research team used a computational approach known as the OptForce algorithm. This led them to enhance certain genes in the pentose phosphate pathway and introduce a gene for geranyl diphosphate synthase from plant *Abies grandis*. An upgraded cyanobacterial strain that generated 6.7 mg/L of limonene as a result showed a 2.3-fold rise in productivity. This approach demonstrates a promising method for using cyanobacteria in the sustainable and photosynthetic production of important compounds like limonene.

On the other hand, a study demonstrated the potential of metabolic engineering in enhancing limonene production in *E. coli*. The researchers engineered the *E. coli* strain to overexpress the key enzymes of the mevalonate pathway, including those for geranyl pyrophosphate (GPP) synthase, limonene synthase, and cytochrome P450 which led to a significant increase in limonene production (nearly 90-fold increase). Consequently, the modified *E. coli* produced limonene at a concentration of 430 mg/L, using 1% glucose over a 72-h period, a significant improvement (Alonso-Gutierrez et al. 2013). In another study, the researchers engineered the yeast *S. cerevisiae* to overexpress the key enzymes of the mevalonate pathway and introduced a limonene synthase gene. The engineered yeast strain produced 23 mg/L of limonene, demonstrating the potential of metabolic engineering in enhancing limonene production (Brennan et al. 2012). In a recent study, researchers used a blend of genetic and metabolic engineering to improve limonene production in the yeast *S. cerevisiae*. They introduced a limonene synthase gene and overexpressed the key enzymes of the mevalonate pathway. Additionally, they deleted the genes associated in the competitive pathways to increase the flux toward limonene production. The altered yeast strain produced 450 mg/L of limonene, a significant improvement compared to previous studies (Zhang et al. 2021).

These approaches have the potential to overcome the current challenges associated with limonene production and pave the way for the commercial production of limonene as a drop-in biofuel. Nonetheless, additional research is necessary to refine these techniques further and address the challenges related to limonene's toxicity to the microbial hosts.

12.6.3 Isobutanol

Isobutanol, a four-carbon alcohol, has emerged as a promising drop-in biofuel due to its higher energy content, improved fuel stability, compatibility with existing infrastructure, and higher-octane rating compared to ethanol. This renewable drop-in biofuel can be produced from biomass, corn, or sugarcane through various fermentation processes. In recent years, genetic and metabolic engineering techniques have been crucial in optimizing isobutanol production, enhancing its viability as a sustainable alternative to conventional gasoline (Gong et al. 2022). By modifying the genetic makeup of bacteria or yeast, key metabolic pathways involved in isobutanol synthesis can be enhanced. This includes overexpressing enzymes such as branched-chain amino acid aminotransferase, 2-keto acid decarboxylase (KDC), and alcohol dehydrogenase (ADH), which are responsible for converting amino acids to 2-keto acids, 2-keto acids to aldehydes, and aldehydes to isobutanol, respectively. Furthermore, by blocking or inhibiting competing pathways, metabolic flux can be redirected to favor the production of isobutanol. A notable case study in isobutanol production involves the application of genetic and metabolic engineering in *Saccharomyces cerevisiae*. In this study by Lee et al., pyruvate, which is produced from glucose via glycolysis, undergoes conversion into 2-ketoisovalerate. This transformation occurs within the mitochondria and is facilitated by the action of enzymes such as acetolactate synthase (Ilv2p), acetohydroxyacid reductoisomerase (Ilv5p), and dihydroxy acid dehydratase (Ilv3p). By overexpressing acetolactate synthase (Ilv2p), acetohydroxyacid reductoisomerase (Ilv5p), and dihydroxyacid dehydratase (Ilv3p) in the cytosol of *S. cerevisiae*, isobutanol synthesis has been found to be increased to 151 mg/L using this approach. Further, branched-chain amino acid aminotransferase (Bat1p) plays a role in converting 2-ketoisovalerate to valine. For effective isobutanol production, it may be necessary to express ILV genes in the cytosol. Cytosolic 2-ketoisovalerate is then transferred to isobutyraldehyde by KDC-like enzymes such as phenylpyruvate decarboxylase (Aro10p), pyruvate decarboxylase (Pdc1p, Pdc5p, Pdc6p), and putative α-ketoisocaproate decarboxylase (Thi3p). The process culminates with the conversion of isobutyraldehyde into isobutanol by the alcohol dehydrogenase enzyme Adh2p (Lee et al. 2012).

In a different study *Saccharomyces cerevisiae* was used to enhance isobutanol production through metabolic engineering. Key enzymes of the valine synthesis pathway, dihydroxy acid dehydratase (Ilv3), acetolactate synthase (Ilv2), and acetohydroxyacid reductoisomerase (Ilv5), were overexpressed in the cytosol. Additionally, mitochondrial valine synthesis was inhibited by disrupting the ILV2 gene. As a result, there was a 22-fold increase in isobutanol production, achieving a yield of 0.22 g/L. Further optimization was achieved by deleting genes of competing pathways, resulting in isobutanol titers of up to 0.56 g/L. Although the removal of the ADH1 gene led to increased glycerol production, the elimination of glycerol formation through the deletion of GPD1 and GPD2 genes resulted in an increase in isobutanol production to 1.32 g/L. Subsequently, by decreasing the production of the by-product isobutyrate via the deletion of the ALD6 gene, further improvements in isobutanol synthesis were achieved. This modification led to an increase in isobutanol production to 2.09 g/L, achieving a yield of 59.55 mg/g of glucose. This marked a more than 200-fold increase compared to the wild type, representing the highest yield obtained with *S. cerevisiae* in shake flask cultures. This represented an increase of over 200 times compared to the wild-type strain, signifying the highest yield achieved with *S. cerevisiae* in shake flask cultures (Wess et al. 2019).

A recent study employed a metabolic engineering method to produce isobutanol from lignocellulosic biomass. This was achieved using a dual clostridial consortium comprising of *Clostridium cellulovorans* and *Clostridium beijerinckii*. Incorporating a different isobutanol pathway into *C. beijerinckii* enabled the engineered strain to produce 194 milligram/liter of isobutanol and 7.16 gram/liter of n-butanol from glucose. Genetic modifications in *C. cellulovorans*, including the overexpression of adhE1, kivD, and yqhD, led to the production of 156 mg/L of isobutanol and 1.81 g/L of n-butanol using alkali-extracted, deshelled corn cobs (AECC) as the carbon source. To further enhance isobutanol production, the two genetically modified strains were combined and optimized as a consortium. Compared to the single culture of recombinant *C. cellulovorans*, the consortium exhibited a remarkable 6.73-fold increase in isobutanol production (Wen et al. 2022). These findings highlight the advancements made in consolidated bioprocessing facilitated by synthetic biology and contribute to developing more efficient and environmentally friendly drop-in biofuel production systems. Furthermore, the genetic and metabolic engineering strategies employed in this case study have significantly enhanced the production of isobutanol from renewable resources.

12.7 Future Perspective

As the field of genetic and metabolic engineering for drop-in biofuel production continues to advance, several challenges and future considerations need to be addressed. These challenges span regulatory and safety aspects, techno-economic analysis, environmental sustainability, integration with existing infrastructure, and future research and development direction. The field of genetic and metabolic engineering has witnessed notable progress in the production of drop-in biofuels. However, it is important to recognize the obstacles that must be overcome and consider future perspectives that can shape this promising area. A key challenge lies in the complexity of cellular metabolic pathways (Nielsen and Keasling 2016). These intricate networks, developed over extended periods of evolution, pose resistance to a redirection of resources, making it challenging to engineer cells for optimal drop-in biofuel production.

Nevertheless, the emergence of metabolic engineering and synthetic biology offers a promising solution for the sustainable generation of these compounds, such as synthetic biology allows for the design and construction of novel genetic circuits and pathways. In contrast, metabolic engineering techniques enable the modification of cellular metabolism to enhance drop-in biofuel production. Moreover, regulatory frameworks and safety considerations serve an essential role in the production and commercialization of drop-in biofuels. These frameworks differ across countries and regions, requiring careful navigation to ensure compliance and safe implementation of genetically modified organisms (GMOs) in drop-in biofuel production. Furthermore, addressing public concerns and fostering acceptance of genetically engineered drop-in biofuels is imperative to promote their widespread adoption (Lee et al. 2008). A further obstacle pertains to the limited accessibility of bioactive natural products, which can pose difficulties, particularly when these products exhibit potential as pharmaceuticals or key compounds. However, this challenge can be tackled by developing yeast cell factories specifically engineered to produce such natural products (Chen et al. 2020). This approach entails altering the cellular processes within yeast to enable the synthesis of bioactive compounds, such as alkaloids, terpenoids, and phenylpropanoids. Researchers can enhance the production of these valuable natural products by employing genetic and metabolic engineering techniques, thereby addressing the scarcity issue. Ethical considerations in the use of synthetic biology present a notable challenge. It is crucial to ensure the implementation of rigorous, reliable, and predictable designs, along with fostering public engagement and adopting a contemporary ethical framework to ensure the ongoing success of synthetic biology. The advent and utilization of CRISPR-Cas9 in genome engineering have undoubtedly marked a significant breakthrough in this domain (Anderson et al. 2012). However, these advancements also bring to the forefront ethical and regulatory concerns that require careful attention and resolution. The future of genetic and metabolic engineering in the production of drop-in biofuels holds excellent promise. One particularly promising direction is the utilization of integrated multispecies microbial communities, as opposed to monocultures, in biodegradation processes (Borchert et al. 2021). This innovative approach, which merges eco-evolutionary principles with applied biotechnology, has the potential to yield more robust and efficient cultures. By leveraging the interactions between different species within these communities, researchers can enhance stability and productivity. Moreover, this approach is anticipated to drive advancements in environmental biotechnology and potentially lead to the identification of new metabolic pathways.

Looking ahead, the potential for applying genetic and metabolic engineering techniques in developing drop-in biofuels is highly promising. Advanced tools and technologies, such as gene editing, high-throughput screening, synthetic biology, and computational modeling, should be further developed and integrated to accelerate progress in drop-in biofuel production. The increasing number of sequenced microbial genomes and metabolic models enables the development of new metabolic routes for generating biofuel (Joshi and Mishra 2022). As understanding of these technologies continues to grow, so will the ability to harness them to produce sustainable and renewable drop-in biofuels.

Future research and development efforts should focus on further refining metabolic pathways, expanding the toolbox of genetic engineering techniques, and exploring emerging technologies. The combination of synthetic biology approaches and artificial intelligence can unlock new possibilities for drop-in biofuel engineering. Using AI algorithms and machine learning models, it is possible to optimize metabolic networks, predict optimal genetic modifications, and design superior drop-in biofuel-producing strains. With continued research and development, it is anticipated to see even more innovative applications of these technologies, fostering high expectations for future advancements in this domain. Despite the challenges faced in the production of drop-in biofuels, their potential benefits are significant. They offer a renewable and potentially more sustainable option compared to fossil fuels, compatibility with existing infrastructure and the potential to stimulate economic growth and create new jobs. However, challenges remain, including the need to optimize the preparation process of drop-in biofuels, increase the efficiency and service life of catalysts for reaction, and reduce the cost of drop-in biofuels. Additionally, drop-in biofuels can play a significant role in decreasing greenhouse gas emissions and fostering environmental sustainability. Ongoing research, technological innovations, and supportive policies are essential to fully realize the potential of drop-in biofuels and accelerate their adoption as a viable and sustainable energy solution.

12.8 Conclusion

In conclusion, recent progress in genetic and metabolic engineering has brought about a revolutionary transformation in the production of drop-in biofuels, presenting a promising path toward achieving a sustainable energy future. By manipulating microbial strains and metabolic pathways in a systematic manner, scientists have unlocked new opportunities for creating drop-in biofuels that are both environmentally friendly and renewable. Although there are still numerous hurdles to overcome, the potential advantages offered by these drop-in biofuels make them an area of research and development with great promise. Through ongoing research, innovation, and the integration of eco-evolutionary principles with applied biotechnology, substantial progress can be expected in this field in the forthcoming years by pushing the boundaries of these technologies and addressing the existing challenges with the help of applied biotechnology, looking forward to a future where drop-in biofuels assume a pivotal role in our energy landscape, facilitating the transition toward sustainability and renewable energy sources.

Acknowledgments This research was funded by the Natural Sciences and Engineering Research Council of Canada (NSERC) CREATE Program, Training in Advanced Biotechnology for Environmental Sustainability (TABES), and we also like to thank James and Joanne Love Chair in Environmental Engineering at York University for the support provided.

References

Alonso-Gutierrez J, Chan R, Batth TS, Adams PD, Keasling JD, Petzold CJ, Lee TS (2013) Metabolic engineering of Escherichia coli for limonene and perillyl alcohol production. Metab Eng 19:33–41. https://doi.org/10.1016/j.ymben.2013.05.004

Anderson J, Strelkowa N, Stan GB, Douglas T, Savulescu J, Barahona M, Papachristodoulou A (2012) Engineering and ethical perspectives in synthetic biology. Rigorous, robust and predictable designs, public engagement and a modern ethical framework are vital to the continued success of synthetic biology. EMBO Rep 13(7):584–590. https://doi.org/10.1038/embor.2012.81

Ashokkumar V, Venkatkarthick R, Jayashree S, Chuetor S, Dharmaraj S, Kumar G, Chen W-H, Ngamcharussrivichai C (2022) Recent advances in lignocellulosic biomass for biofuels and value-added bioproducts – a critical review. Bioresour Technol 344:126195. https://doi.org/10.1016/j.biortech.2021.126195

Bentley FK, Zurbriggen A, Melis A (2014) Heterologous expression of the mevalonic acid pathway in cyanobacteria enhances endogenous carbon partitioning to isoprene. Mol Plant 7(1):71–86. https://doi.org/10.1093/mp/sst134

Birchfield AS, McIntosh CA (2020) Metabolic engineering and synthetic biology of plant natural products—a minireview. Curr Plant Biol 24:100163. https://doi.org/10.1016/j.cpb.2020.100163

Blazeck J, Liu L, Knight R, Alper HS (2013) Heterologous production of pentane in the oleaginous yeast Yarrowia lipolytica. J Biotechnol 165(3–4):184–194. https://doi.org/10.1016/j.jbiotec.2013.04.003

Borchert E, Hammerschmidt K, Hentschel U, Deines P (2021) Enhancing microbial pollutant degradation by integrating eco-evolutionary principles with environmental biotechnology. Trends Microbiol 29(10):908–918. https://doi.org/10.1016/j.tim.2021.03.002

Brennan TCR, Turner CD, Krömer JO, Nielsen LK (2012) Alleviating monoterpene toxicity using a two-phase extractive fermentation for the bioproduction of jet fuel mixtures in Saccharomyces cerevisiae. Biotechnol Bioeng 109(10):2513–2522. https://doi.org/10.1002/bit.24536

Chen R, Yang S, Zhang L, Zhou YJ (2020) Advanced strategies for production of natural products in yeast. iScience 23(3):100879. https://doi.org/10.1016/j.isci.2020.100879

Chisti Y (2007) Biodiesel from microalgae. Biotechnol Adv 25(3):294–306. https://doi.org/10.1016/j.biotechadv.2007.02.001

Chuck CJ, Donnelly J (2014) The compatibility of potential bioderived fuels with jet A-1 aviation kerosene. Appl Energy 118:83–91

Cong L, Ran FA, Cox D, Lin S, Barretto R, Habib N, Hsu PD, Wu X, Jiang W, Marraffini LA, Zhang F (2013) Multiplex genome engineering using CRISPR/Cas systems. Science 339(6121):819–823. https://doi.org/10.1126/science.1231143

Cordova LT, Butler J, Alper HS (2020) Direct production of fatty alcohols from glucose using engineered strains of Yarrowia lipolytica. Metab Eng Commun 10:e00105

Deepanwita B, Ian SY, Xi W, Jinho K, Aparajitha S, Russel M, Yan C, Jennifer WG, Christopher JP, Hector Garcia M, Paul DA, Aindrila M, Joonhoon K, Taek Soon L (2023) Genome-scale and pathway engineering for the sustainable aviation fuel precursor isoprenol production in Pseudomonas putida. *bioRxiv* 2004(2029):538800. https://doi.org/10.1101/2023.04.29.538800

Dusséaux S, Wajn WT, Liu Y, Ignea C, Kampranis SC (2020) Transforming yeast peroxisomes into microfactories for the efficient production of high-value isoprenoids. Proc Natl Acad Sci 117(50):31789–31799. https://doi.org/10.1073/pnas.2013968117

Fan Y, Wu X, Wu X, Li C, Yang Q, Hayat T, Alsaedi A, Wang P, Chen G (2020) A unified ecological assessment of a solar concentrating plant based on an integrated approach joining cosmic exergy analysis with ecological indicators. Renew Sust Energ Rev 129:109934. https://doi.org/10.1016/j.rser.2020.109934

Farooqui SA, Sinha AK, Ray A (2022) Chapter 13 – progress and trends in renewable jet fuels. In: Tuli D, Kasture S, Kuila A (eds) Advanced biofuel technologies. Elsevier, pp 333–364. https://doi.org/10.1016/B978-0-323-88427-3.00011-8

Farrokh P, Sheikhpour M, Kasaeian A, Asadi H, Bavandi R (2019) Cyanobacteria as an eco-friendly resource for biofuel production: a critical review. Biotechnol Prog 35(5):e2835. https://doi.org/10.1002/btpr.2835

Fatma Z, Hartman H, Poolman MG, Fell DA, Srivastava S, Shakeel T, Yazdani SS (2018) Model-assisted metabolic engineering of Escherichia coli for long chain alkane and alcohol production. Metab Eng 46:1–12. https://doi.org/10.1016/j.ymben.2018.01.002

Feng J, Li C, He H, Xu S, Wang X, Chen K (2022) Construction of cell factory through combinatorial metabolic engineering for efficient production of itaconic acid. Microb Cell Factories 21(1):275. https://doi.org/10.1186/s12934-022-02001-1

Gao X, Gao F, Liu D, Zhang H, Nie X, Yang C (2016) Engineering the methylerythritol phosphate pathway in cyanobacteria for photosynthetic isoprene production from CO2 [10.1039/C5EE03102H]. Energy Environ Sci 9(4):1400–1411. https://doi.org/10.1039/C5EE03102H

Goldstein JL, Brown MS (1990) Regulation of the mevalonate pathway. Nature 343(6257):425–430

Gong G, Wu B, Liu L, Li J, Zhu Q, He M, Hu G (2022) Metabolic engineering using acetate as a promising building block for the production of bio-based chemicals. Eng Microbiol 2(4):100036. https://doi.org/10.1016/j.engmic.2022.100036

Heuston S, Begley M, Gahan CG, Hill C (2012) Isoprenoid biosynthesis in bacterial pathogens. Microbiology 158(6):1389–1401

Jakočiūnas T, Bonde I, Herrgård M, Harrison SJ, Kristensen M, Pedersen LE, Jensen MK, Keasling JD (2015) Multiplex metabolic pathway engineering using CRISPR/Cas9 in saccharomyces cerevisiae. Metab Eng 28:213–222. https://doi.org/10.1016/j.ymben.2015.01.008

Joshi S, Mishra S (2022) Recent advances in biofuel production through metabolic engineering. Bioresour Technol 352:127037. https://doi.org/10.1016/j.biortech.2022.127037

Kallio P, Pásztor A, Thiel K, Akhtar MK, Jones PR (2014) An engineered pathway for the biosynthesis of renewable propane. Nat Commun 5(1):4731. https://doi.org/10.1038/ncomms5731

Kalnenieks U, Balodite E, Rutkis R (2019) Metabolic engineering of bacterial respiration: high vs. Low P/O and the Case of Zymomonas mobilis [Review]. Front Bioeng Biotechnol 7:327. https://doi.org/10.3389/fbioe.2019.00327

Kirby J, Geiselman GM, Yaegashi J, Kim J, Zhuang X, Tran-Gyamfi MB, Prahl JP, Sundstrom ER, Gao Y, Munoz N, Burnum-Johnson KE, Benites VT, Baidoo EEK, Fuhrmann A, Seibel K, Webb-Robertson BJM, Zucker J, Nicora CD, Tanjore D, Magnuson JK, Skerker JM, Gladden JM (2021) Further engineering of R. toruloides for the production of terpenes from lignocellulosic biomass [article]. Biotechnol Biofuels 14(1):101. https://doi.org/10.1186/s13068-021-01950-w

Ko JK, Lee S-M (2018) Advances in cellulosic conversion to fuels: engineering yeasts for cellulosic bioethanol and biodiesel production. Curr Opin Biotechnol 50:72–80

Ko JK, Lee JH, Jung JH, Lee S-M (2020) Recent advances and future directions in plant and yeast engineering to improve lignocellulosic biofuel production. Renew Sust Energ Rev 134:110390. https://doi.org/10.1016/j.rser.2020.110390

Lee D, Chen A, Nair R (2008) Genetically engineered crops for biofuel production: regulatory perspectives. Biotechnol Genet Eng Rev 25(1):331–362. https://doi.org/10.5661/bger-25-331

Lee W-H, Seo S-O, Bae Y-H, Nan H, Jin Y-S, Seo J-H (2012) Isobutanol production in engineered Saccharomyces cerevisiae by overexpression of 2-ketoisovalerate decarboxylase and valine biosynthetic enzymes. Bioprocess Biosyst Eng 35(9):1467–1475. https://doi.org/10.1007/s00449-012-0736-y

Lee DS, Fahey DW, Skowron A, Allen MR, Burkhardt U, Chen Q, Doherty SJ, Freeman S, Forster PM, Fuglestvedt J, Gettelman A, De León RR, Lim LL, Lund MT, Millar RJ, Owen B, Penner JE, Pitari G, Prather MJ, Sausen R, Wilcox LJ (2021) The contribution of global aviation to anthropogenic climate forcing for 2000 to 2018. Atmos Environ 244:117834. https://doi.org/10.1016/j.atmosenv.2020.117834

Lichtenthaler H (2000) Non-mevalonate isoprenoid biosynthesis: enzymes, genes and inhibitors. Biochem Soc Trans 28(6):785–789

Liew F, Henstra AM, Köpke M, Winzer K, Simpson SD, Minton NP (2017) Metabolic engineering of clostridium autoethanogenum for selective alcohol production. Metab Eng 40:104–114. https://doi.org/10.1016/j.ymben.2017.01.007

Lin P-C, Saha R, Zhang F, Pakrasi HB (2017) Metabolic engineering of the pentose phosphate pathway for enhanced limonene production in the cyanobacterium Synechocystis sp. PCC 6803. Sci Rep 7(1):17503. https://doi.org/10.1038/s41598-017-17831-y

Lindberg P, Park S, Melis A (2010) Engineering a platform for photosynthetic isoprene production in cyanobacteria, using Synechocystis as the model organism. Metab Eng 12(1):70–79. https://doi.org/10.1016/j.ymben.2009.10.001

Liu S, Zhang M, Ren Y, Jin G, Tao Y, Lyu L, Zhao ZK, Yang X (2021) Engineering Rhodosporidium toruloides for limonene production. Biotechnol Biofuels 14(1):243. https://doi.org/10.1186/s13068-021-02094-7

Luo J, Lehtinen T, Efimova E, Santala V, Santala S (2019) Synthetic metabolic pathway for the production of 1-alkenes from lignin-derived molecules. Microb Cell Factories 18(1):48. https://doi.org/10.1186/s12934-019-1097-x

Lv X, Xu H, Yu H (2013) Significantly enhanced production of isoprene by ordered coexpression of genes dxs, dxr, and idi in Escherichia coli. Appl Microbiol Biotechnol 97(6):2357–2365. https://doi.org/10.1007/s00253-012-4485-2

Lynd LR, Laser MS, Bransby D, Dale BE, Davison B, Hamilton R, Himmel M, Keller M, McMillan JD, Sheehan J, Wyman CE (2008) How biotech can transform biofuels. Nat Biotechnol 26(2):169–172. https://doi.org/10.1038/nbt0208-169

Madhavan A, Jose AA, Binod P, Sindhu R, Sukumaran RK, Pandey A, Castro GE (2017) Synthetic biology and metabolic engineering approaches and its impact on non-conventional yeast and biofuel production [review]. Front Energy Res 5(APR):8. https://doi.org/10.3389/fenrg.2017.00008

Mahmoud SS, Maddock S, Adal AM (2021) Isoprenoid metabolism and engineering in glandular trichomes of Lamiaceae. Front Plant Sci 12:699157. https://doi.org/10.3389/fpls.2021.699157

Majidian P, Tabatabaei M, Zeinolabedini M, Naghshbandi MP, Chisti Y (2018) Metabolic engineering of microorganisms for biofuel production. Renew Sust Energ Rev 82:3863–3885

McNeil BA, Stuart DT (2018) Optimization of C16 and C18 fatty alcohol production by an engineered strain of Lipomyces starkeyi. J Ind Microbiol Biotechnol 45(1):1–14. https://doi.org/10.1007/s10295-017-1985-1

Medipally SR, Yusoff FM, Banerjee S, Shariff M (2015) Microalgae as sustainable renewable energy feedstock for biofuel production. Biomed Res Int 2015:519513. https://doi.org/10.1155/2015/519513

Nielsen J, Keasling JD (2016) Engineering cellular metabolism. Cell 164(6):1185–1197. https://doi.org/10.1016/j.cell.2016.02.004

Pang Y, Zhao Y, Li S, Zhao Y, Li J, Hu Z, Zhang C, Xiao D, Yu A (2019) Engineering the oleaginous yeast Yarrowia lipolytica to produce limonene from waste cooking oil. Biotechnol Biofuels 12(1):241. https://doi.org/10.1186/s13068-019-1580-y

Peralta-Yahya PP, Keasling JD (2010) Advanced biofuel production in microbes. Biotechnol J 5(2):147–162. https://doi.org/10.1002/biot.200900220

Peralta-Yahya PP, Zhang F, del Cardayre SB, Keasling JD (2012) Microbial engineering for the production of advanced biofuels. Nature 488(7411):320–328. https://doi.org/10.1038/nature11478

Pulido P, Perello C, Rodriguez-Concepcion M (2012) New insights into plant isoprenoid metabolism. Mol Plant 5(5):964–967. https://doi.org/10.1093/mp/sss088

Rahmat E, Kang Y (2020) Yeast metabolic engineering for the production of pharmaceutically important secondary metabolites. Appl Microbiol Biotechnol 104(11):4659–4674. https://doi.org/10.1007/s00253-020-10587-y

Reider Apel A, Ouellet M, Szmidt-Middleton H, Keasling JD, Mukhopadhyay A (2016) Evolved hexose transporter enhances xylose uptake and glucose/xylose co-utilization in Saccharomyces cerevisiae. Sci Rep 6(1):1–10

Rodrigues JS, Bourgade B, Galle KR, Lindberg P (2023) Mapping competitive pathways to terpenoid biosynthesis in Synechocystis sp. PCC 6803 using an antisense RNA synthetic tool. Microb Cell Factories 22(1):35. https://doi.org/10.1186/s12934-023-02040-2

Rolf J, Julsing MK, Rosenthal K, Lütz S (2020) A gram-scale limonene production process with engineered Escherichia coli. Molecules 25(8):1881. https://doi.org/10.3390/molecules25081881

Ruza C, Caro-Carretero R (2022) The non-linear impact of financial development on environmental quality and sustainability: evidence from G7 countries. Int J Environ Res Public Health 19(14):8382. https://doi.org/10.3390/ijerph19148382

Saini R, Hegde K, Brar SK, Vezina P (2020) Advanced biofuel production and road to commercialization: an insight into bioconversion potential of Rhodosporidium sp. Biomass Bioenergy 132:105439. https://doi.org/10.1016/j.biombioe.2019.105439

Schirmer A, Rude MA, Li X, Popova E, del Cardayre SB (2010) Microbial biosynthesis of alkanes. Science 329(5991):559–562. https://doi.org/10.1126/science.1187936

Searchinger T, Heimlich R, Houghton RA, Dong F, Elobeid A, Fabiosa J, Tokgoz S, Hayes D, Yu T-H (2008) Use of U.S. croplands for biofuels increases greenhouse gases through emissions from land-use change. Science 319(5867):1238–1240. https://doi.org/10.1126/science.1151861

Song X, Yu H, Zhu K (2016) Improving alkane synthesis in Escherichia coli via metabolic engineering. Appl Microbiol Biotechnol 100(2):757–767. https://doi.org/10.1007/s00253-015-7026-y

Spagnuolo M, Yaguchi A, Blenner M (2019) Oleaginous yeast for biofuel and oleochemical production. Curr Opin Biotechnol 57:73–81. https://doi.org/10.1016/j.copbio.2019.02.011

Wang W, Liu D, Tu Y, Jin L, Wang H (2020) Enrichment of residual carbon in entrained-flow gasification coal fine slag by ultrasonic flotation. Fuel 278:118195. https://doi.org/10.1016/j.fuel.2020.118195

Ward VCA, Chatzivasileiou AO, Stephanopoulos G (2018) Metabolic engineering of Escherichia coli for the production of isoprenoids. FEMS Microbiol Lett 365(10):fny079. https://doi.org/10.1093/femsle/fny079

Wegat V, Fabarius JT, Sieber V (2022) Synthetic methylotrophic yeasts for the sustainable fuel and chemical production. Biotechnol Biofuels Bioprod 15(1):113. https://doi.org/10.1186/s13068-022-02210-1

Wen Z, Tu Z, Al Makishah N, Fang D, Ledesma-Amaro R, Zhu J, Li Q, Li Y, Jin M (2022) Genetic manipulation of a twin clostridia consortium for co-production of n-butanol and Isobutanol by consolidated bioprocessing. ACS Sustain Chem Eng 10(45):14671–14684. https://doi.org/10.1021/acssuschemeng.1c08486

Wess J, Brinek M, Boles E (2019) Improving isobutanol production with the yeast Saccharomyces cerevisiae by successively blocking competing metabolic pathways as well as ethanol and glycerol formation [article]. Biotechnol Biofuels 12(1):173. https://doi.org/10.1186/s13068-019-1486-8

Wichmann J, Baier T, Wentnagel E, Lauersen KJ, Kruse O (2018) Tailored carbon partitioning for phototrophic production of (E)-α-bisabolene from the green microalga Chlamydomonas reinhardtii. Metab Eng 45:211–222. https://doi.org/10.1016/j.ymben.2017.12.010

Wong L, Engel J, Jin E, Holdridge B, Xu P (2017) YaliBricks, a versatile genetic toolkit for streamlined and rapid pathway engineering in Yarrowia lipolytica. Metab Eng Commun 5:68–77. https://doi.org/10.1016/j.meteno.2017.09.001

Wu J, Cheng S, Cao J, Qiao J, Zhao G-R (2019) Systematic optimization of limonene production in engineered Escherichia coli. J Agric Food Chem 67(25):7087–7097. https://doi.org/10.1021/acs.jafc.9b01427

Xu P, Qiao K, Ahn WS, Stephanopoulos G (2016) Engineering <i>Yarrowia lipolytica</i> as a platform for synthesis of drop-in transportation fuels and oleochemicals. Proc Natl Acad Sci 113(39):10848–10853. https://doi.org/10.1073/pnas.1607295113

Xue S-J, Chi Z, Zhang Y, Li Y-F, Liu G-L, Jiang H, Hu Z, Chi Z-M (2018) Fatty acids from oleaginous yeasts and yeast-like fungi and their potential applications. Crit Rev Biotechnol 38(7):1049–1060. https://doi.org/10.1080/07388551.2018.1428167

Yaegashi J, Kirby J, Ito M, Sun J, Dutta T, Mirsiaghi M, Sundstrom ER, Rodriguez A, Baidoo E, Tanjore D, Pray T, Sale K, Singh S, Keasling JD, Simmons BA, Singer SW, Magnuson JK, Arkin AP, Skerker JM, Gladden JM (2017) Rhodosporidium toruloides: a new platform organism for conversion of lignocellulose into terpene biofuels and bioproducts. Biotechnol Biofuels 10(1):241. https://doi.org/10.1186/s13068-017-0927-5

Yang Q, Chen F (2022) Chapter 9 – life cycle assessment of biojet fuels. In: Yousuf A, Gonzalez-Fernandez C (eds) Sustainable alternatives for aviation fuels. Elsevier, pp 215–236. https://doi.org/10.1016/B978-0-323-85715-4.00004-5

Yang J, Xian M, Su S, Zhao G, Nie Q, Jiang X, Zheng Y, Liu W (2012) Enhancing production of bio-isoprene using hybrid MVA pathway and isoprene synthase in E. Coli. PLoS One 7(4): e33509. https://doi.org/10.1371/journal.pone.0033509

Yunus IS, Wichmann J, Wördenweber R, Lauersen KJ, Kruse O, Jones PR (2018) Synthetic metabolic pathways for photobiological conversion of CO(2) into hydrocarbon fuel. Metab Eng 49:201–211. https://doi.org/10.1016/j.ymben.2018.08.008

Zada B, Wang C, Park J-B, Jeong S-H, Park J-E, Birla Singh H, Kim S-W (2018) Metabolic engineering of Escherichia coli for production of mixed isoprenoid alcohols and their derivatives. Biotechnol Biofuels 11:1–12. https://doi.org/10.1186/s13068-018-1210-0

Zhang X, Liu X, Meng Y, Zhang L, Qiao J, Zhao G-R (2021) Combinatorial engineering of Saccharomyces cerevisiae for improving limonene production. Biochem Eng J 176:108155. https://doi.org/10.1016/j.bej.2021.108155

Zhao Y, Yang J, Qin B, Li Y, Sun Y, Su S, Xian M (2011) Biosynthesis of isoprene in Escherichia coli via methylerythritol phosphate (MEP) pathway. Appl Microbiol Biotechnol 90:1915–1922

Zhou J, Wang C, Yang L, Choi E-S, Kim S-W (2015) Geranyl diphosphate synthase: an important regulation point in balancing a recombinant monoterpene pathway in Escherichia coli. Enzym Microb Technol 68:50–55. https://doi.org/10.1016/j.enzmictec.2014.10.005

Chapter 13
Drop-in Biofuel Production: An Environmental Assessment Perspective

Bikash Ranjan Tiwari, Gustavo Amaro Bittencourt, and Satinder Kaur Brar

Abstract Currently fossil-derived fuel is the major input driving the transportation sector. The primary environmental concern associated with this fossil dependency relates to greenhouse gas emission leading to climate change in addition to rapid depletion of natural reserves of petroleum and diesel. An alternative approach to reduce the environmental burdens of the transportation sector is to increase the production and consumption of biomass-derived fuels such as drop-in biofuels. However, despite the change in source of fuel, the process in its entirety is associated with environmental burdens arising from the utilization of materials and energy. Hence, life cycle assessment studies are conducted to quantify the environmental burdens. This chapter reviews the environmental sustainability assessment of various approaches applied for drop-in biofuel derived from biomass. The life cycle stages are majorly categorized into biomass cultivation and harvesting, pretreatment, and biofuel production followed by upgradation for intended use. The biochemical routes are still in the early stages of development, and it is pertinent to discuss the nature of the associated environmental impacts. This could provide an environmental insight into the current and future strategies for drop-in fuel production to derive a comparatively greener fuel as compared to the fossil-derived fuels.

Keywords Drop-in biofuel · Life cycle assessment · Functional unit · Sensitivity analysis · Fermentation

B. R. Tiwari
Institut national de la recherche scientifique – Centre Eau Terre Environnement, Université du Québec, Quebec City, QC, Canada

G. A. Bittencourt
Department of Bioprocess Engineering and Biotechnology, Federal University of Parana, Curitiba, Brazil

S. K. Brar (✉)
Department of Civil Engineering, Lassonde School of Engineering, York University, Toronto, ON, Canada
e-mail: satinder.brar@lassonde.yorku.ca

S. K. Brar et al. (eds.), *The Microbiology of the Drop-in Biofuel Production*, Biofuel and Biorefinery Technologies 15, https://doi.org/10.1007/978-3-031-61637-2_13

13.1 Introduction

According to the targets set out in the Paris Agreement, global emissions have to be controlled to limit the temperature rise to 1.5 °C above pre-industrial thresholds, which would remarkably reduce the worldwide effects of climate change (Pörtner et al. 2019). Across the different sectors in the society that utilize energy and emit greenhouse gases (GHG), the transport sector is responsible for around 23% of the total global emission (~ 34.8 billion tonnes CO_2e per year) (Sandaka and Kumar 2023). The steady increase in the global vehicle population, along with the automation of the motor vehicle industry, has resulted in a significant surge in vehicular carbon emissions. With over 1.2 billion vehicles currently in operation, the transportation sector has undergone rapid progression and transformation (Hardaker et al. 2022). Consequently, there has been a substantial rise in daily petroleum consumption and subsequent per capita emissions. Projections indicate that by 2050, the distance traveled annually in both passenger and freight transport sectors will double, necessitating a significantly higher volume than the current petroleum consumption (Hardaker et al. 2022). In summary, according to the Intergovernmental Panel on Climate Change (IPCC), the global GHG emissions from the transport sector alone would reach 12 giga tonnes of CO_2e by 2050. This has to be addressed worldwide to reduce emissions by 40–70%, to avoid high environmental impacts related to global health, economy, and food security (Pörtner et al. 2019; Sandaka and Kumar 2023; Velvizhi et al. 2023). Additionally, IPCC estimates notable growth in GHG emission for the international aviation industry with an annual increase of 3.4% between 2010 and 2019. It is anticipated that the demand for air passenger travel in 2050 could surpass 10 billion. Subsequently, the annual CO_2 emissions from international aviation are estimated to reach approximately 1.8 billion tons by 2050. It represents a huge challenge to substitute the liquid fuels consumed worldwide, as gasoline and diesel for terrestrial transportation, kerosene by aviation, and many others by marine transportation, for sustainable and low-carbon alternatives.

Considering the high demand for liquid transportation fuels and its sustainability, the development of drop-in biofuels presents advantages in comparison with other potential solutions, such as electric cars and hydrogen fuels (Sandaka and Kumar 2023). Owing to similarity in properties to fossil fuels, drop-in fuels have full compatibility with existing transportation infrastructure, such as conventional engines, pipelines, and fuelling stations, making their integration with existing structures easier, without the need for extensive modifications or development of new infrastructure (Velvizhi et al. 2023). Recently, the interest in gasoline biofuel production has been decreasing, due to the surge and development of electrical light vehicles. However, renewable diesel and jet fuel have received considerable attention predominantly explored by sectors that are challenging to electrify, such as heavy-duty transport, marine transport, and aviation (Santner et al. 2023). Also, drop-in biodiesels can play a crucial role in reducing CO_2 emissions and decarbonizing global transportation, particularly in heavy-duty and marine transport sectors (Liu et al. 2021) (Fig. 13.1).

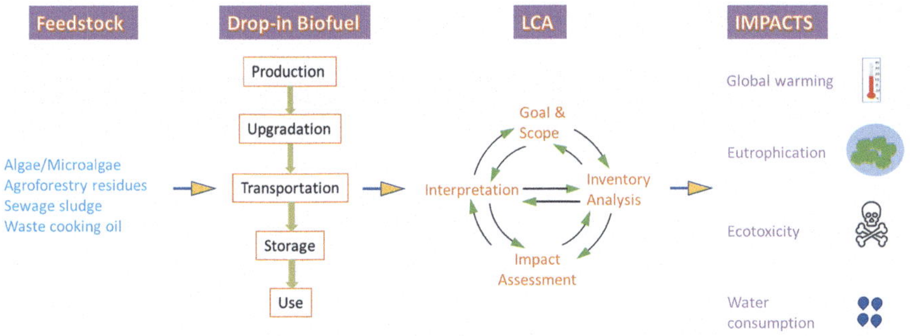

Fig. 13.1 Overview of drop-in biofuels life cycle assessment

However, the environmental burdens associated with the implementation of drop-in biofuels should be evaluated in a wider perspective, for each production process. It is plausible that drop-in biofuel utilization may present benefits over their fossil counterparts considering fuel combustion. However, the overall life cycle of drop-in biofuel production should be evaluated for a range of environmental parameters including those related to land-use change and high carbon footprints, as well as competitiveness with food production (Rathore et al. 2016). Parameters for biofuel drop-in biofuel production such as the feedstock, their extraction and conversion pathways into primary products, and upgrading methodologies, among other steps, need to be evaluated for the complete analysis of carbon footprint and GHG emissions savings. For instance, although first-generation biomass feedstocks (wheat, sugar cane, maize, and vegetable oils) indicate a reduction in GHG emissions, the use of fertilizers and agrochemicals, the promotion of deforestation and resulting loss of biodiversity, and also the use of fossil fuels inside the processes have to be analysed and compared with available conventional technologies in terms of GHG emissions (Kargbo et al. 2021).

The analysis of the entire life cycle of a biofuel production process is internationally recognized by the life cycle assessment (LCA) methodology. LCA is guided by the ISO 14040 series and evaluates the environmental performance of any product or pathway process (Rathore et al. 2016). The ISO 14040 series describes the LCA principles and framework along with the four stages of LCA and the relation between each stage. In addition, it also provides insight into the critical review and limitations of the LCA. The importance of the use of LCA methodology to guide low-carbon biofuel production is related to the consistent level of benchmarks, which enable comparisons between different processes, feedstocks, and technologies, in regard to the complete GHG emissions over their total lifetime, from extraction to production and combustion (Kolosz et al. 2020).

13.2 Life Cycle Assessment Process

LCA is an analytical technique employed for computing and analysing every accounted emission including the GHG balance as an outcome during the entire production process (Tiwari and Brar 2022a). Thus, the production system's energetics, along with environmental effects and impacts, are provided by LCA, offering information on the potential global warming and fossil energy requirement for the specific process of drop-in biofuel production. Other categories of environmental impact are also highlighted, including eutrophication, acidification, and ozone depletion, to address the environmental footprint of such process (Chopra et al. 2020; Kolosz et al. 2020; Panigrahi et al. 2021). The established standard LCA protocol consists of four different methodological phases, according to the "Code of practice" of the Society of Environmental Toxicology and Chemistry (SETAC): (1) goal and scope definition, (2) life cycle inventory, (3) life cycle impact assessment, and (4) life cycle interpretation (Hunkeler et al. 2008; Bhar et al. 2022; Panigrahi et al. 2022).

13.2.1 *Goal and Scope Definition*

The first step in LCA is to determine the goal of the study. The goal involves the purpose to analyse some process (Pradhan et al. 2019). As in the case for a drop-in biofuel process, the goal could be to quantify all the environmental impacts and emissions. This may include materials and energy inputs as well as waste generated, associated with all stages of production, such as cultivation, extraction and transportation of feedstocks, conversion and upgrading pathways of primary products, and waste recycling and demolition of industrial plant for biodiesel production

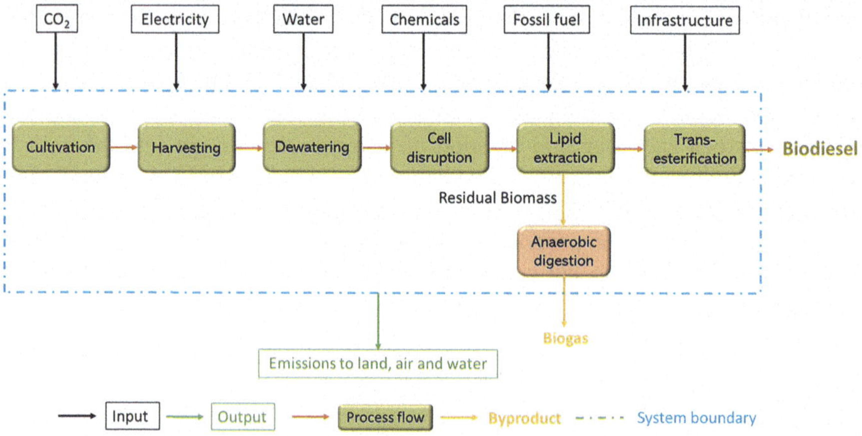

Fig. 13.2 Typical process flow chart for LCA of biodiesel production

(Fig. 13.2). Once the goal is established, the scope outlines the prospective target group and delineates the system being examined by establishing the two most important parameters in LCA investigation: the functional unit and the system boundary.

The functional unit is a central parameter for all LCA studies due to its importance to metrics normalization. It should be measurable and designed according to the defined goal. It provides a quantitative element to be projected across all environmental impacts in the inventory data. It is easy to identify a product by the quantity produced in litres or kg during a process, allowing the comparison of environmental impacts using standard metrics (Gómez and Escobar 2022). For instance, the functional unit of "1 kg of biofuel produced" refers to all the impacts generated in relation to this quantity of biofuel produced (Fazeni-Fraisl and Lindorfer 2023; Kossarev et al. 2023). However, the quantitative functional unit if not properly linked to the final use of the biodiesel may not represent the effectiveness of the product and thus fail to present the true comparison. For example, it may not be always appropriate to compare two different fuels of varying calorific value based solely on their weight or volume. Hence, other strategies can also be applied to design the functional unit as an alternative to the quantity of the biofuel produced. As liquid drop-in biofuels are normally projected to be used in the transport sector as an energy source, energy-based functional units are commonly used, based on one embodied MJ (or GJ) of its product (Kolosz et al. 2020). That was the case in the study of Watanabe et al. (Watanabe et al. 2022a), which used "1 MJ of energy" from the utilization of the evaluated marine drop-in biofuel. The same functional unit was used by Budsberg et al. (2022), in the evaluation of a bio-jet drop-in biofuel. This strategy allows us to evaluate the impacts related directly to the quantity of energy produced, incorporating the different energy densities of each fuel for a fairer comparison between them. Since the drop-in fuel could also be utilized for transportation, another representative parameter for a transportation fuel in addition to the energy units is the distance that can be travelled following its combustion. Functional units such as 1 km driving a passenger vehicle and 1 ton/km have been reported for LCA studies (Deutz et al. 2018; van den Oever et al. 2023).

In a generic process, the production of drop-in biofuels involves biomass cultivation and collection, transportation to facility, biomass pretreatment and conversion to primary products, and its upgrading and downstream processes. This demands many different inputs, such as building construction, agrochemicals and materials for equipment, and fuel for energy and transportation (Kargbo et al. 2021). Hence, the system boundary of an LCA can incorporate different phases of the production process, depending on the goal and scope of the analysis. Different approaches have been used and incorporated, depending on the specific objectives of such study. In general, LCA analyses entail four typical system boundaries. The well-to-gate system boundary is the narrowest approach, while it encompasses processing and refining until the "gate" of the factory, as well as feedstock-related operations—such as cultivation, extraction, and transportation to refinery facility (Fig. 13.3). The well-to-pump approach adds one more boundary level, also taking into account the transport to storage facilities for a wide distribution of biofuel; thus, it incorporates

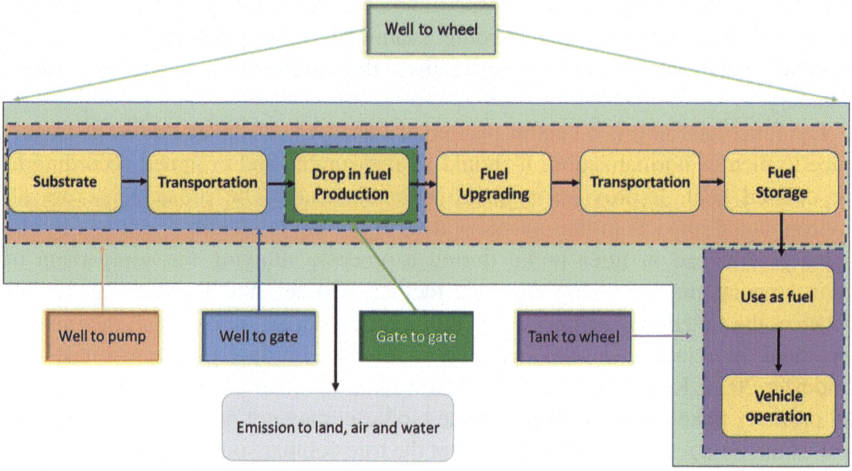

Fig. 13.3 System boundary considerations for drop-in biofuel production

all stages of the production process excluding the biofuel combustion (Budsberg et al. 2022; Fazeni-Fraisl and Lindorfer 2023). Otherwise, the well-to-wake approach, or also called well-to-wheel in liquid fuels contexts, is the wider system boundary, while it encompasses the complete life cycle of the fuel pathway, starting from the feedstock cultivation phase to the analysis of ultimate combustion of the biofuel (Liu and Yang 2020; Zhang et al. 2022). According to Poulikidou et al. (2019), this approach provides the opportunity for a holistic performance assessment of the biofuel, enabling a fairer comparison between different fuel production processes. Due to these varying system boundaries, a notable degree of uncertainty could happen when conducting comparisons between different processes with existing LCA literature.

13.2.2 Inventory Analysis

The inventory analysis is the compilation and quantification of all natural resources and materials consumed and all substances emitted as output during the life cycle, according to the defined system boundary (Angoy et al. 2019). In case of drop-in biofuels production, the inventory analysis includes data related to the different phases of production process. Fertilizers, pesticides, nutrients, and agricultural equipment, as well as fossil fuel demand for transportation to biorefinery facilities, must be accounted for feedstock cultivation phase. Chemicals and energy requirements for conversion and upgrading pathways, such as in the case for cultivation of microbes, pretreatment, fermenters, and unit operations of centrifuge, drying, and heating, among others, must be included (Kargbo et al. 2021). For the generation of data for inventory analysis, best and primary source is typically data collected from real plant operations—including descriptions of technologies, their efficiencies and installed capacities, raw materials, and energy consumption for the life cycle stages (Angoy et al. 2019). Pilot- and lab-scale experiments can also be used; however, in some situations, linear extrapolation for corresponding large-scale applications is not viable, making it necessary to use mathematical modelling techniques to generate the data, or big database software, such as Ecoinvent, which can provide background database (Hischier et al. 2014). To facilitate modelling, evaluation, and optimization of integrated processes across a wide range of industries processes, other simulation software tools are available, like SuperPro and Aspen Plus. Standard chemical equations can be used to derive data from chemical reactions related to emissions or by-products, such as methane and carbon dioxide. Otherwise, it is also feasible to estimate data using mathematical formulas provided in previously published literature. For instance, energy requirements for heating, pumps, stirring, filtration, and centrifugation can be calculated or provided by manufacturers, to express power consumption in terms of kilowatt-hours (kWh) per ton of dry material.

13.2.3 Impact Assessment and Interpretation

Specific software tools are utilized for LCA of drop-in fuels based on different approaches and place of origin (Table 13.1). The most globally used software for LCA studies is SimaPro which utilizes the Ecoinvent database (Tiwari and Brar 2022b). In addition, other databases such as Agri-footprint, EU Danish input–output and the European Life Cycle Database (ELCD), Industry data 2.0, Swiss input output database, and EU and DK input output database are also included. Similar to SimaPro, GaBI has also been commonly used for LCA studies around the world. The open source Life Cycle and Sustainability Assessment software which are freely available for use also garner considerable attention. However, in terms of aviation fuel studies, the GREET (Greenhouse gas Regulated Emissions and Energy use in Transportation) software developed by Argonne National Laboratory, USA, is more

Table 13.1 Software used for transportation fuel LCA studies

Software	Origin	Application	Availability	Website
SimaPro	The Netherlands	Global LCA studies	Commercial	https://simapro.com/
GaBi	Germany	Global LCA studies	Commercial	https://sphera.com/life-cycle-assessment-lca-software/
openLCA	GreenDelta, Germany	Global LCA studies	Free download	https://www.openlca.org/
GREET	Argonne National Laboratory, USA	LCA of aviation fuels	Free download	https://greet.anl.gov/greet/versions.html
GHG Genius	Canada	First-generation aviation fuels	Free download	https://www.ghgenius.ca/index.php/downloads

popular. This database is created on US-based sources for inventory such as electricity generation mix, industrial statistics, transportation modes, and distances (Chandra et al. 2012). Another such LCA tool called GHG genius was developed by Natural Resources Canada primarily used for biomass-derived transportation fuels, mostly for ground transportation (Contadini et al. 2000). This tool has databases which can be used for studies based in USA, India, and Mexico, in addition to Canada.

The evaluation of inventory data involves a series of processing and interpretation steps to assess its impact on the environment which are classified into various categories. ISO 14047 is a standard document which includes various examples to carry out the LCIA phase. The examples are not exclusive but an illustration of ways to apply LCIA phase. It suggests seven impact categories: climate change, acidification, eutrophication, human health toxicity, photochemical oxidation, ecotoxicity, and ozone layer depletion. However, in the case of the drop-in biofuels LCA studies, not all these suggestions are addressed.

Among these categories, the most relevant is the global warming potential (GWP), in the context of climate change, which holds great importance and arises from both natural and human activities, encompassing the direct and indirect emission of carbon dioxide, methane, and nitrous oxide (Gómez and Escobar 2022). Besides, impacts are classified according to their influence on distinct abiotic elements, such as eutrophication in freshwater and marine ecosystems, as well as ecotoxicity in terrestrial, marine, and freshwater environments. The impact categories also focus on the resources as per the assigned system boundary, such as water consumption, land use, mineral resource utilization, and fossil depletion. The impact categories are expressed through midpoint and endpoint indicators. While midpoint indicators represent impacts in the intermediary chain, endpoint indicators reflect the final level of impact and result from the aggregation of different midpoint indicators (Hardaker et al. 2022).

Once the life cycle inventory and the processes within the defined system boundary are translated into corresponding life cycle impacts, the interpretation of the results is carried out. This helps to understand and analyse bottlenecks in the production process. Moreover the comparison with other LCA studies allows us to draw conclusions and recommendations for impact reduction to eventually achieve process sustainability (Hardaker et al. 2022). Sensitivity analysis can be conducted to determine the variation and robustness of results within changes in the input parameters. As certain assumptions made in the inventory can oversimplify the LCA model, some tools can be implemented to incorporate uncertainty degrees in data analysis, such as the Monte Carlo simulation. It provides accurate estimate of uncertainties, integrating it into the inventory and the environmental impact model, leading to a more accurate result interpretation (Hardaker et al. 2022; Tiwari et al. 2023).

13.3 Case Studies and Literature Overview

For an efficient understanding of LCA studies, it is necessary to compare drop-in fuels obtained from various feedstocks and conversion technology pathways with the conventional and widely used petrochemical fuels, in the context of different transport sectors, such as aviation and maritime. As presented in the previous sections, transportation fuel such as aviation fuel is also categorized under drop-in fuel. Utilization of non-fossil environment-friendly bio-based substrate and sustainable pathways/processes presents a huge opportunity for the upcoming generation of researchers and entrepreneurs to invest in alternatives to conventional aviation fuels. Even for the case of reported literature on alternative aviation fuels, LCA studies conducted for biomass-derived aviation fuel constitute the major share. While biomass is a more generalized term for organic material of plant or animal origin, the energy sectors usually categorize them into three different generations of fuels according to their evolution. Table 13.2 shows some recent LCA studies for drop-in biofuel production with different methodologies and approaches. This section will discuss environmental assessment of the drop-in fuels with respect to the biomass, i.e., first generation, second generation, and third generation.

The first-generation biomass consists of substrates which are raw crops and consumed as food. They can be further divided into three categories based on the crops. The first category includes crop-based on sugar such as sugarcane, sorghum, and sugar beet corn. The second category consists of crops based on starch such as corn, barley, and wheat. The final category consists of the oil-based crops such as sunflower, rapeseed, and canola. LCA studies can be conducted on first-generation biofuels considering the production scenarios where the oils, sugar, and starch present in the biomass are converted to fuels through chemical processes or fermentation. The popularity of first-generation biofuels was based on presenting an environment-friendly alternative to conventional fossil fuels. The associated GHG emission from the combustion of first-generation fuels was lower as compared to fossil fuels (Rodionova et al. 2017). This could directly lead to reduction in impacts associated with the midpoint category of GWP. In addition, reduced dependency on fossil fuel would also contribute to saving the natural resources, i.e., fossil fuel and further reflect in fossil resource depletion midpoint category. But cultivation of first-generation biofuels is associated with high agricultural land requirement (Phwan et al. 2018) which can contribute to the land use category in addition to water consumption. Biogenic carbon consideration during fuel combustion is still debatable for GHG emission considerations.

The most popular scenario considers the fact that the carbon dioxide released during combustion is accumulated from the environment. Hence, the GWP attributed to biomass combustion is not accounted for during LCA (Tiwari et al. 2023). However, this should be made clear while framing the goal and the scope of the study. Agricultural practices always require arable land for physical planting and use of fertilizers. Land cultivation can lead to change in carbon stock estimated before and after the cultivation of the crops (Brassard et al. 2021). Similarly, fertilizer

Table 13.2 Summary of studies conducted on LCA of drop-in biofuel

Drop-in fuel	Feedstock	Functional unit	System boundary	Fate of CO_2	Reference
Iso-paraffinic kerosene	Residual woody biomass	1GJ of energy	Well-to-wake	Boiler combustion	(Ganguly et al. 2018)
Drop-in and hydrogen-based fuels	Agricultural and forestry residues		Well-to-wake	Release to environment	(Watanabe et al. 2022b)
Drop-in marine biofuel	Forestry residues	1 MJ of energy	Well-to-pump	Release to environment	(Watanabe et al. 2022a)
Liquid hydrogen and drop-in biofuels	Algae	1 kg of produced biofuel	Well-to-wake	Release to environment	(Kossarev et al. 2023)
Marine biofuel	Agroforestry residues	1 GJ of fuel	Well-to-wake	Release to environment	(Tanzer et al. 2019)
Naphtha, jet fuel and diesel	Red maple wood	1 MJ of energy	Well-to-wake	Extracted from the system	(Olcay et al. 2018)
Bio-jet fuel	Poplar wood	1 MJ of energy	Well-to-gate	Release to environment	(Budsberg et al. 2022)
Isobutene (IBN)	Sugars and cereal straw	1 kg of bio-based IBN	Well-to-gate	Release to environment	(Fazeni-Fraisl and Lindorfer 2023)
Drop-in bio-jet fuel	Poplar wood	1 MJ of jet fuel	Well-to-wake	Release to environment	(Budsberg et al. 2016)
Marine biofuels	Hydrothermal liquefaction (HTL) of sewage sludge	1 kg of treated sludge	Just treatment and upgrading	Release to environment	(Lozano et al. 2022)
Oxymethylene ether (OME)	Formaldehyde (FA), methanol, CO_2, and H_2	1 km driving a passenger vehicle, and 1 kg of OME produced	Well-to-wake	OME production	(Deutz et al. 2018)
High-cetane hydrocarbons and C6–C11 alcohols	Syngas from bark chips gasification	1 ton/km	Well-to-wake	Release to environment	(van den Oever et al. 2023)
Renewable diesel, renewable gasoline, renewable jet fuel	Microalgae (*Scenedesmus* sp.)	1 MJ of usable energy	Well-to-pump	Biomass assimilation	(Bessette et al. 2018)
Bio-jet fuel	Waste cooking oil	1 MJ of bio-jet fuel produced	Well-to-wake	Release to environment	(Zhang et al. 2022)
2-ethylhexanol (2-EH)	Forest residues (softwood)	1 MJ of biomass-based 2-EH	Well-to-wake	Release to environment	(Poulikidou et al. 2019)

(continued)

Table 13.2 (continued)

Drop-in fuel	Feedstock	Functional unit	System boundary	Fate of CO_2	Reference
Bio-jet fuel	Microalgae (*Nannochloropsis* sp.) and Jatropha	1 MJ of usable energy	Well-to-wake	Release to environment	(Liu and Yang 2020)
Diesel and marine drop-in renewable fuel	Forestry residues	2000 barrel/d of fuel	Well-to-gate	CO_2 sequestration	(Jensen et al. 2017)
Hydrocarbons, suitable for upgrading to drop-in biofuels	Sawdust, bark, and logging residues	1 MJ of fuel	Well-to-wake	Release to environment	(Stigsson et al. 2022)
Renewable diesel and biodiesel	Microalgae	1 MJ of delivered fuel product	Well-to-pump	Recirculated for biomass growth	(Borkowski et al. 2012)

production is energy and material intensive and may lead to emission to land and water often included in eutrophication and ecotoxicity categories. The comparison of first-generation biomass revealed an associated GWP emission in the range of −27.0 to 19.7, 47.5 to 117.5, and 11.7 to 89.8 gCO_2e per MJ middle distillate (diesel and jet fuels) for sugarcane, corn grain, and switchgrass (Chandra et al. 2012). However, the conventional process for middle distillate production leads to 90 gCO_2e per MJ. This is attributed to the overall environmental land footprint emissions associated with arable area used for the biomass growth which subsequently increases the GHG emission. Apart from the environmental point of view, the biggest controversy surrounding the first-generation biofuel is the issue of competition with food supply and food security (Rodionova et al. 2017). Hence, the focus shifted to alternative non-edible substrates which have either comparable or lower environmental burdens as compared to first-generation biomass.

The need to avoid the debate surrounding the substrate intended for food and fuel led to the rise of second-generation biomass and corresponding transportation fuel. Hence, second-generation fuels are derived from non-food-based crops and agricultural residues (Chandra et al. 2012). The second-generation biomass scores over the first-generation biomass from the sustainability point of view considering that no waste is generated during their utilization rather the waste is utilized (Saini et al. 2015). This subsequently reduces the emissions and environmental burdens associated with the transportation, disposal, and management of the wastes. As opposed to the first-generation biomass, land use category is avoided. Despite these benefits, second-generation biomass is often energy and material intensive owing to the pretreatment step (Brassard et al. 2021; Bhar et al. 2022). Pretreatment step may include utilization of chemicals such as acids and base or heat, etc. This is reflected in the environmental categories of GWP, eutrophication, and ecotoxicity. The use of heat or electricity if fossil-derived can further increase GHG emission. However, if the source of electricity is renewable, the contribution from energy use may be comparably lower (Tiwari and Brar 2022a). Similarly, the exposure of water sources to the chemical discharge such as acid and base from the process may lead to change in the surrounding ecosystem and affect the inhabiting species. This is usually reflected in the endpoint categories of the ecosystem for the ReCiPe method. The damage to the ecosystem is expressed in terms of potentially disappearing fraction of species per m^2 of land area per year (Chopra et al. 2020). Comparative LCA for second-generation fuel derived from switchgrass, Jatropha, and Salicornia revealed GWP of −222.7, −70.5, and − 105.3 gCO_2e per MJ (Chandra et al. 2012). However, when land use changes were included the highest environmental footprint of 698 g CO_2e per MJ was estimated. Cellulosic biomass was found to contribute around 10 g CO_2e per MJ which is considerably lower than fuel produced from corn stover (Chandra et al. 2012). Apart from the agricultural and forestry residues, the organic fraction of municipal solid waste is also utilized for drop-in fuel production. However, the municipal solid waste is a mixture of food waste, garden waste, paper, etc. which may lead to issues during the substrate utilization. Further, this variability may lead to reduced yield of the drop-in fuel which may be reflected in the cumulative impacts generated per litre or kg of fuel. Hence, the process yield associated with the substrate is also a defining parameter to reduce the associated environmental impacts of GWP, eutrophication, ecotoxicity, and resource depletion (Tiwari et al. 2023).

Ganguly et al. (2018) evaluated the emissions associated with the production of iso-paraffinic kerosene (IPK) from residual woody biomass, for conventional kerosene substitution in aircrafts, comparing both by means of the Ecoinvent database, in a well-to-wake system boundary approach. The authors observed that the overall GWP impact associated with the drop-in biofuel was 22.1% of that associated with the baseline petroleum-based jet fuel, which represents a 77.9% reduction in overall GHG emissions. Not including the biogenic CO_2 emissions, which are considered carbon neutral, the impact on GWP from fuel combustion, associated with the production of 1 GJ of energy, was considered 19.4 kg of CO_2e (Ganguly et al. 2018). In the context of maritime transport sector, Watanabe et al. (2022b) evaluated different technologies pathways for the conversion of agricultural and forest residues, such as fast pyrolysis, hydrothermal liquefaction, and gasification, to Fischer–Tropsch fuels or bio-synthetic natural gas. The authors reported that biofuels produced from gasification (to Fischer–Tropsch fuels or bio-synthetic natural gas) had the lowest environmental impacts between the pathways analysed. In terms of 100-year GWP, average results were between 5 and 16 g CO_2e/ MJ, with a small uncertainty range. This uncertainty was due to the absence of dependency on the electricity from the grid, which was a dominant factor in the Monte Carlo uncertainty analysis (Watanabe et al. 2022b).

Tanzer et al. (2019) evaluated an integrated screening model to compare the technological, economic, and environmental performance of 33 drop-in marine biofuel blend stock supply chains, with nine different agroforestry residues and three thermochemical technologies. Although all biofuels showed an improvement in life cycle emissions comparing with heavy fuel oil, forestry residue emissions depended highly on biogenic carbon dioxide accounting, which makes biofuels from biomass wastes a better option, due to a decrease in life cycle GHG emissions of 40–100 kg/GJ. This shows that the accounting method for estimating biogenic GHG emissions substantially influences the resulting estimated emissions, as well as the delay of carbon payback. Agricultural wastes also have showed higher GHG savings when compared to first-generation sugar feedstocks (Fazeni-Fraisl and Lindorfer 2023). The study evaluated different designed scenarios for a comparative environmental sustainability assessment of bio-based isobutene production with fossil isobutene, which is a widely used liquid fuel in the industry (Fazeni-Fraisl and Lindorfer 2023). It was reported that savings in GWP for first-generation isobutene are up to 68%, comparing with conventional isobutene process, while for second-generation isobutene, incorporating the energy from lignin combustion into the process chain, was 105%.

The limitations arising from the yield and biomass to fuel conversion led to the evolution of the third generation of biofuels. Apart from agricultural waste residues, algae and microalgae have been preferred choices for drop-in biofuel production [11, 15, 29, 32]. Algae is a popular choice for biofuels since they have a high lipid content of around 70% (dry weight). Hence, it is possible that algae have the ability to generate ten times more fuel as compared to some of the previous generation fuels (Chandra et al. 2012). It was reported that while the total energy consumption for aviation fuel production reduced from around 2–3 $MJ \cdot MJ^{-1}$ for increased lipid content of around 15–28%, the fossil fuel consumption and GWP increased in the range of 0.3–0.7 $MJ \cdot MJ^{-1}$ and 17–51 g $CO_2e \cdot MJ^{-1}$, respectively (Chandra et al. 2012). Since algae are grown in water streams, the environmental footprint associated with agricultural land requirement is considerably lower than the first-generation biomass. Moreover, algae can grow in waste stream as well as brackish water which reduced the environmental burdens on utilization of freshwater sources for drop-in fuel production (Suali and Sarbatly 2012).

One of the major factors leading to the high score in sustainability is the ability of alga to utilize carbon dioxide for its growth. This reduces the score of GHG emission. Moreover, if flue gas released from thermal power plants is utilized, the process becomes more environment friendly. A study reported that although energy consumption from coal-derived power plant led to higher environmental footprint, the CO_2 that was utilized from the plant led to emission savings of around 86 g CO_2e per MJ. Hence, the final GWP was reduced to around 32 g CO_2e per MJ. This could be further reduced to 19 g CO_2e per MJ if clean flue gas is utilized [46]. Contrary to the popular belief, it was found that for a long-range aircraft show, the adverse environmental impacts associated with hydrogen and biofuel were higher in comparison to the conventional aircraft fuel (Kossarev et al. 2023). While the environmental impacts for hydrogen-powered aircraft increased by ~15%, the increase was significantly higher at 235% and 548% for algae produced with hydrogenated vegetable oil and oil-rich biomass, respectively. For all the cases, the increase was due to the inherent consumption of energy for the production process. Another study reported opposing results where algae-derived biofuel accounted for the least GHG emission of 33.7 gCO_2e per MJ (Liu and Yang 2020). This was considerably lower than coal-based (182.8 gCO_2e per MJ) and petroleum-based jet fuel (87 gCO_2e per

MJ). The reduction in algae-based jet fuel was owing to the introduction of a refining step which resulted in a reduction of around 27% of GHG emissions. Borkowski et al. compared the environmental burdens for algae-derived renewable biodiesel (hydrogenation and decarboxylation-based) and biodiesel (transesterification-based) fuel (Borkowski et al. 2012). When the cultivation and harvesting conditions are similar, the benefits of renewable biodiesel were reported to be marginally higher than biodiesel in terms of energy consumption and GHG emission. Hence, algae-derived renewable biodiesel was the preferred alternative for future research and development.

13.4 Future Perspectives

It is a common practice to conduct LCA for individual process pathways for biofuel production. Although it provides valuable information regarding the specific pathways, the comparison with existing alternatives can elevate the outcome of the study. This is intended to identify and compare the necessary emissions and impacts associated with the current and proposed fuels. Moreover, individual LCA studies should be extended to comparative LCA where the goal should be to compare the environmental benefits and adverse impacts of two or more competitive technologies for biofuel production. As previously mentioned, the outcome of an LCA study depends on various considerations such as functional unit, system boundary, energy consideration, and finally the source of primary and secondary inventory data. During the LCA, it is possible that certain assumptions are undertaken to fulfil the lack of data by citing various literature studies or following standard ideal equations. Considering this disparity, it becomes an arduous job to compare outcomes of two or more different studies when the criteria for evaluation are different. The comparison will be holistic and more deterministic in order to select the best biofuel or technology if a proper code or criteria for evaluation is applicable to all studies conducted for LCA. Hence, it is recommended that a uniform code should be designed and followed for future LCA studies. Since drop-in biofuel is still at a developmental stage, it is understood that most of the studies reported are conducted on either lab-scale or pilot-scale. In order to provide the most relevant information with respect to future implementation, it is important that the technology is scaled to the real field-scale which replicates/resembles the industrial scale production. For instance, fermentation is the main pathway for biochemical conversion of substrate to drop-in fuels. However, industrial scale production for drop-in fuel production is mostly limited to first-generation feedstocks. It is expected that LCA based on material and energy inventories designed for large-scale implementation will provide a clearer picture as compared to a lab-scale design. Hence, drop-in biofuel production models used for LCA should be upscaled from lab-scale to field-scale by following specific design and empirical consideration. It is true that most LCA studies take into consideration the material and energy resources involved in cultivation, production, and transportation. However, infrastructure-related data are often neglected which should be included in order to provide the complete life cycle scenario including activities such as demolition, disposal, and recycling of construction materials. Since the feedstock is the most important material input for drop-in biofuel which strongly affects the environmental sustainability criteria, it is proposed to opt for waste-derived substrate for biofuel production. This could potentially reduce the GHG emission in comparison to first-generation biofuels. A final parameter which must be included in LCA studies is the time period consideration, say 30 years or 100 years for drop-in biofuel production and use which would subsequently generate environmental impacts. For example, for a biorefinery supposed to be operative for 50 years, the accumulated impacts will be different if the time period under consideration is increased or reduced. Currently, only the GWP100 estimates the GHG emissions for a standard time horizon of 100 years. Hence, the impacts should be expressed either considering the rate of emission or a standard time should be fixed for impact estimation.

13.5 Conclusions

LCA studies reveal that drop-in fuel could lower the carbon footprint by almost half as compared to their fossil counterpart. However, commercialization and upscaling of technologies such as fermentation will eventually provide a clear picture of the sustainability. The environmental impacts are also expected to reduce with implementation of waste to bioenergy recovery route for biofuel production. Since utilization of waste is already associated with avoided emission in the form of landfilling or GHG emission, this will eventually project beneficial impacts on the production process. Global warming or climate change is the broadly accepted representation of adverse impacts on the environment; most LCA studies are limited to estimation of GWP for drop-in fuels. However, this limits the ability of LCA for a comprehensive evaluation of other impacts with respect to ecotoxicity, resource consumption, particulate emission, etc. Hence, in addition to achieving the primary goal of estimating the carbon footprint, LCA studies on drop-in biofuels should widen their scope to include other characterization factors for other impact categories relating to soil, air, water, human health, and the flora and fauna.

Acknowledgements The authors would like to acknowledge the support of NSERC CRD grants and James and Joanne Love Chair in Environmental Engineering.

References

Angoy A, Brianceau S, Chabrier F et al (2019) 17 – microwave technology for food applications. In: Chemat F, EBT-GFPT V (eds) . Academic press, pp 455–498

Bessette AP, Teymouri A, Martin MJ et al (2018) Life cycle impacts and techno-economic implications of flash hydrolysis in algae processing. ACS Sustain Chem Eng 6:3580–3588. https://doi.org/10.1021/acssuschemeng.7b03912

Bhar R, Tiwari BR, Sarmah AK et al (2022) A comparative life cycle assessment of different pyrolysis-pretreatment pathways of wood biomass for Levoglucosan production. Bioresour Technol 356:127305

Borkowski MG, Zaimes GG, Khanna V (2012) Integrating LCA and thermodynamic analysis for sustainability assessment of algal biofuels: comparison of renewable diesel vs. biodiesel. In: IEEE international symposium on sustainable systems and technology (ISSST). IEEE. https://doi.org/10.1109/ISSST.2012.6227997

Brassard P, Godbout S, Hamelin L (2021) Framework for consequential life cycle assessment of pyrolysis biorefineries: a case study for the conversion of primary forestry residues. Renew Sust Energ Rev 138:110549. https://doi.org/10.1016/j.rser.2020.110549

Budsberg E, Crawford JT, Morgan H et al (2016) Hydrocarbon bio-jet fuel from bioconversion of poplar biomass: life cycle assessment. Biotechnol Biofuels 9:1–13. https://doi.org/10.1186/s13068-016-0582-2

Budsberg E, Parker N, Bandaru V et al (2022) Hydrocarbon bio-jet fuel from bioconversion of poplar biomass: life cycle assessment of site-specific impacts. Forests 13:1–26. https://doi.org/10.3390/f13040549

Chandra R, Takeuchi H, Hasegawa T (2012) Methane production from lignocellulosic agricultural crop wastes: a review in context to second generation of biofuel production. Renew Sust Energ Rev 16:1462–1476. https://doi.org/10.1016/j.rser.2011.11.035

Chopra J, Tiwari BR, Dubey BK, Sen R (2020) Environmental impact analysis of oleaginous yeast based biodiesel and bio-crude production by life cycle assessment. J Clean Prod 271:122349. https://doi.org/10.1016/j.jclepro.2020.122349

Contadini JF, Moore RM, Sperling D, Sundaresan M (2000) Life-cycle emissions of alternative fuels for transportation: dealing with uncertainties. SAE Tech Pap. https://doi.org/10.4271/2000-01-0597

Deutz S, Bongartz D, Heuser B et al (2018) Cleaner production of cleaner fuels: wind-to-wheel-environmental assessment of CO2-based oxymethylene ether as a drop-in fuel. Energy Environ Sci 11:331–343. https://doi.org/10.1039/c7ee01657c

Fazeni-Fraisl K, Lindorfer J (2023) Comparative life cycle assessment of first- and second-generation bio-isobutene as a drop-in substitute for fossil isobutene. Biofuels Bioprod Biorefin 17:207–225. https://doi.org/10.1002/bbb.2444

Ganguly I, Pierobon F, Bowers TC et al (2018) 'Woods-to-wake' life cycle assessment of residual woody biomass based jet-fuel using mild bisulfite pretreatment. Biomass Bioenergy 108:207–216. https://doi.org/10.1016/j.biombioe.2017.10.041

Gómez IDL, Escobar AS (2022) The dilemma of plastic bags and their substitutes: a review on LCA studies. Sustain Prod Consum 30:107–116. https://doi.org/10.1016/j.spc.2021.11.021

Hardaker A, Styles D, Williams P et al (2022) A framework for integrating ecosystem services as endpoint impacts in life cycle assessment. J Clean Prod 370:133450. https://doi.org/10.1016/j.jclepro.2022.133450

Hischier R, Achachlouei MA, Hilty LM (2014) Evaluating the sustainability of electronic media: strategies for life cycle inventory data collection and their implications for LCA results. Environ Model Softw 56:27–36. https://doi.org/10.1016/j.envsoft.2014.01.001

Hunkeler D, Lichtenvort K, Rebitzer G (2008) Environmental life cycle costing. CRC Press

Jensen CU, Guerrero JKR, Karatzos S et al (2017) Hydrofaction™ of forestry residues to drop-in renewable transportation fuels. Elsevier Ltd.

Kargbo H, Harris JS, Phan AN (2021) "Drop-in" fuel production from biomass: critical review on techno-economic feasibility and sustainability. Renew Sust Energ Rev 135:110168. https://doi.org/10.1016/j.rser.2020.110168

Kolosz BW, Luo Y, Xu B et al (2020) Life cycle environmental analysis of 'drop in' alternative aviation fuels: a review. Sustain Energy Fuels 4:3229–3263. https://doi.org/10.1039/C9SE00788A

Kossarev K, Scholz AE, Hornung M (2023) Comparative environmental life cycle assessment and operating cost analysis of long-range hydrogen and biofuel fueled transport aircraft. CEAS Aeronaut J 14:3–28. https://doi.org/10.1007/s13272-022-00627-w

Liu Z, Yang X (2020) Refining drop-in jet fuel coupling GHGs reduction in LCA with airworthiness in aero-engine and aircraft. Catal Today 353:260–268. https://doi.org/10.1016/j.cattod.2018.04.049

Liu Y, Cruz-Morales P, Zargar A et al (2021) Biofuels for a sustainable future. Cell 184:1636–1647. https://doi.org/10.1016/j.cell.2021.01.052

Lozano EM, Løkke S, Rosendahl LA, Pedersen TH (2022) Production of marine biofuels from hydrothermal liquefaction of sewage sludge. Preliminary techno-economic analysis and life-cycle GHG emissions assessment of Dutch case study. Energy Convers Manag X 14:100178. https://doi.org/10.1016/j.ecmx.2022.100178

Olcay H, Malina R, Upadhye AA et al (2018) Techno-economic and environmental evaluation of producing chemicals and drop-in aviation biofuels via aqueous phase processing. Energy Environ Sci 11:2085–2101. https://doi.org/10.1039/c7ee03557h

Panigrahi S, Sharma HB, Tiwari BR et al (2021) Insight into understanding the performance of electrochemical pretreatment on improving anaerobic biodegradability of yard waste. Renew Energy 180:1166–1178. https://doi.org/10.1016/j.renene.2021.08.123

Panigrahi S, Tiwari BR, Brar SK, Dubey BK (2022) Thermo-chemo-sonic pretreatment of lignocellulosic waste: evaluating anaerobic biodegradability and environmental impacts. Bioresour Technol 361:127675

Phwan CK, Ong HC, Chen W-H et al (2018) Overview: comparison of pretreatment technologies and fermentation processes of bioethanol from microalgae. Energy Convers Manag 173:81–94. https://doi.org/10.1016/j.enconman.2018.07.054

Pörtner H, Roberts D, Masson-Delmotte V et al (2019) IPCC intergovernmental panel on climate change. Switz, Geneva

Poulikidou S, Heyne S, Grahn M, Harvey S (2019) Lifecycle energy and greenhouse gas emissions analysis of biomass-based 2-ethylhexanol as an alternative transportation fuel. Energy Sci Eng 7:851–867. https://doi.org/10.1002/ese3.315

Pradhan S, Tiwari BR, Kumar S, Barai SV (2019) Comparative LCA of recycled and natural aggregate concrete using particle packing method and conventional method of design mix. J Clean Prod 228:679–691. https://doi.org/10.1016/j.jclepro.2019.04.328

Rathore D, Nizami A-S, Singh A, Pant D (2016) Key issues in estimating energy and greenhouse gas savings of biofuels: challenges and perspectives. Biofuel Res J 3:380–393. https://doi.org/10.18331/BRJ2016.3.2.3

Rodionova MV, Poudyal RS, Tiwari I et al (2017) Biofuel production: challenges and opportunities. Int J Hydrog Energy 42:8450–8461. https://doi.org/10.1016/j.ijhydene.2016.11.125

Saini JK, Saini R, Tewari L (2015) Lignocellulosic agriculture wastes as biomass feedstocks for second-generation bioethanol production: concepts and recent developments. Biotech 5:337–353. https://doi.org/10.1007/s13205-014-0246-5

Sandaka BP, Kumar J (2023) Alternative vehicular fuels for environmental decarbonization: a critical review of challenges in using electricity, hydrogen, and biofuels as a sustainable vehicular fuel. Chem Eng J Adv 14:100442. https://doi.org/10.1016/j.ceja.2022.100442

Santner P, Chanquia SN, Petrovai N et al (2023) Biocatalytic conversion of fatty acids into drop-in biofuels: towards sustainable energy sources. EFB Bioeconomy J 3:100049. https://doi.org/10.1016/j.bioeco.2023.100049

Stigsson C, Furusjö E, Börjesson P (2022) A model of an integrated hydrothermal liquefaction, gasification and Fischer-Tropsch synthesis process for converting lignocellulosic forest residues into hydrocarbons. Bioresour Technol 353:126070. https://doi.org/10.1016/j.biortech.2021.126070

Suali E, Sarbatly R (2012) Conversion of microalgae to biofuel. Renew Sust Energ Rev 16:4316–4342. https://doi.org/10.1016/j.rser.2012.03.047

Tanzer SE, Posada J, Geraedts S, Ramírez A (2019) Lignocellulosic marine biofuel: Technoeconomic and environmental assessment for production in Brazil and Sweden. J Clean Prod 239:117845. https://doi.org/10.1016/j.jclepro.2019.117845

Tiwari BR, Brar SK (2022a) Chapter 11 – a life cycle assessment perspective to conventional and modular wastewater treatment. In: Kaur Brar S, Kumar P, Cuprys ABT-MTA for DW and W (eds) . Elsevier, pp 187–205

Tiwari BR, Brar SK (2022b) A life cycle assessment perspective to conventional and modular wastewater treatment. In: Modular treatment approach for drink water and wastewater, pp 187–205

Tiwari BR, Bhar R, Dubey BK et al (2023) Life cycle assessment of microbial 2,3-Butanediol production from Brewer's spent grain modeled on pinch technology. ACS Sustain Chem Eng 11(22):8271–8280. https://doi.org/10.1021/acssuschemeng.3c00616

van den Oever AEM, Costa D, Messagie M (2023) Prospective life cycle assessment of alternatively fueled heavy-duty trucks. Appl Energy 336:120834. https://doi.org/10.1016/j.apenergy.2023.120834

Velvizhi G, Nair R, Goswami C et al (2023) Carbon credit reduction: a techno-economic analysis of "drop-in" fuel production. Environ Pollut 316:120507. https://doi.org/10.1016/j.envpol.2022.120507

Watanabe MDB, Cherubini F, Cavalett O (2022a) Climate change mitigation of drop-in biofuels for deep-sea shipping under a prospective life-cycle assessment. J Clean Prod 364:132662. https://doi.org/10.1016/j.jclepro.2022.132662

Watanabe MDB, Cherubini F, Tisserant A, Cavalett O (2022b) Drop-in and hydrogen-based biofuels for maritime transport: country-based assessment of climate change impacts in Europe up to 2050. Energy Convers Manag 273:116403. https://doi.org/10.1016/j.enconman.2022.116403

Zhang Z, Wei K, Li J, Wang Z (2022) Life-cycle assessment of bio-jet fuel production from waste cooking oil via hydroconversion. Energies 15:6612. https://doi.org/10.3390/en15186612

Chapter 14
Currently and Future Legislation in Drop-in Biofuel Production

Gaurav Bhardwaj, Lachi Wankhede, Rahul Saini, Carlos S. Osorio-Gonzalez, and Satinder Kaur Brar

Abstract The search for sustainable energy has led to the development of biofuels as a promising alternative to fossil fuels. Drop-in biofuels show significant progress by being compatible with existing fuel infrastructure. However, moving from the conception of biofuels to commercializing drop-in variants faces numerous challenges mainly due to complex legislation. This chapter addresses these challenges by providing a detailed analysis of the legislative frameworks that influence the biofuel sector. It begins by exploring the historical development of biofuel laws, transitioning into an examination of the current regulatory environment. Key challenges and ambiguities within these frameworks are then scrutinized. The chapter further discusses emerging trends and factors that will shape future legislation, supported by global, Canadian, and emerging economies' case studies. Policy recommendations, based on these analyses, are presented to address the identified issues and facilitate sustainable biofuel development. The chapter concludes with insights on future directions and the imperative for interdisciplinary research and dynamic policy approaches in navigating the evolving biofuel landscape.

Keywords Drop-in biofuel legislation · Sustainable biofuel policies · Biofuel regulatory frameworks · Future biofuel market trends · Bioenergy law and commercialization

14.1 Introduction

Drop-in biofuels represent a significant advancement in renewable energy, providing a sustainable alternative to fossil-based fuels. Unlike conventional biofuels, drop-in biofuels are fully compatible with the existing petroleum infrastructure, eliminating the need for specialized storage, distribution, or engine modifications (Demirbas

G. Bhardwaj · L. Wankhede · R. Saini · C. S. Osorio-Gonzalez · S. K. Brar (✉)
Department of Civil Engineering, Lassonde School of Engineering, York University, Toronto, ON, Canada
e-mail: satinder.brar@lassonde.yorku.ca

S. K. Brar et al. (eds.), *The Microbiology of the Drop-in Biofuel Production*, Biofuel and Biorefinery Technologies 15, https://doi.org/10.1007/978-3-031-61637-2_14

2011). This compatibility makes them highly attractive for immediate implementation in the current energy landscape (Ponte and Birch 2014).

The global significance of drop-in biofuels is increasingly recognized due to their potential in addressing climate change, enhancing energy security, and promoting economic sustainability (Hill et al. 2006). These biofuels provide a promising solution that corresponds with international commitments to decrease greenhouse gas emissions (Naik et al. 2010). These biofuels, derived from sustainable sources such as lignocellulosic biomass (Balat and Balat 2009), waste material (Zhang et al. 2016), and algae (Chisti 2007), offer a greener alternative to fossil fuels. They contribute to a closed carbon cycle by utilizing renewable resources that absorb carbon dioxide, thereby mitigating climate impacts (William Stafford et al. 2017; Stafford et al. 2019), and have a higher energy density than fossil fuels, allowing efficient energy production. Furthermore, their use of diverse, renewable feedstocks not only diminishes reliance on finite fossil fuels, enhancing energy independence, but also brings economic benefits, especially in rural areas rich in biomass, by creating new markets and job opportunities. This multifaceted approach positions drop-in biofuels as a key element in the transition to sustainable energy production (Koh and Ghazoul 2008; Prasad et al. 2019). Furthermore, drop-in biofuels have led to decreased emissions, enhancing air quality and addressing the effects of climate change (Cherubini 2010). Also, drop-in biofuels are in line with various Sustainable Development Goals, specifically SDG 13 and SDG 7. However, it is crucial to note that the sustainability of biofuels is contingent upon responsible production practices to minimize the negative impacts on other SDGs (Nazari et al. 2021). It is essential to acknowledge that the sustainability of drop-in biofuels is challenging. Land-use change, competition with food production, and environmental degradation must be carefully managed to ensure that biofuels contribute to a sustainable future (Fletcher Jr. et al. 2011).

In summary, drop-in biofuels hold considerable promise for the transition toward a more sustainable energy landscape. However, widespread adoption and commercialization of drop-in biofuels are intricately linked to a complex regulatory landscape. This landscape encompasses various laws, policies, and guidelines that both facilitate and constrain the development and deployment of biofuels. The regulatory framework shapes emerging technologies, such as drop-in biofuels. Legislative measures not only establish guidelines for the production, distribution, and utilization of biofuels but also significantly impact investment decisions and research priorities (Bacovsky et al. 2013). In this context, it is imperative to understand the legislative framework governing drop-in biofuels.

Legislation plays a pivotal role in shaping the development, commercialization, and sustainability of drop-in biofuels, acting as both a catalyst and, at times, a hindrance. Regulatory frameworks enforce the use of sustainable feedstocks and low-emission production processes, crucial for ensuring the environmental benefits of biofuels and aligning with global efforts to mitigate climate change (Stoeglehner and Narodoslawsky 2009). Economic viability is significantly influenced by legislative actions such as subsidies, tax incentives, and trade policies, which can either facilitate or impede the commercialization of biofuels (Sorda et al. 2010).

Fig. 14.1 Evolution and landscape of global drop-in biofuel legislation: From historical roots to future prospects

Additionally, legislation spurs innovation and technological advancements in the biofuel sector by influencing intellectual property rights and providing research grants and support for public–private partnerships (Caspeta and Nielsen 2013). However, the complexity and inconsistency of regulations across jurisdictions can create barriers, with stringent sustainability criteria and complex permitting processes potentially escalating costs and impeding scalability (Shao 2019; Li et al. 2021). Understanding this regulatory landscape is essential for stakeholders, including researchers, policymakers, and industry leaders, as it serves as a framework guiding sustainable development and climate change mitigation efforts in the biofuel industry.

This chapter offers an extensive exploration of the legislative frameworks worldwide that influence drop-in biofuels. Through a detailed analysis of existing and prospective legal structures, and by identifying influential regulatory bodies, it provides a critical assessment of how these laws bolster or restrict the growth of drop-in biofuels. The chapter also proposes policy recommendations to nurture the development of this sector. Its aim is to furnish a thorough overview that is pertinent to a wide range of stakeholders, including researchers, policymakers, and industry professionals. This comprehensive approach not only contributes significantly to the discourse on sustainable energy but also bridges the gap between regulatory theory and practice. Figure 14.1 encapsulates this narrative, visually summarizing the chapter's objectives and the global scope of legislative impacts on drop-in biofuels. It serves as a guide, highlighting key points and case studies that illustrate the practical implications of legislation on biofuel initiatives, thereby underscoring the chapter's commitment to providing actionable insights and fostering a more sustainable and equitable biofuel industry.

14.2 Historical Context of Global Biofuel Legislation

14.2.1 Early Regulations in Various Countries and their Impact on Biofuel Development

The historical trajectory of biofuel legislation is a tapestry of diverse regulatory approaches shaped by varying national priorities, technological capabilities, and environmental concerns. Early regulations were often driven by the need for energy security and the desire to promote domestic agricultural industries (Ragauskas et al. 2006; Sorda et al. 2010).

The Energy Policy Act of 1992 in the USA was an early and important legislation that promoted the use of biofuels, particularly by requiring federal fleets to start using alternative fuels (Yacobucci 2013). This led to two significant periods of innovation in the biofuel industry: one from 1995 to 2000, characterized by extensive patenting by first-generation (food-based) biofuel companies, and another from 2005 to 2010, marked by a second wave of innovation from leading firms and biotechnology companies focusing on second-generation biofuels (Kessler and Sperling 2016). These surges were influenced by the legislative landscape, including mandates and subsidies, which acted as drivers of technological advancement. However, early regulations also posed challenges. For example, advanced biofuel production facilities in the USA often lack historical air pollutant emission data, complicating the obtaining of the necessary air emission permits for construction and operation (Eberle et al. 2017). This highlights the importance of regulatory frameworks for addressing innovation and environmental concerns. In the European Union, the Biofuels Directive 2003 was a significant law that established projected goals for biofuel use. It aimed to reach a 5.75% share of biofuels in the overall transport fuel market by 2010 (Hinkle and Rosencranz 2008). Subsequently, the Renewable Energy Directive in 2009 mandated a target of achieving a 10% share of renewable energy in the transportation sector by 2020 (Scarlat et al. 2011). These directives significantly accelerated biofuel research and market development in Europe but also led to controversies regarding the sustainability of biofuels, particularly concerning indirect land-use change and food security (Edwards and Marelli 2010). Brazil, a pioneering country in the biofuel sector, introduced the Proálcool Program (Brazilian Alcohol Program) in 1975, which aimed to replace gasoline with ethanol produced from sugarcane. This program had a profound impact on boosting Brazil's ethanol industry and setting a precedent for other countries to follow.

Early biofuel regulations in Canada, such as the Ethanol Expansion Program launched in 2003, aimed to boost ethanol production through capital assistance to producers (Stefanelli et al. 2019). The 2010 Renewable Fuels Regulations subsequently required that gasoline contain an average renewable content of 5%. These early regulations laid the groundwork for Canada's biofuel industry by initially increasing its production capacity and later providing a stable biofuel market, thereby attracting further investment and innovation in the sector. In the context of developing countries, biofuel legislation has often been influenced by a combination of factors such as energy security, rural development, and foreign exchange savings and has had mixed outcomes. For instance, India launched its Ethanol Blended Petrol (EBP) Programme in 2003 to promote ethanol as a biofuel (Balat 2011). Similarly, African countries like Ethiopia and Mozambique have initiated biofuel programs to diversify their energy portfolios and encourage rural development (Jumbe et al. 2009). In Zambia, the early evolution of the biofuel sector was examined to assess the country's preparedness to leverage the benefits of emerging industries while mitigating social and environmental costs (German and Schoneveld 2012). Similarly, in Tanzania, policy challenges related to biofuel development were identified as critical for realizing the expected benefits for local communities (Sosovele 2010). China, another key player, introduced the Renewable Energy Law in 2005, laying the foundation for biofuels development, including biodiesel and ethanol (Qiu et al. 2010). The law aimed to reduce dependency on fossil fuels and mitigate environmental pollution but has faced challenges related to feedstock availability and land-use conflicts (Zhong et al. 2010).

Moreover, early biofuel mandates have been criticized for causing higher food prices and lower consumer welfare, raising concerns regarding crop-based biofuel production's sustainability (Claassen et al. 2017). This underscores the need for carefully designed policies to balance the competing interests of energy production, environmental sustainability, and social welfare. In summary, these early regulations had a dual impact. On the one hand, they provided the necessary impetus for research and development activities, leading to technological advancements in biofuel production processes (Demirbas 2011). On the other hand, they also led to unintended consequences, such as land-use changes and increased food prices, sparking debates on the sustainability of biofuels (Searchinger et al. 2008).

14.2.2 Evolution of Legislation in Response to Environmental Concerns and Technological Advancements

The legislative landscape governing biofuels has evolved significantly due to technological advancement and environmental concerns. This evolution has been marked by a shift from focusing on market development and energy security to a more nuanced approach incorporating sustainability criteria.

Biofuel policies originally stemmed from the desire to ensure energy security and support domestic industries. For instance, the development of Germany's biofuel industry can be attributed to interventions at both national and supranational levels in the market (Kaup and Selbmann 2013). However, as the industry matured, the focus shifted to addressing environmental sustainability. For instance, the World Trade Organization's (WTO) legal framework has become more receptive to ecological sustainability concerns over time (Daugbjerg and Swinbank 2016). Environmental concerns have further shaped legislation. Indirect land-use change (ILUC) has been incorporated into several regulatory frameworks to account for the broader environmental impact of biofuel production (Plevin et al. 2010). This reflects a growing awareness of the need to consider the entire life cycle of biofuels, from feedstock production to end use, in regulatory policies (Edwards and Marelli 2010).

Technological progress has also been crucial in influencing biofuel regulations. The US biofuel sector saw two major periods of innovation driven by legislative requirements and financial support, which spurred technological development (Kessler and Sperling 2016). These advancements have necessitated legislation updates to accommodate new technologies and production methods (Costantini et al. 2015). Moreover, the rise of second-generation biofuels and advanced production technologies such as pyrolysis has led to re-evaluating existing policies. These technological shifts can potentially ease some environmental and societal costs associated with first-generation biofuels (Koh and Ghazoul 2008). Technological advancements in enzymatic hydrolysis, gasification, and synthetic biology have also influenced legislative agendas. These technologies can produce drop-in biofuels from non-food feedstocks, mitigating some sustainability concerns associated with first-generation biofuels (Lynd et al. 2008).

In light of these developments, the US Renewable Fuel Standard, first established in 2005 and later expanded in 2007, incorporated provisions for advanced biofuels, including drop-in biofuels, which meet the specific greenhouse gas emission reduction criteria (Schnepf and Yacobucci 2012). Similarly, the Renewable Energy Directive of the European Union of 2009 set sustainability criteria for biofuels and was later amended in 2015 to include provisions for advanced biofuels made from waste and residues (Scarlat et al. 2015). In summary, the evolution of biofuel legislation has been a dynamic process influenced by technological advancements and environmental concerns. This evolution reflects the growing recognition of the need for a balanced approach that considers both the opportunities and challenges the biofuel industry presents.

14.3 Current Regulatory Framework

The global biofuel market operates within a complex web of international laws and treaties that aim to harmonize standards, promote sustainable practices, and facilitate trade. The Paris Agreement stands out as a significant global framework. While not specifically centered on biofuels, it establishes the wider setting for renewable energy endeavors, including biofuels. Its objective is to restrict global warming to below 2 °C (Oberthür and Bodle 2016).

The International Energy Agency (IEA) also plays a significant role through its Bioenergy Technology Collaboration Program, which fosters international cooperation in research and development of bioenergy, including biofuels (IEA 2017). Additionally, the World Trade Organization (WTO) has implications for the biofuel market through its agreements on tariffs and trade, which can affect the competitiveness of biofuels in the global market (Matsushita et al. 2015). The Roundtable on Sustainable Biomaterials (RSB) offers another critical framework that provides an international standard for the sustainability of biofuels, including criteria for greenhouse gas emissions, land rights, and water use (RSB 2016). This standard is often cited in national regulations and is a benchmark for sustainable biofuel production. The Food and Agriculture Organization has created the Bioenergy and Food Security Criteria and Indicators, offering policy recommendations to safeguard food security while promoting biofuel production (FAO 2012). With this global context established, examining specific national frameworks that have influenced the biofuel industry is pertinent. One such framework that has had a significant impact, domestically and globally, is that of the USA.

14.3.1 United States

The Renewable Fuel Standard (RFS) in the USA, a pivotal policy in the nation's energy landscape, began under the Energy Policy Act of 2005, aiming to blend 7.5 billion gallons of renewable fuel into gasoline by 2012. The Energy Independence and Security Act of 2007 expanded the program, setting a goal of 36 billion gallons of renewable fuel by 2022, and focusing on advanced biofuels made from non-food sources.

In 2023, the Environmental Protection Agency (EPA) announced new targets for these years, extending the program's scope beyond the original 2022 deadline. These targets are especially focused on advanced biofuels, which are more environmentally friendly and sustainable. For example, the target for advanced biofuels is set to rise from 5.94 billion Renewable Identification Numbers (RINs) in 2023 to 7.33 billion RINs by 2025. Other categories like cellulosic biofuel and biomass-based diesel also have higher targets, indicating a broad push for more diverse and sustainable fuel options (US EPA 2023).

In addition to RFS, the US Department of Energy funds research and development in next-generation biofuels. Various tax incentives and loan guarantees are available to stimulate investments in biofuel production and infrastructure development. Public–private partnerships complement these financial mechanisms to accelerate the commercialization of next-generation biofuels—similarly, the US Department of Agriculture (USDA) also plays a role in various programs to support biofuel feedstock development and infrastructure. For instance, the Biomass Crop Assistance Program (BCAP) provides financial assistance to farmers and landowners to cultivate biomass feedstock for biofuel production (USDA 2015).

State-level policies further augment federal initiatives. Several states have their own biofuel mandates and incentives, creating a multilayered policy landscape that drives biofuel development nationwide. An instance of this is the impact of regulations like California's Low Carbon Fuel Standard in driving the adoption of drop-in biofuels. The LCFS requires a decrease in the carbon intensity of transportation fuels and specifically allows for drop-in biofuels as a way to comply with these requirements (California Air Resources Board 2023). The USA is a global leader in next-generation biofuels, including drop-in biofuels. The existing policies are designed to boost domestic production and position the USA as a key player in the worldwide biofuel market, both producers and exporters.

14.3.2 European Union

The European Union's Renewable Energy Directive (RED) stands as a pivotal policy framework shaping the region's energy landscape. Initially introduced in 2009, the RED set a goal for the EU to source 20% of its energy from renewables by 2020, a target aimed at reducing carbon emissions and enhancing energy security (Scarlat et al. 2015). In 2018, recognizing the increasing feasibility and importance of renewable energy, the RED was amended to include the concept of "advanced biofuels," which are produced from waste and residues rather than food crops, raising the target to 32% by 2030. This revision acknowledged the significant strides in renewable energy, particularly the cost-effectiveness of onshore wind energy. Further developments saw the European Commission propose an increase to 40% by 2030 as part of its broader climate action plan in 2021. Following this, the REPowerEU Plan in 2022 suggested an even more ambitious target of 45%, reflecting the EU's commitment to addressing climate change and energy challenges. In 2023, an agreement was reached to aim for at least 42.5%, with aspirations to reach 45%. This progression underscores the EU's dedication to a sustainable energy future, continually adapting its policies to meet the urgent demands of environmental conservation and energy security. The RED has thus been a key driver in the EU's transition toward renewable energy, influencing not just energy production but also economic and social development within the region (European Commission 2023).

The Fuel Quality Directive (FQD) complements RED by setting sustainability criteria for biofuels, including greenhouse gas emission reductions, and it has been instrumental in guiding the sustainability aspects of drop-in biofuels (Valin et al. 2015). The EU launched the Horizon 2020 initiative, offering financial support for research and development across a range of industries, such as biofuels (CINEA 2020). Additionally, individual member states have policies aligning with EU directives but may include specific biofuel incentives or mandates. For example, Germany's Biofuel Quota Act and the UK's Renewable Transport Fuel Obligation (RTFO) are national-level policies that affect drop-in biofuels.

However, the EU faces challenges such as varying member-state policies and market fragmentation. To address these issues, the EU aims for international collaboration in biofuel development and encourages intra-EU cooperation and global partnerships. The overarching goal is to position the EU as a leader in sustainable biofuel production, including drop-in biofuels, and contribute to international efforts to mitigate climate change.

14.3.3 Canada

The complex interplay between federal and provincial policies shapes Canada's biofuel development landscape. These policies serve as the cornerstone for developing, commercializing, and utilizing biofuels, including drop-in biofuels.

14.3.3.1 Federal Policies

The Canadian government has proactively established a regulatory framework to promote biofuel development and commercialization. Federal mandates for biofuel blending have been instrumental in driving this industry forward. These mandates have created a guaranteed market for biofuel producers and have encouraged investment in biofuel technologies. Two crucial federal initiatives have been launched: the Ethanol Expansion Program and ecoENERGY. These programs have received funding of $78 million and $1.5 billion, respectively, to support the biofuel industry (Saini et al. 2021). The ecoENERGY for Biofuels program, which operated from 2008 to 2017, provided financial incentives to produce renewable fuels, including advanced biofuels, which can serve as drop-in replacements for fossil fuels (Govt. of Canada 2013).

In response to the evolving landscape post-COVID-19, the Canadian government updated the Clean Fuel Standard (CFS) to support the biofuel sector further. The latest Clean Fuel Regulations aim to deliver up to 26.6 million tonnes of emission reduction annually by 2030, reflecting a significant step in Canada's climate action plan (Govt. of Canada 2023). These regulations, which build on the success of the long-standing federal renewable fuel regulations, introduce a more flexible approach in reducing the emission intensity of fuels and encouraging a broader range of cleaner fuels and technologies. The first compliance period for these regulations runs from July 2023 to December 2023, with the provision that regulated parties will not need to submit compliance credits until July 2024. This transitional phase is designed to ease the industry into new standards with renewable fuels that comply with existing provincial and federal fuel-blending requirements recognized under the CFS. The government anticipated a surplus of credits in the early years, which is expected to keep the costs for companies needing to acquire credits relatively low (Govt. of Canada 2023). Furthermore, the Renewable Fuels Regulations under the Canadian Environmental Protection Act (CEPA), established in 2010, continue to mandate a minimum amount of renewable fuel in gasoline and diesel fuel, sustaining the biofuel market. The adjustments to the CFS post-COVID-19 highlight Canada's ongoing commitment to biofuels as a critical component of its strategy for a cleaner and more sustainable energy future.

14.3.3.2 Provincial Policies

Provincial regulations in Canada work alongside federal efforts and tailor them to fit local circumstances and objectives. For instance, British Columbia's Regulation on Renewable and Low Carbon Fuels Requirements continues to require a decrease in the carbon emissions of transportation fuels, promoting the use of biofuels that can be seamlessly integrated (Ministry of Energy 2023). In Quebec, a draft regulation is in progress that requires distributors to incorporate 15% low-carbon content in gasoline and 10% in diesel fuel by 2030, representing a significant increase from current levels and underscoring Quebec's commitment to reduce emissions from transportation fuels (Gouvernement du Québec 2023).

Ontario announced plans to increase the amount of renewable energy in regular-grade gasoline to 15% by 2030, up from the current 10%. This progressive policy move is in the province's "Made-In-Ontario Environment Plan." Ontario's broader goal is to reduce greenhouse gas emissions by 30% by the same year. The increase in ethanol content is expected to lead to an annual reduction of up to one megaton of greenhouse gas emissions, marking a substantial contribution to the province's environmental objectives (Sherrard 2020). This regulatory change is also positioned as a stimulus for economic recovery post-COVID-19, potentially creating new jobs and supporting the biofuel and agriculture sectors. The introduction of this new standard will be gradual, with increases to 11% by 2025, 13% by 2028, and finally reaching the 15% target by 2030 (Environmental Registry of Ontario 2018). Ontario's Ethanol Growth Fund and Quebec's 2030 Energy Policy are other provincial initiatives that affect the drop-in biofuel sector. For instance, Ontario has moved toward incorporating 10% ethanol in gasoline, a step applauded by Renewable Industries Canada (Renewable Industries Canada 2022). These policies aim to stimulate investment in renewable fuel production and infrastructure, thereby facilitating the market entry of advanced biofuels, including drop-in variants.

14.3.3.3 Comparison with US and EU Frameworks

The Canadian, US, and EU frameworks for drop-in biofuels each present unique characteristics shaped by their respective policy objectives, market conditions, and technological advancements.

- **Regulatory mandates**: While all three jurisdictions have mandates for renewable fuel content, the scope and stringency differ. The US Renewable Fuel Standard (RFS) is perhaps the most comprehensive, with specific categories and quotas for advanced biofuels, including drop-in biofuels (Schnepf and Yacobucci). The EU's Renewable Energy Directive (RED) also includes progressive biofuel provisions but focuses more on sustainability criteria (Scarlat et al. 2015). Canada's Renewable Fuels Regulations under CEPA are less prescriptive but serve a similar purpose of creating a biofuel market (Govt. of Canada 2010).

- **Financial incentives**: Canada's ecoENERGY for Biofuels program and the US Department of Agriculture's various programs offer financial incentives for biofuel production. However, the EU's Horizon 2020 program is more research-oriented, primarily funding innovation and technology development (Govt. of Canada 2013).
- **Provincial/state-level policies**: Canada and the USA both have regional or state-level regulations in place. For example, British Columbia has the Renewable and Low Carbon Fuel Requirements Regulation, while California implemented the Low Carbon Fuel Standard to decrease transportation fuel's carbon intensity (Ministry of Energy 2023). The EU, being a supranational entity, does not have an equivalent level of governance, but member states often enact national policies that align with EU directives.

Compared to the USA and the EU, Canada's biofuel policies are relatively nascent but rapidly evolving. However, Canada is making strides in policy development, focusing on reducing GHG emissions and attracting private-sector investment for commercial fuel production.

14.3.4 Emerging Economies

Emerging economies such as Brazil, India, and China are becoming increasingly significant players in the global biofuel market. Combining energy security concerns, environmental objectives, and economic development goals shapes their regulatory frameworks.

- **Brazil**: Brazil has been a pioneer in the biofuel sector, particularly with its Proálcool Program initiated in 1975, which focused on ethanol production from sugarcane. This program focused on ethanol production from sugarcane and has been instrumental in shaping Brazil's energy matrix (Saravanan et al. 2020). The RenovaBio program, launched in 2017, to promote producing advanced biofuels, including drop-in biofuels, by setting decarbonization targets for the fuel sector (Grangeia et al. 2022).
- **India**: India initiated its biofuel journey with the Ethanol Blended Petrol Programme in 2002, aiming for a 5% blending mandate. However, the program faced implementation challenges due to feedstock shortages (Saravanan et al. 2020). In India, the regulatory framework for drop-in biofuels is shaped by the National Policy on Biofuels, established in 2018 by the Ministry of Petroleum and Natural Gas and amended in 2022. This policy aims to reduce dependence on imported petroleum by promoting domestic biofuel production, and it has made

significant strides by advancing the deadline for achieving a 20% bioethanol blend in petrol from 2030 to 2025–2026 and broadening the range of eligible feedstocks for biofuel production. Complementing this policy, the "Roadmap for Ethanol Blending in India 2020–2025" provides a detailed strategy for meeting these targets, addressing logistical, infrastructural, and financial aspects. These initiatives mark India's proactive approach to enhancing energy security, stimulating economic growth, and mitigating environmental impacts through increased use of sustainable biofuels (IEA 2022).
- **China**: China first implemented its biofuel policy with the Pilot Testing Program for bioethanol in 2002, which included the adoption of E10 standards for fuel ethanol. This commitment to biofuels was reinforced by the Renewable Energy Law in 2006 and the Medium and Long-Term Development Plan for Renewable Energy in 2007 (Su et al. 2015). China's 13th Five-Year Plan for Bioenergy identifies biofuels as a critical area for development and sets targets for biofuel production and utilization (Govt. of China 2016). The plan also emphasizes the importance of advanced biofuels, including drop-in biofuels, in achieving China's energy and environmental goals.

In conclusion, the regulatory landscape for drop-in biofuels is a complex tapestry woven from various national and international policies. Whether it is the comprehensive mandates of the US RFS, the sustainability focuses of the EU's RED, Canada's blend of federal and provincial initiatives, or the burgeoning frameworks in emerging economies like Brazil, India, and China, each contributes uniquely to the global biofuel narrative. These frameworks dictate the production and utilization of drop-in biofuels and influence research, development, and commercialization efforts. As we transition to the next section, we will delve into the role of governmental and non-governmental organizations, which serve as critical agents in shaping, implementing, and overseeing these diverse regulatory landscapes.

14.4 Role of Governmental and Non-governmental Organizations

14.4.1 Governmental Organizations

The global landscape for drop-in biofuels is significantly influenced by the roles and responsibilities of international agencies, which serve as critical factors in shaping policy frameworks, fostering technological advancements, and promoting environmental sustainability.

- **International Energy Agency (IEA)**

The International Energy Agency (IEA) is a cornerstone for international cooperation in the energy sector, including biofuels. Established in 1974, the IEA's Bioenergy Technology Collaboration Program has been instrumental in fostering research and development in bioenergy technologies. The program facilitates international projects, workshops, and publications that aim to accelerate the commercialization of bioenergy technologies, including drop-in biofuels. The IEA's latest World Energy Outlook reports and technology roadmaps remain essential resources for biofuel policy and market analysis, with an increased emphasis on aligning bioenergy with climate change targets (IEA 2017).

- **United Nations Environment Programme (UNEP)**

The United Nations Environment Programme (UNEP) ensures that biofuel development aligns with global environmental objectives. Established in 1972, UNEP has been actively involved in assessing the environmental impacts of biofuels through its various publications and initiatives. UNEP's role in assessing the environmental impact of biofuels has expanded, with greater focus on aligning biofuel development with the Sustainable Development Goals and the Paris Agreement. UNEP continues to work on sustainability criteria to mitigate the environmental risks of biofuel production, including more stringent guidelines on land use and biodiversity conservation (Environment 2017).

- **National Agencies**

National agencies are critical in implementing and overseeing biofuel policies, including drop-in biofuels. These agencies are tasked with many responsibilities, ranging from policy formulation and implementation to research and development and market facilitation.

Canada: Natural Resources Canada (NRCan) and Environment and Climate Change Canada (ECCC) remain pivotal in Canada's biofuel sector. Recent developments include updated mandates for renewable fuel content and increased funding for biofuel research and development (Govt. of Canada 2013).

USA: The Environmental Protection Agency (EPA) has updated its Renewable Fuel Standard targets, reflecting new objectives in sustainable biofuel production and usage (Schnepf and Yacobucci 2012).

European Union: The Directorate-General for Energy now focuses on integrating biofuels within the broader framework of the European Green Deal, with updated directives and guidelines for member states (Scarlat et al. 2015).

14.4.2 Non-governmental Organizations

Non-governmental organizations (NGOs) influence biofuel, contributing to policy discourse, public awareness, and technical advancements. These organizations often act as intermediaries between the public, policymakers, and industry, advocating for sustainable and equitable biofuel practices.

14.4.2.1 Global NGOs

- **World Wildlife Fund (WWF)**: The World Wildlife Fund (WWF) advocates sustainable biofuel production. Through its various publications and campaigns, the WWF emphasizes the need for biofuels to meet stringent sustainability criteria to mitigate potential negative impacts on biodiversity and land use (WWF 2013).
- **Greenpeace**: Greenpeace has been critical of biofuel policies that do not adequately address environmental concerns, such as deforestation and water usage. The organization often releases reports and engages in advocacy to push for more stringent regulations and sustainability criteria for biofuels, including drop-in biofuels (Greenpeace UK 2007).

14.4.2.2 Canadian NGOs

- **Pembina Institute**: The Pembina Institute is a Canadian NGO focusing on sustainable energy solutions. It has published several reports on the potential of biofuels, including drop-in biofuels, to contribute to Canada's renewable energy portfolio. The institute also engages policymakers to advocate for sustainable biofuel policies (Todd 2018).
- **Clean Energy Canada**: Canada's shift toward renewable energy, such as biofuels, is advancing with the efforts of Clean Energy Canada. The organization provides research, conducts public engagement campaigns, and collaborates with various stakeholders to promote policies that encourage the development and adoption of sustainable biofuels.

14.5 Regulatory Challenges and Criticisms

14.5.1 Global Analysis of Gaps, Ambiguities, and Contradictions in Existing Laws

The regulatory landscape for drop-in biofuels is fraught with challenges from various factors, including technological limitations, environmental concerns, and economic considerations. This section aims to comprehensively analyze these challenges, highlighting the gaps, ambiguities, and contradictions in existing laws.

- **Gaps in legislation**: One of the critical gaps in current biofuel legislation across various regions is the absence of wide-ranging sustainability standards. While frameworks like the European Union's Renewable Energy Directive have begun to bridge this gap by establishing criteria that biofuel production must meet to be sustainable, such standards are not universally adopted. For example, environmental safeguards to prevent ecosystem degradation, social policies to ensure fair labor practices, and economic strategies to maintain market stability without heavy subsidies are often missing in the mandates of several US states. Furthermore, a lifecycle assessment to evaluate the overall carbon footprint of biofuels is not a consistent requirement. This lack of uniformity in sustainability criteria leads to a patchwork of regulations that fail to comprehensively address biofuel production's global nature and its potential impacts. Without these criteria, there is a risk that biofuel policies may inadvertently support practices that are detrimental to the environment, are socially unjust, or are economically unsustainable, thus undermining the holistic objectives of such legislation (Scarlat et al. 2015).

- **Ambiguities in legislation**: Ambiguities in biofuel legislation frequently stem from imprecise or unspecified terms within legal texts. Words such as "renewable biomass," "biomass sustainability," "low carbon fuel standards," and "greenhouse gas lifecycle emissions" are often employed without universally accepted definitions, leading to varied interpretations and applications among different regions. For instance, "renewable biomass" could refer to a wide range of organic materials, but it is unclear which materials qualify without a specific definition. Similarly, "biomass sustainability" may include factors like land use and energy balance, yet legislation does not always clearly outline these components. The term "advanced biofuels" exemplifies this issue, as it is commonly used without a clear delineation, resulting in inconsistent categorization and enforcement across jurisdictions. This lack of clarity can create confusion and hinder the effective implementation and enforcement of biofuel policies, as stakeholders may not have a uniform understanding of the legislative requirements or standards (Valin et al. 2015).
- **Contradictions in legislation**: Contradictions in biofuel legislation can arise when different levels of government or various agencies within the same government have conflicting policies. The US federal government encourages the use of biofuels through the Renewable Fuel Standard, while certain state-level regulations, like California's Low Carbon Fuel Standard, may place additional demands that conflict with federal directives (California Air Resources Board 2023).
- **Criticisms**: The biofuel sector, including drop-in biofuels, has faced criticisms on several fronts. Environmental organizations frequently highlight the possible adverse effects of biofuel manufacturing on land utilization, water supplies, and biodiversity (Greenpeace UK 2007). Additionally, there are economic criticisms related to the cost-effectiveness of biofuels as a renewable energy source, particularly when compared to other renewable technologies like wind and solar power (Todd 2018).

14.5.2 Specific Challenges and Criticisms in the Canadian Context and Emerging Economies

14.5.2.1 Canadian Context

The landscape for drop-in biofuels in Canada is fraught with technical, economic, and policy challenges. These challenges necessitate a comprehensive and nuanced understanding of effective policy formulation and implementation.

- **Lack of policy mechanisms**: The most pressing issue is the inadequacy of existing policy frameworks to address the complexities associated with drop-in biofuels. The absence of holistic policy mechanisms for energy and sustainability, including gaps in addressing greenhouse gas (GHG) emissions, creates a precarious foundation for the sector's growth.
- **Political interests**: The influence of political interests, mainly the federal government's diminished focus on renewable energy, has shifted the responsibility to provinces like Ontario. This political landscape complicates the national strategy for renewable energy, including drop-in biofuels.
- **Economic viability**: Another criticism in the Canadian context is the economic viability of biofuels, including drop-in biofuels. Despite federal and provincial incentives, production costs often remain high, raising questions about long-term sustainability without government subsidies (Todd 2018).
- **Regulatory fragmentation**: One of the primary challenges in Canada is the fragmentation of biofuel regulations across federal and provincial jurisdictions. National regulations like the Renewable Fuels Regulations establish a foundation, while local policies such as British Columbia's Renewable and Low Carbon Fuel Requirements Regulation can add further complications (Ministry of Energy 2023).
- **Integration into existing infrastructure**: While drop-in biofuels are compatible with existing distribution systems, their integration requires significant changes in land use, production facilities, and transportation logistics, as well as issues related to resource ownership.
- **Upstream disruptions**: Changes in upstream activities, such as land use and logistics for biomass transportation, have far-reaching implications. These disruptions can affect regional economies and land-use paradigms, necessitating re-evaluating current policies.
- **The materiality of biomass**: The unique nature of biomass as a feedstock introduces complexities that are markedly different from fossil fuels. The requirement for expansive land areas for biomass collection and subsequent transportation to processing plants adds layers of complexity and public visibility to energy production activities.

14.5.2.2 Emerging Economies

The challenges and criticisms in drop-in biofuels are not confined to developed nations; emerging economies like Brazil, India, and China also grapple with many issues that impede the effective implementation of biofuel policies. In emerging economies, the biofuel sector confronts unique challenges that stem from rapid development and resource constraints. Limited availability of sustainable feedstock, infrastructure inadequacies, and technological limitations are expected. Countries like Brazil grapple with sustainable feedstock sourcing amid concerns over land use and environmental degradation (Grangeia et al. 2022). India contends with inconsistent feedstock quality and a deficit in production and storage infrastructure, which hampers its biofuel blending targets (Saravanan et al. 2020). While making strides in policy measures, China still faces challenges in local implementation and competition from fossil fuels. These barriers highlight the need for tailored approaches to support the biofuel industry's growth within these diverse economic landscapes (Govt. of China 2016).

14.6 Emerging Trends and Future Legislation

14.6.1 Technological Innovations

Technological innovations in the biofuel sector are rapidly evolving, offering new opportunities and challenges for legislation. These advancements not only influence the economic viability of biofuels but also have the potential to reshape existing regulatory frameworks.

- **Synthetic biology and metabolic engineering**: Advancements in synthetic biology and metabolic engineering have enabled the development of microorganisms capable of converting various feedstocks into drop-in biofuels more efficiently (Peralta-Yahya et al. 2012). These innovations could necessitate updates to existing regulations concerning genetically modified organisms (GMOs) and their use in biofuel production.
- **Catalytic conversion technologies**: New catalytic conversion technologies are emerging that can convert biomass into drop-in biofuels using more efficient and less energy-intensive processes (Huber et al. 2006). These technologies could influence emissions standards and sustainability criteria in future legislation.
- **Waste-to-fuel technologies**: Using waste materials, such as agricultural residues and municipal solid waste, as feedstocks for drop-in biofuels is gaining traction (Kumar et al. 2009). This trend could lead to legislative changes incentivizing waste-to-fuel technologies, potentially impacting waste management policies.
- **Integrated biorefineries**: The concept of integrated biorefineries, which produce a range of products including biofuels, chemicals, and materials, is becoming

more prevalent (Cherubini 2010). This could lead to more comprehensive legislation addressing biofuels and the entire value chain of biomass utilization.

14.6.2 Environmental Concerns

As the global community becomes increasingly aware of the environmental impacts of energy production, upcoming legislation is expected to place greater emphasis on sustainability. This trend is evident both globally and in the Canadian context, where new laws and amendments aim to address the environmental aspects of biofuel production and utilization.

14.6.2.1 Global Legislation

- **Circular economy principles**: Adopting circular economy principles in biofuel legislation is gaining momentum. The European Union, for example, is contemplating revisions to the Renewable Energy Directive with the aim of encouraging the utilization of waste and by-products as sources for biofuel (European Commission 2020).
- **Carbon pricing mechanisms**: Global initiatives like carbon pricing and cap-and-trade systems are increasingly integrated into biofuel policies. These mechanisms could provide economic incentives for producing and utilizing more sustainable drop-in biofuels (World Bank 2020).

14.6.2.2 Canadian Legislation

- **Clean Fuel Standard:** Canada's forthcoming Clean Fuel Standard seeks to decrease the carbon footprint of various fuels, including biofuels. This legislation is expected to include provisions that promote the use of sustainable feedstocks and advanced biofuel technologies (Govt. of Canada 2022).
- **Provincial sustainability mandates**: Several Canadian provinces are considering or have already implemented sustainability mandates for biofuels. British Columbia is planning to revise its Low Carbon Fuel Requirements Regulation by implementing more rigorous sustainability standards for biofuels (Ministry of Energy 2023).

14.6.3 Commercialization and Market Dynamics

A complex interplay of market dynamics and regulatory frameworks influences the commercialization of drop-in biofuels. As the sector matures, future legislation is expected to focus on market regulation to facilitate the large-scale adoption of these biofuels.

14.6.3.1 Global Legislation

- **Trade policies**: Global trade policies will likely play an increasingly significant role in commercializing drop-in biofuels. Tariffs, import/export restrictions, and trade agreements could be revised to encourage or restrict the global trade of biofuels and their feedstocks (UNCTAD 2019). By 2032, the international trade of biodiesel is projected to decrease to 11% of total production, while ethanol trade will drop slightly to 7% (OECD 2023).
- **Quality standards**: International quality standards for drop-in biofuels are expected to be developed to facilitate market acceptance. Such standards could be integrated into legislation to ensure product consistency and reliability. In the USA, the Renewable Fuel Standard (RFS) aims to increase the ethanol blend rate to 11% by 2032 despite decreasing transportation fuel usage (OECD 2023). The European Union's Renewable Energy Directive (RED II) seeks to expand the renewable energy target to 32% by 2030, with specific biofuel feedstock like palm oil being reclassified due to sustainability concerns.
- In emerging economies, national policies, such as Brazil's RenovaBio, India's National Biofuel Policy, and China's Five-Year Plans, are evolving to better integrate drop-in biofuels through incentives like decarbonization credits, blending mandates, and market-driven mechanisms, reflecting a concerted effort toward a sustainable energy future. Brazil's National Energy Policy Council plans to return to a biodiesel blending target of 15% by 2026, up from the current 10%. India is accelerating ethanol production with a goal of E20 by 2025, despite feedstock limitations, and anticipates a significant increase in the share of ethanol produced from sugarcane. China's biofuel policy remains in flux, with the ethanol blending rate expected to reach 1.7% by 2032, a slight increase to support growth in ethanol consumption over the next decade (Govt. of China 2016; Grangeia et al. 2022).

14.6.3.2 Canadian Legislation

- **Investment incentives**: Canada is expected to introduce new laws or amend existing ones to provide investment incentives for biofuel production facilities, including drop-in biofuels. These could include tax credits, grants, or low-interest loans to stimulate domestic production (Govt. of Canada 2021).
- **Market quotas**: Future Canadian legislation may include market quotas for drop-in biofuels, similar to the Renewable Fuels Regulations, to ensure a minimum market share and stimulate demand (Govt. of Canada 2013).

Fig. 14.2 Shaping the future of biofuel legislation: The convergence of technology, environment, and market dynamics

The complex relationship between technology, environmental priorities, and market trends in shaping future biofuel policies can be explained with the help of Fig. 14.2. It showcases how recent innovations in biofuel technology, the necessity for environmental protection, and the realities of market economics are critical drivers for new legislation. This visual overview serves as a gateway to understanding the multifaceted factors that legislators must consider supporting the sustainable growth of the biofuel industry.

14.7 Case Studies

Case studies serve as invaluable tools for understanding the real-world implications of legislation on drop-in biofuel projects. By examining specific examples, this section aims to elucidate how various legislative frameworks have either facilitated or hindered the development and commercialization of drop-in biofuels (Table 14.1).

Table 14.1 Global perspectives on biofuel development: Case studies of policy impact and industry success

Introduction	Key policies and their impact	Conclusion
Global case study: Neste's renewable diesel		
Due to favorable legislation and policies, Neste, a Finnish renewable diesel producer, has become a global leader. This case study explores these contributing factors.	• EU's Renewable Energy Directive (RED): The RED's mandatory renewable energy targets have enabled Neste to become a significant player in the European renewable diesel market • California's Low Carbon Fuel Standard: This standard provides financial incentives for low-carbon fuels, benefiting Neste by allowing it to sell credits (California Air Resources Board 2023). • Environmental Policy Design: Neste aligns with policies aimed at sustainable development and climate change mitigation, positioning itself as a leader in eco-innovation (Zafar et al. 2020). • Education and research: Investment in R&D has enhanced Neste's biofuel production efficiency, giving it a competitive edge • Trade openness: Neste has capitalized on technology transfer and economies of scale, contributing to its growth (Ben Jebli and Ben Youssef 2015).	Neste's success is attributed to a synergy of supportive legislations, financial incentives, and a focus on sustainability and innovation. Policies like the EU's RED and California's Low Carbon Fuel Standard have been instrumental, while emphasis on R&D and trade openness has furthered Neste's impact in the renewable energy sector
Canadian case study: Enerkem Alberta Biofuels		
Enerkem, based in Edmonton, Alberta, converts municipal waste into fuels like methanol and ethanol. This study examines how legislation, policy, and funding have facilitated this success	• Legislative Support: Enerkem benefited from federal funding via Natural Resources Canada's ecoENERGY for Biofuels program, which aims to incentivize renewable energy projects (Canada 2018). • Policy framework: Supportive waste management and renewable energy policies have enabled Enerkem to establish its waste-to-chemicals facility.	Enerkem's success in waste-to-fuel conversion is attributed to legislative incentives, a supportive policy framework, and adequate funding. These elements collectively foster innovation and sustainability in renewable energy ventures

(continued)

Table 14.1 (continued)

Introduction	Key policies and their impact	Conclusion
	• Funding support: Federal funding provided the necessary capital for R&D, infrastructure, and operations, enabling Enerkem's ongoing innovation and expansion	
Brazil: Raízen's sugarcane ethanol		
This study explores how Raízen, a leading sugarcane ethanol producer in Brazil and a joint venture between Shell and Cosan, has capitalized on Brazil's RenovaBio program to expand its operations. The focus is on the role of biofuel legislation and policies in facilitating Raízen's growth	• Legislative Support: Brazil's biofuel policies, particularly the RenovaBio program, have provided Raízen with incentives like decarbonization credits, fostering its growth (Zaman Sajid 2021) • Strategic partnerships: The joint venture with Shell and Cosan has enabled Raízen to leverage resources and expertise, enhancing its production capacity • R&D investment: Raízen's investment in advanced agricultural practices has increased its production efficiency and reduced its environmental impact	Raízen's success is attributed to supportive Brazilian biofuel policies, strategic partnerships, and R&D investments. The case underscores the importance of legislation and policy in fostering biofuel industry growth and sustainability
India: Praj Industries' second-generation ethanol		
Praj Industries, an Indian biofuel company, has successfully navigated India's National Policy on Biofuels and emerged as an industry leader. This case study explores how Praj Industries aligns with critical success factors and India's biofuel policy to achieve its success	• Identification of critical success factors: Praj Industries' growth can be attributed to its focus on critical success factors such as technological advancements, feedstock availability, and policy support, as identified in a study by Borse and Sheth (2017) • Alignment with National Policy on Biofuels: The company's success aligns with India's National Policy on Biofuels, mainly focusing on second-generation ethanol projects utilizing agricultural residues. • Technological advancements and innovation: Praj Industries has invested in R&D to develop advanced	Praj Industries' success is attributed to its alignment with India's biofuel policy, technological innovation, and active policy advocacy. The company addresses critical success factors, positioning itself as a leader in the sustainable development of the biofuel industry

(continued)

Table 14.1 (continued)

Introduction	Key policies and their impact	Conclusion
	biofuel production technologies, positioning it as an industry leader • Policy advocacy and collaboration: The company actively engages in policy advocacy and collaborates with stakeholders, influencing the formulation of supportive biofuel policies	
China: Sinopec's algal biofuel		
Sinopec is researching algal biofuels, aligning with China's 13th and 14th Five-Year Plans focusing on advanced biofuels and carbon reduction	• Research and collaboration: Sinopec collaborates with experts in catalysis to design efficient biofuel catalysts. The study aims to unlock the biofuel potential of microalgae, a lipid-rich biomass. • Policy alignment: The project aligns with China's Five-Year Plans, which aim for advanced biofuels and carbon peaking by 2030 (Govt. of China 2016).	China's biofuel policies support Sinopec's research and focus on overcoming scalability and commercial viability challenges in algal biofuels

14.8 Policy Recommendations

The preceding sections have elucidated the complexities and challenges of the existing legislative frameworks governing drop-in biofuels. Based on these insights, this section aims to provide policy recommendations to facilitate drop-in biofuels' sustainable development and commercialization.

14.8.1 Global Recommendations

For global and emerging economies alike, it is essential to establish a uniform standard for sustainability criteria to ensure the environmental and social integrity of biofuels. Enhanced international cooperation to incentivize research and development is also crucial.

- **Standardization of sustainability criteria**: A global standard for sustainability criteria should be developed to ensure that biofuels, including drop-in biofuels, meet uniform environmental and social benchmarks.

- **Incentivizing research and development**: International collaborations should be encouraged to accelerate research and development in drop-in biofuels. Funding mechanisms, such as grants and tax incentives, could be established globally.

  Specifically for emerging economies:

- Brazil should expand RenovaBio to include drop-in biofuels, promoting diversity in its energy mix.
- India must streamline its biofuel supply chain and set clear directives for drop-in fuels.
- In its Five-Year Plans, China should articulate explicit commercialization strategies for drop-in biofuels, emphasizing market incentives and collaboration between public and private sectors.

14.8.2 Recommendations for Canada

- **Strengthen investment incentives**: Canada should expand its investment incentives, such as grants and low-interest loans, to stimulate the domestic production of drop-in biofuels.
- **Update provincial mandates**: Provincial governments should update their biofuel mandates to include specific provisions for drop-in biofuels, thereby ensuring a more comprehensive approach to renewable fuel adoption.

14.9 Future Perspectives

The work presented in this chapter underscores the need for continued interdisciplinary research that bridges technological innovation and policy development. For researchers, this means delving into how sustainability metrics can be better integrated into legislative frameworks to support the comprehensive advancement of drop-in biofuels. Policymakers, meanwhile, are encouraged to draw from the insights and recommendations of this chapter to craft adaptable and forward-looking policies that can keep pace with rapid technological and market changes. For industry stakeholders, there is a clear call to engage more deeply with policymakers and researchers to ensure that the commercialization of drop-in biofuels is both innovative and sustainable. Looking ahead, public–private partnerships have significant potential to play a transformative role in this sector, driving both technological and commercial progress.

14.10 Conclusion

This chapter has provided a detailed examination of the legislative frameworks that govern the field of drop-in biofuels. It has articulated the complexities and challenges inherent in this domain and identified the dynamic trends shaping it. The discussion has been enriched by case studies from around the world, including those from global leaders, Canada, and emerging economies, to illustrate the diverse impacts of these frameworks on the development and commercialization of drop-in biofuels. From these analyses, a series of policy recommendations has emerged. These recommendations aim to tackle the challenges and capitalize on the opportunities for sustainable growth within the sector, offering a pathway toward a more robust and environmentally responsible future for drop-in biofuels.

Acknowledgments This research was funded by the Natural Sciences and Engineering Research Council (Discovery Grant 23451, NSERC Alliance Option-2 Grants, ALLRP 571066 – 21). We would also like to thank James and Joanne Love, Chair in Environmental Engineering, for their financial support.

References

Bacovsky D, Ludwiczek N, Ognissanto M, Woergetter M (2013) Status of advanced biofuels demonstration facilities in 2012. A report to IEA Bioenergy task 39

Balat M (2011) Production of bioethanol from lignocellulosic materials via the biochemical pathway: a review. Energy Convers Manag 52:858–875. https://doi.org/10.1016/j.enconman.2010.08.013

Balat M, Balat H (2009) Recent trends in global production and utilization of bio-ethanol fuel. Appl Energy 86:2273–2282. https://doi.org/10.1016/j.apenergy.2009.03.015

Ben Jebli M, Ben Youssef S (2015) Output, renewable and non-renewable energy consumption and international trade: evidence from a panel of 69 countries. Renew Energy 83:799–808. https://doi.org/10.1016/j.renene.2015.04.061

Borse P, Sheth A (2017) Technological and commercial update for first- and second-generation ethanol production in india. In: Chandel AK, Sukumaran RK (eds) Sustainable biofuels development in india. Springer International Publishing, Cham, pp 279–297

California Air Resources Board (2023) Low carbon fuel standard | California Air Resources Board

Canada NR (2018) Ecoenergy for biofuels program lessons learned assessment. https://natural-resources.canada.ca/21100. Accessed 28 Aug 2023

Caspeta L, Nielsen J (2013) Economic and environmental impacts of microbial biodiesel. Nat Biotechnol 31:789–793. https://doi.org/10.1038/nbt.2683

Cherubini F (2010) The biorefinery concept: using biomass instead of oil for producing energy and chemicals. Energy Convers Manag 51:1412–1421. https://doi.org/10.1016/j.enconman.2010.01.015

Chisti Y (2007) Biodiesel from microalgae. Biotechnol Adv 25:294–306. https://doi.org/10.1016/j.biotechadv.2007.02.001

CINEA (2020) European climate, infrastructure and environment executive agency. https://cinea.ec.europa.eu/index_en. Accessed 16 Nov 2023

Claassen R, Langpap C, Wu J (2017) Impacts of federal crop insurance on land use and environmental quality. Am J Agric Econ 99:592–613. https://doi.org/10.1093/ajae/aaw075

Costantini V, Crespi F, Martini C, Pennacchio L (2015) Demand-pull and technology-push public support for eco-innovation: the case of the biofuels sector. Res Policy 44:577–595. https://doi.org/10.1016/j.respol.2014.12.011

Daugbjerg C, Swinbank A (2016) Three decades of policy layering and politically sustainable reform in the European Union's agricultural policy. Governance 29:265–280. https://doi.org/10.1111/gove.12171

Demirbas A (2011) Competitive liquid biofuels from biomass. Appl Energy 88:17–28. https://doi.org/10.1016/j.apenergy.2010.07.016

Eberle A, Bhatt A, Zhang Y, Heath G (2017) Potential air pollutant emissions and permitting classifications for two biorefinery process designs in the United States. Environ Sci Technol 51:5879–5888. https://doi.org/10.1021/acs.est.7b00229

Edwards R, Marelli L (2010) Indirect land use change from increased biofuels demand. Comparison of models and results for marginal biofuels production from different feedstocks

Environment UN (2017) UNEP Annual Report 2016. In: UNEP – UN Environment Programme. http://www.unep.org/resources/unep-annual-report-2016. Accessed 28 Aug 2023

Environmental Registry of Ontario (2018) Preserving and protecting our environment for future generations: a made-in-Ontario environment plan | Environmental Registry of Ontario. https://ero.ontario.ca/notice/013-4208. Accessed 29 Nov 2023

European Commission (2020) A new circular economy action plan

European Commission (2023) Renewable energy directive. https://energy.ec.europa.eu/topics/renewable-energy/renewable-energy-directive-targets-and-rules/renewable-energy-directive_en. Accessed 28 Aug 2023

FAO (2012) Bioenergy | energy | food and agriculture organization of the United Nations

Fletcher RJ Jr, Robertson BA, Evans J, Doran PJ, Alavalapati JR, Schemske DW (2011) Biodiversity conservation in the era of biofuels: risks and opportunities. Front Ecol Environ 9:161–168. https://doi.org/10.1890/090091

German LA, Schoneveld G (2012) Biofuel investments in sub-Saharan Africa: a review of the early legal and institutional framework in Zambia. Rev Policy Res 29:467–491. https://doi.org/10.1111/j.1541-1338.2012.00572.x

Gouvernement du Québec (2023) Ethanol and biodiesel. In: Gouvernement du Québec. https://www.quebec.ca/en/agriculture-environment-and-natural-resources/energy/energy-production-supply-distribution/bioenergy/ethanol-biodiesel. Accessed 15 Nov 2023

Govt. of Canada (2013). https://canadagazette.gc.ca/rp-pr/p1/2013/2013-05-18/html/reg3-eng.html. https://canadagazette.gc.ca/rp-pr/p1/2013/2013-05-18/html/reg3-eng.html. Accessed 28 Aug 2023

Govt. of Canada E and CC (2010) Renewable fuels regulations. https://www.canada.ca/en/environment-climate-change/services/managing-pollution/energy-production/fuel-regulations/renewable.html. Accessed 28 Aug 2023

Govt. of Canada E and CC (2022) Clean fuel standard regulatory design. https://www.canada.ca/en/environment-climate-change/services/managing-pollution/energy-production/fuel-regulations/clean-fuel-regulations/regulatory-design.html. Accessed 28 Aug 2023

Govt. of Canada E and CC (2023) Clean fuel regulations: recap of June 2023 media technical briefing. https://www.canada.ca/en/environment-climate-change/news/2023/06/clean-fuel-regulations-recap-of-june-2023-media-technical-briefing.html. Accessed 15 Nov 2023

Govt. of Canada NR (2021) Canada invests over $960-million in renewable energy and grid modernization projects. https://www.canada.ca/en/natural-resources-canada/news/2021/06/canada-invests-over-960-million-in-renewable-energy-and-grid-modernization-projects.html. Accessed 28 Aug 2023

Govt. of China (2016) Outline of the 13th five-year plan for national economic and social development of the People's Republic of China_Rolling News_Chinese Government Website. https://www.gov.cn/xinwen/2016-03/17/content_5054992.htm. Accessed 28 Aug 2023

Grangeia C, Santos L, Lazaro LLB (2022) The Brazilian biofuel policy (RenovaBio) and its uncertainties: an assessment of technical, socioeconomic and institutional aspects. Energy Convers Manag: X 13:100156. https://doi.org/10.1016/j.ecmx.2021.100156

Greenpeace UK (2007) Biofuels: green dream or climate change nightmare? | Greenpeace UK. https://storage.googleapis.com/gpuk-archive/blog/climate/biofuels-green-dream-or-climate-change-nightmare-20070509.html. Accessed 28 Aug 2023

Hill J, Nelson E, Tilman D, Polasky S, Tiffany D (2006) Environmental, economic, and energetic costs and benefits of biodiesel and ethanol biofuels. Proc Natl Acad Sci USA 103:11206–11210. https://doi.org/10.1073/pnas.0604600103

Hinkle J, Rosencranz A (2008) Jon Birger Skjærseth and Tora Skodvin, climate change and the oil industry: common problem, varying strategies. Int Environ Agreements 8:91–94. https://doi.org/10.1007/s10784-007-9056-0

Huber GW, Iborra S, Corma A (2006) Synthesis of transportation fuels from biomass: chemistry, catalysts, and engineering. Chem Rev 106:4044–4098. https://doi.org/10.1021/cr068360d

IEA (2017) Technology roadmap – delivering sustainable bioenergy—analysis. In: IEA. https://www.iea.org/reports/technology-roadmap-delivering-sustainable-bioenergy. Accessed 28 Aug 2023

IEA (2022) National policy on biofuels (2022 amendment)—policies. In: IEA. https://www.iea.org/policies/17006-national-policy-on-biofuels-2022-amendment. Accessed 16 Nov 2023

Jumbe CBL, Msiska FBM, Madjera M (2009) Biofuels development in sub-Saharan Africa: are the policies conducive? Energy Policy 37:4980–4986. https://doi.org/10.1016/j.enpol.2009.06.064

Kaup F, Selbmann K (2013) The seesaw of Germany's biofuel policy—tracing the evolvement to its current state. Energy Policy 62:513–521. https://doi.org/10.1016/j.enpol.2013.08.024

Kessler J, Sperling D (2016) Tracking U.S. biofuel innovation through patents. Energy Policy 98: 97–107. https://doi.org/10.1016/j.enpol.2016.08.021

Koh LP, Ghazoul J (2008) Biofuels, biodiversity, and people: understanding the conflicts and finding opportunities. Biol Conserv 141:2450–2460. https://doi.org/10.1016/j.biocon.2008.08.005

Kumar A, Jones DD, Hanna MA (2009) Thermochemical biomass gasification: a review of the current status of the technology. Energies 2:556–581. https://doi.org/10.3390/en20300556

Li G, Lu S, Shao S, Yang L, Zhang K (2021) Do environmental regulations hamper small enterprises' market entry? Evidence from China. Bus Strateg Environ 30:252–266. https://doi.org/10.1002/bse.2619

Lynd LR, Laser MS, Bransby D, Dale BE, Davison B, Hamilton R, Himmel M, Keller M, McMillan JD, Sheehan J, Wyman CE (2008) How biotech can transform biofuels. Nat Biotechnol 26:169–172. https://doi.org/10.1038/nbt0208-169

Matsushita M, Schoenbaum TJ, Mavroidis PC, Hahn M (2015) The World Trade Organization: law, practice, and policy, 3rd edn. Oxford University Press, Oxford, New York

Ministry of Energy M and LCI (2023) Renewable & low carbon fuel requirements regulation – province of British Columbia. https://www2.gov.bc.ca/gov/content/industry/electricity-alternative-energy/transportation-energies/renewable-low-carbon-fuels. Accessed 28 Aug 2023

Naik SN, Goud VV, Rout PK, Dalai AK (2010) Production of first and second generation biofuels: a comprehensive review. Renew Sust Energ Rev 14:578–597. https://doi.org/10.1016/j.rser.2009.10.003

Nazari MT, Mazutti J, Basso LG, Colla LM, Brandli L (2021) Biofuels and their connections with the sustainable development goals: a bibliometric and systematic review. Environ Dev Sustain 23:11139–11156. https://doi.org/10.1007/s10668-020-01110-4

Oberthür S, Bodle R (2016) Legal form and nature of the Paris outcome. Clim Law 6:40–57. https://doi.org/10.1163/18786561-00601003

OECD (2023) OECD-FAO agricultural outlook 2023–2032. OECD

Peralta-Yahya PP, Zhang F, del Cardayre SB, Keasling JD (2012) Microbial engineering for the production of advanced biofuels. Nature 488:320–328. https://doi.org/10.1038/nature11478

Plevin RJ, O'Hare M, Jones AD, Torn MS, Gibbs HK (2010) Greenhouse gas emissions from biofuels' indirect land use change are uncertain but may be much greater than previously estimated. Environ Sci Technol 44:8015–8021. https://doi.org/10.1021/es101946t

Ponte S, Birch K (2014) Guest editorial. Environ Plann Econ Space 46(2):271–279. https://doi.org/10.1068/a46296

Prasad S, Venkatramanan V, Kumar S, Sheetal KR (2019) Biofuels: a clean technology for environment management. In: Shah S, Venkatramanan V, Prasad R (eds) Sustainable green technologies for environmental management. Springer, Singapore, pp 219–240

Qiu H, Huang J, Yang J, Rozelle S, Zhang Y, Zhang Y, Zhang Y (2010) Bioethanol development in China and the potential impacts on its agricultural economy. Appl Energy 87:76–83. https://doi.org/10.1016/j.apenergy.2009.07.015

Ragauskas AJ, Williams CK, Davison BH, Britovsek G, Cairney J, Eckert CA, Frederick WJ, Hallett JP, Leak DJ, Liotta CL, Mielenz JR, Murphy R, Templer R, Tschaplinski T (2006) The path forward for biofuels and biomaterials. Science 311:484–489. https://doi.org/10.1126/science.1114736

Renewable Industries Canada (2022) RICanada welcomes the finalization of Canada's clean fuel regulations • renewable industries Canada. In: Renewable Industries Canada. https://ricanada.org/ricanada-welcomes-the-finalization-of-canadas-clean-fuel-regulations/. Accessed 16 Nov 2023

RSB (2016) RSB

Saini R, Osorio-Gonzalez CS, Brar SK, Kwong R (2021) A critical insight into the development, regulation and future prospects of biofuels in Canada. Bioengineered 12:9847–9859. https://doi.org/10.1080/21655979.2021.1996017

Zaman Sajid (2021) Sustainability | free full-text | historical analysis of the role of governance systems in the sustainable development of biofuels in Brazil and the United States of America (USA). https://www.mdpi.com/2071-1050/13/12/6881. Accessed 30 Aug 2023

Saravanan AP, Pugazhendhi A, Mathimani T (2020) A comprehensive assessment of biofuel policies in the BRICS nations: implementation, blending target and gaps. Fuel 272:117635. https://doi.org/10.1016/j.fuel.2020.117635

Scarlat N, Dallemand J-F, Skjelhaugen OJ, Asplund D, Nesheim L (2011) An overview of the biomass resource potential of Norway for bioenergy use. Renew Sust Energ Rev 15:3388–3398. https://doi.org/10.1016/j.rser.2011.04.028

Scarlat N, Dallemand J-F, Monforti-Ferrario F, Nita V (2015) The role of biomass and bioenergy in a future bioeconomy: policies and facts. Environ Dev 15:3–34. https://doi.org/10.1016/j.envdev.2015.03.006

Schnepf R, Yacobucci B (2012) Renewable fuel standard (Rfs): overview and issues

Searchinger T, Heimlich R, Houghton RA, Dong F, Elobeid A, Fabiosa J, Tokgoz S, Hayes D, Yu T-H (2008) Use of U.S. croplands for biofuels increases greenhouse gases through emissions from land-use change. Science 319:1238–1240. https://doi.org/10.1126/science.1151861

Shao A (2019) Chapter 3 – global market entry regulations for nutraceuticals, functional foods, and dietary/food/health supplements. In: Bagchi D (ed) Nutraceutical and functional food regulations in the United States and around the world (third edition). Academic Press, pp 37–45

Sherrard A (2020) Ontario to boost renewable blending in gasoline to 15 percent by 2030. In: Bioenergy International. https://bioenergyinternational.com/ontario-to-boost-renewable-blending-in-gasoline-to-15-percent-by-2030/. Accessed 15 Nov 2023

Sorda G, Banse M, Kemfert C (2010) An overview of biofuel policies across the world. Energy Policy 38:6977–6988. https://doi.org/10.1016/j.enpol.2010.06.066

Sosovele H (2010) Policy challenges related to biofuel development in Tanzania. Afr Spectr 45: 117–129. https://doi.org/10.1177/000203971004500105

Stafford W, Lotter Adrian, Brent A, von Maltitz, G (2017) Working paper: biofuels technology. In: UNU WIDER. http://www.wider.unu.edu/publication/biofuels-technology. Accessed 29 Aug 2023

Stafford WHL, Lotter GA, von Maltitz GP, Brent AC (2019) Biofuels technology development in southern Africa. Dev South Afr 36:155–174. https://doi.org/10.1080/0376835X.2018.1481732

Stefanelli RD, Walker C, Kornelsen D, Lewis D, Martin DH, Masuda J, Richmond CAM, Root E, Tait Neufeld H, Castleden H (2019) Renewable energy and energy autonomy: how indigenous peoples in Canada are shaping an energy future. Environ Rev 27:95–105. https://doi.org/10.1139/er-2018-0024

Stoeglehner G, Narodoslawsky M (2009) How sustainable are biofuels? Answers and further questions arising from an ecological footprint perspective. Bioresour Technol 100:3825–3830. https://doi.org/10.1016/j.biortech.2009.01.059

Su Y, Zhang P, Su Y (2015) An overview of biofuels policies and industrialization in the major biofuel producing countries. Renew Sust Energ Rev 50:991–1003. https://doi.org/10.1016/j.rser.2015.04.032

Todd L (2018) Pembina Institute Report to Donors 2018

UNCTAD (2019) Trade and development report 2019

US EPA O (2023) Regulations and volume standards for renewable fuel standards. https://www.epa.gov/renewable-fuel-standard-program/regulations-and-volume-standards-renewable-fuel-standards. Accessed 16 Nov 2023

USDA (2015) USDA announces restart of biomass crop assistance program for renewable energy. https://www.usda.gov/media/press-releases/2015/06/01/usda-announces-restart-biomass-crop-assistance-program-renewable. Accessed 28 Aug 2023

Valin H, Peters D, Berg M, Frank S, Havlík P, Forsell N, Hamelinck C, Pirker J, Mosnier A, Balkovič J, Schmidt E, Dürauer M, Fulvio FD (2015) The land use change impact of biofuels consumed in the EU: quantification of area and greenhouse gas impacts

World Bank W (2020) State and trends of carbon pricing 2020. World Bank, Washington, DC

WWF (2013) WWF analysis: searching for sustainability—comparative analysis of certification schemes for. https://wwf.panda.org/wwf_news/?212775/WWF-analysis-Searching-for-Sustainability%2D%2DComparative-analysis-of-certification-schemes-for-biomass-used-for-the-production-of-biofuels. Accessed 28 Aug 2023

Yacobucci B (2013) Fuel ethanol: background and public policy issues

Zafar MW, Shahbaz M, Sinha A, Sengupta T, Qin Q (2020) How renewable energy consumption contribute to environmental quality? The role of education in OECD countries. J Clean Prod 268:122149. https://doi.org/10.1016/j.jclepro.2020.122149

Zhang Z, O'Hara IM, Mundree S, Gao B, Ball AS, Zhu N, Bai Z, Jin B (2016) Biofuels from food processing wastes. Curr Opin Biotechnol 38:97–105. https://doi.org/10.1016/j.copbio.2016.01.010

Zhong C, Cao Y-X, Li B-Z, Yuan Y-J (2010) Biofuels in China: past, present and future. Biofuels Bioprod Biorefin 4:326–342. https://doi.org/10.1002/bbb.207

Chapter 15
The Future of the Microbial-Based Drop-in Biofuel Commercialization

Luiz Alberto Junior Letti, Susan Grace Karp, Leonardo Wedderhoff Herrmann, Maria Clara Manzoki, Thamarys Scapini, Walter José Martínez Burgos, and Carlos Ricardo Soccol

Abstract This chapter delves into the prospect of commercializing drop-in biofuels. Recognizing the need to transition away from fossil fuels to cleaner sources of energy, it highlights the significance of global goals for mitigating climate change and achieving sustainability targets. In this context, drop-in biofuels emerge as a promising solution. However, their commercialization has not yet reached maturity when compared to classical methods, such as hydrothermal liquefaction, mainly due to technological bottlenecks and especially due to economic feasibility. On the other hand, many companies are interested in the development of drop-in biofuels and are investing in new technical solutions, focusing on cost reduction. The chapter also details the importance of regulations and standardization to lower costs and establish a comprehensive value chain encompassing biofuel production, transportation, and commercialization. Notable advancements have been made in the aviation sector, such as the ASTM D7566 standard. Similarly, the transportation sector has made significant progress through initiatives like California's Low Carbon Fuel Standard (LCFS). Policy and legislation are also detailed. Various policies have been implemented to incentivize biofuel usage across different levels, from production to distribution. Also, this chapter discusses commercialization policies, approaching subjects such as trade tariffs, storage and logistic guarantees, and international trade policies related to biofuel utilization. Examples of public policies include Brazil's Proálcool and RENOVABIO programs, the United States Energy Policy Act and Energy Dependency Security Act, Canada's Clean Fuel Regulation, and the European Union's Renewable Energy Directive. Sustainability aspects are also considered during the discussion of new policies which could enhance the future commercialization of drop-in biofuels. Public acceptance is another key aspect to be taken into account when prospecting the future of biofuels. In this regard, three

L. A. J. Letti · S. G. Karp · L. W. Herrmann · M. C. Manzoki · T. Scapini · W. J. M. Burgos · C. R. Soccol (✉)
Department of Bioprocess Engineering and Biotechnology, Federal University of Paraná, Curitiba, Paraná, Brazil
e-mail: soccol@ufpr.br

S. K. Brar et al. (eds.), *The Microbiology of the Drop-in Biofuel Production*, Biofuel and Biorefinery Technologies 15, https://doi.org/10.1007/978-3-031-61637-2_15

major obstacles (price, performance of biofuels, and understanding of their use) are detailed. Finally, the chapter briefly presents the current scenarios of drop-in biofuels suppliers, highlighting the current well-developed infrastructure for some biofuels (such as bioethanol and biodiesel) and the most recent advances for other sectors, such as aviation fuels.

Keywords Drop-in biofuels · Sustainability · Regulation · Policies · Social acceptance

15.1 Introduction

Concerns about climate change and the need to reduce greenhouse gas emissions have driven the search for attractive solutions in the energy sector. In this context, biofuels emerge as a promising alternative to replace fossil fuels and contribute to the mitigation of environmental impacts. However, despite advances in the production of conventional biofuels such as ethanol and biodiesel, the basket of next-generation biofuels, known as "drop-in biofuels," presents distinct challenges. These biofuels are designed to be compatible with the existing infrastructure and can be used as direct substitutes for conventional fuels, without the need to modify engines or supply logistics.

Since drop-in biofuels are functionally equivalent to petroleum-based fuels, they are considered an important pillar to achieve the goals of reduction in greenhouse gases and climate change (Detsios et al. 2023). According to industry developments, two drop-in biofuel alternatives dominate the market and are the main source of investment in the sector, renewable diesel and sustainable aviation fuels (SAF), since these are high-emission sectors that are difficult to decarbonize, making it essential to replace fossil fuels with low-carbon ones (Bloomberg NEF's 2023; IEA Bioenergy 2023b).

Currently, most commercial processes for drop-in biofuel production are based on oleochemical lipid hydrotreatment technologies (hydrotreated vegetable oils or hydrotreated esters and fatty acids) (IEA Bioenergy 2023b). However, global decarbonization goals drive investments and research in the field of drop-in biofuels, demonstrating an expanding trend in the sector, with other technological routes being developed for commercialization, with biorefinery facilities being designed and built (IEA Bioenergy 2023b; Detsios et al. 2023). Nonetheless, the development of drop-in biofuels has been slow and only limited volumes of production and blending are available in large-scale plants (IEA Bioenergy 2023b). For example, advanced drop-in biofuels (derived from lignocellulosic biomass, such as agricultural and forestry) have not yet reached commercial maturity due to significant technical challenges associated with driving processes and the high capital costs of converting feedstock into the bioproducts (van Dyk et al. 2019; van der Kroft and Pruyn 2021).

This scenario also makes the price uncompetitive compared to prices of petroleum products (van Dyk et al. 2019). However, it does not undervalue the importance of drop-in biofuels for large-scale development and deployment to achieve a cleaner circular economy with lower impacts on the environment. Currently, companies interested in drop-in biofuels development have focused on achieving technologies that enable coprocessing while overcoming the associated challenges (e.g., operational, product quality, corrosion, etc.), including reactor modifications, use of unit operations for pretreatment and protective beds, and specific catalysts (IEA Bioenergy 2023b).

Considering previous sections of this book have already detailed these technological challenges, this chapter will focus on the potential for the commercialization of drop-in biofuels soon. The key hurdles that must be overcome will be highlighted, encompassing policy, regulatory frameworks, market dynamics and public acceptance, and the scenario for some scale suppliers of drop-in biofuels.

15.2 Regulation and Standards

Standardization and regulation of drop-in biofuel production are imperative criteria for the viability of large-scale commercialization, and they directly impact the establishment of drop-in biofuel refineries and the effectiveness of commercialization at scale. Among the current regulations that nominate drop-in biofuels, from a collaborative approach to evaluate and approve alternative fuels, the aviation community has more consolidated processes, given the results already presented in established processes at scale based on the American Society for Testing and Materials (ASTM) International consensus (Rumizen 2021; van Dyk et al. 2019).

The investments in drop-in biofuels in the aviation sector—called SAF—reflect the creation of regulations that encompass new manufacturing processes. Globally, the aviation industry has designated the ASTM D7566 standard as the governing specification in the production of drop-in aviation biofuels, which has been incorporated into existing specifications (ASTM D1655—conventional jet fuel). These processes for conversion of feedstocks to SAF are evaluated by ASTM International which authorizes the conversion processes, possible feedstocks, and maximum blending ratio with conventional fuel (ASTM D7566-22 2022). Until April 2023, nine conversion processes are authorized, including biomass, natural gas, coal, recycled oils, residual biomass from ethanol and biobutanol production, and algae as possible feedstocks. Processes from companies such as Shell, Virent, and Swedish Biofuels are also being evaluated (ICAO 2023). The use of drop-in aviation biofuels has properties and composition similar to conventional fuels, enabling seamless insertion into the industry's existing and well-established fuel supply infrastructure (Rumizen 2021).

Also in the transportation sector, the USA is leading the market investments for the development of drop-in biofuels, mainly due to the establishment of regulations in California (LCFS) for reducing Greenhouse Gases (GHG) emissions on roads (California Code of Regulations 2014). The impact of that program positively reached the country, encouraging other initiatives to be developed that can contribute to the reduction of environmental impacts. Although the program is not aimed at the production of drop-in biofuels, the regulation emphasizes the use of "low-carbon fuel," allowing the entry of stakeholders and the expansion of non-petrochemical-based industries. In this regard, a recent study demonstrated that drop-in biofuels can be low-carbon substitutes for gasoline and diesel, considering the context of California (Taptich et al. 2018). Under this scenario, simulations based on small-scale data show that California can achieve (and even exceed) the 2030 GHG emissions goals using biofuels (Taptich et al. 2018). A major challenge considering California's scenario, and one that can be extrapolated to other locations, the GHG reduction goal with exclusive drop-in biofuel supply (as discussed by Taptich et al. 2018) is demonstrated to be possible by simulation data but would require the use of biomass resources outside the State, since there is a limited supply of the right type of biomass for large-scale production (Bloomberg NEF's 2023; Taptich et al. 2018). One strategy would be the participation of technologies that enable high yield with the exploration of a wider variety of biomasses (Bloomberg NEF's 2023).

Considering the future of drop-in biofuels, research from stakeholders expects the investment in drop-in biofuels to grow, making it feasible to develop more specific regulations for their commercialization. Countries such as the USA, the Netherlands, Canada, China, and Singapore are highlighted in this scenario (Bloomberg NEF's 2023), although there are not yet specific regulations for drop-in biofuels that do not include the production and regulation of biomass-derived biofuels (e.g., ethanol). Incentives and regulations may further boost this market. Although a great innovation, the difficulty of creating wide-ranging regulations for drop-in biofuels makes the processes bureaucratic since each process needs to be evaluated separately by communities of experts in each field, like the example for SAF, which arises as a major challenge to the advances of drop-in biofuels in different fields, especially in developing countries. However, the demand for drop-in biofuels is guaranteed and growing, especially in the aviation sector (Bloomberg NEF's 2023). Since the global scenario is constantly changing and new approaches to reduce environmental impacts and climate change are being found, it is expected that commercialization will be expanded and consolidated on public and regulatory policies in force for each new technology and process that emerges within the drop-in biofuels. It is already possible to notice a movement toward studies and scale-up, aiming to understand how the operation and how the insertion of drop-in biofuels can occur within different sectors, and this movement can guide future advances. An important example is the research programs conducted in Switzerland, organized by the federal government, to conduct priority research for 2021–2024 on biomass that includes drop-in biofuels, aiming at making the most efficient technologies for the insertion of these circularity models in the country's economy (IEA Energy 2021; Swiss Federal Office of Energy 2017). Prospected results may be crucial for the development of regulations for the commercialization of drop-in biofuels.

15.3 Policy and Legislation

Drop-in biofuels have many environmental advantages over fossil fuels, but given their limitations in techno-economic terms, all routes to drop-in biofuel production will require policy support to develop to their full potential (IEA Energy 2019).

Global agreements are important drivers of the development of national and regional policies, which are then directed by precise targets. One of the most significant and global policies established today is, undoubtedly, the Paris Agreement, sealed in 2015 and implemented in 2016. It is an agreement between 194 parties (today) that aims to guide all nations by long-term goals such as (i) reducing global greenhouse gas emissions to limit the global temperature increase in this century to 2 degrees Celsius (limit the increase even further to 1.5 degrees); (ii) review countries commitments every 5 years; (iii) provide funding to developing countries to mitigate climate change and enhance abilities to adapt to climate impacts (United Nations 2015). One of the envisaged actions to achieve these goals is related to the establishment of a technological framework for developing biofuels and applying the technology on an industrial scale.

Since the Paris Agreement proposes monitoring the evolution of each country every 5-year cycle, national climate action plans of this same duration are proposed by each country. Those are known as Nationally Determined Contributions (NDCs). Each NDC is meant to be more ambitious over the cycles, and recently the COP27 suggested a revision of the NDCs 2030 targets due to the accelerated necessity of limiting global warming to 1.5 °C. Until today, the impacts of the Paris Agreement have led to local carbon neutrality targets, low-carbon solutions, and new markets worldwide, creating changes in the power and transport sectors and many new business opportunities (United Nations 2023). To accomplish the NDC targets and internal goals, biofuel policies have been developed and implemented in most countries aiming to support the production and utilization of biofuels. These policies focus on different stages of the biofuel chain: its production, commercialization, and use by consumers. Advanced biofuel investments are capital intensive and involve substantial risks, notably the technology selection, feedstock availability, and the financial viability gap in the implementation of the technology (ADB 2020). In Fig. 15.1, most of the strategic policies are implemented worldwide to overcome those challenges.

The International Civil Aviation Organization (ICAO) also plays an important role in encouraging the development of drop-in biofuels. ICAO has committed to reduce the greenhouse gas emissions of the aviation sector by 50% until 2050 over 2005 levels. This reduction is strongly related to the expansion of the use of SAF since this drop-in biofuel is equivalent to the commonly utilized fossil jet fuel but much less hazardous to the environment (Ringsred et al. 2021).

In the following, an overview of the biofuel policies of some of the countries/regions with significant potential in biofuel development is provided. It is important to notice that drop-in biofuels policies specifically are still extremely scarce: normally no differentiation from the conventional biofuels is made. The most highlighted strategies or limitations of each of these regions are exemplified.

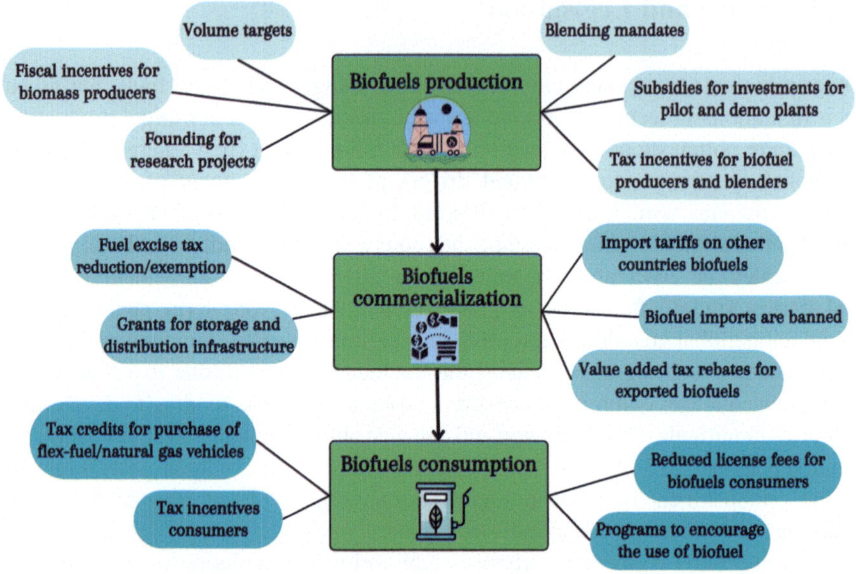

Fig. 15.1 Types of policies applied worldwide in the production, commercialization, and consumption of biofuels

15.3.1 Brazil

Brazil, a major sugarcane producer, has a history concerning bioenergy. In 1975, the National Alcohol Program (Proálcool) was created, motivated by the first petroleum crisis. This program was active until the decade of 1980 and contributed widely to the development and valorization of biofuels in the country (Cortez 2016). In 2017, the National Biofuels Policy (RENOVABIO) was signed into law (Federal Law N° 13.576) (IEA Bioenergy 2023a). This policy has the goal of promoting the national biofuel industry, reducing greenhouse gas emissions, and improving the life-cycle energy efficiency of biofuels. It establishes national carbon intensity targets, which are distributed to all fuel distributors, proportionally to their participation in the fossil fuel market. Financial penalties are applied for those who do not compromise the targets. Besides, decarbonization credits (CBIO) are issued from the commercialization of biofuel and serve as a financial asset, traded on the stock exchange (ANP 2020).

Blending mandates are also applied in Brazil: the country's National Council for Energy Policy (CNPE) elevated the level of biodiesel in commercial diesel from 10% (B10) to 12% (B12), with a progressive increase in mandates to B15 by 2026 (Ministério da Agricultura e Pecuária 2023). It is also mandatory that gasoline be blended with 27% of ethanol, with provisions of increasing the amount to 30% (CNN 2023).

15.3.2 United States of America (USA)

In the USA, the Energy Policy Act of 2005 aimed to promote biofuel research and development through economic incentives, including grants, income tax credits, subsidies, and loans. Two years later, in 2007, the Energy Independence and Security Act (EISA) expanded the Renewable Fuel Standard to increase biofuel production and target values of production from different biofuel types such as cellulosic biofuel and advanced biofuels derived from feedstocks other than cornstarch. EISA also provides tax credits, cash awards, grants, subsidies, and loans for research and development, biorefineries that displace more than 80% of the fossil fuels used in operations, and commercial applications of cellulosic biofuel (EPA 2023).

The Renewable Fuel Standard (RFS) specifies the use of biofuels in the U SA and thereby guides nearly half of all global biofuel production, being the largest biofuel program in the world (Lark et al. 2022). More recently, in August of 2022, the USA approved the Inflation Reduction Act (IRA) which includes incentives for biodiesel, SAF, and advanced fuels as well as support for up to 75% of the costs of biofuels infrastructure (IEA Bioenergy 2022). The IRA represents the most significant federal commitment to low-carbon biofuels since the RFS expansion in 2007.

15.3.3 Canada

Canada has committed to reduce emissions by 40–45% from 2005 levels by 2030 and reach net-zero emissions by 2050. Since transportation accounts for 25% of Canadian GHG emissions, meeting those goals is strictly related to the use of biofuels and especially advanced ones. In 2022, the Canadian Government published the Clean Fuel Regulations (CFR), which replaced the federal Renewable Fuels Regulations. This last policy focuses on following the biofuel sustainability criteria aligned with North American requirements, on the opt-in eligibility for sustainable aviation fuel, and on adopting measures to ensure that clean fuels used in Canada guarantee the protection of ecosystems and create purchasing opportunities for consumers (Government of Canada 2022; Advanced Biofuels Canada 2022).

15.3.4 European Union (EU)

The European Union first introduced its Renewable Energy Directive (RED) in 2009. It is an incentive program focused on biofuels, which aimed to achieve a 20% usage of renewable energy by 2020, particularly in the transportation sector, where the target was 10% (European Commission 2009). More recently, in 2018, RED II raised the target for the renewable energy sector to 32% and for the transportation sector, up to 14% by 2030. In 2015, the EU also started to show noticeable concern about establishing ceilings for biofuels based on food sources, therefore adding targets for advanced biofuels. The reason for that is the acknowledgment of the impacts caused by the indirect land-use change (iLUC) which is the possibility of agricultural expansion in regions previously not cultivated, due to the increased demand for food-based biofuels. Nowadays, biofuels based on food sources are holding at 7%, with a flexibility of 1%, and advanced biofuels went to 3.5% (European Commission 2018).

15.3.5 India

India's National Policy on Biofuels from 2018 set ambitious targets for biofuels in the country. By 2030, India expects a blending of 20% ethanol and 5% biodiesel, both biofuels being produced from non-food feedstock. Those measures aim to balance food security with climate security and will directly accelerate the development of advanced biofuels in the country in a decade (Pavlenko and Searle 2019; Ministry of Petroleum and Natural Gas 2018). In this scenario, India is working on the exploration of sustainable next-generation feedstocks like algae biomass and aquatic plants (India Energy and Resources Institute 2023).

15.3.6 Africa

Biofuels are still in the early stages of development in most African countries, but some of them already have established policies for biofuel development. It is the case of South Africa, Mozambique, Zambia, Malawi, and Zimbabwe, the last two having implemented the blending of bioethanol with petrol up to 20%. Fiscal incentives for biofuels or disincentives for fossil fuels are seen in those countries (Stafford et al. 2019), although the demand for transport fuels is low due to low vehicle ownership per capita. The lack of infrastructure in rural areas and water scarcity present additional challenges to the expansion of biofuel production (Barahira et al. 2021). Immature technology can be a barrier to producing conventional and advanced biofuels on a large scale in Africa, and strong policies are needed. South Africa being the country with the most potential, technology transfer with other African countries can be easy afterward (Barahira et al. 2021).

Upscaling drop-in biofuels worldwide will depend on well-formulated drop-in biofuel support policies, and, nowadays, most biofuel policies do not differentiate drop-in biofuels from conventional biofuels. Although significant technical progress has been made in developing and improving the various routes to drop-in fuels, the establishment of supportive policies directed specifically toward their industrial production has not kept pace (IEA Bioenergy 2014).

Among the current policies in place, volumetric mandates for biofuels are interesting, but alone will not be enough if climate change mitigation is the goal (IEA Energy 2019). Low carbon fuel standards are a better strategy for promoting drop-in biofuel production since it rewards those biofuels that provide the greatest emission reductions (accountable by their carbon intensity using tools such as life-cycle assessment—LCA) (IEA Energy 2019). Besides, policies that enhance drop-in biofuel R&D should be envisaged, with the opportunity of the establishment of policies between developed and developing countries to exchange feedstock and biofuels technological know-how (Das and Gundimeda 2022). Policies that encourage the production and sustainability of the feedstocks/intermediates (for example lipids and bio crudes) utilized in the drop-in production will be needed (IEA Energy 2019).

Concerning the aviation and marine transportation sectors, since they are reliant on lower carbon intensity fuels such as drop-in biofuels, it may require specific policies to support the commercialization and consumption of these biofuels. Those sectors must be included in the development of drop-in biofuel-promoting strategies, tax incentives, and biofuel mandates (IEA Bioenergy 2014). All policies must be elaborated based on solid studies and life-cycle analysis considering all the aspects involved in all drop-in biofuels routes, in a way to wisely promote drop-in commercialization in an economically viable, environmentally clean, and land-preservation way. Since the primary source of cost in drop-in biofuel production is feedstock cost, using waste biomass as a feedstock (second-generation biofuels) would be the best option to reduce costs and GHG emissions (Kargbo et al. 2021).

15.4 Social Aspects and Public Acceptance

Besides contemplating all the quality control standards and policies for its production and commercialization, the drop-in fuels market must also consider the public acceptance and general aspects of its application in society. At first glance, this kind of combustible presents ecological features compared to oil-derived ones, due to the bio-based feedstock used in its production. Alternative fuel research and share in the market are constantly growing, and renewable energy sources favor the current progress toward a circular bioeconomy, reducing pollutant emissions considerably (Mohr et al. 2015; Maggio et al. 2019; Carus and Dammer 2018; IEA Energy 2019).

As mentioned previously, several current policies are being applied toward a more sustainable and greener future. The United Nations Organization, for instance, established the Sustainable Development Goal (SDG) number 7, specifically Sect. 7.2, which focused on enhancing the renewable energy share in the world's energetic matrix. Associated with that, other goals are addressing the reduction of environmental impact (SDG numbers 6, 13, 14, and 15) and responsibility for sustainable production, growth, and consumption (SDG numbers 9, 11, and 12) (FAO 2016). These societal changes are being encouraged through SDG since 2016, yet date from the Conference on the Human Environment in Stockholm in 1972, after realizing the prejudice against oil-derived fuels (Londoño-Pulgarin et al. 2021). Therefore, an initial acceptance of the drop-in biofuels by the public would occur, considering the eco-friendly approach to its production.

However, at least three major obstacles hinder the commercialization of drop-in biofuels, which are (i) the selling prices and affordability of the material; (ii) its energetic performance for daily uses; and (iii) the public knowledge and comprehension of its use. The oil-based fuels and other fossil carbon energy sources are deeply internalized in modern society, from each car, bus, or transportation system to house heating or industrial production purposes. In that aspect, subtle changes in prices and efficiency directly hinder the life of several people (Londoño-Pulgarin et al. 2021).

International price fluctuations are especially severe, as they represent disputes and tensions over the oil energy sources. In daily life, it affects the availability of fuel for regular purposes. For instance, in January 2021, the gasoline average price in the USA was US$ 2.34 per gallon (US$ 0.62 per liter), and in January 2023, it reached US$ 3.33 per gallon (US$ 0.88 per liter). This indicates an increase of over 1.4 times in the prices, and considering an average engine efficiency of 10 km per liter and 20 km driven per day, it would represent an increase of around US$ 190,00 in fuels per year (EIA 2023). However, in most European countries, such as Germany, France, and the UK, the prices reach US$ 2.00 per liter, indicating over US$ 430,00 increase in this fuel consumption per year (Global Petrol Price 2023). These fluctuations disturb not only the gasoline but also all fuels, affecting the day-to-day life of regular people. In this aspect, it is desirable that the drop-in fuels present as low prices as possible, covering the production costs yet maintaining affordable conditions for the consumers.

Additionally, not only fuel is desirable to be inexpensive but also offers highly energetic performance. Whenever selecting the liquid to fill the tank, the client balances the value and the distance that the vehicle car reaches with the bought volume. The energetic efficiency of the fuel is associated with its chemical composition, estimating that huge carbon and hydrogen rates while the oxygen remains low usually generate more energy during combustion (Maggio et al. 2019; IEA Energy 2019). Another interesting parameter is the octane number, which indicates the resistance to compression without detonating, affecting mainly gasoline engines. Comparing ethanol and gasoline performance, for example, both have similar octane numbers that indicate they can be used in regular engines, however, gasoline presents higher energy density as its chemical composition only contains carbon and hydrogen, not oxygen. Therefore, gasoline usually allows the automobile to reach longer distances with the same initial volume, around 35% more than ethanol, and this comparison also extends to iso-propanol (27% more) and n-butanol (22% more) (Li et al. 2016). It would be interesting for the drop-in fuel to provide similar characteristics to the gasoline energetic yield, yet not its emission, to reduce the amount of fuel volume necessary for daily use.

One more detail to take into consideration is the general people's knowledge about fuel. The concept of drop-in biofuels is that they are functionally equivalent to petroleum fuels and fully compatible with existing petroleum infrastructure.

Therefore, every engine and machine would work normally as it was using oil-based fuels and should use the same gas station's available structure (IEA Energy 2019). This demonstrates that drop-in fuels do not require vehicle modifications or special care during fueling up compared to compressed natural gas (CNG), such as moving away from the car, or even electricity, which requires special stations. However, as a completely new product, the public would naturally suspect if it worked or not in regular engines. The apprehensiveness in applying those combustibles and damaging the equipment, especially among non-specialist people, would be an obstacle to their use. This indicates that the proper advertisement of the product could affect the overall acceptance. One interesting strategy is instructing the applications, characteristics, and origins of the drop-in fuels to the public through common media. Examples of cars, buses, and automobiles working only on these biofuels would increase societal approval, especially with popular model cars.

Society is already directing toward the use of alternative fuels to replace oil-based ones, especially after reaching such a pollution rate (Mohr et al. 2015). The arrival of a new bio-based renewable drop-in fuel would certainly be attractive for people; however, it is important to take into account that not only the environmental effects are relevant, but also the final effect on human well-being and energy availability. Therefore, the social aspects of a new combustible being inserted in the energetic matrix must be considered while designing a more sustainable future.

15.5 Commercial-Scale Suppliers of Drop-in Biofuels

The commercialization of drop-in biofuels depends on the availability of industrial facilities to supply products to the market according to the increasing demand. In this scenario, some drop-in biofuels have being produced at commercial scale for decades, and others are still depending on scale-up and start-up of new facilities. The production of bioethanol and biodiesel is well-established worldwide. The USA is the largest producer of bioethanol globally, using corn as the primary feedstock. Dry mill and wet mill processes are employed, with dry mill plants being more prevalent. The largest bioethanol producers in the USA include Archer Daniels Midland (ADM) (ADM 2023b), POET (POET 2023), Valero (Valero 2023), and Green Plains (Green Plains 2023).

In second position is Brazil, a leader in bioethanol production using sugarcane as the feedstock. The country has established an extensive bioethanol industry, with a significant portion of its transportation fuel being composed of ethanol. Key players in Brazil's bioethanol sector include Raízen (Raízen 2022) and São Martinho (São Martinho 2023). Raízen and Granbio also produce second-generation ethanol from sugarcane bagasse and straw (GranBio 2023; Raízen 2022). In the European Union (EU), bioethanol production predominantly relies on wheat, corn, and sugar beet as feedstocks. Several countries, including Germany, France, and Spain, have well-established bioethanol industries. Key companies in Europe include CropEnergies

(CropEnergies 2019), Tereos (Tereos 2023), and Abengoa (Abengoa Bioenergía 2006).

In Asia, Thailand is a major producer of bioethanol utilizing cassava as its primary feedstock. Major producers are Ubon Bio Ethanol, E85, and Mitrphol Biofuel, the last one using molasses as a feedstock (Krungsri Research 2021). In India, ethanol is produced from sugarcane molasses and lignocellulosic residues, the major producer being Shree Renuka Sugars (Groww 2023). Europe is a prominent producer of biodiesel, with Germany, France, and Spain leading the industry. In the European Union (EU), biodiesel is primarily derived from rapeseed oil, which is abundant in the region. Other feedstocks used in European biodiesel production include soybean oil, palm oil, and waste cooking oil. Key players in the European biodiesel market include Neste (Neste 2023a), ADM (ADM 2023a), and TotalEnergies (Total Energies 2023a).

The USA is another significant producer of biodiesel, with soybean oil being the primary feedstock. Other feedstocks used in the USA include animal fats, corn oil, and used cooking oil. Biodiesel production facilities are spread across various states, including Iowa, Illinois, and Texas. Major players in the US biodiesel industry include the Renewable Energy Group (REG) (Renewable Energy Group 2023), ADM (ADM 2023a), and World Energy (World Energy 2023). South American countries, such as Brazil and Argentina, are major producers of biodiesel. In Brazil, soybean oil is the primary feedstock, while Argentina primarily utilizes soybean oil and animal fats. Brazil has implemented a biodiesel blending mandate, further driving the growth of the industry. Key companies in the South American biodiesel market include Olfar (Olfar 2021), Granol, Cargill (Cargill 2023), and Louis Dreyfus (LDC 2023).

China, Indonesia, Malaysia, and India have also experienced significant growth in biodiesel production. Palm oil is the dominant feedstock in these regions due to its abundant availability. The governments of these countries have implemented policies and incentives to promote biodiesel use, addressing environmental concerns and boosting the domestic agricultural sector. Important producers are Beijing Sanju Environmental Protection (Statista 2023), PT Pertamina (Biofuels International 2020), and Vance Bioenergy (Vance Bioenergy 2023).

The commercial-scale production of SAF is more recent as compared to other drop-in biofuels. The USA, Europe, and the Asia-Pacific region are at the forefront of SAF production, with several key players investing in scaling up production capacity. These facilities employ various technologies, including hydrotreatment, Fischer–Tropsch synthesis, and alcohol-to-jet processes. The USA has made substantial progress in developing SAF production facilities. Companies like Fulcrum BioEnergy (Fulcrum Bioenergy 2023), World Energy (World Energy 2023), and Velocys (Velocys 2023) have established commercial-scale plants that convert municipal solid waste and woody biomass into SAF. These facilities contribute to reducing greenhouse gas emissions and promoting a circular economy by diverting waste from landfills.

Europe is a pioneer in the production and utilization of SAF. France and the UK have incorporated the ExxonMobil SAF in their fuel blends for lower-emission aviation (Exxon Mobil 2023). Companies such as Neste (Neste 2023b), Total Energies (Total Energies 2023b), and SkyNRG (SkyNRG 2023) have invested in the construction of SAF refineries, utilizing feedstocks like used cooking oil, vegetable oils, and agricultural residues. Furthermore, several European airports have implemented blending mandates, requiring a percentage of SAF in their fuel supply. The Asia-Pacific region has shown significant potential for SAF production. Companies like Neste (Neste 2023b) and China Petroleum & Chemical Corporation (SINOPEC 2023) have initiated projects to manufacture SAF from waste oils, cooking oil, and algae. Several airlines in the region have also started testing and using SAF in their operations, driving market growth and stimulating further investments (Exxon Mobil 2023).

At present, however, SAF production costs are higher than those of conventional jet fuels due to various factors, including feedstock availability, production scale, and infrastructure. Continued technological advancements, supportive policies, and economies of scale are vital to reducing costs and enhancing competitiveness.

15.6 Future Perspectives

Technological advancements are crucial for the widespread commercialization of drop-in biofuels. While existing processes, particularly hydrotreatment technologies, have enabled limited-scale production, but further research and development efforts are required to overcome technical challenges and reduce production costs. The emergence of alternative technological routes, such as biorefinery facilities, demonstrates an expanding trend in the sector and offers promising prospects for scaling up drop-in biofuel production and commercialization. Addressing technical challenges, including driving processes and capital costs associated with feedstock conversion, will be essential for achieving commercial maturity.

Policy and regulatory frameworks play a vital role in shaping the future of drop-in biofuels. Several initiatives (such as ASTM D7566 California's LCFS) have paved the way for greater adoption and commercialization in specific sectors. Moreover, governments worldwide have implemented policies to incentivize biofuel usage, reduce trade barriers, and establish comprehensive value chains. However, the future and the maturity for commercialization of drop-in biofuels strongly rely on the continued collaboration between governments, industry stakeholders, and research institutes. These collaborations are essential to refine and expand the policies, ensuring their effectiveness in driving commercialization.

We have also demonstrated that market dynamics, including pricing competitiveness and market demands, are critical factors that influence the future of drop-in biofuels. While current prices may not be as competitive as petroleum products, it is essential to recognize the long-term environmental benefits and the role drop-in biofuels play in achieving a cleaner circular economy. Ongoing efforts by companies to develop technologies for coprocessing, reactor modifications, pretreatment, and catalyst improvements demonstrate the industry's commitment to overcoming challenges and reducing production costs. Continued investment, research, and collaboration will help unlock the potential of drop-in biofuels as a viable and economically feasible alternative to fossil fuels.

Another key aspect of the future of drop-in biofuels which we have addressed is public acceptance. Price, performance, and public understanding of biofuel usage are the three major obstacles that need to be overcome to garner broader public support. Educating the public about the benefits and potential of drop-in biofuels, dispelling misconceptions, and ensuring transparency in communication will contribute to increased acceptance and adoption. In terms of drop-in biofuels suppliers, we have found the current landscape showcases a well-developed infrastructure for certain biofuels like bioethanol and biodiesel. However, significant progress is being made in other sectors, such as aviation fuels and renewable diesel, where investment and research are concentrated. Scaling up production capacity and diversifying feedstock sources will be crucial for meeting growing demand and achieving commercial viability.

15.7 Conclusions

The commercialization of drop-in biofuels holds great potential for addressing the pressing challenges of transitioning to cleaner and more sustainable sources of energy. While significant progress has been made in the production and adoption of conventional biofuels like ethanol and biodiesel, drop-in biofuels, which can be seamlessly integrated into existing infrastructure without requiring modifications, present unique challenges and opportunities. In this chapter, we have explored various aspects related to the future commercialization of drop-in biofuels, including a brief overview of technological bottlenecks and a more detailed discussion of policy, regulatory frameworks, specific legislation, and also public acceptance.

In conclusion, the commercialization of drop-in biofuels represents a promising solution for transitioning to cleaner energy sources and mitigating climate change. Although challenges exist, significant advancements have been made in technological innovation, policy frameworks, and market dynamics. Overcoming these hurdles will require continued collaboration, research, and investments from governments, industry stakeholders, and research institutions. By addressing technical limitations, refining policy frameworks, fostering public acceptance, and ensuring competitive market dynamics, drop-in biofuels can realize their full potential as a viable and economically feasible alternative to fossil fuels. Continued efforts to reduce production costs, expand production capacity, and diversify feedstock sources will be essential in driving the commercialization of drop-in biofuels. This will pave the way for a sustainable future powered by clean and renewable energy sources.

References

Abengoa Bioenergía (2006) Bioenergy. https://www.abengoa.com/export/sites/abengoa_corp/resources/pdf/en/gobierno_corporativo/informes_anuales/2006/2006_Volume1_AnnualReport_Bioenergy.pdf
ADB (2020) India: promoting advanced biofuels through high technology report. https://www.adb.org/projects/54222-001/main
ADM (2023a) Biodiesel. 2023. https://www.adm.com/en-us/products-services/industrial-biosolutions/products/biodiesel/
ADM (2023b) Cleaner-burning biofuel. 2023. https://www.adm.com/en-us/products-services/industrial-biosolutions/products/ethanol/
Advanced Biofuels Canada (2022) Advanced biofuels Canada applauds release of final clean fuel regulations. https://advancedbiofuels.ca/advanced-biofuels-canada-applauds-release-of-final-clean-fuel-regulations/
ANP (2020) Renovabio. https://www.gov.br/anp/pt-br/assuntos/renovabio
ASTM D7566-22 (2022) Standard specification for aviation turbine fuel containing synthesized hydrocarbons. https://www.astm.org/d7566-22.html
Barahira DS, Okudoh VI, Eloka-Eboka AC (2021) Suitability of crop residues as feedstock for biofuel production in South Africa: a sustainable win-win scenario. J Oleo Sci 70(2):213–226. https://doi.org/10.5650/jos.ess20288
Biofuels International (2020) Indonesia's state oil company to produce B100 biodiesel next Year. 2020. https://biofuels-news.com/news/indonesias-state-oil-company-to-produce-b100-biodiesel-next-year/
Bloomberg NEF's (2023) Global drop-in fuels market to see a flood of new entrants. https://about.bnef.com/blog/global-drop-in-fuels-market-to-see-a-flood-of-new-entrants/
California Code of Regulations (2014) California's low carbon fuel standard (LCFS) program. http://carules.elaws.us/code/t.17_d.3_ch.1_subch.10_art4_subart1_sec.95303
Cargill (2023) Biodiesel. 2023. https://www.biodieselbr.com/usinas_brasil/fabrica/cargill
Carus M, Dammer L (2018) The circular bioeconomy—concepts, opportunities, and limitations. Ind Biotechnol 14(2):83–91. https://doi.org/10.1089/ind.2018.29121.mca
CNN (2023) Governo Estuda Aumentar Mistura Do Etanol Na Gasolina de 27% Para 30%. April 2023. https://www.cnnbrasil.com.br/economia/governo-estuda-aumentar-mistura-do-etanol-na-gasolina-de-27-para-30/
Cortez LAB (2016) Proálcool 40 Anos: Universidade e Empresas, 40 Anos de Ciência e Tecnologia Para Etanol Brasileiro, 1st edn. Blucher, São Paulo
CropEnergies (2019) Ethanol: sustainable, climate-friendly and versatile – especially as a petrol substitute. 2019. https://www.cropenergies.com/en/products/ethanol
Das P, Gundimeda H (2022) Is biofuel expansion in developing countries reasonable? A review of empirical evidence of food and land use impacts. J Clean Prod 372(October):133501. https://doi.org/10.1016/j.jclepro.2022.133501
Detsios N, Maragoudaki L, Atsonios K, Grammelis P, Orfanoudakis NG (2023) Design considerations of an integrated thermochemical/biochemical route for aviation and maritime biofuel production. Biomass Convers Bior. https://doi.org/10.1007/s13399-023-03754-4
EIA (2023) Gasoline and diesel fuel update. https://www.eia.gov/petroleum/gasdiesel/
EPA (2023) Economics of biofuels. https://www.epa.gov/environmental-economics/economics-biofuels#:~:text=U.S.%20policy%20approaches%20to%20support%20biofuel%20production,-The%20Energy%20Policy&text=EISA%20expanded%20the%20Renewable%20Fuel,from%20feedstocks%20other%20than%20cornstarch
European Commission (2009) European Commission renewable Energy directive 2009. https://eur-lex.europa.eu/legal-content/EN/ALL/?uri=CELEX:32009L0028
European Commission (2018) Revised renewable energy directive (EU) 2018/2001 (RED II). https://eur-lex.europa.eu/legal-content/EN/TXT/?uri=uriserv:OJ.L_.2018.328.01.0082.01.ENG&toc=OJ:L:2018:328:TOC

Exxon Mobil (2023) Aviation: lower emissions at high altitude. 2023. https://corporate.exxonmobil.com/what-we-do/lower-emission-transportation/aviation

FAO (2016) The 17 goals for sustainable development. https://sdgs.un.org/goals

Fulcrum Bioenergy (2023) Our fuel process: from your trash can to the fly. 2023. https://www.fulcrum-bioenergy.com/our-fuel-process

Global Petrol Price (2023) Gasoline prices, litre, 29-May-2023. https://www.globalpetrolprices.com/gasoline_prices/

Government of Canada (2022) Clean fuel regulations. https://laws-lois.justice.gc.ca/eng/regulations/SOR-2022-140/index.html

GranBio (2023) BioFlex. 2023. https://www.granbio.com.br/en/our-technology/bioflex/

Green Plains (2023) Biofuels: providing clean, renewable fuel to keep America running. 2023. https://gpreinc.com/what-we-do/biofuel-production/

Groww (2023) Shree Renuka Sugars. 2023. https://groww.in/stocks/shree-renuka-sugars-ltd.

ICAO (2023) Global framework for aviation alternative fuels. 2023. https://www.icao.int/environmental-protection/GFAAF/Pages/default.aspx

IEA Bioenergy (2014) The potential and challenges of drop-in biofuels. https://task39.sites.olt.ubc.ca/files/2014/01/Task-39-drop-in-biofuels-report-summary-FINAL-14-July-2014-ecopy.pdf

IEA Bioenergy (2022) Inflation reduction act 2022: sec. 22003 biofuel infrastructure and agriculture product market expansion. https://www.iea.org/policies/16289-inflation-reduction-act-2022-sec-22003-biofuel-infrastructure-and-agriculture-product-market-expansion

IEA Bioenergy (2023a) National biofuels policy. https://www.iea.org/policies/2475-national-biofuels-policy

IEA Bioenergy (2023b) Recent progress in the production of low carbon intensive drop-in fuels. https://www.ieabioenergy.com/blog/publications/recent-progress-in-the-production-of-low-carbon-intensive-drop-in-fuels/

IEA Energy (2019) Drop-in biofuels: the key role that co-processing will play in its production. https://www.ieabioenergy.com/blog/publications/new-publication-drop-in-biofuels-the-key-role-that-co-processing-will-play-in-its-production/

IEA Energy (2021) Implementation of bioenergy in Switzerland. https://www.ieabioenergy.com/wp-content/uploads/2021/11/CountryReport2021_Switzerland_final.pdf

India Energy and Resources Institute (2023) The energy and resources institute Mission and goals. https://www.teriin.org/dbt-teri-centre-excellence-advanced-biofuels-and-bio-commodities

Kargbo H, Harris JS, Phan AN (2021) 'Drop-in' fuel production from biomass: critical review on techno-economic feasibility and sustainability. Renew Sust Energ Rev 135(January):110168. https://doi.org/10.1016/j.rser.2020.110168

Krungsri Research (2021) Industry outlook 2021–2023: Ethanol Industry. 2021. https://www.krungsri.com/en/research/industry/industry-outlook/Energy-Utilities/Ethanol/IO/io-ethanol-21

Lark TJ, Hendricks NP, Smith A, Pates N, Spawn-Lee SA, Bougie M, Booth EG, Kucharik CJ, Gibbs HK (2022) Environmental outcomes of the US renewable fuel standard. Proc Natl Acad Sci 119(9):e2101084119. https://doi.org/10.1073/pnas.2101084119

LDC (2023) Bioenergy. 2023. https://www.ldc.com/us/en/product/bioenergy/

Li Y, Meng L, Nithyanandan K, Lee TH, Lin Y, Lee C-f F, Liao S (2016) Combustion, performance and emissions characteristics of a spark-ignition engine fueled with isopropanol-n-butanol-ethanol and gasoline blends. Fuel 184(November):864–872. https://doi.org/10.1016/j.fuel.2016.07.063

Londoño-Pulgarin D, Cardona-Montoya G, Restrepo JC, Muñoz-Leiva F (2021) Fossil or bioenergy? Global fuel market trends. Renew Sust Energ Rev 143(June):110905. https://doi.org/10.1016/j.rser.2021.110905

Maggio G, Nicita A, Squadrito G (2019) How the hydrogen production from RES could change energy and fuel markets: a review of recent literature. Int J Hydrog Energy 44(23):11371–11384. https://doi.org/10.1016/j.ijhydene.2019.03.121

Ministério da Agricultura e Pecuária (2023) CNPE Aprova Aumento Para 12% Da Mistura de Biodiesel Ao Diesel a Partir de Abril. 2023. https://www.gov.br/agricultura/pt-br/assuntos/noticias/cnpe-aprova-aumento-para-12-da-mistura-de-biodiesel-ao-diesel-a-partir-de-abril

Ministry of Petroleum and Natural Gas (2018) National policy on biofuels. https://vikaspedia.in/energy/policy-support/renewable-energy-1/biofuels/national-policy-on-biofuels-2018#:~:text=The%20Policy%20expands%20the%20scope,for%20human%20consumption%20for%20ethanol

Mohr SH, Wang J, Ellem G, Ward J, Giurco D (2015) Projection of world fossil fuels by country. Fuel 141(February):120–135. https://doi.org/10.1016/j.fuel.2014.10.030

Neste (2023a) Neste MY Renewable Diesel. 2023. https://www.neste.com/products/all-products/renewable-road-transport/neste-my-renewable-diesel#c75ce717

Neste (2023b) Sustainable air travel: SAF. 2023. https://www.neste.com/products/all-products/saf#c75ce717

Olfar (2021) Novo Líder: Olfar Se Torna a Maior Fabricante de Biodiesel Do Brasil. 2021. https://www.biodieselbr.com/noticias/usinas/info/novo-lider-olfar-se-torna-a-maior-fabricante-de-biodiesel-do-brasil-020321

Pavlenko N, Searle S (2019) The potential for advanced biofuels in India: assessing the availability of feedstocks and deployable technologies. Int Counc Clean Transp (ICCT)

POET (2023) Bioethanol. 2023. https://poet.com/bioethanol/

Raízen (2022) Our business: renewables. 2022. https://www.raizen.com.br/en/our-business/renewables#ethano

Renewable Energy Group (2023) Helping lead the way toward a lower carbon future. 2023. https://www.regi.com/products/endura-fuels

Ringsred A, van Dyk S, John (Jack) Saddler. (2021) Life-cycle analysis of drop-in biojet fuel produced from British Columbia forest residues and wood pellets via fast-pyrolysis. Appl Energy 287(April):116587. https://doi.org/10.1016/j.apenergy.2021.116587

Rumizen MA (2021) Qualification of alternative jet fuels. Front Energy Res 9(November):760713. https://doi.org/10.3389/fenrg.2021.760713

São Martinho (2023) Ethanol. 2023. https://www.saomartinho.com.br/show.aspx?idCanal=rhw93OXFh1cptbt5Xx8owA==&linguagem=en

SINOPEC (2023) Institutional web page. 2023. http://www.sinopec.com/listco/en/about_sinopec/

SkyNRG (2023) Fueling a new era of progress in aviation. 2023. https://skynrg.com/

Stafford WH, Lee GA, Lotter GP, von Maltitz, and Alan Colin Brent. (2019) Biofuels technology development in southern Africa. Dev South Afr 36(2):155–174. https://doi.org/10.1080/0376835X.2018.1481732

Statista (2023) Leading biodiesel producers in China in 2022, by estimated production capacity (in million liters). 2023. https://www.statista.com/statistics/1285565/largest-producers-of-biodiesel-by-capacity-china/

Swiss Federal Office of Energy (2017) Bioenergy research programme." https://www.bfe.admin.ch/bfe/en/home/research-and-cleantech/research-programmes/bioenergy.html

Taptich MN, Scown CD, Piscopo K, Horvath A (2018) Drop-in biofuels offer strategies for meeting California's 2030 climate mandate. Environ Res Lett 13(9):094018. https://doi.org/10.1088/1748-9326/aadcb2

Tereos (2023) Bioethanol. 2023. https://tereos.com/en/activities-and-products/product-lines/bioethanol/

Total Energies (2023a) Biofuels. 2023. https://miningsolutions.totalenergies.com/renewables-and-low-carbon-energies/biofuels

Total Energies (2023b) SAF (Sustainable Aviation Fuel). 2023. https://aviation.totalenergies.com/en/fuels-and-services-aviation/sustainable-aviation-fuel

United Nations (2015). Adoption of the Paris agreement. https://unfccc.int/sites/default/files/english_paris_agreement.pdf

United Nations (2023) The Paris agreement – what Is the Paris Agreement? https://unfccc.int/process-and-meetings/the-paris-agreement

Valero (2023) Ethanol: from corn to fuel and livestock feed. 2023. https://www.valero.com/renewables/ethanol

van der Kroft DFA, Pruyn JF (2021) A study into the availability, costs and GHG reduction in drop-in biofuels for shipping under different regimes between 2020 and 2050. Sustain For 13(17): 9900. https://doi.org/10.3390/su13179900

van Dyk S, Jianping S, Mcmillan JD, Jack (John) Saddler. (2019) Potential synergies of drop-in biofuel production with further co-processing at oil refineries. Biofuels Bioprod Biorefin 13(3): 760–775. https://doi.org/10.1002/bbb.1974

Vance Bioenergy (2023) Vance biodiesel. 2023. https://www.vancebioenergy.com/product_biodiesel.php

Velocys (2023) Institutional web page. 2023. https://velocys.com/

World Energy (2023) Institutional web page. https://www.worldenergy.net/

SPRINGER NATURE

GPSR Compliance

The European Union's (EU) General Product Safety Regulation (GPSR) is a set of rules that requires consumer products to be safe and our obligations to ensure this.

If you have any concerns about our products, you can contact us on ProductSafety@springernature.com

In case Publisher is established outside the EU, the EU authorized representative is:

Springer Nature Customer Service Center GmbH
Europaplatz 3
69115 Heidelberg, Germany

The manufacturer's authorised representative in the EU is Springer
Nature Customer Service Centre GmbH, Europaplatz 3, 69115 Heidelberg,
Germany. If you have any concerns regarding our products, please
contact ProductSafety@springernature.com

Printed and bound by CPI Group (UK) Ltd, Croydon, CR0 4YY
01/10/2026
02225904-0018